DIGITAL TECHNOLOGY

Second Edition

GERALD E. WILLIAMS, P.E.

Riverside City College

SCIENCE RESEARCH ASSOCIATES, INC.
Chicago, Palo Alto, Toronto, Henley-on-Thames, Sydney
A Subsidiary of IBM

Acquisition Editor	*Alan W. Lowe*
Project Editor	*Ann Wood*
Compositor	*Bi-Comp, Inc.*
Illustrator	*John Foster*
Designer	*Carol Harris*

Library of Congress Cataloging in Publication Data

Williams, Gerald Earl, 1931–
Digital technology.

Includes index.
1. Digital electronics. I. Title.
TK7868.D5W5 1981 621.3819′582 81-8870
ISBN 0-574-21555-7 AACR2

Printed in the United States of America.

10 9 8 7 6 5 4 3 2

DEDICATION

With the pride that only a satisfied parent can know, I dedicate this book to two of the finest young men I know—my sons Geoffrey and Kelly, who have filled my life with the magic and beauty of love.

PREFACE

The main objective we established for this second edition was to update the coverage by incorporating new technology that has come on the industrial scene since the first edition. As a result of our ongoing assessment of industrial practices, you will find many more C-MOS devices; new coverage of Programmable Logic Arrays (PLAs): the new memory technologies, such as magnetic bubble and charge-coupled devices; and expanded coverage of the low-power Schottky subfamily. Also included are many new challenging student problems.

The general overall presentation remains the same. Primarily intended for a one-term first course in general digital theory, this text includes the latest digital-logic "building-blocks" and the associated techniques, circuits, and systems. Real-world integrated circuits are depicted and referenced throughout.

Coverage of programmable logic and bus-system organization concepts gives the student not only a solid grounding in modern digital integrated circuits, but also a foundation for future study of microprocessors. The concluding chapter, which gives a general orientation to the microprocessor scene, forms a bridge between hard-wired logic systems and software-oriented microprocessor systems.

The prerequisites are an understanding of electronics fundamentals, including semiconductor theory. Although aimed at electronics technology programs in colleges and technical institutes, the first edition proved popular as well in physics and electrical-engineering departments, especially for technology upgrading purposes.

An associated laboratory manual is also available. Although its organization correlates with the textbook, the lab manual is not tightly tied to the text and can be used independently to support any lab program with a modern integrated-circuit approach.

In preparing the second edition, great care was taken to mitigate the "glitches" that can occur because of the degree of complexity inherent in a digital book. (As technically oriented individuals we all know what happens when technical complexity increases.) We established special check procedures and reused originals from the prior edition, wherever pages remained essentially unchanged. Nevertheless, the reality is that "glitches" will appear, as they do in almost all technical books. If you run across any, let us know. Once an error has been validated, the SRA policy has always been to make a correction in the next printing. We welcome your feedback.

ACKNOWLEDGMENTS

I would like to thank the following people for their invaluable help on this second edition: My wife Patty, whose industry and talent kept wolves and creditors from the door and gave me a heady taste of creative freedom: Susan Radice for her expert manuscript typing; Richard and Annette Burchell for their many valuable ideas and contributions; Donna Maskell Stits for her excellent proofreading; Gerald Nolan for his help with the sections on C-MOS; Glen Graham for his help with Analog-to-Digital converters; Ann Wood of SRA for her diligent production editing; and William H. Hudelson, Jr. for his photographic work.

Because an author's greatest asset is a team of knowledgeable and thorough reviewers, I wish to thank the following users of the first edition for their constructive comments:

William C. Callahan, State Area Voc-Tech School, Pulaski, Tennessee
Bill Chow, Texas State Technical Institute, Waco
Pearley Cunningham, Community College of Allegheny County, South Campus
Jim Emerson, Wentworth Institute
Glenn F. Goff, Idaho State University
Steve Harrington, Heald Institute of Technology, Santa Clara Campus
Russ Heiserman, Oklahoma State University
James M. Huddleston, The Pennsylvania State University, York Campus
Robert Laursen, Parkland College
James Mumaw, Terra Technical College
Delbert R. Newman, Indiana State University
Louis Tornillo, Palomar College
Neal D. Voke, Triton College

Last, because his is a hard act to follow, my thanks to Alan W. Lowe of Science Research Associates—every inch the competent professional editor, with a patience that fans of Job could not help but admire.

Gerald E. Williams

CONTENTS

CHAPTER 1

THE LANGUAGE OF LOGIC

Learning Objectives. *Upon completing this chapter you should know:*

1. *How the binary number system works.*
2. *What a logic diagram is and why it is used.*
3. *What a truth table is, how it is constructed, and how it is used.*
4. *What a timing diagram is and its use.*
5. *What a Boolean equation is and how it is used.*
6. *What an AND gate is, its characteristics, and its truth table.*
7. *What an OR gate is, its characteristics, and its truth table.*
8. *What NAND and NOR gates are, their characteristics, and truth tables.*
9. *The properties of an inverter.*
10. *The basic Boolean laws.*
11. *DeMorgan's laws and the system of logic duals.*
12. *How bubble notation works.*
13. *In a limited way, how gates can be combined to perform a logical task.*

Digital circuits, no matter how complex, are composed of a small group of identical building blocks. These blocks are either basic gates or special circuits such as Schmitt triggers, special memory cells, and other structures for which gates are less suitable. The vast majority of the functional building blocks are gates or combinations of gates. A flip-flop, for example, can be considered as a functional block, but it too is composed of standard gates interconnected within the package.

A counter is a more complex functional unit than a flip-flop and is composed of flip-flops and control gates connected inside the package to perform a particular counting function. The flip-flops are made up of combinations of basic gates.

At a much higher level of organization, the microprocessor is the central processing unit of a computer in a package using thousands of gates, flip-flops, and memory cells. Even the memory cells are modified versions of basic gates.

In this chapter we will examine the functional (logical) properties of basic gates. In the next chapter we will examine the electronic properties of the most important gate circuits. In subsequent chapters we will see how these basic gates can be integrated into larger systems to perform all of the computing, counting, and control functions required by computers as well as by smaller-scale digital systems.

In most electronics systems the schematic diagram is our most valuable symbolic tool. We also use block diagrams to explain systems behavior. In logic systems the schematic diagram is often an unsatisfactory form of communication because of the large number of individual circuits involved and because modern integrated circuits are fairly complex. Schematic diagrams of digital systems would be difficult to follow and to draw simply because of the vast number of components involved. There are only a half dozen or so basic circuits for which we normally draw schematics. In this chapter we will be concerned with logic gates, but later we will encounter some building blocks that are not made up of basic gates alone.

BOOLEAN ALGEBRA

Boolean algebra is a two-value symbolic logic system. This symbolic tool allows us to design complex logic systems with the certainty that they will carry out their function exactly as intended. Conversely, we can use Boolean algebra to determine exactly what function an existing logic circuit was intended to perform. A very important aspect of the Boolean system is that it formalizes the difficult problem of getting the job done with an absolute minimum of hardware.

In applied Boolean, the symbols and equations of Boolean algebra are integrated with conventional electronic symbols. Digital waveforms called *timing diagrams* and digital block diagrams called *logic diagrams* are liberally mixed with their symbolic (Boolean) counterparts.

BINARY ARITHMETIC

Ordinary algebra is a generalization of our human system of arithmetic. Machine arithmetic is accomplished in a two-value (binary) number system, but Boolean algebra is not a generalization of the binary number system. The symbols 0, 1, +, and · are used in both systems but with totally different meanings in each system. Table 1-1 illustrates some important differences between them. There are more differences than similarities, and binary numbers and Boolean algebra must be treated as entirely different entities.

The use in one machine of two different systems having the same symbols but different meanings sounds confusing. In practice, both counting in binary and arithmetic in binary are performed by Boolean logic circuits. The actual circuits are strictly Boolean logic circuits, although the input to the circuit or the output from it may be binary. Binary numbers and arithmetic may be thought of as being *external* to

Table 1-1 Numerical Addition and Logical Addition Compared

A	+ B	= Sum	
0	0	0	
0	1	1	
1	0	1	
1	1	0	→ 1 (carry)

a. Numerical addition table
(The + symbol is pronounced plus.)

A	+ B	= Sum	
0	0	0	
0	1	1	
1	0	1	
1	1	1	→ (no carry exists)

b. Logical (Boolean) addition table
(The + symbol is pronounced or.)

the electronic logic circuitry. Because of the above-mentioned distinction, there is little real confusion generated.

Although we will not get into binary arithmetic until a later chapter, we need to understand right away how the basic number system works. We will use binary numbers to enumerate things, to count entries in Boolean truth tables, and so on.

1-1 The Binary Number System

The binary system contains only two symbols, 0 and 1, but any number that can be written in the decimal system can also be written in the binary system. The binary system is structurally similar to our decimal system.

In the decimal system we have the symbols for weight digits, 0, 1, 2, 3, 4, 5, 6, 7, 8, and 9, and the position of a digit in the number determines its value. We call the position values ones, tens, hundreds, thousands, and so on. Moving a given digit one position to the left increases its value by a factor of 10:

Hundreds	*Tens*	*Units*	*Position value*
5	5	5	Decimal number

In the units position the digit 5 represents the decimal number 5 (5 units), in the tens position it represents 50 (5 × 10), and in the hundreds position it represents 500 (5 × 100).

In the binary system each position to the left increases the value of the symbol by a factor of 2. Thus the position values in the binary system are ones, twos, fours, and so on:

Eights	*Fours*	*Twos*	*Ones*	*Position value*
1	1	1	1	Binary number

The digit 1 in the ones column represents 1, in the twos column it represents 2, in the fours column it stands for 4, and so on. Thus the number 1111 in the example would be equal to $(1 \times 8) + (1 \times 4) + (1 \times 2) + (1 \times 1)$, or decimal 15. The binary number 101 would be equal to $(1 \times 4) + (0 \times 2) + (1 \times 1)$, or decimal 5.

The principles involved in converting binary numbers into our usual decimal numbers are illustrated by the following examples:

Example 1

2^4	2^3	2^2	2^1	2^0	Exponent value
16	8	4	2	1	Column value
1	0	1	0	0	Binary number
(1×16) +	(0×8) +	(1×4) +	(0×2) +	(0×1)	Decimal value
16	+ 0	+ 4	+ 0	+ 0	= 20 (decimal)

Example 2

8	4	2	1	Column value
1	0	1	1	Binary number
8 +	0 +	2 +	1 =	11 (decimal)

Example 3

	16	8	4	2	1	Column heading	
(1)	0	0	1	1	0	= 4 + 2 + 0	= 6 (decimal)
(2)	0	1	0	0	1	= 8 + 0 + 0 + 1	= 9 (decimal)
(3)	1	0	0	0	1	= 16 + 0 + 0 + 0 + 1	= 17 (decimal)
(4)	1	1	0	1	0	= 16 + 8 + 0 + 2 + 0	= 26 (decimal)

HOW TO CONVERT DECIMAL NUMBERS INTO BINARY NUMBERS

The following example shows how to convert decimal numbers into their binary equivalents.

Example Write the decimal number 43 as its equivalent binary number.
Procedure

1. Set up the column values as shown.

6	5	4	3	2	1	0	Column number
64	32	16	8	4	2	1	Decimal value of each column
0	1	0	1	0	1	1	Binary number

2. Examine each column in turn starting with column 6 (the 64s column).

Column 6: 64. This is larger than 43, and we enter a 0 in column 6.
Column 5: 32. This is less than 43, and we enter a 1 in column 5.
Column 4: 16, and 16 + 32 = 48. This is larger than 43, and we enter a 0 in column 4.
Column 3: 8, and 32 + 8 = 40. This is less than 43, and we enter a 1 in column 3.
Column 2: 4, 40 + 4 is more than 43, and we enter a 0 in column 2.
Column 1: 2, and 40 + 2 = 42. This is less than 43, and we enter a 1 in column 1.
Column 0: 1, 42 + 1 = 43. We enter a 1 in column 0, and the job is done.

Thus decimal 43 = 101011 in binary.

Table 1-2 lists the decimal numbers 0 through 7 and their binary equivalents.

1-2 Logic Symbols, Functions, and Conventions

THE LOGIC DIAGRAM

Gates are the basic universal logic building blocks. Each gate has a special symbol that represents the circuitry in the gate. These logic symbols will be combined to form block diagrams called *logic diagrams*.

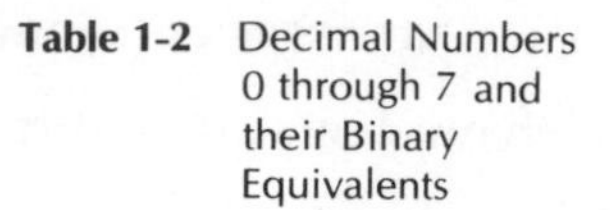

Table 1-2 Decimal Numbers 0 through 7 and their Binary Equivalents

	Binary		
Decimal	4	2	1
0	0	0	0
1	0	0	1
2	0	1	0
3	0	1	1
4	1	0	0
5	1	0	1
6	1	1	0
7	1	1	1

Although the logic diagram will be our primary method of symbolizing logic systems, there are also some supplementary forms of notation that serve to interpret the operation of logic circuits and systems.

While the logic diagram indicates how the various gates are interconnected, the supplementary forms tell us exactly what the particular logic circuit is intended to do and when.

THE TRUTH TABLE

One of the best statements of exactly what a particular circuit does (and does not do) is called a *truth table*. The truth table is a formal specifications sheet that describes the exact circuit behavior for every possible set of conditions. Logic circuits normally have a number of inputs and one or more outputs.

Each input signal can be only one of two possible values designated as 1 or 0 in a logic discussion and as *high* or *low* when electronic conditions are being emphasized. All input signals are usually independent of each other so that 2^n possible combinations of input 0's and 1's exist. The base of 2 is used because there are two possible input conditions, 0 or 1. The exponent (n) is the number of independent inputs to a given circuit. Each output can produce one of two values (0 or 1) and is a function of the various combinations of input signals. The truth table, then, shows all possible combinations (2^n) of input conditions and the resultant condition of each output for all the input conditions.

The truth table is a complete statement of logic circuit functions because no condition can exist that is not included in the truth table.

THE BOOLEAN EQUATION

Boolean algebra is a special logical algebra that provides the same information as the truth table in the form of equations. It has specific rules for manipulating equations to discover several alternate logic structures that will perform the desired function. In addition, Boolean notation, in conjunction with a special kind of truth table, provides a tool to insure that any given logical requirements are met with an absolute minimum number of gates.

THE TIMING DIAGRAM

The logic diagram shows how the logic gates are interconnected to form the specific logic system; truth tables and Boolean equations are used to describe what the gates and the system accomplish. The timing diagram tells us when each gate or each part of the system responds with respect to a standard timing signal called the *clock* signal.

These four systems—the logic diagram, the truth table, the Boolean equation, and the timing diagram—provide the methods of describing and explaining logic systems on paper and make up the symbolic *language of logic*.

ELECTRONIC GATES

Logic circuits are composed of combinations of high-speed electronic switches called *gates*. These gates are the electronic equivalent of simple conventional switches connected in series or parallel. Various systems combine groups of these series and parallel switches. We will use simple switches only for the purpose of explaining the basic logic functions and combinations.

In most digital circuits we cannot use mechanical switches (or relays) because they are much too slow. Electronic switches have been designed that can switch in much less than a microsecond (10^{-6} sec) and, in most cases, in a few nanoseconds (10^{-9} sec). To give you some idea of how fast this is, light travels at a speed of 186,000 miles per second or 982,080,000 feet per second. In 1 microsecond light will cover a distance of 982 feet, and in 1 nanosecond it will cover a little under 1 foot (0.982 ft). Many modern electronic switches are capable of switching in the time it takes light to travel the length of a man's stride.

Input voltages used to represent logic 0's and 1's in electronic gates can be any pair of distinctly different voltage levels. However, in practice they have been well standardized, eliminating much of the confusion that existed at one time. There are two logic conventions in use: *positive logic* and *negative logic*. In positive logic a logic 1 is represented by the most positive of the two levels, and a logic zero is represented by the less positive level. In current transistor-transistor-logic (TTL) technology positive logic universally uses +5 volts for a logic 1 and zero volts for logic 0. The less common emitter-coupled logic family uses positive logic but a −0.8 volt = logic 1 and −1.9 volts = logic 0. The level −0.8 volt is more positive than −1.9 volts.

In the *P*-channel field-effect family of logic circuits the output voltages are negative for both levels, as they are in emitter-coupled logic. The absolute voltage levels are somewhat more variable and will be examined later. The logic is positive.

In the *N*-channel field-effect family the levels are +5 volts = logic 1 and 0 volts = logic 0. These are the same levels used in the TTL family.

Since nearly all modern logic circuits use positive logic, negative logic devices are becoming increasingly rare. In negative logic, the most negative of the two voltage levels represents a logic 1 and the less negative voltage represents a logic 0. These two definitions are fairly recent and other definitions may sometimes be found in older texts.

GATE-CIRCUIT CONVENTIONS

a. A 1 on the input of an electronic gate can also be designated *High, Hi,* or *H*.
b. A 0 on the input of an electronic gate can also be designated *Low, Lo,* or *L*.
c. A 1 on the output of an electronic gate can also be designated by *High, Hi, or H,* and a 0 can be represented by *Low, Lo,* or *L*.
d. Inputs will be labeled *A, B, C,* and so forth, and the output of a gate will be called *f* (for *function*).

SWITCHING CIRCUIT CONVENTIONS

a. An open switch is designated 0. If some switch A is open, $A = 0$.
b. A closed switch is designated 1. If some switch A is closed, $A = 1$.
c. If lamp f is lit, we designate that lit condition as a 1.
d. If lamp f is dark, we designate it by: $f = 0$.

BOOLEAN ALGEBRA CONVENTIONS

There are only three basic logical functions in digital circuits and Boolean algebra: AND, OR, and NOT. Variables in Boolean algebra correspond to the input conditions on a gate (or gates). They are designated A, B, C, and so on and can have only a value of either 0 (low) or 1 (high). Boolean expressions are normally written as equations with an f used (generally by itself) on the left side of the equal sign. The f corresponds to the output function of a gate (or system of gates) and can have either one of two values, 0 or 1. (See Figure 1-1b and c.)

Equations consist of variables joined by operators: (·) the Boolean *product* symbol and (+) the Boolean *sum* symbol, along with an equal sign and the f (function). For example, $f = (A \cdot B) + (A \cdot C)$ is read as f equals (A *and* B) *or* (A *and* C).

It is important to understand that the (·) and (+) symbols do not describe the same operation as ordinary arithmetic multiplication and addition.

The (·) symbol is called AND. The equation $f = A \cdot B$ is read as f equals A AND B. The (+) symbol is called the OR symbol. The equation $f = A + B$ is read as f equals A OR B.

THE COMPLEMENT

The complement or NOT function is the third Boolean operator. A bar over a variable, constant, or operator (· or +) indicates a NOT or complemented variable, constant, or operator.

$$\text{If } A = 1, \bar{A} \text{ (not } A) = 0.$$
$$\text{If } B = 0, \bar{B} \text{ (not } B) = 1.$$

The complement bar may be read in two ways:

$\bar{A}$ is read as NOT A, or as A NOT

SUMMARY

1. The AND function (symbolizing the AND gate):

$$f = AB \text{ or}$$
$$f = A \cdot B$$
$$\text{Read as: } f = A \text{ AND } B$$

2. The OR function (symbolizing the OR gate):

$$f = A + B$$
$$\text{Read as: } f = A \text{ OR } B$$

3. The *complement* or NOT function (symbolizing the *inverter*) is indicated by a bar over a variable (or larger segments of an equation). For example, $\bar{A}$ would be read as NOT A, or A NOT.

There are no squares, square roots, or any other roots or powers in Boolean algebra nor are there any fractions or division operations.

There are only two possible values for constants and two possible values for variables, 0 or 1.

CONSTRUCTING TRUTH TABLES

Columns

The truth table will have the following columns: one for each variable (A, B, C, D, and so on), one or more f columns, and one m column. The m column simply numbers the rows, always beginning with 0.

Rows

The rows contain every possible combination of 0's and 1's, with each row containing one combination. The number of rows on a given truth table depends on the number of variables involved in the corresponding Boolean equation. The number of rows $= 2^n$, where n is the number of variables.

Example In the equation $f = (A \cdot B) + (A \cdot C)$ there are three variables: A, B, and C. The number of rows is as follows: $\mathbf{n} = 2^n$ where $\mathbf{n} = 3$, and $\mathbf{n} = 2^3 = 8$ rows. These would be numbered 0 through 7.
We will use these conventions throughout the text.

Table 1-3 Constructing a Truth Table

Decimal	4	2	1 ←Binary column headings	
m	A	B	C	f
0	0	0	0	
1	0	0	1	
2	0	1	0	
3	0	1	1	
4	1	0	0	
5	1	0	1	
6	1	1	0	
7	1	1	1	
	Binary numbers (A, B, C)			↑ Output column

Procedure (See Table 1-3)

1. Enter the (2^n) decimal numbers in the m column in Table 1-3. Start numbering with zero.
2. For each row, write the binary equivalent of the decimal number in the m column.
3. The f column will be used to specify the behavior of the particular logic circuit under consideration.

This procedure satisfies two important requirements: (1) it ensures against duplication and omissions of possible combinations of zeros and ones in the table; and (2) it provides a standard form for all truth tables. This makes it possible to compare all entries in a pair of truth tables by examining only the f column.

1-3 The AND Gate

FUNCTIONAL DEFINITION

The AND gate will produce a 1 (high) output *if and only if* there is a logical 1 (high) on *all* inputs at the same time.

The Boolean representation of the AND gate is $f = A \cdot B$ and is read $f = A$ AND B.

The AND gate is the electronic equivalent of series-connected switches. Figure 1-1 shows the AND switching circuit, the AND gate symbol, two forms of truth table, and the equation for a two-input and a three-input AND function. The AND gate can have any number of inputs.

Example To start a car with an automatic transmission, you must put the shift lever out of gear (*neutral* or *park*) AND turn the key to start the car. The conditions are: key turned to "start" AND "shift lever out of gear" equals start. Both conditions must exist at the same time if the car is to start. The equation for this example is as follows: start = neutral · key on.

TIMING DIAGRAMS

Digital circuits are normally driven by pulses derived from a master oscillator called a *clock*. Diagrams of these pulses in different parts of the system are called *timing diagrams*. Timing diagrams are often a convenient way to illustrate graphically the behavior of a digital circuit. The timing diagram for a two-input AND gate is illustrated in Figure 1-2.

The input and output conditions are shown on the timing diagram in the same order as they are listed on the truth table.

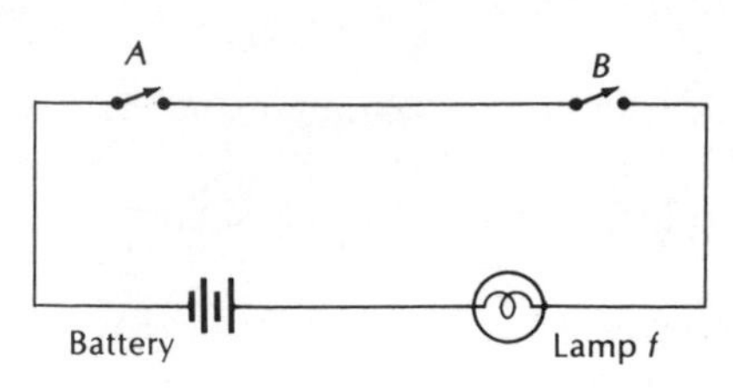

a. The two-variable AND switching circuit

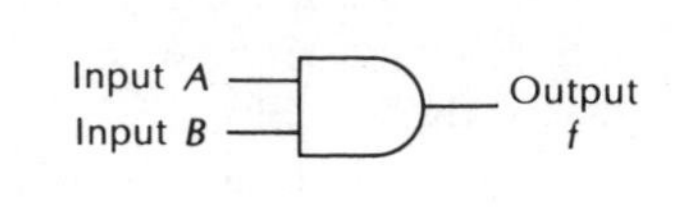

b. The AND gate symbol

c. Equation: $f = (A \cdot B)$

m	Switch A	Switch B	f (lamp)
0	open	open	not lit
1	open	closed	not lit
2	closed	open	not lit
3	closed	closed	lit

d. The two-variable switching-circuit truth table

m	A	B	f
0	0	0	0
1	0	1	0
2	1	0	0
3	1	1	1

$f = (A \cdot B)$

e. The standard truth table

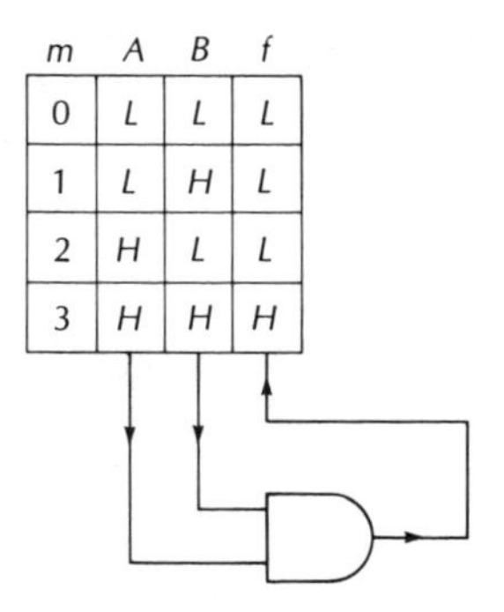

m	A	B	f
0	L	L	L
1	L	H	L
2	H	L	L
3	H	H	H

f. Truth table illustrating
gate input-output conditions

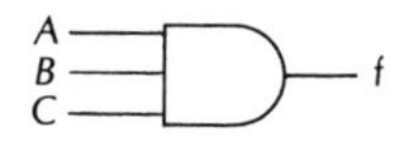

g. The three-input AND gate

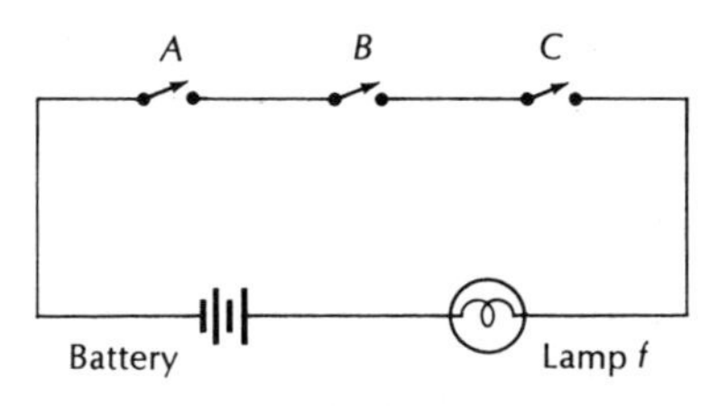

h. A three-switch AND circuit

Figure 1-1 AND Circuit, Gate, and Truth Tables

m	A	B	C	f
0	0	0	0	0
1	0	0	1	0
2	0	1	0	0
3	0	1	1	0
4	1	0	0	0
5	1	0	1	0
6	1	1	0	0
7	1	1	1	1

Equation:
$f = A \cdot B \cdot C$

i. The three-input AND gate truth table

m	A	B	C	f
0	L	L	L	L
1	L	L	H	L
2	L	H	L	L
3	L	H	H	L
4	H	L	L	L
5	H	L	H	L
6	H	H	L	L
7	H	H	H	H

j. The three-input AND gate truth table in terms of High and Low input levels

Figure 1-1 continued

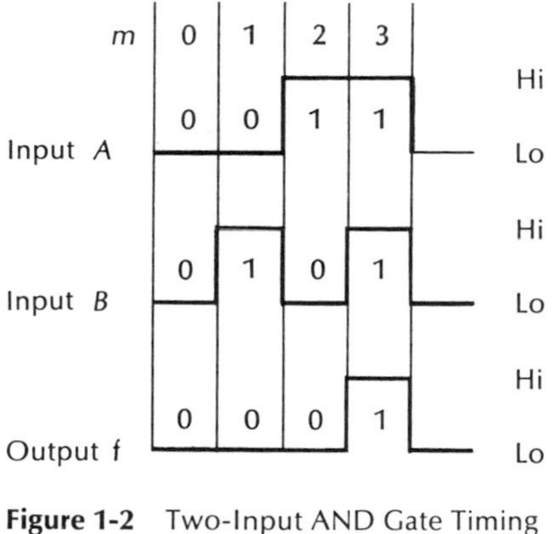

Figure 1-2 Two-Input AND Gate Timing Diagram

1-4 The OR Gate

The OR* gate is the electronic equivalent of switches connected in parallel. Figure 1-3 shows the OR switching circuit, the OR gate symbols, the truth tables, and the equations. The truth table indicates that a logic 1 (high) on one or more inputs at the same time will produce a 1 (high) output. The OR gate will output a 0 *only* when *all* inputs are 0 (low) at the same time. The OR gate can have any number of inputs.

* Sometimes called an *inclusive OR*.

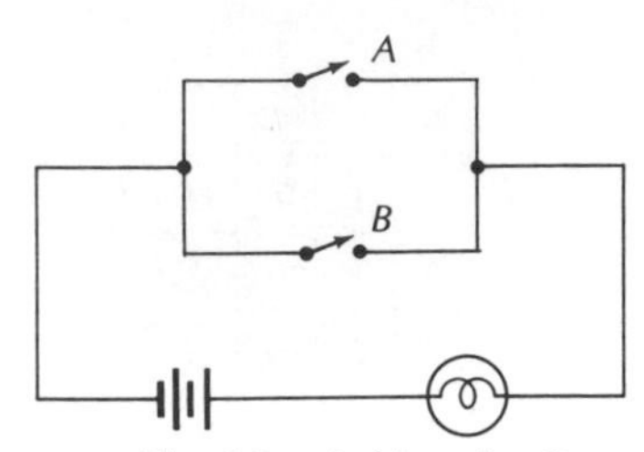

a. The OR switching circuit

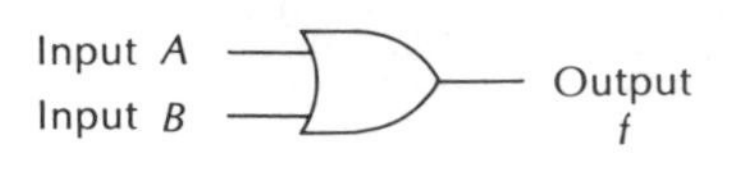

b. The two-input OR gate symbol

m	A	B	f
0	0	0	0
1	0	1	1
2	1	0	1
3	1	1	1

c. Standard truth table

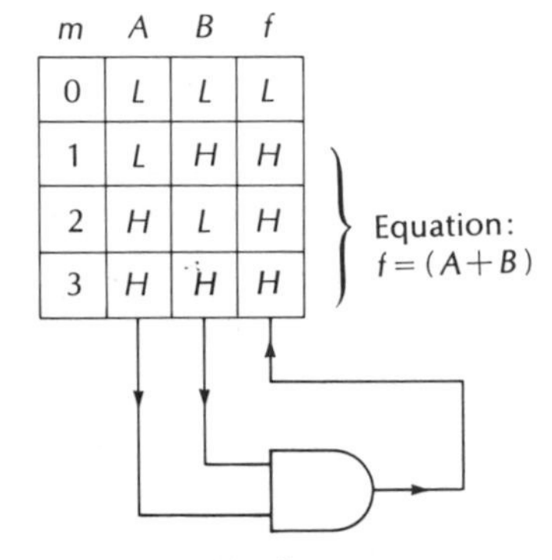

m	A	B	f
0	L	L	L
1	L	H	H
2	H	L	H
3	H	H	H

d. Truth table illustrating gate input-output conditions

Figure 1-3 The OR Function

Figure 1-4 shows a four-input OR switching circuit, gate symbol, truth table, and equation. Figure 1-5 is the timing diagram for a two-input OR gate. The extension wings in Figure 1-4c are sometimes used to prevent the input lines from appearing too crowded on the drawing.

The AND gate symbol may also be extended in a similar fashion as shown in Figure 1-6. These extension lines can also have a special symbolic meaning as an indication that the gate shown is actually constructed of more than one gate. When more inputs are required than are available in a readily obtainable gate package, several gates may be connected to function as a single basic gate. Only experience can tell you whether this meaning is intended in a given situation.

Example You and an associate have a joint checking account. Either of you can write checks without consulting the other. The bank will honor your signature *or* your associate's. If both signatures appear on the check, the bank will still honor the check.

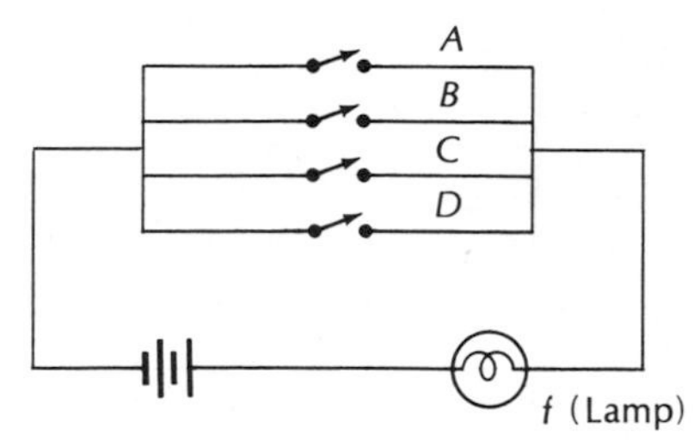

a. The switching equivalent

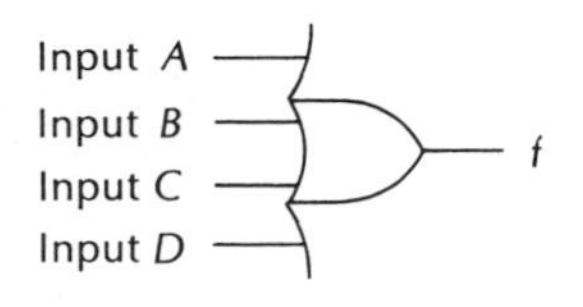

c. Four-input OR gate; alternate symbol

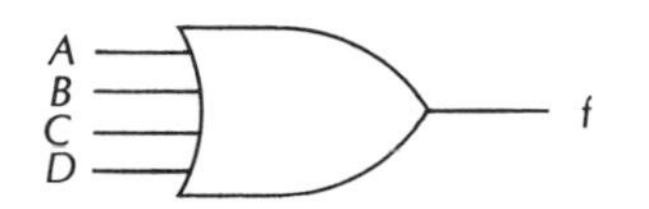

b. The four-input OR gate symbol

Equation: $f = A + B + C + D$

	A	B	C	D	f
m_0	0	0	0	0	0
m_1	0	0	0	1	1
m_2	0	0	1	0	1
m_3	0	0	1	1	1
m_4	0	1	0	0	1
m_5	0	1	0	1	1
m_6	0	1	1	0	1
m_7	0	1	1	1	1
m_8	1	0	0	0	1
m_9	1	0	0	1	1
m_{10}	1	0	1	0	1
m_{11}	1	0	1	1	1
m_{12}	1	1	0	0	1
m_{13}	1	1	0	1	1
m_{14}	1	1	1	0	1
m_{15}	1	1	1	1	1

d. Truth table

Figure 1-4 The Four-Input OR Gate

1-5 The Inverter

The inverter, the simplest of the three gates, has one input and one output. The output is always the complement of the input. The complement of a 1 is 0; the complement of 0 is 1. A 1 (high) at the input of an

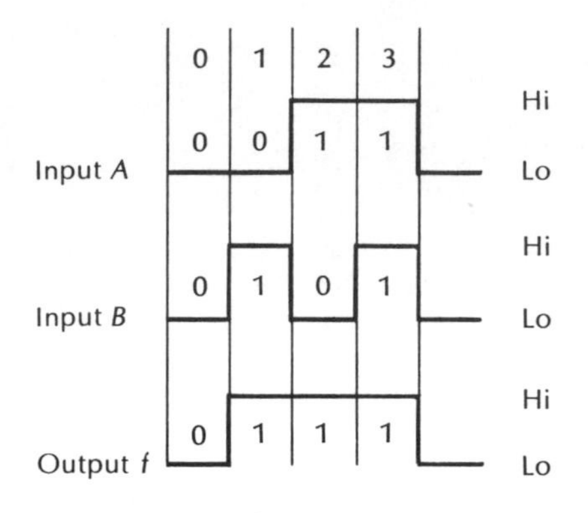

Figure 1-5 Two-Input OR Gate Timing Diagram

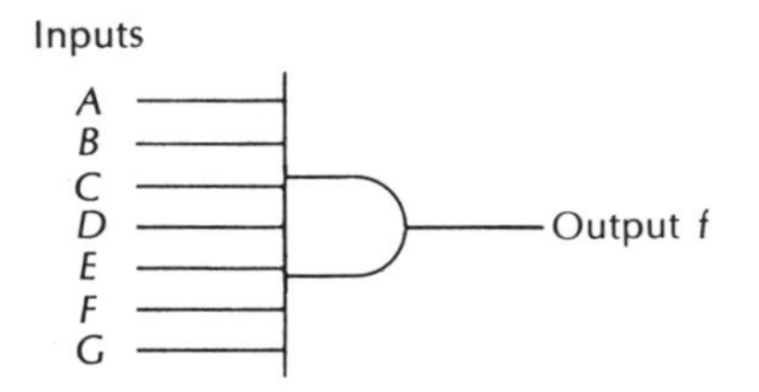

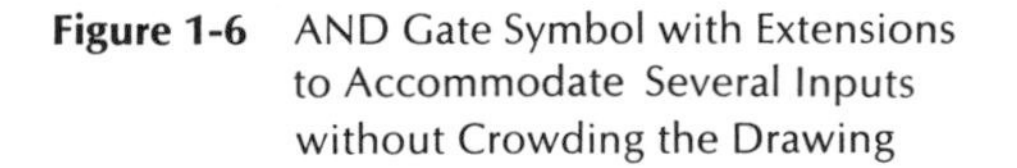

Figure 1-6 AND Gate Symbol with Extensions to Accommodate Several Inputs without Crowding the Drawing

inverter produces a 0 (low) at the inverter output. A 0 (low) input produces a 1 (high) output.

Notice that Figure 1-7 shows two different symbols, one with a circle on the input and the other with the circle on the output. The circle, often called a *bubble,* is always a part of the inverter symbol, and identifies the symbol as an inverting amplifier.

In the case of multiple inverters in series: (1) An *even* number of inverters yields the same results as no inverters at all, and (2) an *odd* number of inverters is equivalent to a single inverter. (See Fig. 1-8.) Figure 1-9 shows the inverter timing diagram.

1-6 Basic Boolean Operations

Because Boolean equations are such a common and useful form for describing digital circuits, it is important to be familiar with the basic

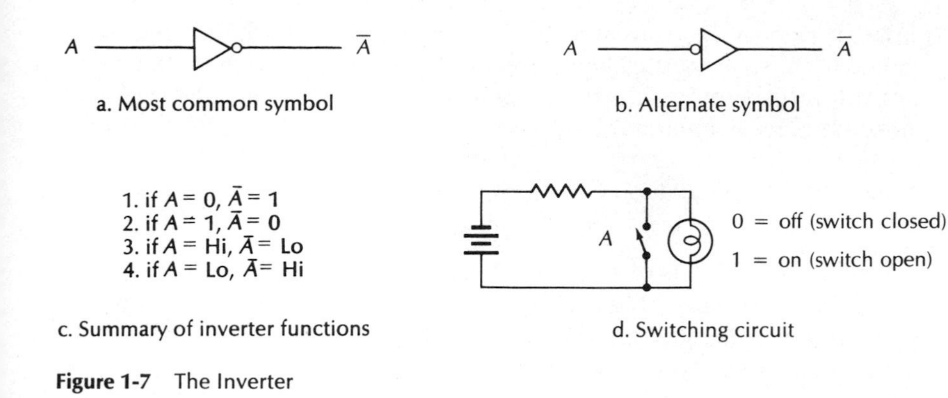

Figure 1-7 The Inverter

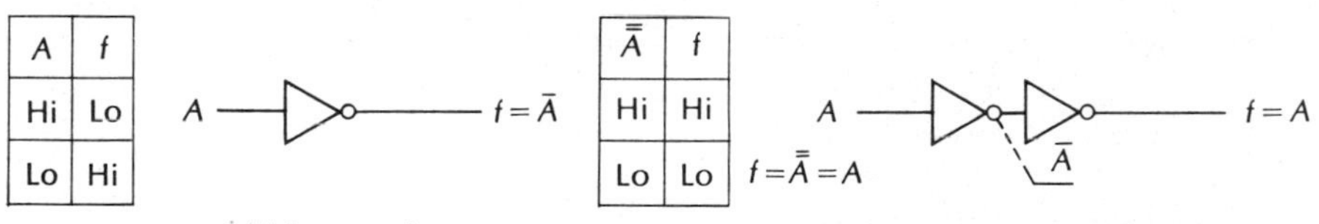

b. Using two inverters

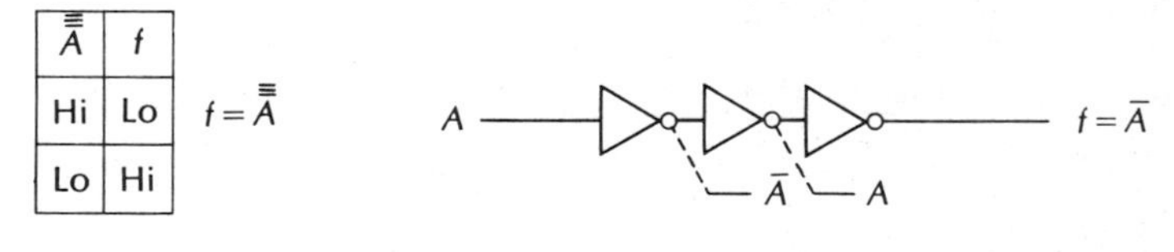

c. Using three inverters

Figure 1-8 Inverter Truth Tables and Diagrams

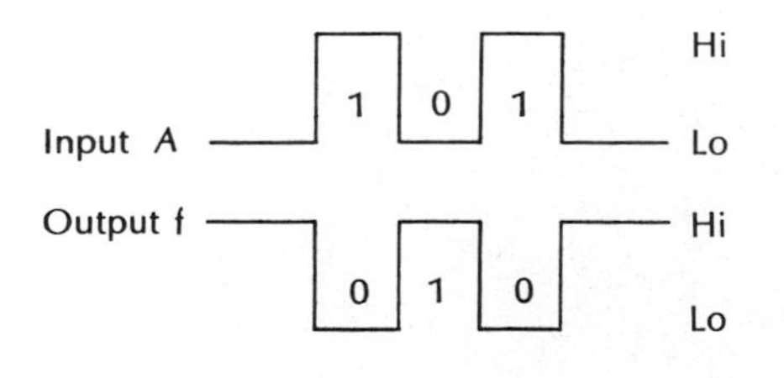

Figure 1-9 Inverter Timing Diagram

laws of Boolean algebra and how they are applied to digital circuits. Some of these properties have already been briefly examined, but now we will formalize the already familiar ones and examine some that we have not yet encountered in this text.

1. The AND operation equation:

$$f = A \cdot B$$

   This operation is often called the *Boolean product*. The AND function is unique to Boolean algebra and should not be confused with binary arithmetic multiplication. Table 1-4 shows the truth tables for Boolean product combinations.

2. The OR operation equation:

$$f = A + B$$

   The OR operation is referred to as a *Boolean sum*. Again, this is a unique Boolean function not to be confused with arithmetic addition. Figure 1-11 shows the electronic implementation of the OR

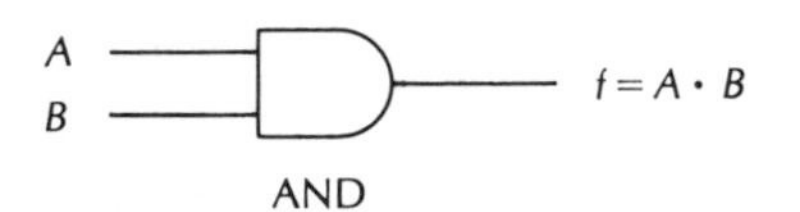

Figure 1-10 Electronic Implementation of *A* AND *B*

Table 1-4 Tables of Boolean Product Combinations

	A	*B*	*f*
0	0	0	0
1	0	1	0
2	1	0	0
3	1	1	1

a. Operations *A* AND *B*

	AND		*f*
0	*A*	*A*	*A*
1	1	*A*	*A*
2	*A*	1	*A*
3	0	*A*	0
4	*A*	0	0

b. Mixed variables and constants

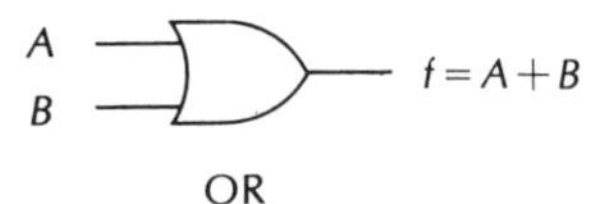

Figure 1-11 Electronic Implementation of the OR Function

functions and Table 1-5 provides truth tables of Boolean sum combinations.

3. The NOT operation and the laws of complementation:

   First law of complementation:

$$\text{If } A = 0, \overline{A} = 1$$
$$\text{If } A = 1, \overline{A} = 0$$

   Second law of complementation:

$$A \cdot \overline{A} = 0$$

   Third law of complementation:

$$A + \overline{A} = 1$$

   Law of double complementation:

$$\overline{\overline{A}} = A$$

   Figure 1-12 illustrates the electronic interpretation of the laws of complementation.

1-7 The Commutative Laws

There are two commutative laws in Boolean algebra, one for the logical AND function and one for the logical OR function. Both commutative

Table 1-5 Tables of Boolean Sum Combinations

	A	B	f
0	0	0	0
1	0	1	1
2	1	0	1
3	1	1	1

a. Operations *A* OR *B*

	OR		f
0	A	A	A
1	A	1	1
2	A	0	A

b. Mixed variables and constants

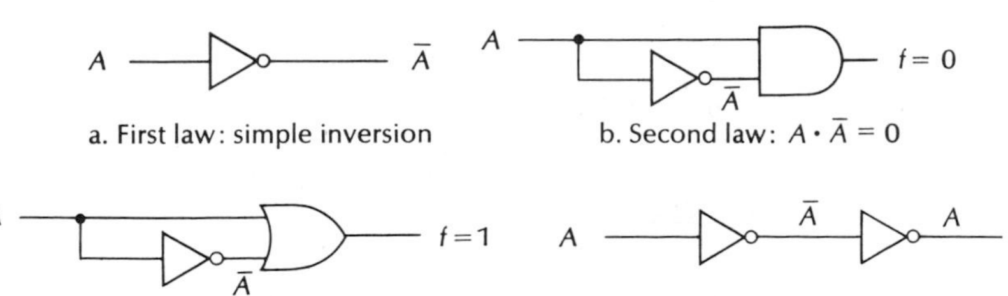

Figure 1-12 Electronic Interpretation of the Laws of Complementation

laws state that order is not important, that A may be ORed or ANDed to B, or B ORed or ANDed to A with the same result.

Commutative law for the OR function:

$$A + B = B + A$$

Commutative law for the AND function:

$$A \cdot B = B \cdot A$$

1-8 The Associative Laws

The associative laws apply when three or more variables (or constants) are to be combined by the AND or OR function. The associative laws state that the variables may be combined in any order without changing the outcome.

Associative law for the OR function:

$$A + (B + C) = C + (A + B)$$

Associative law for the AND function:

$$A \cdot (B \cdot C) = C \cdot (A \cdot B)$$

1-9 The Distributive Laws

In addition to specifying rules for grouping and multiplying out, the distributive laws also lead to a kind of factoring that will prove to be of value later on in this course.

First distributive law:

$$A \cdot (B + C) = (A \cdot B) + (A \cdot C)$$

Second distributive law:

$$A + (B \cdot C) = (A + B) \cdot (A + C)$$

1-10 The Laws of Tautology

First law of tautology:

$$A \cdot A = A$$

because: If $A = 1$, 1 AND 1 = 1, and
If $A = 0$, 0 AND 0 = 0

Second law of tautology:

$$A + A = A$$

because: If $A = 1$, 1 OR 1 = 1, and
If $A = 0$, 0 OR 0 = 0

When constants are involved:

$$A \cdot 1 = A$$
$$A \cdot 0 = 0$$
$$A + 1 = 1$$
$$A + 0 = A$$

1-11 The Laws of Absorption

First law of absorption:

$$A \cdot (A + B) = A$$

Second law of absorption:

$$A + (A \cdot B) = A$$

In both cases in Figure 1-13, a 1 on A will produce a 1 at f. B can be either 1 or 0 without affecting the output at f. If there is a 0 on A, the output will be 0 for either $B = 0$ or $B = 1$. The B has no influence over the output result and can be dropped from the system.

1-12 DeMorgan's Laws

DeMorgan's laws describe the dual nature of Boolean algebra. When DeMorgan's laws are applied to the implementation of Boolean functions using electronic gates, they can be interpreted to mean: *Any Boolean function can be accomplished using either AND gates and inverters or OR gates and inverters.*

Boolean expressions are often written initially in a form that implies the use of all three gate forms (AND, OR, Invert) to implement the function. DeMorgan's laws provide the basis for the use of most modern electronic gate families. These commercial gate families are based on, for example, an AND gate and a built-in inverter. In this example the OR gate would exist in the family but would probably be more expensive and not as frequently used.

DeMorgan's laws imply that the OR gate would not actually be necessary and that any circuit requirements could be met using only the AND-inverter gate package.

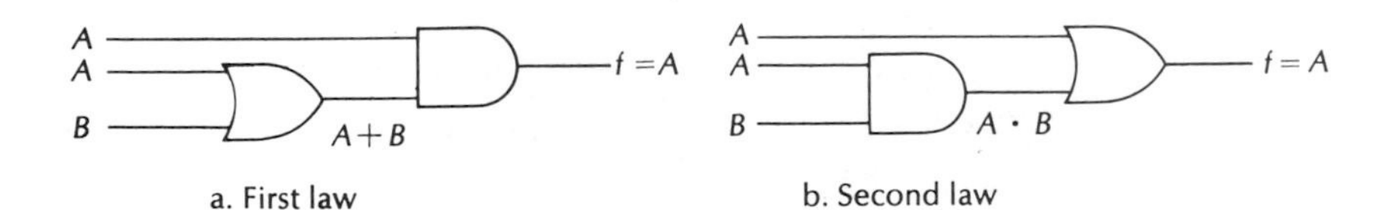

Figure 1-13 Electronic Interpretation of the Laws of Absorption

DEMORGAN'S LAW: CASE 1

$$\overline{A \cdot B} = \overline{A} + \overline{B}$$

Figure 1-14 shows the gate circuits corresponding to the expression on each side of equation 1. The truth tables shown are identical. Because every possible condition is included on the truth table, identical truth tables provide *proof* that the two circuits are functionally identical.

The truth table entries in Figure 1-14a show all of the possible conditions of 0's and 1's for inputs A and B. Under the heading "$A \cdot B$, output," the entries define the functional operation of any AND gate.

The bar over $A \cdot B$ ($\overline{A \cdot B}$) means that the entire expression is complemented by "passing" it through an inverter. The complete logic function $\overline{A \cdot B}$ requires the use of an AND gate and an inverter as shown in Figure 1-14a.

The rightmost column in the truth table, the result of the complete function, is the defined function. The column adjacent (to the left) of the $\overline{A \cdot B}$ column is shown here for the purpose of explanation only. It would not normally be included because it merely represents one of the steps in obtaining the desired function $f = \overline{A \cdot B}$.

Figure 1-14b shows the logic diagram for $f = \overline{A} + \overline{B}$. The truth table shows the headings of the two leftmost columns as $\overline{A}$ and $\overline{B}$. Each 1 appearing in the truth table in part a of this figure, columns A and B, shows up as a zero in the table in part b. Each 0 in the truth table in part a (columns A and B) shows up in part b as a 1. The f column defines the functional behavior of any OR gate for all possible combinations of 0's and 1's for variables A and B.

These two configurations are called *logic duals* because, although they are structurally different, they perform identical functions.

DEMORGAN'S LAW: CASE 2

$$f = \overline{A + B} = \overline{A} \cdot \overline{B}$$

Figure 1-14, Case 2a shows the truth table and logic diagram for this pair of duals. In both cases of DeMorgan's laws the following applies:

Inverted Output Positive Logic	*Inverted Input Logic Dual*
$f = \overline{A \cdot B}$	$f = \overline{A} + \overline{B}$
$f = \overline{A + B}$	$f = \overline{A} \cdot \overline{B}$

Case 1

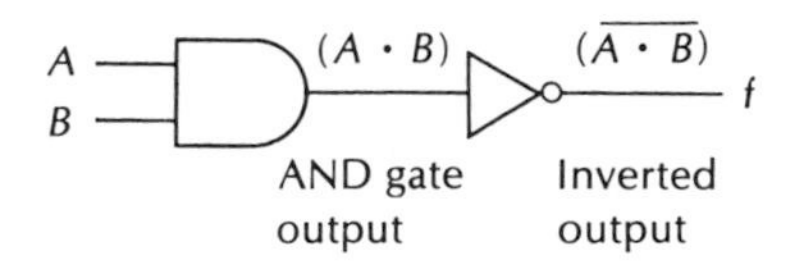

a. Equation: $f = \overline{A \cdot B}$
(Read as f equals A AND B, NOT)

m	A	B	$A \cdot B$ output	f inverted to $\overline{A \cdot B}$
0	0	0	0	1
1	0	1	0	1
2	1	0	0	1
3	1	1	1	0

Truth table for $f = \overline{A \cdot B}$

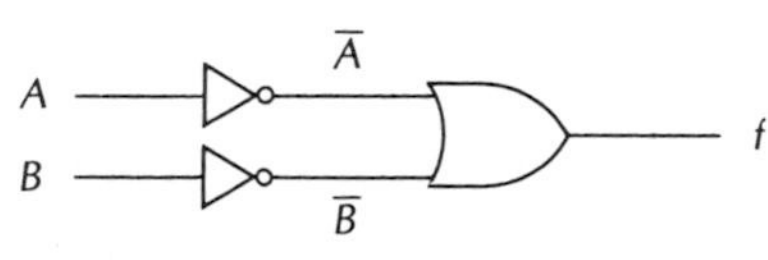

Logic diagram

Equation: $f = \overline{A} + \overline{B}$
(Read as f equals not A AND not B)

b. The logic dual of $\overline{A \cdot B}$

m	$\overline{A}$	$+\overline{B}$	f
0	1	1	1
1	1	0	1
2	0	1	1
3	0	0	0

Truth table for $f = \overline{A} + \overline{B}$

Case 2

$A+B$

A

B

$f = \overline{A+B}$

Logic diagram

a. Equation: $f = \overline{A+B}$

m	A	B	$A+B$	$f = \overline{A+B}$
0	0	0	0	1
1	0	1	1	0
2	1	0	1	0
3	1	1	1	0

Truth table

$\overline{A}$

A

B

$\overline{B}$

$f = \overline{A} \cdot \overline{B}$

Logic diagram

b. Equation: $f = \overline{A} \cdot \overline{B}$

m	$\overline{A}$	$\overline{B}$	f
0	1	1	1
1	1	0	0
2	0	1	0
3	0	0	0

Truth table (inverted)

Figure 1-14 De Morgan's Laws: Case 1 and Case 2

SUMMARY OF FUNDAMENTAL LAWS

1. Laws of tautology
   (1) $A \cdot A = A$
   (2) $A + A = A$
   Constants
   (1) $A \cdot 1 = A$
   (2) $A \cdot 0 = 0$
   (3) $A + 1 = 1$
   (4) $A + 0 = A$
2. Laws of complementation
   (1) $A \cdot \overline{A} = 0$
   (2) $A + \overline{A} = 1$
   (3) $\overline{\overline{A}} = A$ (double complement)
3. DeMorgan's laws
   (1) $\overline{A \cdot B} = \overline{A} + \overline{B}$
   (2) $\overline{A + B} = \overline{A} \cdot \overline{B}$
4. Commutative laws
   (1) $A \cdot B = B \cdot A$
   (2) $A + B = B + A$
5. Distributive laws
   (1) $A \cdot (B + C) = (A \cdot B) + (A \cdot C)$
   (2) $A + (B \cdot C) = (A + B) \cdot (A + C)$
6. Associative laws
   (1) $A \cdot (B \cdot C) = C \cdot (A \cdot B)$
   (2) $A + (B + C) = C + (A + B)$
7. Laws of absorption
   (1) $A \cdot (A + B) = A$
   (2) $A + (A \cdot B) = A$

1-13 The NAND Gate

Two of the most popular modern gate structures are the NAND (not AND) and the NOR (not OR) gates. Any digital circuit can be constructed using only NAND gates or only NOR gates.

The NAND gate is an AND gate with a built-in inverter in the output line. Figure 1-15 shows the NAND gate symbol, its AND-inverter equivalent circuit, and how the law of the double complement can be used to convert a NAND gate into an AND gate. The bubble on the output represents the built-in inverter and is part of the NAND sym-

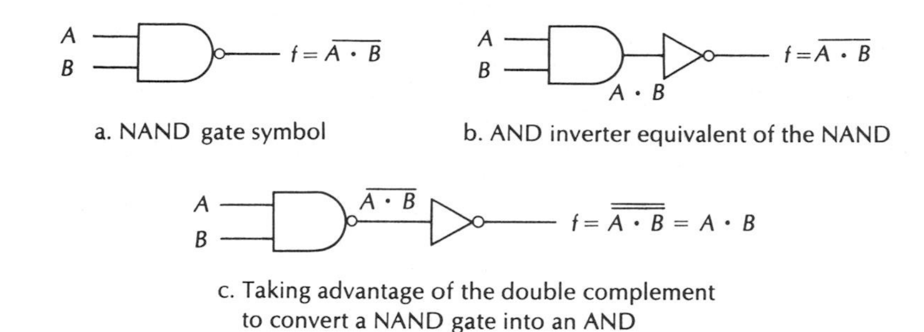

Figure 1-15 The NAND Gate

bol. The AND function is not directly accessible in the NAND gate package.

In Figure 1-15b, notice that the equivalent circuit (and equation) is identical to Figure 1-14, Case 1a.

1-14 The NOR Gate

The NOR gate is an OR gate with a built-in inverter in the output line. The NOR gate is a NOT-OR gate. The OR function is not available from a NOR gate. Figure 1-16 shows the NOR gate symbol, its OR-inverter equivalent circuit, and how the law of the double complement can be used to obtain the OR function from a NOR gate. The bubble on the output is an integral part of the NOR gate symbol.

Notice that the NOR equivalent circuit (Fig. 1-16b) is identical to the Case 2a logic diagram and equation in Figure 1-14. (Note: Properly connected, either the NAND gate or the NOR gate can serve as an inverter. Most logic families that feature NAND or NOR gates as the

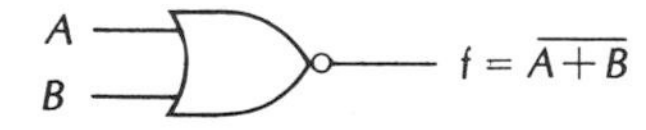

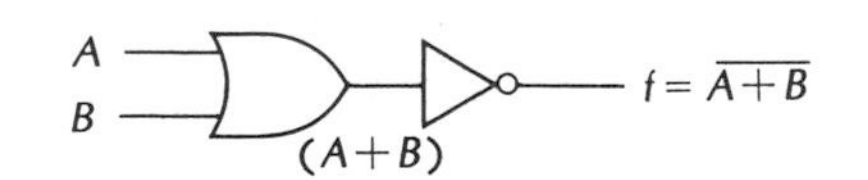

a. NOR gate symbol

b. The OR — inverter equivalent of the NOR

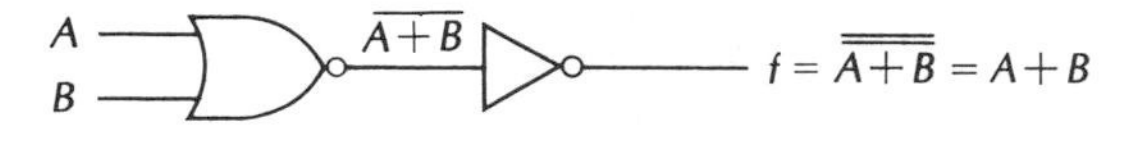

c. Taking advantage of the double complement law to convert a NOR gate into an OR

Figure 1-16 The NOR Gate

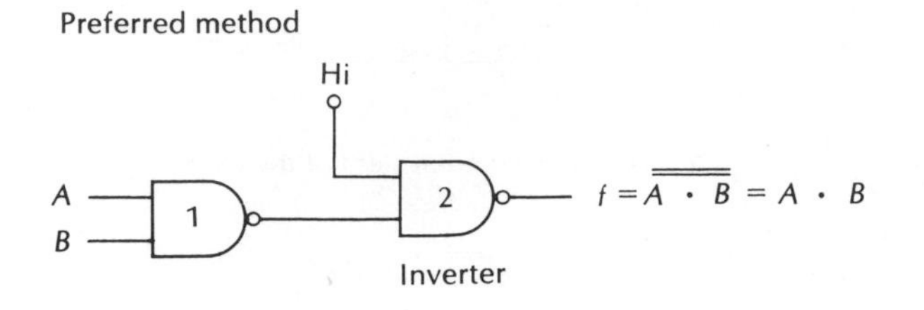

Note: The High (Hi) on the upper leg of gate 2 is normally a fixed voltage power supply line, in most cases the same line that supplies power to the gates.

Alternate method

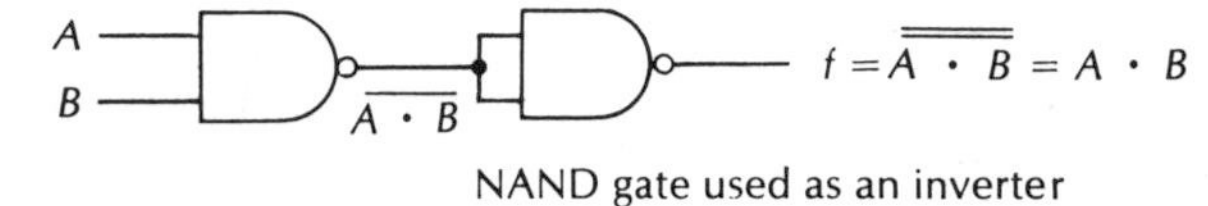

NAND gate used as an inverter

a. $(A \cdot B)$ synthesized with NAND gates

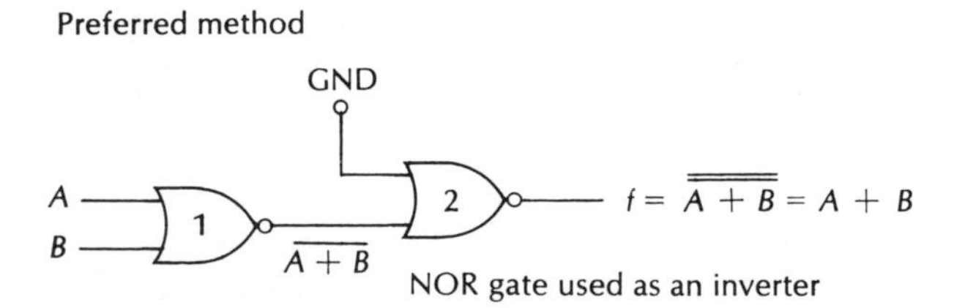

Note: The GND (Lo) on the upper leg of gate 2 is normally a fixed voltage power supply line, in most cases the same line that supplies power to the gates.

Alternate method

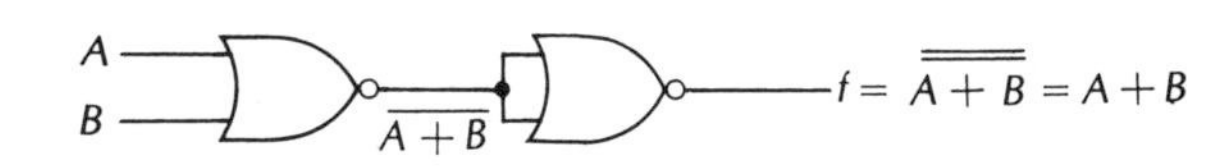

NOR gate used as an inverter

b. $(A+B)$ synthesized with NOR gates

Figure 1-17 Synthesized Gates

primary gate form also provide inverters for use when it is more convenient to use them.)

1-15 Using NAND and NOR Gates as Inverters

The drawings in Figure 1-17 indicate methods for using NAND and NOR gates as inverters. It is important that unused inputs on any gate be tied to some low-impedance source—generally to the positive side of the power supply or ground. An open input circuit *sees* a nearly infinite driving impedance and is an invitation for noise or unwanted pulses to sneak in. A circuit may function in a lab with an open OR gate input, for example, but may become erratic in the field where electrical noise levels are higher.

Figure 1-18 summarizes NAND and NOR gates and their various equivalent forms.

1-16 Bubble Notation

The use of bubble notation makes logic diagrams less cluttered, easier to read, and easier to draw.

SOME RULES FOR BUBBLE NOTATION

1. The bubble on the output of a gate is a part of that particular symbol and the indicated inverter is built into the gate.
2. The input bubbles do *not* indicate whether the inverters are internal to the gate or connected externally. In general, for basic gates the inverters are connected externally and are *not* a part of the gate circuit. The best interpretation of an input bubble is to consider that input as a *low* active input. It takes a zero (low) instead of a one (high) on a bubbled input leg to activate that input. Some logic circuits, particularly flip-flops, do have built-in low active inputs.

1-17 DeMorgan's Law and Logic Gate Equivalents

The equivalent circuits in Figure 1-19 are based on the two cases of DeMorgan's law:

$$\text{Case 1: } \overline{A \cdot B} = \bar{A} + \bar{B}$$
$$\text{Case 2: } \overline{A + B} = \bar{A} \cdot \bar{B}$$

Figure 1-19 shows the equivalent logic gates (logic duals) based on the two cases of DeMorgan's law. The symbols differ from Figure 1-13 in that inverters are symbolized by circles on the inputs and outputs of

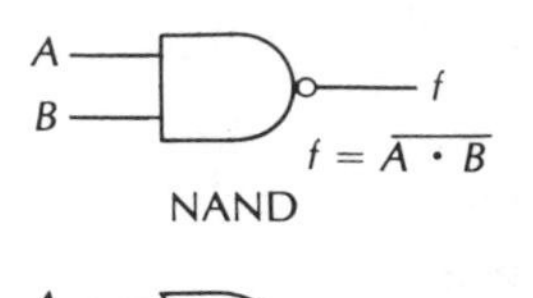

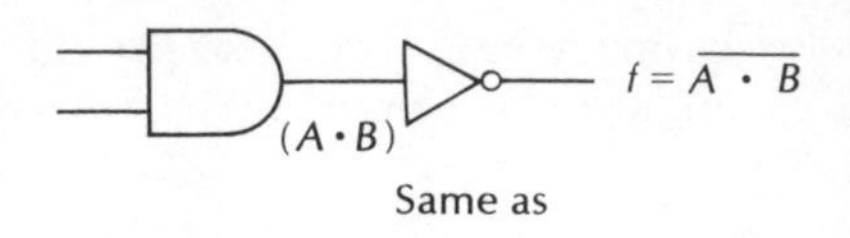

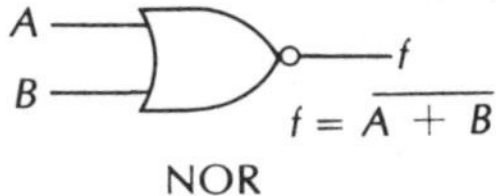

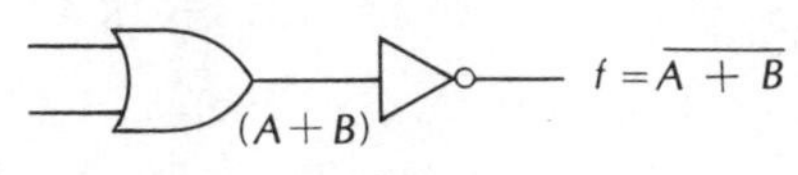

PART 1
Defined

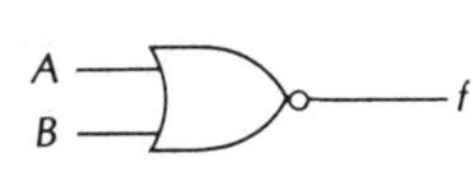

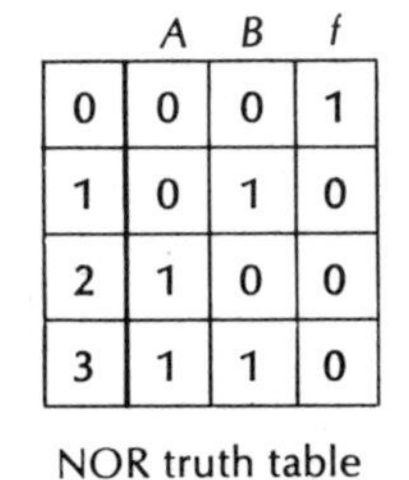

	A	B	f
0	0	0	1
1	0	1	0
2	1	0	0
3	1	1	0

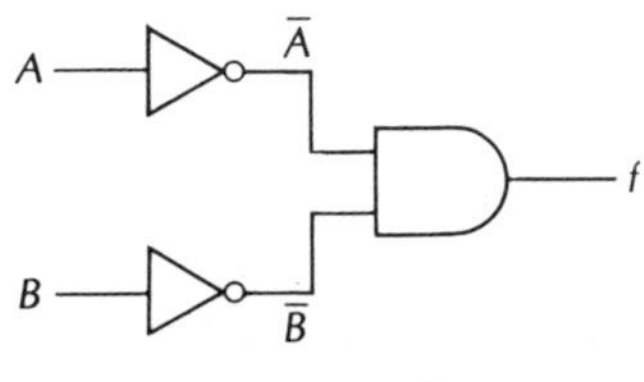

a. NOR gate NOR truth table b. AND equivalent

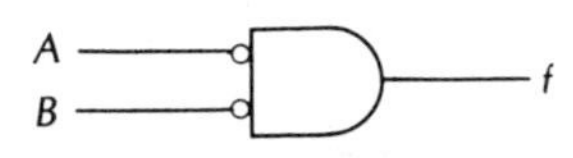

c. AND equivalent

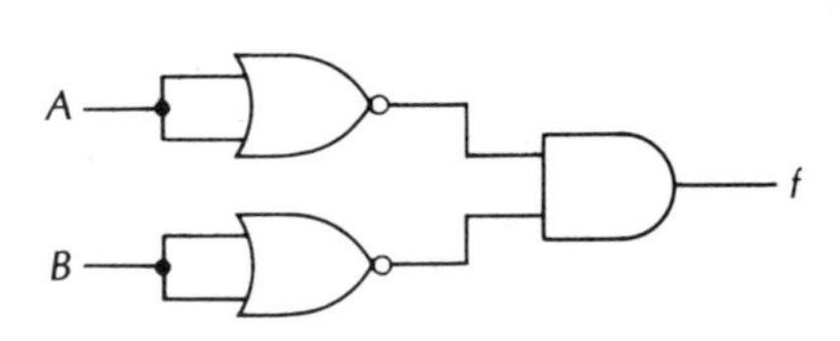

	A	B	f
0	0	0	1
1	0	1	0
2	1	0	0
3	1	1	0

d. AND equivalent AND equivalent truth table

PART 2

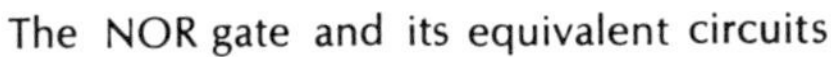

The NOR gate and its equivalent circuits

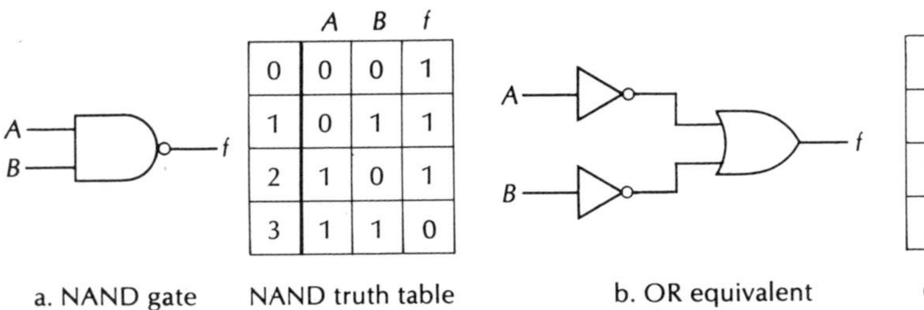

	A	B	f
0	0	0	1
1	0	1	1
2	1	0	1
3	1	1	0

	A	B	f
0	0	0	1
1	0	1	1
2	1	0	1
3	1	1	0

a. NAND gate NAND truth table b. OR equivalent OR equivalent truth table

Figure 1-18 NOR and NAND Gate Circuits and Equivalents

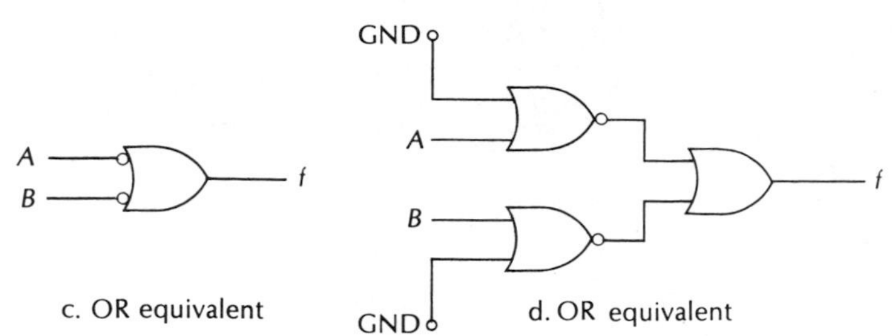

c. OR equivalent

d. OR equivalent

PART 3

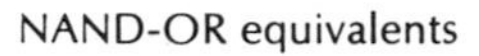
NAND-OR equivalents

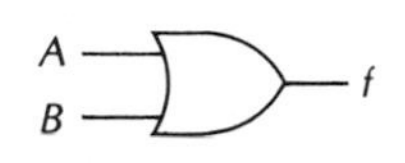

a. OR gate

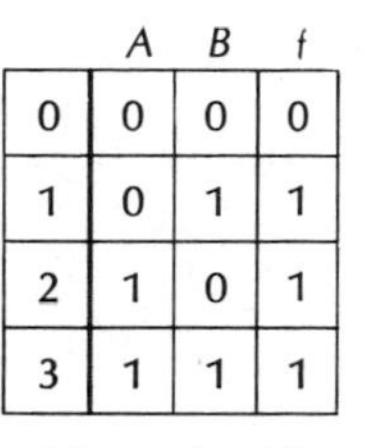

	A	B	f
0	0	0	0
1	0	1	1
2	1	0	1
3	1	1	1

OR truth table

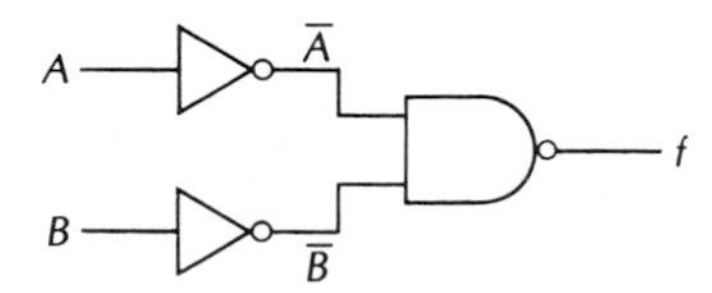

b. NAND equivalent

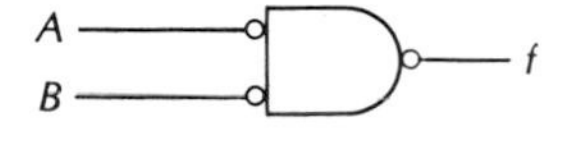

c. NAND equivalent

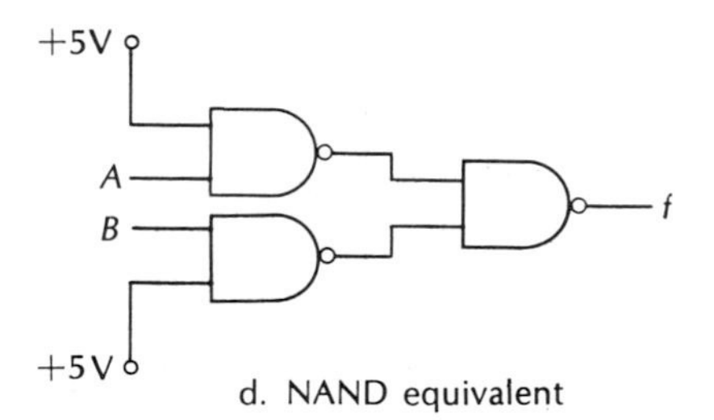

d. NAND equivalent

	A	B	f
0	0	0	0
1	0	1	1
2	1	0	1
3	1	1	1

NAND equivalent truth table

PART 4

The OR-NAND equivalents

Figure 1-18 continued

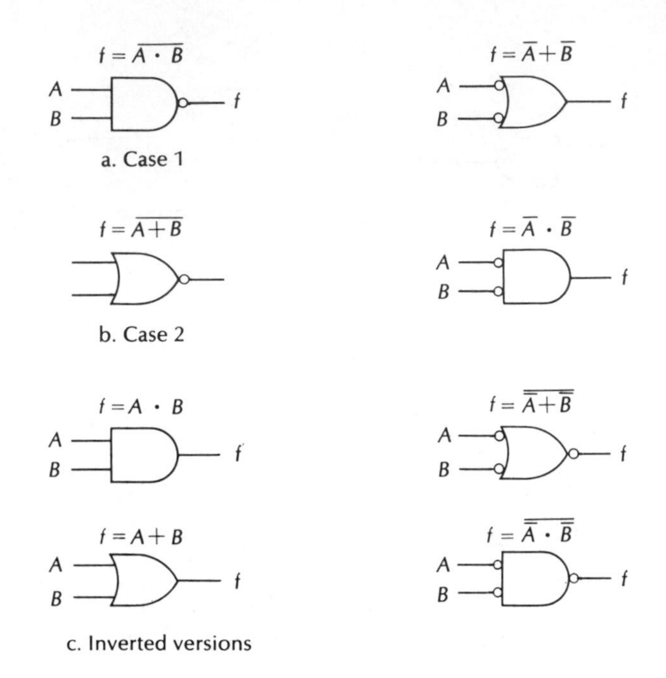

Figure 1-19 DeMorgan's Law Gate Equivalents

the gates. This method of presentation is called *bubble notation,* derived from MIL-STD 806B and The American National Standards Institute (ANSI) y32.14.

1-18 Introduction to Gates and Pulses

Most logic systems are pulse-operated. Combinations of pulses arriving at the inputs of a logic circuit produce a string of output pulses according to the operating rules for the particular circuit. In pulse operation both high and low levels and timing are important considerations. Figure 1-20 illustrates pulse operation of four basic gates.

1-19 A Simple Logic Circuit Example

The simple logic circuit in Figure 1-21 makes the "decision" to sound or not to sound a warning buzzer. The buzzer will sound when:

The seatbelt is *not* fastened	AND	
The ignition switch is *on*	AND	
The gearshift is in *reverse*	OR	*drive*

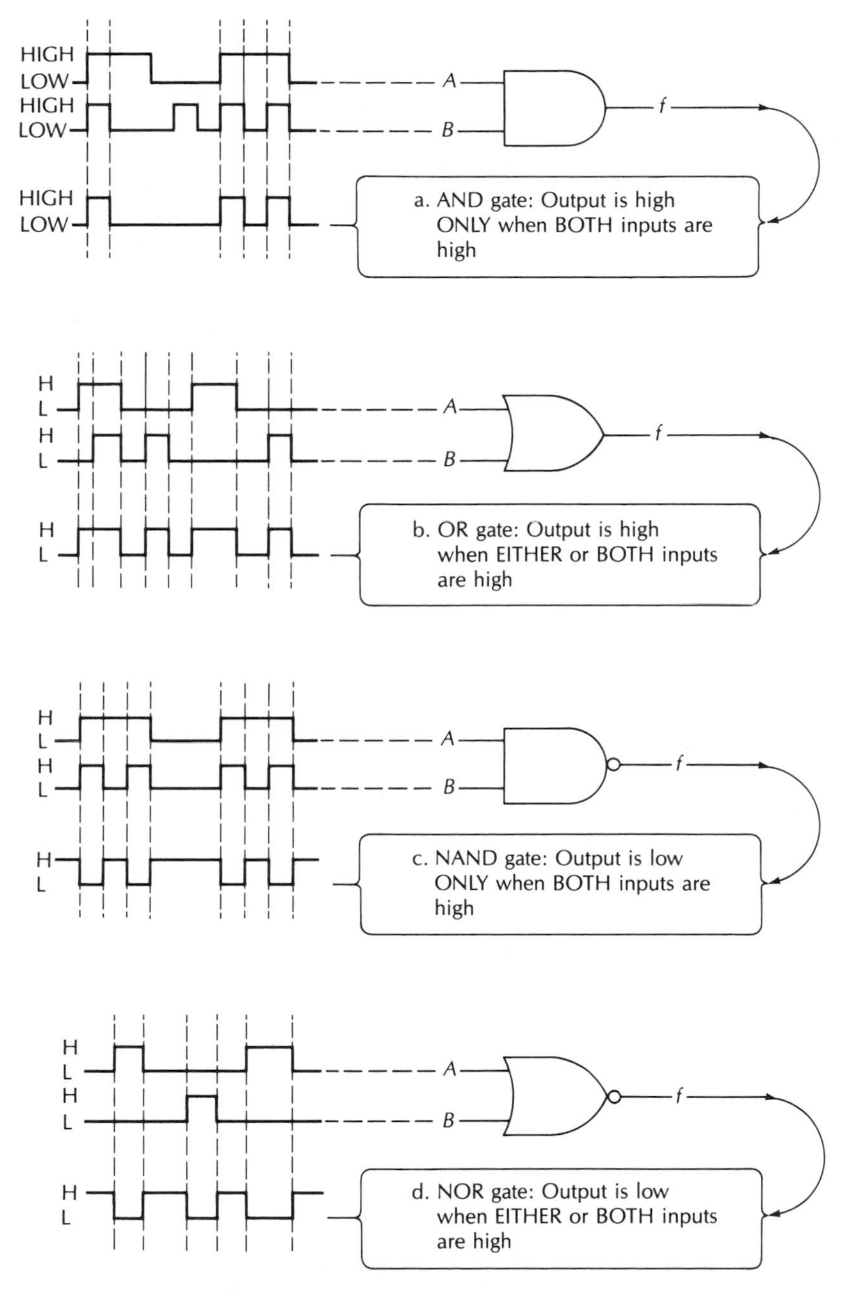

Figure 1-20 Gates and Pulses

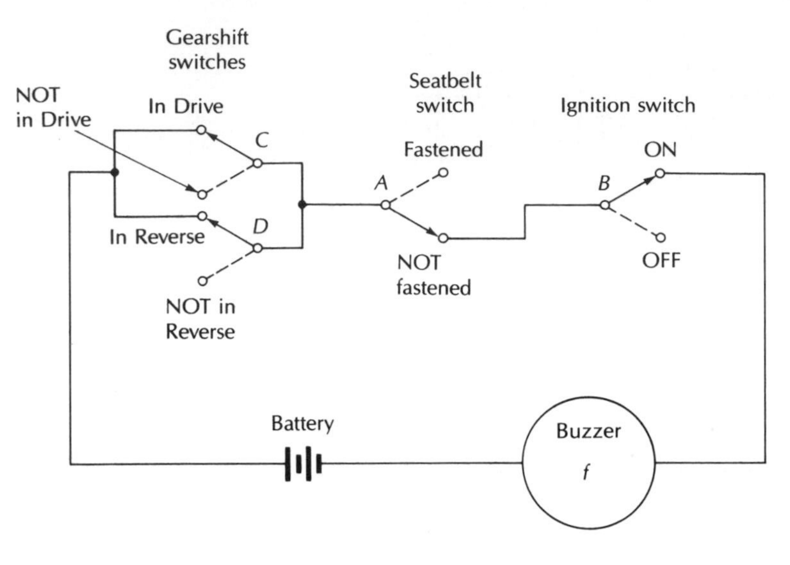

a. Switching diagram

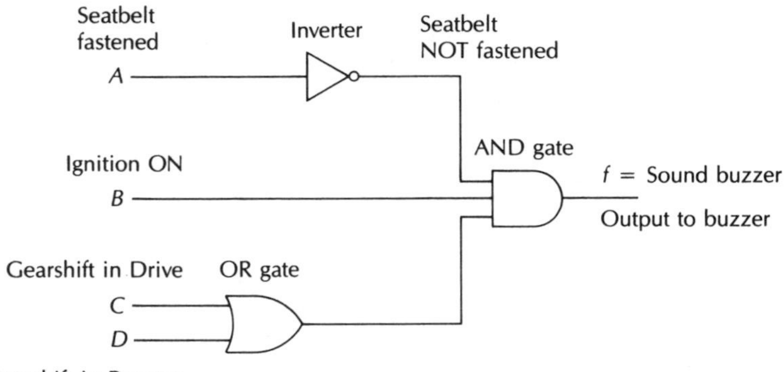

b. Logic diagram

Sound buzzer (f) = ignition ON (B) AND seatbelt not fastened ($\overline{A}$)
AND [gearshift in drive (C) OR in reverse (D)]

c. The Boolean equation in words

$$f = \overline{A} \cdot B \cdot (C + D)$$

d. The equation in symbols

Figure 1-21 Simple Logic Circuit Example

Conventions:

Buzzer sounds; $f = 1$
Closed switch = 1
Open switch = 0
A = Seatbelt switch
B = Ignition switch
C = Drive switch
D = Reverse switch

ADVANCED TOPICS

The following Boolean techniques are not likely to be needed on a daily basis, but there will be times when you will not be able to solve the problem at hand without resorting to them.

1-20 Demorganizing an Equation

In more complex systems it sometimes becomes difficult to determine the exact function of the system without resorting to the formal technique of *demorganizing* equations. The problem arises because entire Boolean terms or even complete equations are often inverted. When a minterm-form equation is inverted it becomes a maxterm-form equation. When a maxterm-form equation is inverted it becomes a minterm-form equation.

The most common form of Boolean equations is the minterm form, often called the *sum-of-products* form. For example, $f = (A \cdot B \cdot C) + (A \cdot \bar{B} \cdot \bar{C})$ is a minterm equation where variables within a term are ANDed and individual terms are ORed together. The maxterm form, also called the *product-of-sums,* takes the form $f = (\bar{A} + \bar{B} + \bar{C}) \cdot (\bar{A} + B + C)$, where the variables within the terms are ORed and individual terms ANDed together. (Section 3-5 provides a further discussion of the two basic Boolean equation forms).

Demorganization is simply an easy way to determine what happens to equations and parts of equations when they are passed through an inverter.

The Rules

1. Complement each variable, including the f, on the left-hand side of the equation.
2. Change all AND operators to OR operators and all OR operators to AND operators.
3. Clear all double complements.

Example Demorganize the equation:

$$f = (A \cdot B \cdot \bar{C}) + (A \cdot \bar{B} \cdot C) + (\bar{A} \cdot \bar{B} \cdot \bar{C})$$

Complementing both sides and changing operators, we have:

$$\bar{f} = (\bar{A} + \bar{B} + \bar{\bar{C}}) \cdot (\bar{A} + \bar{\bar{B}} + \bar{C}) \cdot (\bar{\bar{A}} + \bar{\bar{B}} + \bar{\bar{C}})$$

Since $\bar{\bar{A}} = A$, we can clear the equation of double complements and write the equation as:

$$\bar{f} = (\bar{A} + \bar{B} + C) \cdot (\bar{A} + B + \bar{C}) \cdot (A + B + C)$$

Example Demorganize the equation:

$$\bar{f} = (\bar{A} + B + \bar{C}) \cdot (\bar{A} + B + C)$$

Complementing and changing operators:

$$\bar{\bar{f}} = (\bar{\bar{A}} \cdot \bar{B} \cdot \bar{\bar{C}}) + (\bar{\bar{A}} \cdot \bar{B} \cdot \bar{C})$$

Clearing the equation of double complements:

$$f = (A \cdot \bar{B} \cdot C) + (A \cdot \bar{B} \cdot \bar{C})$$

This is the finished minterm-form equation.

1-21 Complement-Bar Notation

Because of the dual nature of Boolean algebra, complement bars can be used as an instruction to complement an entire term or equation as well as individual variables. Some examples of complement-bar notation, coupled with a little practice, will provide you with ample facility to handle this kind of notation.

Example The long complement bar in this example is an implicit instruction to demorganize the equation.

Given the following equation, eliminate the long complement bar by performing the indicated operation (demorganization):

$$\overline{f = (A + \bar{B} + C) \cdot (\bar{A} + B + \bar{C})}$$

Complementing all variables and changing operational signs, we get:

$$\bar{f} = (\bar{A} \cdot \bar{\bar{B}} \cdot \bar{C}) + (\bar{\bar{A}} \cdot \bar{B} \cdot \bar{\bar{C}})$$

Clearing the double complements, we get:

$$\bar{f} = (\bar{A} \cdot B \cdot \bar{C}) + (A \cdot \bar{B} \cdot C)$$

Here we have interpreted the complement bar over an operational sign as conforming to the rule:

$$\overline{+} = \cdot \text{ and } \overline{\cdot} = + \text{ (not OR = AND, not AND = OR)}$$

This may be considered a general rule when the operational sign is part of a term or an equation.

Example Demorganize the equation:

$$f = \overline{(\bar{A} \cdot B \cdot C) + (A \cdot \bar{B} \cdot C)}$$

Complementing all variables and changing operational signs, we get:

$$\bar{f} = (\bar{\bar{A}} + \bar{B} + \bar{C}) \cdot (\bar{A} + \bar{\bar{B}} + \bar{C})$$

Clearing the double complements yields:

$$\bar{f} = (A + \bar{B} + \bar{C}) \cdot (\bar{A} + B + \bar{C})$$

Example This example illustrates the case where a long complement bar exists over a single term. The complemented term is demorganized, but the rest of the equation is not altered.

Given the following equation, remove the long complement bar by performing the operations indicated:

$$f = (\bar{A} \cdot \bar{B} \cdot \bar{C}) + (\overline{A + \bar{B} + C})$$

Complementing the indicated variables and changing operational signs, we get:

$$f = (\bar{A} \cdot \bar{B} \cdot \bar{C}) + (\bar{A} \cdot \bar{\bar{B}} \cdot \bar{C})$$

Clearing the double complement, we get:

$$f = (\bar{A} \cdot \bar{B} \cdot \bar{C}) + (\bar{A} \cdot B \cdot \bar{C})$$

Example This example covers the case where both of the previous cases are included in a single equation.

Given the following equation, remove the long complement bars by performing the indicated demorganizations:

$$f = \overline{(\bar{A} \cdot B \cdot C) + (\overline{A + B + C})}$$

Removing the longer bar, we get:

$$\bar{f} = \overline{(\bar{A} \cdot B \cdot C)} \cdot \overline{(\overline{A + B + C})}$$

Clearing the remaining long complement bars yields:

$$\bar{f} = (A + \bar{B} + \bar{C}) \cdot (A + B + C)$$

Problems

1. Why are schematics used infrequently in describing logic systems?
2. Match the following to items (1) through (4) below:
   a. Boolean equations
   b. Schematic diagrams
   c. Logic diagrams
   d. Timing diagrams
   e. Truth table

   (1) Tells *how* logic elements are interconnected
   (2) Tells *when* a gate is expected to operate
   (3) Provides a tool for describing logic circuit organization and discovering other equivalent organizations
   (4) Specifies all possible operating conditions for a given logic circuit

3. Describe the AND function in words.
4. Write the Boolean equation for the AND function, draw the AND gate symbol, and write the truth table for the AND function.
5. Write the Boolean equation, the truth table, and draw the logic symbol for the OR function.
6. Describe the OR gate function in words.
7. Explain the NOT operation. What is the logic gate called that performs the NOT function?
8. Draw the logic symbol for a NAND gate, and draw a logic diagram of its equivalent using:
   a. AND's inverters
   b. OR's inverters
   c. OR's inverters with bubble notation
   d. OR's, NOR's
9. Draw the logic symbol for a NOR gate, and draw a logic diagram of its equivalent using:
   a. OR's inverters
   b. AND's inverters
   c. AND's inverters with bubble notation
   d. AND's, NAND's
10. Write the two cases of DeMorgan's law.
11. Write the truth table for a NAND gate.
12. Given the diagram in Figure 1-22, draw the output (f) pulse waveform.
13. Given the diagram in Figure 1-23, draw the output (f) pulse waveform.
14. Given the diagram in Figure 1-24, draw the output (f) pulse waveform.
15. Given the diagram in Figure 1-25, draw the output (f) pulse waveform.

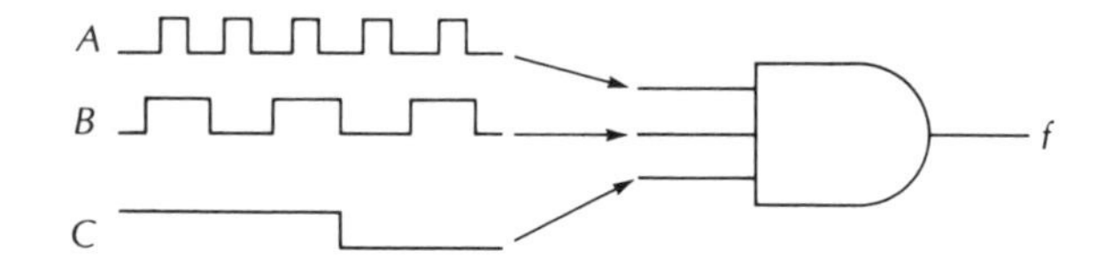

Figure 1-22 Diagram for Problem 12

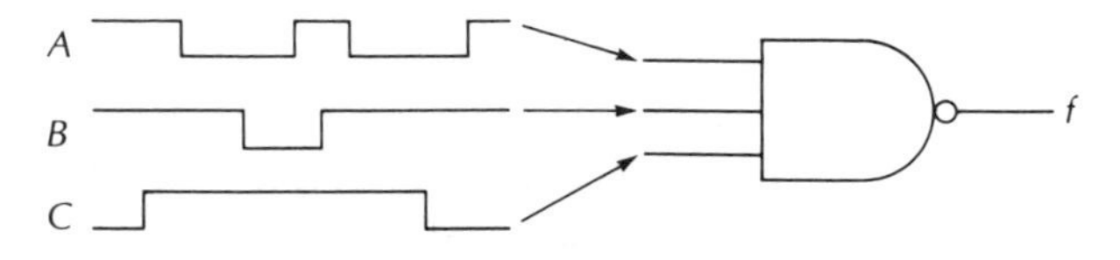

Figure 1-23 Diagram for Problem 13

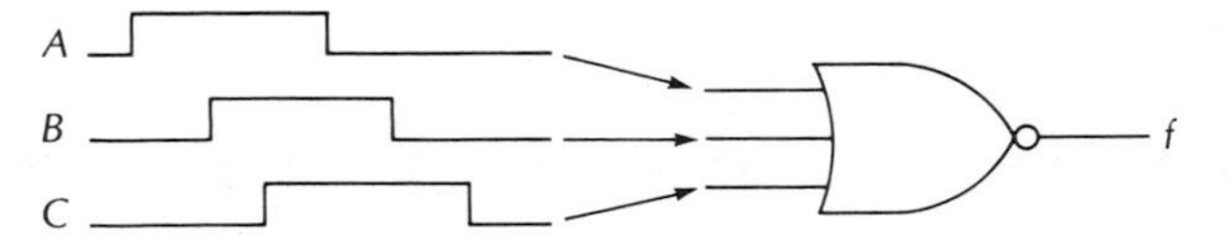

Figure 1-24 Diagram for Problem 14

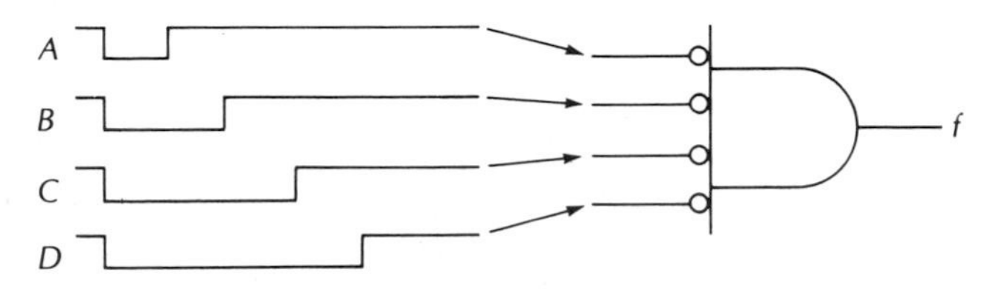

Figure 1-25 Diagram for Problem 15

Problems for Advanced Topics

Demorganize the following equations:

16. $f = (A + \bar{B} + C) \cdot (\bar{A} + B + \bar{C}) \cdot (\bar{A} + \bar{B} + C)$
17. $\bar{f} = (A + \bar{B} + C) \cdot (\bar{A} + \bar{B}) \cdot (\bar{A} + \bar{B} + C)$
18. $f = (A + B + C) \cdot (\bar{A} + \bar{B} + \bar{C})$
19. $\bar{f} = (\bar{A} + \bar{B} + \bar{C}) \cdot (\bar{A} + B + C) \cdot (A + B + \bar{C})$
20. $f = (A \cdot \bar{B} \cdot C) + (A \cdot \bar{B} \cdot \bar{C}) = f$
21. $\bar{f} = (A \cdot B \cdot C) + (\bar{A} \cdot \bar{B} \cdot \bar{C}) + (A \cdot \bar{B} \cdot C)$
22. $f = (\bar{A} \cdot \bar{B}) + (B \cdot \bar{C}) + (A \cdot \bar{B} \cdot C)$

Remove all complement bars that cover more than a single variable:

23. $f = \overline{(A \cdot \bar{B}) + (A \cdot \bar{C})}$
24. $f = \overline{(A + \bar{\bar{B}}) \cdot (\bar{A} + \bar{\bar{C}})}$
25. $f = \overline{\overline{(A + B)} \cdot (A \cdot \bar{\bar{B}})}$
26. $f = \overline{\overline{(\bar{\bar{A}} + \bar{\bar{B}})} + (\bar{A} \cdot B \cdot C)}$
27. $f = \overline{(A \cdot B \cdot C) + \overline{\overline{(\bar{A} + B + \bar{\bar{C}})}}}$
28. $f = \overline{\overline{(A \cdot \bar{\bar{B}} \cdot C)} + (A + \bar{B})}$
29. $f = \overline{(\bar{A} \cdot \bar{B})} + (A \cdot \bar{C}) + \overline{(A \cdot \bar{B} \cdot C)}$
30. $f = \overline{\overline{(\bar{A} \cdot \bar{B})} + (A \cdot \bar{C}) + \overline{(A \cdot \bar{B} \cdot C)}}$

CHAPTER 2

INTEGRATED LOGIC CIRCUITS

Learning Objectives. *Upon completing this chapter you should know:*

1. *How to compare speed and power consumption among the various logic circuits.*
2. *The logic levels and power supply requirements of each logic circuit.*
3. *Which class of applications involve which logic circuit forms.*
4. *The meaning of the terms fan-out, fan-in, compatibility, dynamic and static logic, LSI, MSI, SSI, speed, Schottky, noise immunity, and interfacing.*
5. *How to identify each of the logic circuits presented in the chapter.*
6. *The data manual symbols.*
7. *The basic rules for power supply decoupling.*
8. *What dynamic and static testing are.*
9. *How to define storage time.*
10. *What factors limit logic circuit speed.*
11. *What a totem-pole circuit is and why it is used.*
12. *The most important operating parameters of TTL, MOS, C-MOS, and MTL.*
13. *The theory of operation for each of the major logic circuits.*

In most applied digital logic we are not too concerned about the internal circuitry of a gate package. We are far more involved with input and output characteristics, operating levels, gate propagation times, loading rules, and so on. Still, there is some justification for spending time in examining the internal schematics of typical gate structures, because many of the input-output and transfer characteristics are dictated by internal circuitry.

This chapter is concerned with the practical considerations necessary to make real circuits work. It also briefly covers the theory of operation of basic integrated circuit gates.

2-1 Integrated Circuits

Nearly all modern logic circuits and subsystems are monolithic* integrated circuits. Transistors, diodes, resistors, and small capacitors are

* Circuits are fabricated on a slab of semiconductor material by selectively altering the conductivity (at the molecular level) of the semiconductor material. Various conductivity levels correspond to the transistor elements, base collector, etc.

formed on small chips of silicon—from about 0.1 inch to 0.5 inch on a side (dimensions approximate). Individual components are interconnected by aluminum or gold wiring patterns that resemble ordinary printed circuit wiring.

Here we will examine briefly how integrated circuits are fabricated and assembled. This is not intended to be a comprehensive coverage and will necessarily be simplified.

The procedure begins with the development of the artwork for fabricating glass photographic masks. Each mask controls the areas on the silicon wafer (substrate) where various levels of *doping* (defined in next paragraph) are required for transistor bases, collectors, field effect channels, resistors, and so on.

Each mask pattern is developed photographically on a silicon wafer. The wafer is then heated to the point where it almost melts but where surface tension still holds it in a solid form. Impurity elements are diffused into the silicon like butter diffusing into hot toast. The process of adding controlled amounts of certain impurity elements to the silicon crystal is called *doping*. Several masks are used to control the areas of diffusion, one mask for each doping level. One set of masks is designed for a circuit consisting of anywhere from a few to thousands of gates connected in an array ranging from a few individual gates to a complete microcomputer. (The term microcomputer refers to the machine's physical size not its computing capabilities.) Figure 2-1 shows integrated circuit (IC) transistor cross sections.

A silicon ingot some two inches (more or less) in diameter is sliced into wafers a few mils thick. Each wafer will eventually have as many as several hundred complete *independent* circuits on it. At the end of the diffusing and interconnecting wiring operations, the wafer is diced by

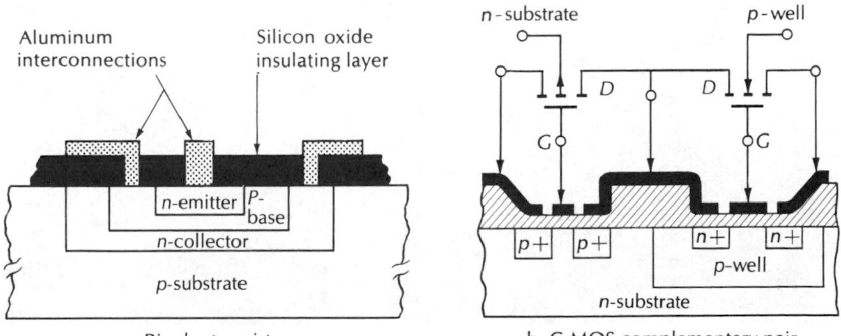

a. Bipolar transistor

b. C-MOS complementary pair

Figure 2-1 Cross-Sections of Bipolar and C-MOS Integrated Circuit Transistors

laser cutting, scribing and breaking, or by a diamond saw. This yields up to several hundred chips, each a complete integrated circuit. Each die or chip is then mounted on a header similar to that shown in Figure 2-2. Tiny hairlike gold wires that connect input, output, and power pads on the chip are welded (under a microscope) to pins on the header. Each chip and its gold leads are molded in a plastic block (or housed in some other case). The header strip is then cut apart, the leads formed, and the type number stamped on the case. Figure 2-3 shows the most common case style. There are, of course, a number of quality control and testing steps involved in producing the highly reliable integrated circuits currently available.

Individual integrated circuits are then mounted on etched circuit boards.

2-2 Integrated Circuit Classification

There are two basic types of integrated circuits; one is based on bipolar transistors and the other is built around *metal-oxide-semiconductor* (MOS) field effect transistors. In both technologies transistors or parts of transistors are used as diodes, resistors, and capacitors. The amount of functional capability on a given chip is generally defined by the

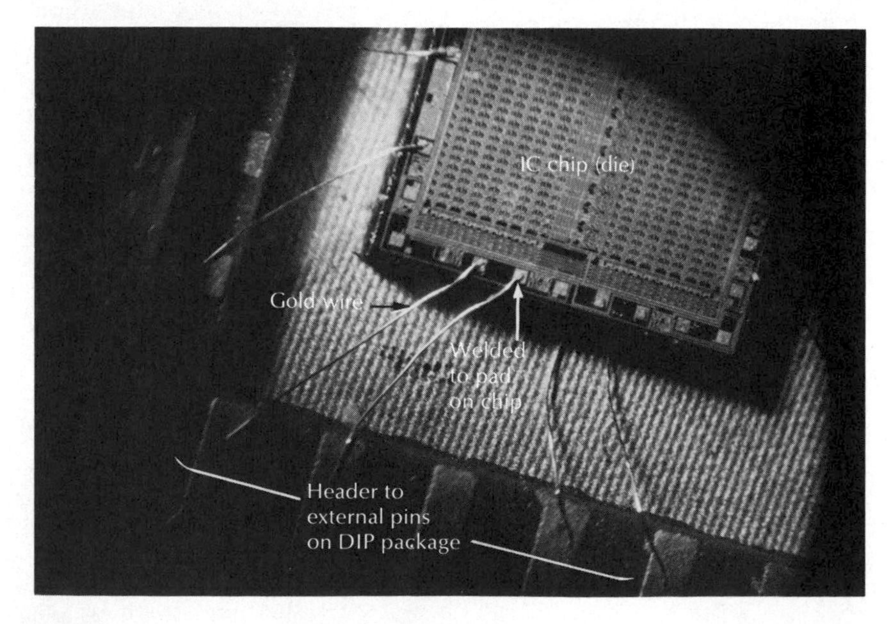

Figure 2-2 IC Chip (Die) Connected to Header

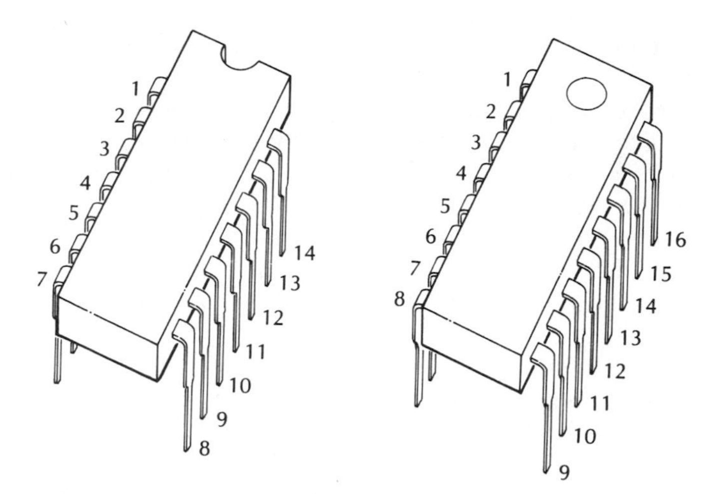

Figure 2-3 The Dual In-line Package (DIP)

number of operational gates on the chip even though the number of components may differ from one kind of gate to another.

SSI

The acronym SSI stands for *small-scale integration* and identifies integrated circuit packages that contain less than twelve logic gates.

MSI

The acronym MSI, *medium-scale integration,* identifies logic packages containing more than 12 but less than 100 logic gates.

LSI

Large-scale integration is defined as any integrated circuit with more than 100 gates in a single package. In many cases an LSI chip may contain several thousand gates.*

An important consideration about any logic family is that all its logic circuits—gates, flip-flops, counters, and so on—are compatible. This means that the output(s) of any gate in the family can be connected to the input(s) of any other gate (or more complex structures) without

* The terms very large (VLSI), ultra large (ULSI), and super large (SLSI) scale integration appear in current literature, but at this writing none of these terms have been defined by recognized standards groups.

elaborate buffering circuits, voltage level changers, or other interface circuitry.

Each of the popular families provides a remarkably complete set of logic building blocks, and because of the concept of compatibility, members of each family can be assembled in almost any combination required, as long as a few simple rules are followed. Whenever circuits using one family are connected to circuits based on a different family, some problems often arise requiring that extra circuitry be added to bridge the gap. This extra hardware is called *interface* circuitry.

Digital integrated circuits (except for some special LSI circuits) are divided into logic families, each of which is based on one particular type of transistor circuit. The same basic circuit is used for all gates, inverters, and flip-flops. MSI and LSI circuits of considerable complexity are composed of interconnected arrays of standard gates.

All devices in a given logic family use the same logic levels and operate from the same power supply voltage, and the output of one device can supply the proper amount of voltage and current to drive the input of another. Because each logic family uses a different circuit, any two given logic families are generally not totally compatible with each other. In addition to the basic logic elements and MSI or LSI circuits, each logic family contains special circuits, such as level translators (for interfacing to the circuits of another logic family), signal conditioning circuits (Schmitt triggers and multivibrators), display driver circuits (for interfacing a logic circuit to visual display), and other circuits for special applications.

The compatible logic families based on bipolar technology cover most SSI and MSI logic needs. MOS technology is more often the basis of LSI systems, such as memories, microprocessors, calculators, digital clocks (the time-of-day kind), and other specialized products. These circuits are not defined as families. Complementary MOS (C-MOS) is used for a fairly complete family in direct competition with bipolar logic circuits. C-MOS is relatively slow compared with bipolar circuits, but it is fairly compatible with both large-scale MOS circuits and bipolar logic.

2-3 Pulse Parameters

Because logic circuits are most often pulse-operated, we must have some pulse measurement criteria and standard terminology if we are to deal effectively with them. In this section we will examine standard pulse terminology and measurement parameters.

FREQUENCY AND PERIOD

Clock signals (and other pulses) have certain defined parameters to describe their dimensions in amplitude and time. These parameters are illustrated in Figure 2-4.

Frequency

The *frequency* of pulses is a measure of how many pulses occur in a given time period. The measurement is generally given in megahertz (MHz).

Period

The *period* is defined as the time between two adjacent pulses, and its base unit is the second. It is dependent upon the frequency and bears the mathematical relationship: $T = 1/F$, where T = time between pulses and f = the frequency of pulses. If we rearrange the equation in terms of F, we get $F = 1/T$. If we know the period, we can always find the frequency by taking the reciprocal; conversely, if we know the frequency, we can take the reciprocal to determine the period.

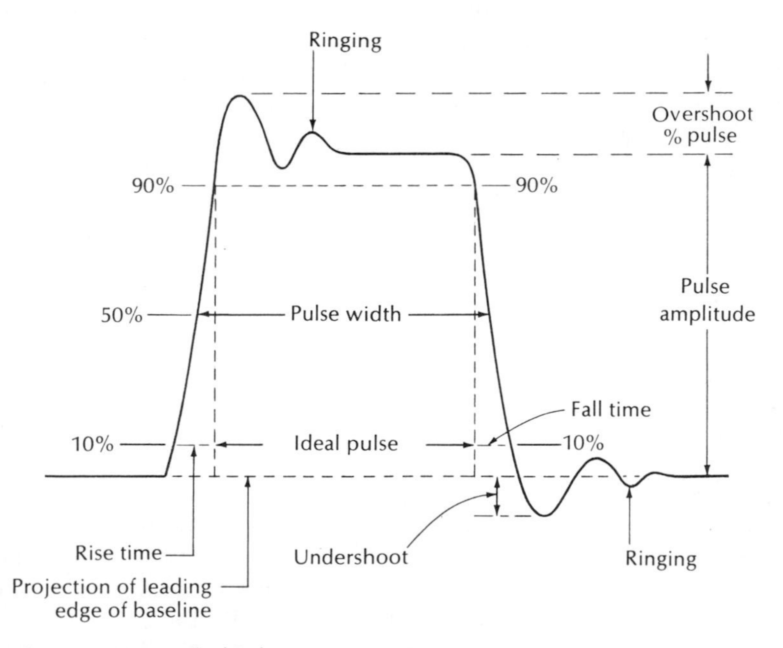

Figure 2-4 Non-Ideal Pulse

Rise and Fall Times

Rise time is generally defined as the time required for a pulse voltage to rise from 10 percent above the (*ideal* pulse) zero volt state to 90 percent of the ideal pulse maximum amplitude. The 10 percent at the beginning of the pulse's rise and the 10 percent at the end of the rise are practical considerations, allowing for measurements in spite of *ringing, overshoot,* and *undershoot* (defined below). Therefore, 10 percent above the zero volts base line and 10 percent below the maximum pulse amplitude has been allowed to insure meaningful measurements. *Fall time* is the time (in nanoseconds) required for the pulse voltage to fall from 90 percent of its maximum steady-state voltage to 10 percent of maximum amplitude.

Pulse Width

Pulse width is the time interval, measured at 50 percent above the ideal baseline—zero. Some companies may have different in-house standards.

Ringing, Undershoot, and Overshoot

At the end of a sudden transition (from low to high or from high to low), distributed capacitances and inductances form a resonant circuit that produces a decaying sinusoidal oscillation, which lasts for a time from less than half a cycle to several cycles. The added amplitude is called *overshoot* on the upward transition. At the bottom of the transition, from high to low, ringing may occur again. When the tank circuit energy is dissipated in approximately half a cycle, it is said to be *critically damped,* and the portion of the first half (ringing) cycle is called *undershoot*. Undershoot and overshoot are measured as a fraction of the maximum steady-state pulse amplitude. Ringing may also be classified by its *self-resonant frequency*. The frequency of the ringing is important only in the sense that it may offer some clue as to the elements causing it.

Pulse Amplitude

Pulse amplitude is the maximum steady-state pulse height. The actual measurement is taken by projecting the leading edge baseline to the trailing edge of the pulse and measuring from the projected baseline to the top of the trailing edge of the pulse. The measurement is taken in this way because any overshoot (or ringing) will occur at the end of the leading edge high-going transition. This tends to obscure the actual pulse amplitude at the top of the leading edge. By the beginning of the pulse downward transition (trailing edge), overshoot and ringing energy should have been dissipated and the steady-state pulse amplitude established. At the end of the pulse fall to zero, undershoot (or ringing) can obscure the baseline for the trailing edge of the pulse. Logic circuits can tolerate a reasonable amount of pulse imperfection and still operate properly. As a result, we are usually not too concerned about the absolute values of these pulse parameters. However, in difficult troubleshooting problems these parameters can be crucial. The system manufacturer will generally provide information about maximum allowable overshoot, ringing, undershoot, rise time, and so on.

2-4 Propagation Delay Time

A pulse cannot pass through a gate instantly. It requires a small but finite time period called the *propagation delay time*. The propagation delay is a direct consequence of the rise and fall times shown in Figure 2-4.

The rise time delay can be primarily attributed to gate input capacitance and resistance and their attendant *R-C* time constant. The fall time delay also has an *R-C* time constant component, but it may also involve a phenomenon known as *storage time*. We will look at these problems in greater detail later in the chapter.

Figure 2-5 illustrates the relationship between rise and fall times and propagation delay. Fortunately, because of their switching nature, gates do not amplify waveform defects from a previous gate. The waveform distortions at the output of a gate are only those produced by that gate and are relatively independent of the shape of the input waveform.

The distinction between rise/fall times and propagation delays is that propagation delays are cumulative from gate to gate while rise and fall times are not. As illustrated in Figure 2-5, rise and fall waveform slopes are not passed along, but the time delays they produce are passed along. Propagation delays are additive.

2-5 Diode Logic

Electronic logic gates must not only perform their respective logic functions but also have certain electrical characteristics. High-speed switching capability is one of these characteristics, but there are less obvious but equally important ones.

First, the gate must switch in a positive fashion; it must be on or off. There must be no intermediate condition even if the circuit is operating well below normal performance standards.

Second, the gate must have a dead-band between the on and off voltage levels. The dead-band allows the input voltage to change over a small range without causing the gate to change states from on to off or

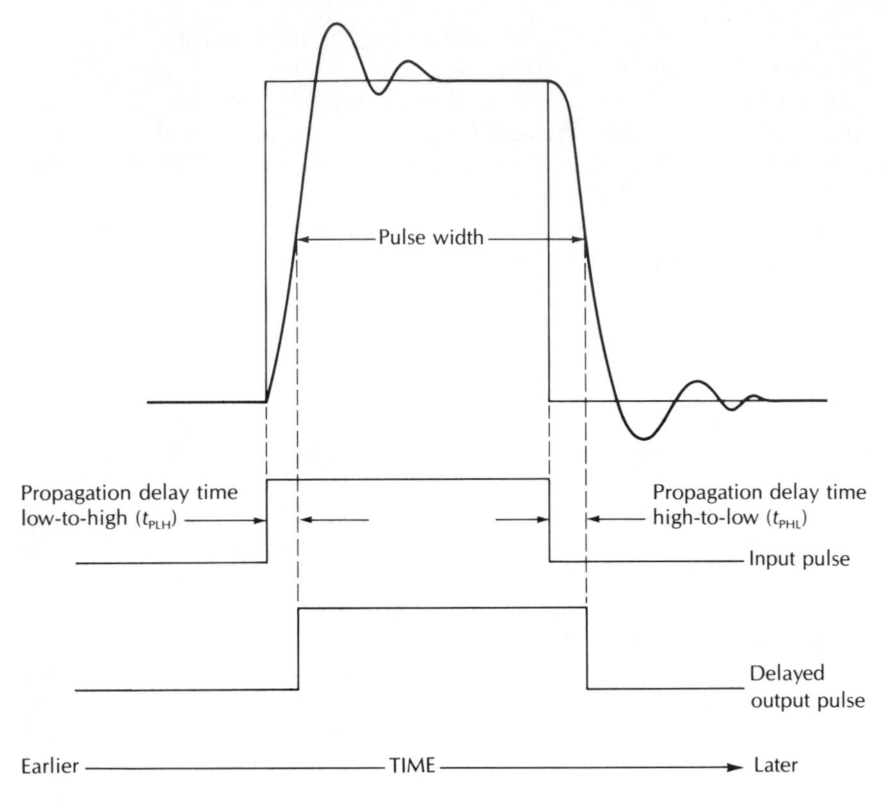

Figure 2-5 Propagation Delay

from off to on. The dead-band offers a safe area so voltage levels need not be impossibly precise. It also allows a certain amount of electrical noise to enter the gate without switching it.

Third, the several inputs to the gate must be isolated from each other. A particular voltage level applied to one input must not be reflected onto other input lines.

Table 2-1 Diode AND Gate Truth Table

Input A	Input B	Output
GND	GND	≈GND
GND	+5V	≈GND
+5V	GND	≈GND
+5V	+5V	≈+5V

The diode conduction curve in Figure 2-6 reveals why the silicon junction diode is a nearly ideal device for logic gate application. The diode has a very high resistance in the reverse-bias direction and a low resistance in the forward-bias direction. The diode has a built-in dead-band in the form of a 0.6 V junction potential. Even the steep non-linear forward conduction is important, as we will see shortly.

HOW THE DIODE *AND* GATE WORKS

Figure 2-7 defines the electrical conditions for each entry in the AND gate truth table (Table 2-1). For this discussion:

$$\text{Logical } 1 = 4.4 \text{ to } 5 \text{ V}$$
$$\text{Logical } 0 = 0 \text{ to } 0.6 \text{ V}$$

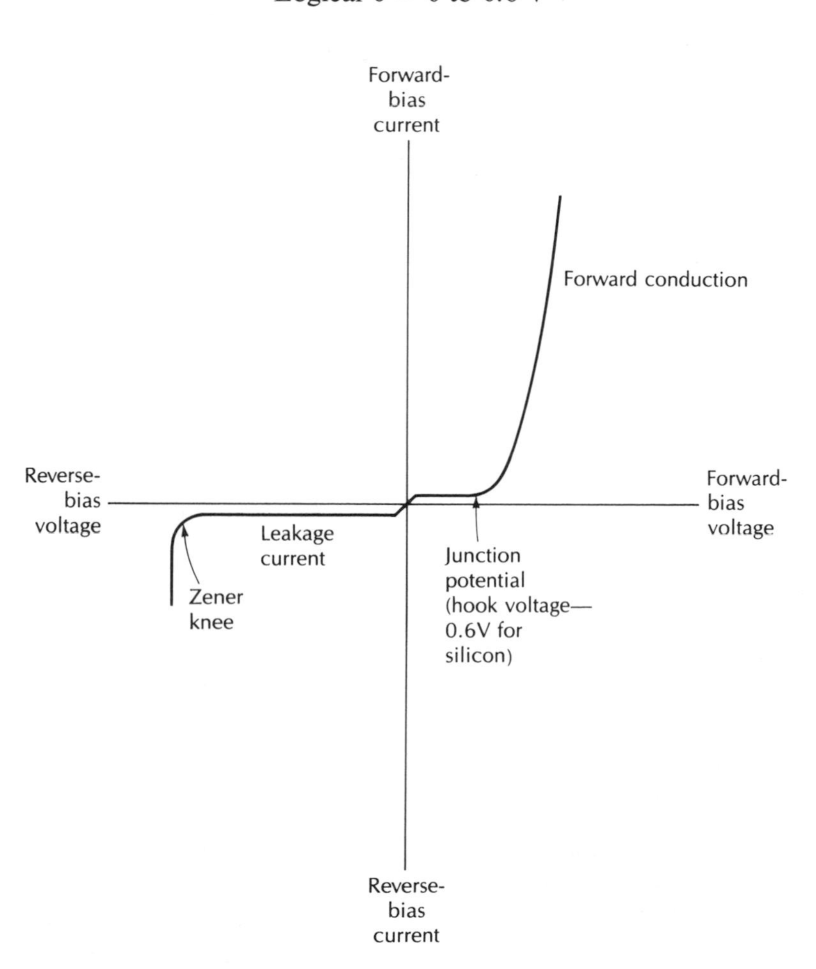

Figure 2-6 The Diode Conduction Curve

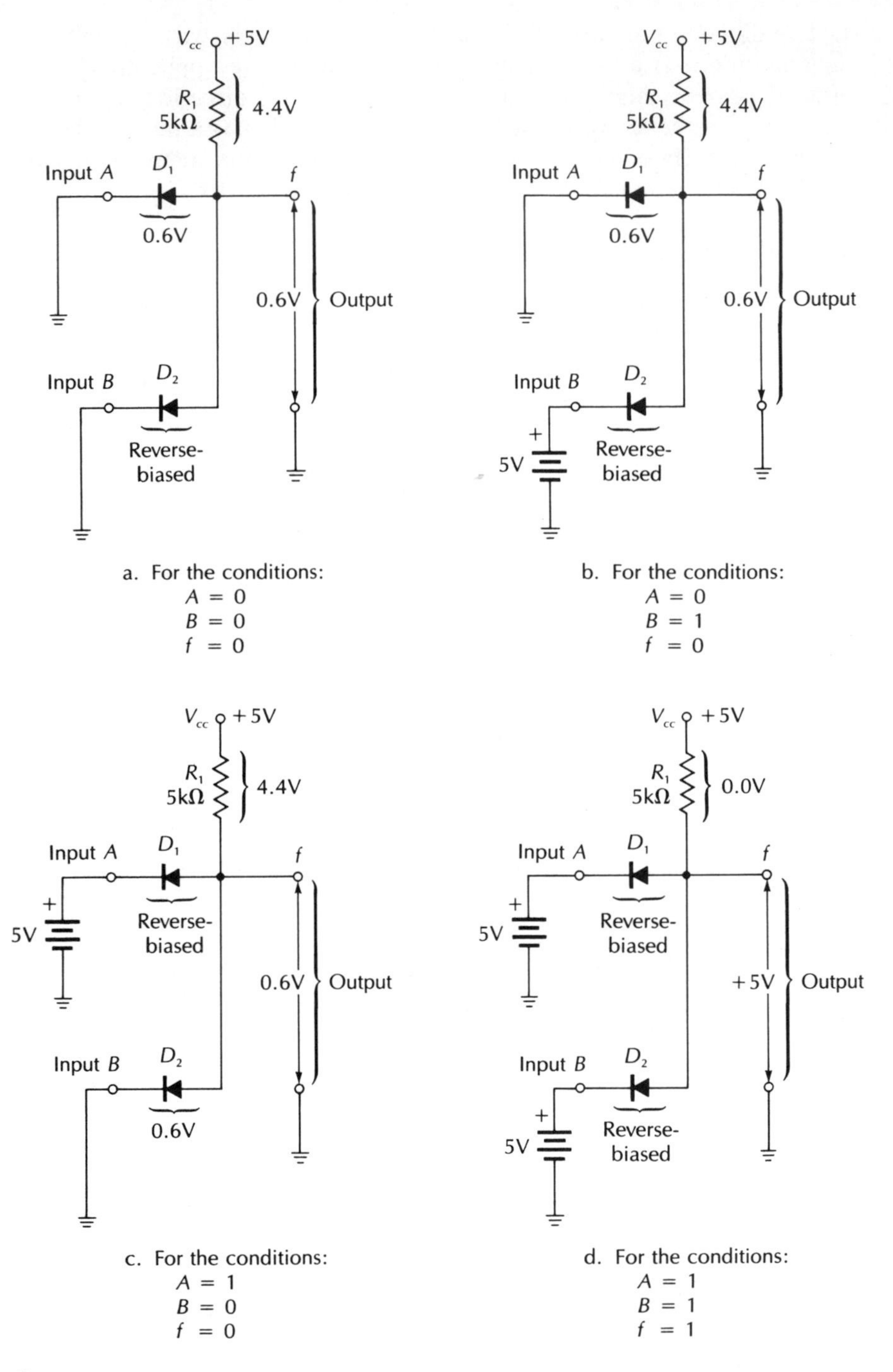

Figure 2-7 Diode AND Gate

Please follow Figure 2-7 as we discuss each set of logic conditions.

1. See Figure 2-7a.

$$A = 0 \text{ (ground)}$$
$$B = 0 \text{ (ground)}$$

Diode D_1 is forward-biased (positive anode) and draws nearly 1 mA of current through R_1. This current causes 4.4 of the 5-volt V_{cc} to be dropped across R_1. The output is effectively in parallel with D_1 (and D_2), and the output voltage is 0.6 V, the diode junction potential (0.6 volts is a logical zero).

Diode D_2 appears to be as much forward-biased as diode D_1, but it is labeled *reverse-biased* on Figure 2-7a. The explanation for this seeming contradiction lies in the steep non-linear conduction curve in Figure 2-6. The curve indicates very large changes in current for very small changes in voltage. When diodes are connected in parallel, one of them will invariably start to conduct before the others. If the forward-bias voltage increases, the first diode to start conducting rides up the conduction curve, demanding much more current and preventing the other parallel diodes from starting to conduct. This phenomenon, called *current hogging,* is usually a problem when diodes are connected in parallel. In the case of the diode logic circuit, current hogging serves an important function: it ensures that no more than one of the gate input diodes will be forward-biased at any given time. All of the other input diodes are effectively open circuits and thus the input lines are isolated from each other.

2. See Figure 2-7b.

$$A = 0 \text{ (GND)}$$
$$B = 1 \text{ (+5 V)}$$

Diode D_1 is forward-biased and the current through R_1 drops 4.4 V across it. The output voltage is the same as the D_1 junction voltage, 0.6 V (logical zero).

Diode D_2 has +5 V on both anode and cathode, and is therefore intrinsically reverse-biased. Diode D_2 is effectively an open circuit.

3. See Figure 2-7c.

$$A = 1 \text{ (+5 V)}$$
$$B = 0 \text{ (GND)}$$

The conditions here are the same as those in Figure 2-7b except that D_2 is now forward-biased and D_1 is an open circuit. The output voltage is 0.6 V, a logical zero.

4. See Figure 2-7d.

$$A = 1\ (+5\ \text{V})$$
$$B = 1\ (+5\ \text{V})$$

In this case, both diodes are reverse-biased. There is no current flow through either diode and therefore no current flow through R_1. The output voltage rises to approximately V_{cc} (+5 V). The gate output is a logical one.

2-6 The Diode OR Gate

Diode gates are not often used in a discrete form except in an array of diode gates called a *matrix*. Diode logic does form part of the internal circuitry in many commercial integrated circuits. Diode matrices are usually used in conjunction with TTL and other integrated circuit gates. The input circuit in TTL integrated circuits is essentially a diode AND gate, so TTL output circuits are designed to provide proper drive currents for diode AND gates. Please look at the diode AND gate in Figure 2-7b.

Notice that a high (+5 V) on an input reverse-biases the diode. The source of the +5 volts needs to supply a few microamperes (mA) of diode leakage current for a logic 1 (+5 V). Now, notice that a low (ground) forward-biases the diode and about 1 mA flows to ground. If the output circuit of a gate is taking that diode to ground, the output circuitry must be able to carry that 1 mA of current to ground.

SOURCE AND SINK CURRENTS

1. The current that flows when the output of a logic gate is high is called the *source current.*
2. The current that flows when the output of a logic gate is low is called the *sink current.*
3. TTL gates can source only a few hundred microamperes.
4. Standard TTL (7400/5400) series gates can sink 16 mA.
5. High-power (54H/74H) series gates can sink 20 mA.
6. Low-power (54L/74L) series gates can sink 2 mA.
7. Low-power Schottky (54LS/74LS) series gates can sink 4 mA (54LS) or 8 mA (74LS).
8. TTL gates can sink 16 mA.

Figure 2-8 shows the positive-logic diode OR gate circuit. The circuit is not compatible with TTL drive currents so the circuit is not often used. A driving gate for the circuit in Figure 2-8 would have to be capable of sourcing about 1 mA (see Fig. 2-9).

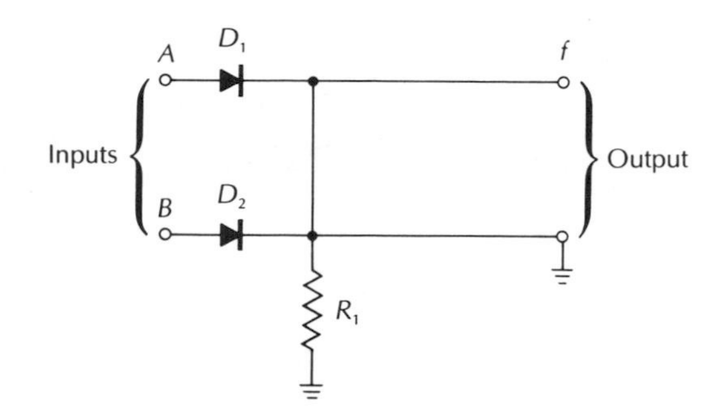

Figure 2-8 Diode OR Gate

Because the OR gate in Figure 2-8 is not compatible with TTL, we must take a different approach if we need an OR gate matrix that is compatible with TTL logic. We can take advantage of the concept of logic duals or of DeMorgan's laws, from which the duals were derived. (If you will refer back to Figure 1-14, you will find a summary of the logic duals.) In this case we can use the positive-logic diode AND gate (Fig. 2-7), and by adding some inverters we can get a TTL-compatible diode OR gate. Figure 2-9 shows how.

2-7 Transistor-Transistor Logic (TTL)

THE EVOLUTION OF TTL LOGIC

The TTL circuit is the culmination of several years of development and improvement in the electrical characteristics of logic gates.

Diode logic (a pre-integrated circuit form) has excellent switching characteristics, good isolation among inputs, a small input capacitance, and some inherent noise immunity. The fact that no amplification takes place makes each driven gate a heavy load on the driving gate.

Adding an amplifier stage to the diode logic circuit yielded the basic DTL diode-transistor-logic circuit in Figure 2-10. DTL gates are still found in digital systems, but they are being replaced by TTL in most new equipment.

The basic TTL circuit is shown in Figure 2-11. The introduction of the multi-emitter transistor considerably reduced the cost and chip space required for an integrated circuit logic gate as well as providing performance superior to that of DTL.

The TTL circuit still retains the excellent diode switching properties, because each emitter forms a junction diode with the single collector of

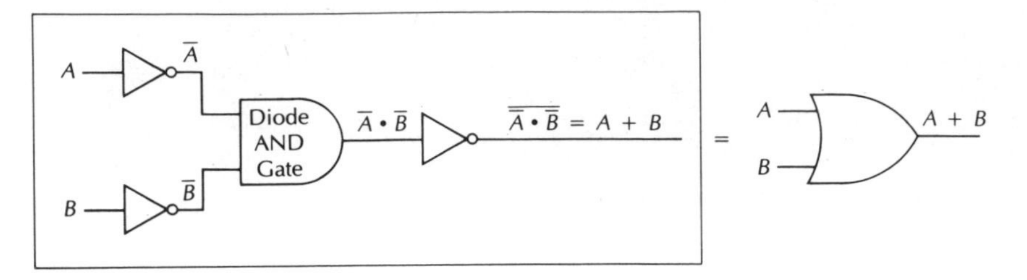

a. Logic circuit and equations

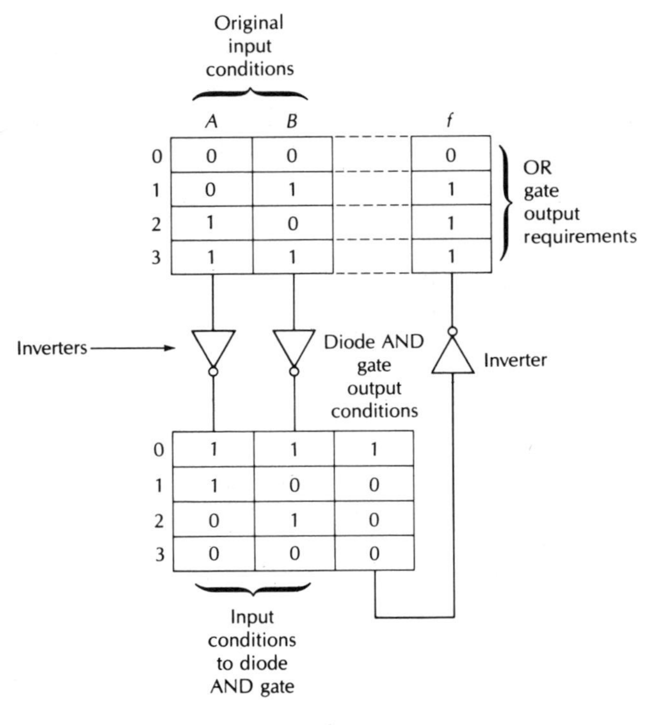

b. Truth tables

Figure 2-9 Converting the Diode AND Gate into an OR Gate

the multi-emitter transistor. The collector-base junction voltage drop provides additional noise immunity by increasing the input switching threshold voltage.

One other refinement has been added to TTL circuits in the form of a push-pull (generally called a *totem-pole configuration*) output stage

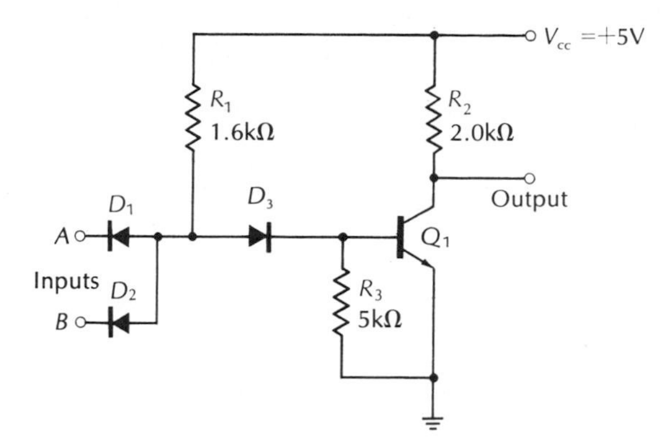

Figure 2-10 DTL NAND Gate Schematic Diagram

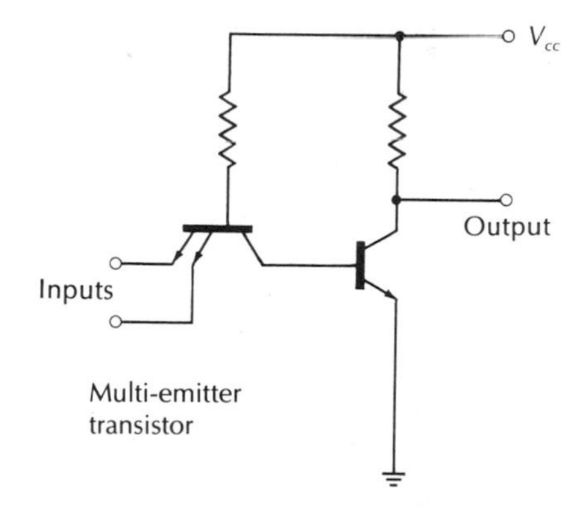

Figure 2-11 (Simplified) Basic TTL Circuit

(see Fig. 2-12). This output stage provides a low impedance drive to other gate inputs, allowing one gate to drive ten or more gate inputs. In addition, switching time is significantly reduced and made more predictable. There are two principal factors that tend to limit switching speed in a transistor—input capacitance and carrier storage within the depletion zone at the transistor collector-base junction.

CAPACITANCE

There is always some input capacitance that has to be charged before the transistor can switch. Part of the capacitance is simply stray-wiring capacitance, but generally more important is the collector-base junction capacitance when the transistor is turned off. The capacitance is the result of a reverse-biased junction that consists of two areas contain-

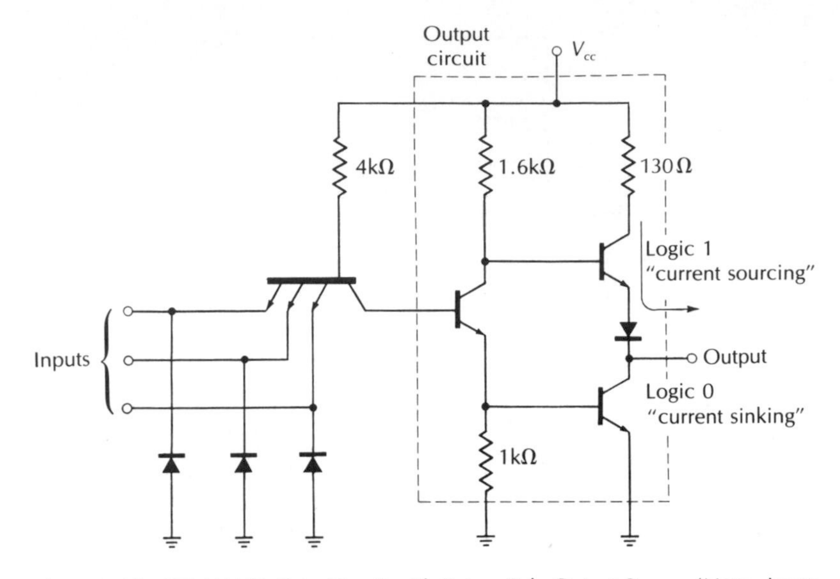

Figure 2-12 TTL NAND Gate Circuit with Totem-Pole Output Stage. (Note: Input diodes protect against negative input voltages.)

ing an abundance of free carriers separated by a layer (the depletion layer) that has been swept clear of carriers and behaves as a dielectric. The base-collector junction, in the reverse-biased condition, behaves as a capacitor between collector and base. Proper design of the transistor can minimize this capacitance, but there is another factor involved. The transistor does have voltage gain during the switching transition; and, because of the *Miller effect,* this small interelectrode capacitance appears to the driving gate as the voltage gain—interelectrode capacitance *product*. It might appear that voltage gain is a negligible factor because the steady-state input and output voltages are so nearly equal, a fact that would seem to imply near unity voltage gain. However, when we talk about propagation delays, we are actually referring to the time the transistor takes to slew from the *off* state to the *on* state and from the *on* state to the *off* state. During these changes in conduction states there is voltage gain involved, and the interelectrode capacitance is effectively amplified.

STORAGE TIME

Once the transistor is driven into a full *on* condition, called *saturation,* the collector-base junction is forward-biased and the depletion zone is saturated with carriers. When the time comes to switch the transistor from full conduction (saturation) into the *off* (no conduction) state, a finite time is required to sweep these carriers out of the depletion zone. The time required to clear these carriers out of the depletion zone is called *storage time,* which contributes to the overall gate propagation delay time.

The two factors, interelectrode capacitance and carrier storage, increase the rise and fall times resulting in the propagation delay illustrated in Figure 2-5.

THE IMPORTANCE OF THE TOTEM-POLE OUTPUT STAGE

On the rising part of the drive pulse one of the two totem-pole transistors provides a large charging current to charge input capacitances rapidly. On the falling part of the drive pulse the other transistor provides a low-impedance capacitor discharge path that also helps clear stored charges more rapidly. This action is known as *active pull-up* and *pull-down*. Pull-up involves a resistor or transistor that pulls the output *up* towards V_{cc}. Pull-down involves a device that provides a low-impedance path, pulling the output down towards ground.

THE 54/74 TTL FAMILY

There are two basic series of TTL logic—the 5400 series, which was originally designed to meet military requirements, and the 7400 series commercial variety. The 5400 series has an operating temperature range of from −55°C to +125°C and conforms to military documentation, testing, and reliability requirements. The 7400 series commercial grade TTL has a temperature range of from 0°C to 70°C. A military grade package can always be used as a direct replacement for its commercial equivalent although such a substitution is generally prohibitively expensive. A typical catalog number might read as 54/7404, indicating that the 5404 military grade and the 7404 commercial grade are both available. All manufacturers use the same numbers for equivalent devices, although a particular manufacturer may use a one- or two-letter prefix company symbol or an *in-house* number in addition to usual 7400 or 5400 series numbers. For example, the SN 7475 and 9375/7475 are both quad latches and are direct substitutes for one another.

TTL logic is the most widely used SSI and MSI logic family. Nearly every major manufacturer has a TTL product line, and most common TTL integrated circuits are produced by a number of companies. The TTL product line consists of the following subfamilies:

Standard TTL
Low-power TTL
High-speed TTL
Schottky-clamped TTL
Low-power Schottky TTL

VARIATIONS ON THE BASIC TTL CIRCUIT

The basic TTL circuit is the most common and popular of bipolar logic forms, but there are several variations of the basic circuit to satisfy special logic needs. All of the variations, however, have the following properties in common:

1. Supply voltage: $V_{cc} = 5.0$ V
2. Noise immunity: 1.0 V
3. Fan-in per gate input: 1 unit
4. Fan-out: 10

2-8 Operating Theory and Characteristics of the 7400 NAND Gate

We will examine the 7400 two-input NAND gate because it is the basis for the entire TTL family and the most often used member of the family. The NAND gate has a positive (high) output when either input is grounded (low). The output is low (ground) only when both inputs are high (positive).

The following are the NAND rules:

Output	*Input Conditions*
Positive (high)	Either or both inputs grounded (low)
Grounded (low)	Both inputs positive (high)

INPUT AND OUTPUT LEVELS

Input levels typically range from an absolute minimum of 2 V (for a logic 1 input) to a typical maximum 3.5 V. The inputs can tolerate voltages of up to 5 V, but negative voltages can damage the gate and must be avoided. An input left unconnected behaves as though there were 2 to 5 V being applied. However, inputs should not be left open in practice because it allows noise injection and erratic operation.

When an input is grounded, approximately 1.6 milliamperes (mA) of current flows along the input to ground path. If there is an appreciable resistance in the input to ground path, a voltage drop will be developed that can prevent the input from pulling near enough to ground for reliable operation. The resistance from input to ground should not exceed 500 Ω. The maximum voltage for the input to be effectively grounded is 0.8 V and is normally closer to 0.2 V.

Output levels range typically from 0.2 V for a low (ground) to 3.5 V for a high. These output levels satisfy the input requirements of less than 0.8 V for a low and greater than 2 V for a high. The output of a TTL gate can be (and almost always is) directly connected to the inputs of other TTL gates.

When the output is driving the input of another TTL gate at the low (0.2 V) level, it must be capable of sinking 1.6 mA. Standard TTL can sink 16 mA and can drive 10 standard TTL inputs.

When the output is driving the input of another TTL gate at the high (positive) level, input diodes in the driven TTL gate are reverse-biased. The driving gate need supply only a small leakage current while holding the voltage level at 2.4 V or greater (typically closer to 3.3 V). Table 2-2 summarizes input-output conditions.

THEORY OF OPERATION

Figure 2-13 shows the schematic diagram for the 7400 standard TTL two-input NAND gate.

Both Inputs Unconnected

Assume that both inputs are open (not connected to anything). There is no current flowing out of either of the emitters in the dual emitter transistor, Q_1. There is no transistor action and the collector-base junction is forward-biased. Current flows from the +5 V V_{cc} source, through the forward-biased junction, and into the base of Q_2. This turns Q_2 and Q_3 on. Heavy conduction through Q_2 shunts nearly all available current away from the base of Q_4. Q_4 turns off. The *output* drops to below 0.6 V, a logic low level. In this condition the output can sink up to 16 mA. For this reason TTL is often called *current-sinking logic*.

Table 2-2 Summary of Standard TTL Input-Output Conditions

	Input Logic 0 (Low)	Input Logic 1 (High)	Output Logic 0 (Low)	Output Logic 1 (High)
Minimum	0.0V	2.0V	0.0V	2.4V
Typical	0.6V	3.3V	0.2V	3.3V
Maximum	0.8V	5.0V	0.4V	3.6V

a. Voltages

	Input Logic 0 (Low)	Input Logic 1 (High)	Output Logic 0 (Low)	Output Logic 1 (High)
Typical	1.6mA	40μA	Sinks 16mA	Sources 400μA

b. Currents

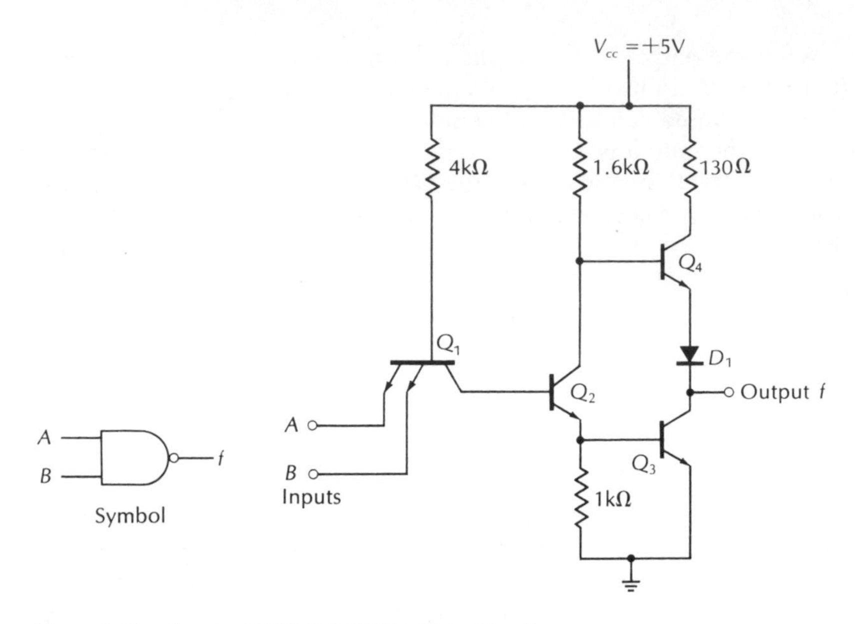

Figure 2-13 Standard 7400 TTL NAND Gate Circuit

Both Inputs Connected to Positive

Now assume that *both* inputs are connected to a voltage between 2 and 5 V. This condition simply drives the base-emitter junction deeper into reverse bias. The results are the same as when both inputs are left unconnected. Remember that even though the same results can be had by either unconnected inputs or inputs connected to positive, leaving inputs unconnected is poor practice because of the possibility of noise pickup.

One Input Grounded and One Positive

In this case the grounded emitter biases the emitter junction forward. Q_1 now behaves like a transistor and a large base current flows through the 4 kΩ base resistor, out of the emitter to ground. This pulls Q_1's collector to near ground potential, turning Q_2 off. The collector voltage of Q_2 rises to approximately V_{cc} turning Q_4 on. The output goes high. The voltage output is about 3.3 V because of the drop across the 130 Ω resistor, the collector-emitter drop of Q_4, and the diode (D_1) junction potential.

Both Inputs at Ground

When both inputs are connected to ground, both emitter-base junctions of Q_1 are potentially forward-biased. But because the two junctions are never absolutely identical, one will conduct more heavily than the other and hog the available current, holding the other junction in a reverse-bias condition. Only one of the input junctions is forward-biased at any given time, and one is all that is required.

So far we have paid very little attention to Q_1 as a transistor. For the most part it functions as a diode array, and calling it a transistor is more a matter of structural fabrication than function. When all emitters are high, the hFE (current gain) is less than unity.

If you carefully inspect the low to high (input) voltages, you will see that total current drain from the power supply varies during each phase of operation. This fact is important because these current variations can be reflected down both the V_{cc} and ground lines. The particular phase in which Q_2, Q_3, and Q_4 are all on, followed within fractions of a nanosecond by Q_4's turning off, generates a current spike of about ten times the normal current.

Table 2-3a summarizes the 7400 operation; Part b shows common symbols as they are generally found in the manufacturer's data manual. Typical values for TTL are also indicated in part c.

2-9 Decoupling

The current spike generated by the on-to-off or off-to-on switching transition of a TTL gate is necessary for fast operation, but if these spikes are allowed to propagate down the V_{cc} line, they can cause the system to malfunction.

To prevent current spikes from being reflected down the V_{cc} line, decoupling capacitors are used to store energy to supply the brief (spike) current demand. These capacitors are essential in all but the simplest TTL systems.

Actual values are less critical than the way in which they are distributed on the circuit board. For example, a single 0.2 microfarad (μF) capacitor would be less effective than four 0.05 μF capacitors properly distributed. Further, since electrolytic capacitors and wound mylar capacitors tend to be highly inductive, wound mylar capacitors should be avoided in favor of disc capacitors. Electrolytic capacitors should be of the tantalum type and bypassed by a 0.05 to 0.1 μF disc.

The following represent typical decoupling capacitor usage:

1. One 0.01 to 0.1 μF disc from V_{cc} to ground for every four small-scale IC's
2. One 0.01 to 0.1 μF disc for each pair of MSI IC's
3. One tantalum 10 μF 10 V capacitor where the V_{cc} line enters the circuit board
4. A 0.01 to 0.1 μF capacitor near any IC package that is more than 7.5 centimeters (about 3 inches) from the nearest decoupling capacitor

Table 2-3 Table of TTL Gate Operating Characteristics

Input A	Input B	Output
Unconnected	Unconnected	Logic 0 Low (ground)
+	+	Logic 0 Low (ground)
+	0 Gnd	Logic 1 High (+)
0 Gnd	+	Logic 1 High (+)
0 Gnd	0 Gnd	Logic 1 High (+)

a. Summary of 7400 NAND Gate Operation

Definition of Terms:

V_{cc} Power supply voltage: 5.0V ±10% (±5% for 5400 series)

V_{IH} High-level input voltage: voltage required for logic 1 at an input. It is a guaranteed minimum of 2.0V.

V_{IL} Low-level input voltage: voltage required for a logic 0 at an input. It is a guaranteed maximum of 0.8V.

V_{OH} High-level output voltage: voltage level output from an output in the logic 1 state. it is a guaranteed minimum of 2.4V.

V_{OL} Low-level output voltage: voltage level output from an output in the logical 0 state. It is a guaranteed maximum of 0.4V.

I_{IH} High level input current: the current flowing into an input when a logic 1 voltage is applied to that input

I_{IL} Low-level input current: the current flowing from an input when a logic 0 voltage is applied to that input

I_{OH} High level output current: the current flowing from the output while the output voltage is at logic 1

I_{OL} Low-level output current: the current flowing into an output, while the output voltage is at logic 0

Logic High State		Logic Low State	
V_{IH}	must be 2.0V or greater	V_{IL}	must not exceed 0.8V
I_{IH}	will not exceed 40 μA	I_{IL}	will source at least 1.6mA
V_{OH}	will be 2.4 V or greater	V_{OL}	will not exceed 0.4V
I_{OH}	will source at least 400 μA	I_{OL}	will sink at least 16mA

b. TTL Characteristics

Table 2-3 continued

Circuit Form	V_{CC}	Typical Power Dissipation per Gate	Fan-out	Propagation Delay per Gate	Immunity to External Noise	Clock Rate
Standard TTL (NAND)	5V	12mW	10	10ns	1V Guaranteed 0.4V	35MHz
High-speed TTL (NAND)	5V	22mW	10	6ns	Guaranteed 0.4V	50MHz
Low-power TTL (NAND)	5V	1mW	10	33ns	Guaranteed 0.4V	3MHz
Schottky TTL (NAND)	5V	20mW	10	3ns	0.4V	125MHz
Low-power Schottky TTL (NAND)	5V	2mW	10	10ns	0.4V	45MHz

c. Table of TTL Gate Operating Characteristics

All leads should be kept as short as possible and capacitors should be placed close to the IC packages.

2-10 Noise Immunity

TTL circuits generally are considered to be immune to noise of 1 V in amplitude, although as much as 1.5 V of noise is seldom a problem. The

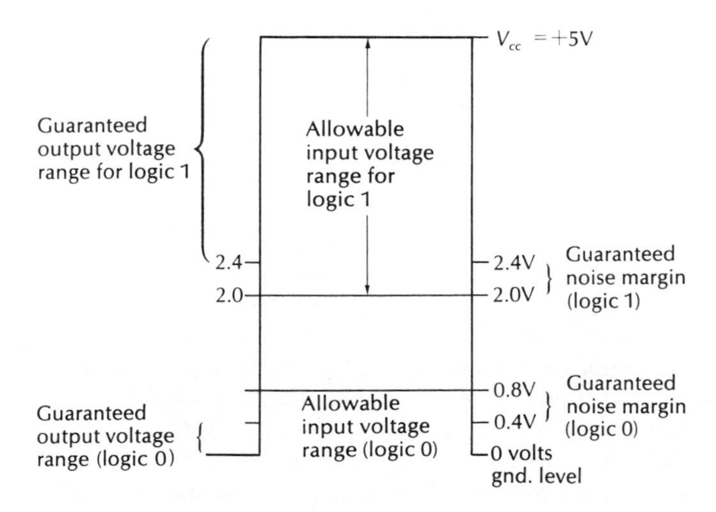

Figure 2-14 TTL Switching Levels and Noise Immunity

guaranteed DC noise margin is 0.4 V at all temperatures within the operating range. Because of the high switching speed of TTL circuits, most noise pulses are so slow by comparison that considering them to be DC is generally a realistic approach. Figure 2-14 illustrates TTL switching and noise immunity levels.

TTL SUBFAMILIES

In addition to the standard TTL family, there are four (common) distinct subfamilies: high-speed TTL, low-power TTL, Schottky-clamped TTL, and special TTL gates.

2-11 High-Speed TTL (High Power)

The high-speed TTL circuit uses a Darlington driver which improves the current drive to the totem-pole output to increase switching speed. The resistor values are lower than in the standard TTL and, consequently, the high-speed TTL has a high power consumption, 22 milliwatts (mW), as compared to the standard TTL's 12 mW. The high-speed TTL has a 6 nanoseconds (ns) delay time and a flip-flop operating speed of 50 megahertz (MHz). The standard TTL has a propagation delay time of 10 ns and a flip-flop speed of 35 MHz.

The high-speed unit provides roughly twice the speed of standard TTL at the expense of approximately twice the power consumption. The fan-out of the high-speed TTL is 10, but its fan-in is about 1.3. A standard TTL gate can drive no more than 7 high-speed TTL gate inputs. High-speed TTL is gradually being replaced by the more recent Schottky TTL gates. Figure 2-15 shows the high-speed TTL circuit. A 74H04 is the number designation for the high-speed version of the standard 7404.

2-12 The Low-Power TTL

The circuit for the low-power TTL is essentially the same as for the standard TTL except that resistor values have been increased, reducing the power consumption from 12 mW to 1 mW, with a decrease in speed from 10 ns for the standard TTL to 33 ns for the low-power version. Number designations containing an *L* indicate low power; for example, 74L04 is the low-power version of the standard 7404.

The low-power TTL gate has a fan-out of 10. It will drive ten low-power TTL inputs but only one standard TTL input. The low-power TTL gate is meeting considerable competition from complementary MOS (C-MOS) devices.

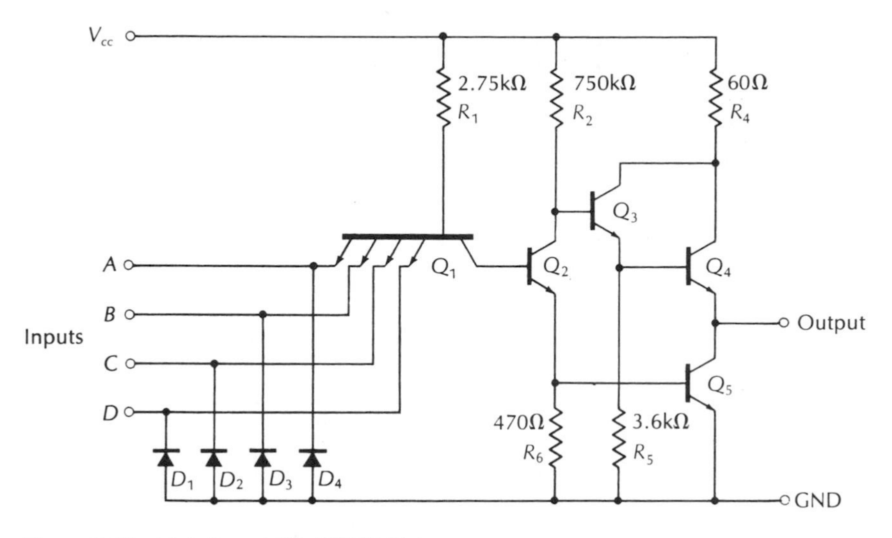

Figure 2-15 High-Speed TTL NAND Gate

2-13 The Schottky-Clamped TTL

The Schottky-clamped TTL is the fastest member of the TTL family. With a 3 ns propagation time and a 125 MHz clocking rate, it is a serious rival to the fastest available logic (ECL) family with its 1 to 4 ns propagation times.

The Schottky TTL also compares favorably with ECL (emitter-coupled logic) in power consumption and cost, as well as being compatible with the other members of the TTL family. The secret of this circuit's high speed is the elimination of storage-time delay by clamping the collector of each transistor to its base with a Schottky barrier diode. This prevents the collector-base junction from going into saturation in which the junction becomes heavily forward-biased, drawing carriers into the depletion zone that must be cleared before the transistor can be turned off. The use of a clamping diode between collector and base to prevent saturation is not a new idea, but ordinary junction diodes also have some storage problem. The Schottky diode has no storage problems and virtually eliminates transistor carrier storage when used as a collector-base clamp. Figure 2-16 shows a Schottky-clamped transistor, the special symbol for a Schottky transistor, and the schematic diagram of the Schottky TTL *NAND* gate. The very popular low-power Schottky circuit features the same speed as the standard TTL, but with a power requirement of only 2 mW per gate as opposed to the 12 mW for a standard TTL gate. Table 2-4 compares the several TTL subfamilies with respect to speed and power consumption.

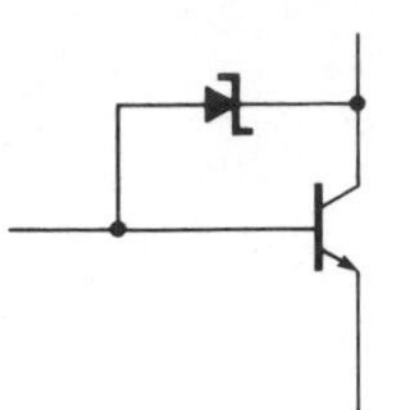

a. Transistor and Schottky barrier diode clamp

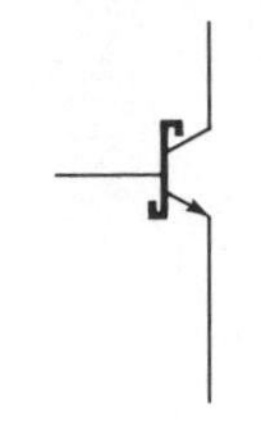

b. Symbol for transistor with Schottky barrier diode clamp

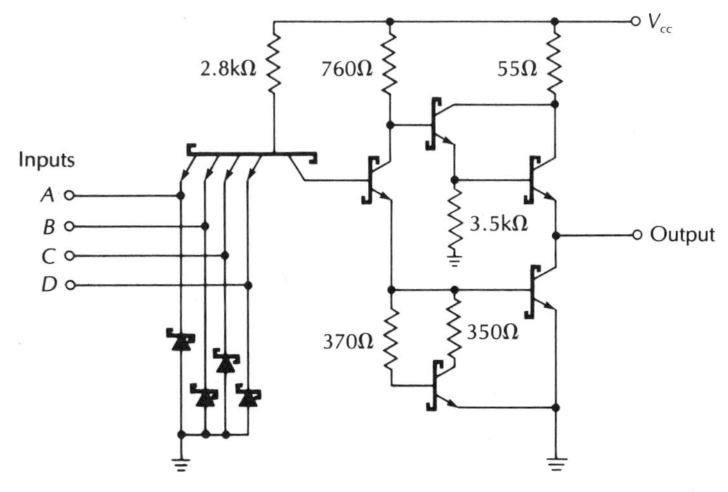

c. Schematic diagram

Figure 2-16 The Schottky Diode Clamped TTL NAND Gate

Table 2-4 Comparison of TTL Subfamilies

TTL Subfamily	Gate Power (milliwatts)	Propagation Time (nanoseconds)	Upper Counting Frequency Limit (megahertz)
Standard	10	10	35
High-speed	22	6	50
Low-power	1	33	3
Schottky	19	3	125
Low-power Schottky	2	10	45

2-14 Special TTL Gates

OPEN COLLECTOR GATES

Most TTL gates use a totem-pole output stage. The outputs can be tied together only via the inputs of another gate. They cannot be tied directly together because if one output goes high while another goes low, the direct connection will yield an indeterminate output level. In addition, gate damage is a very likely result.

Some TTL gates are available using a single output transistor with an uncommitted collector. Figure 2-17 is a typical example of an *open collector* gate.

Outputs of open collector devices can be connected together to a common external load resistor. For example, all of the outputs of a 7405 Hex inverter (six inverters in a package) can be connected in this way to get a six-input NOR gate (Fig. 2-18e).

If two 7401 open-collector two-input NAND gates are connected as shown in Figure 2-18d, the result is not a four-input NAND gate but rather the following logic function:

$$f = \overline{(A \cdot B) + (C \cdot D)}$$

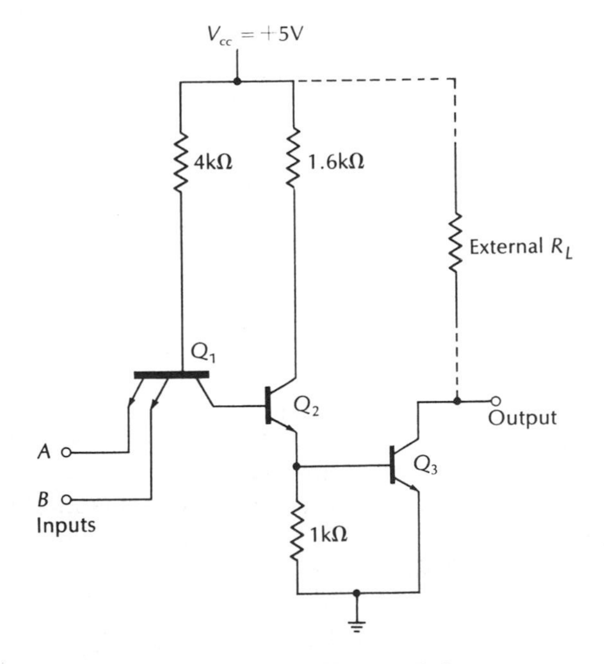

Figure 2-17 Open Collector TTL NAND Gate

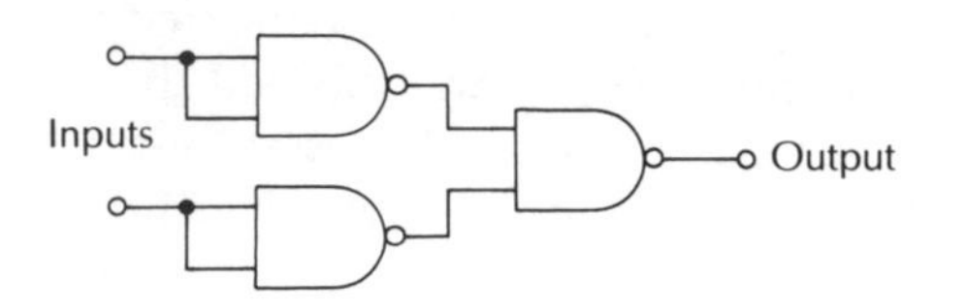

a. Conventional (ORed) circuit

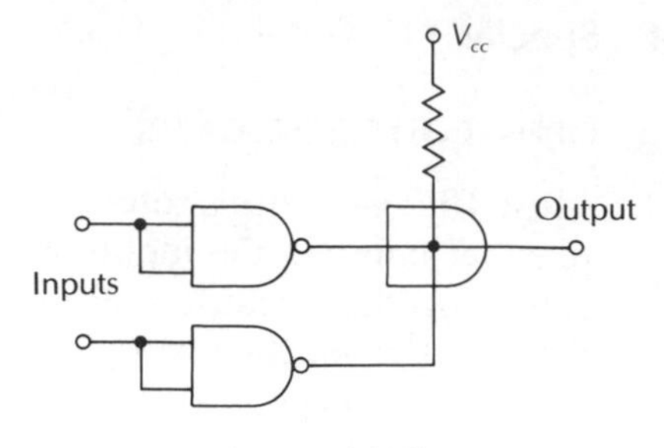

b. Wired AND

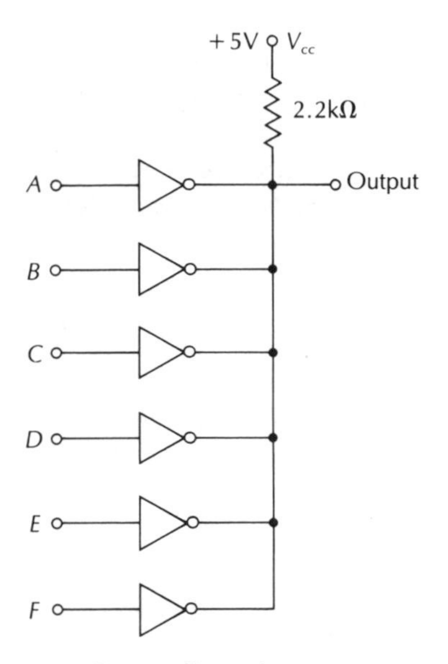

c. Open collector inverters wired as 6-input NOR gate

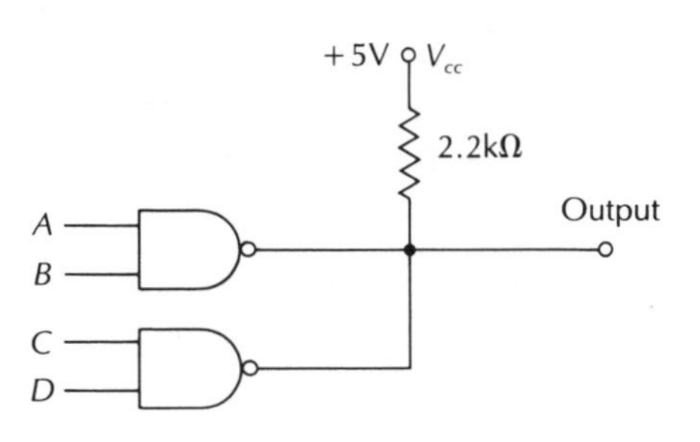

d. Circuit for $f = \overline{(A \cdot B) + (C \cdot D)}$

Figure 2-18 Hard-Wired Open Collector Gates

The wired output configuration is not very popular with designers for several reasons: open collector circuits are much slower than totem-pole output devices; the noise problem in open collector circuits is greater than in totem-pole devices; and troubleshooting hard wired groups of open collector gates is next to impossible without serious damage to the circuit board.

AND GATES

An AND gate can easily be obtained by inverting the output of a NAND gate. Combining two gates adds additional propagation delay, power consumption, and stray wiring capacitance. By building the inverter into the IC, we can form an AND gate that adds only about 4 ns delay and 5 mW of power consumption and eliminates external wiring problems. Input-output characteristics are the same as for the NAND gate.

NOR GATES

A similar approach could be taken to fabricate OR or NOR by combining NAND's on a chip but the total propagation delay would be greater than would be acceptable if such a gate were to be used with other gates having much shorter propagation times. Instead, the NOR gate uses a slightly different circuit but does have the same basic input and output circuits to insure compatibility with the rest of the TTL family.

SCHMITT-TRIGGER NAND GATES

Because many data sources are comparatively slow and may produce waveforms that are not acceptable to high-speed logic circuits, a NAND gate with a *Schmitt trigger* is available. The 5413/7413 Schmitt NAND accepts almost any input waveform and produces an output waveform fully compatible with the requirements of TTL circuits. A Schmitt trigger inverter is also available. The Schmitt trigger is a threshold-controlled regenerative switch. It is particularly useful when a slowly changing waveform, such as a sine wave or sawtooth, must be reshaped into a fast-rising pulse to be compatible with high-speed logic systems. The important characteristics of the Schmitt trigger are the lower-trip-level (LTL), the upper-trip-level (UTL), and the difference between the LTL and UTL voltages, called the *hysteresis* voltage.

An ordinary (TTL) gate has a conduction threshold level of about 0.8 V, at which point the transistor begins operating in its linear region where the gain is high. During the transistor's transition through the linear region, it is very sensitive to noise or signal amplitude variations.

The normal logic pulse drives the transistor through the sensitive region in a very short time and provides a constantly rising drive signal. A slow rising signal tends to remain for a comparatively long time in this sensitive high-gain condition where even small noise levels or signal amplitude variations can cause the gate to jitter—to fall in and out of the on state. The system could well interpret these variations as several input pulses instead of one or reject the input signal as no pulse at all.

The Schmitt trigger provides a regenerative action that drives the transistor rapidly through the linear region, once the action has been initiated by the input signal. In addition, it provides two threshold levels—one to initiate the trigger's *turn-on* condition and the other, a significantly lower level, to initiate the regenerative *turn-off* action. The regenerative action causes a fast-rising (or falling) drive that simulates the rise of a normal digital pulse. The hysteresis makes the system insensitive to noise and small signal amplitude variations that could otherwise cause false triggering. Figure 2-19 shows typical Schmitt trigger waveforms. In this case we will use V_{t+} for the positive-going trip level and V_{t-} for the negative or downward-going trip level. This convention is becoming more common in TTL systems than the notation UTL and LTL.

Figure 2-20 shows the symbolic representations for a Schmitt trigger input. In Figure 2-20b the Schmitt trigger symbol is shown in conjunction with a NAND gate, indicating a gate with a built-in Schmitt trigger. The 7413 is a dual four-input NAND gate with Schmitt action. It will respond to very low rates of change but not to a discrete DC level. Typical TTL Schmitt devices have a hysteresis voltage of about 0.8V.

The Schmitt trigger is essentially a two-gate, two-state multivibrator of the general form shown in Figure 2-21.

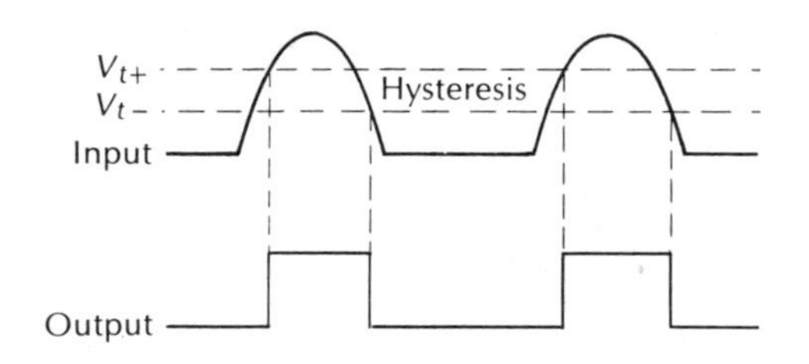

a. Schmitt input and output waveforms

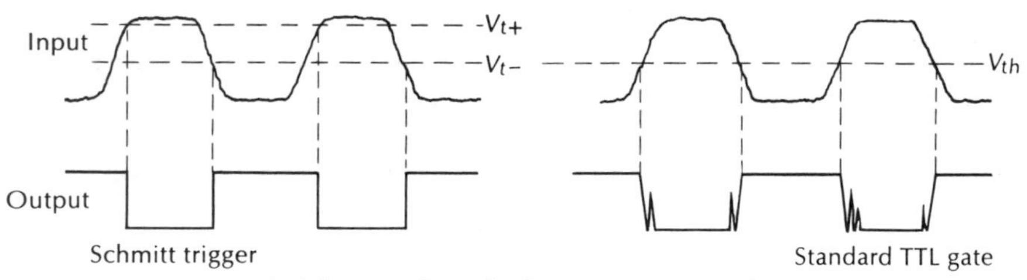

b. Schmitt and standard TTL gate compared

Figure 2-19 Schmitt Trigger Waveforms

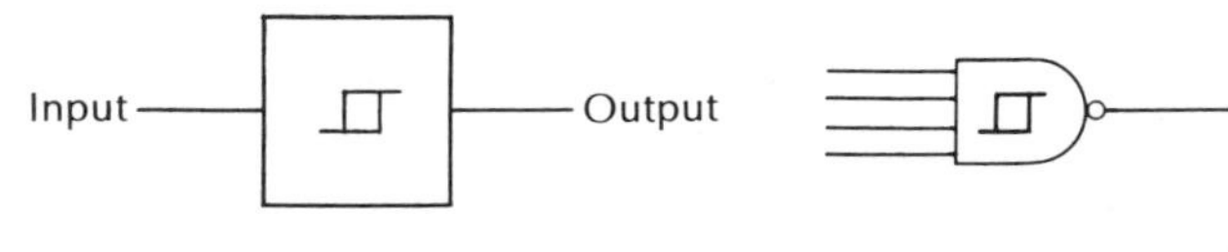

a. General Schmitt trigger symbol b. Gate with Schmitt action

Figure 2-20 Schmitt Trigger Symbols

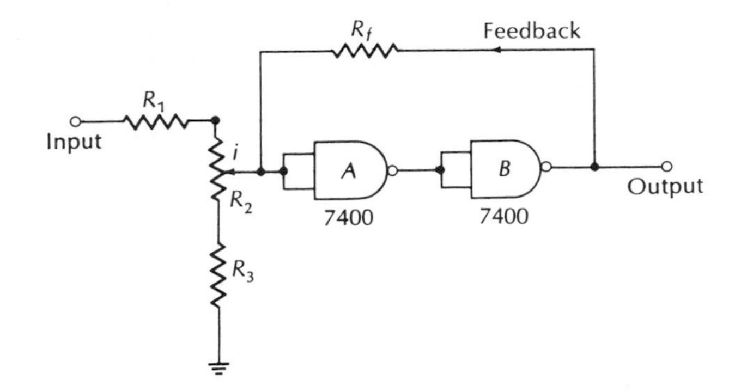

Figure 2-21 Two-Gate, Two-State Multivibrator

The circuit operates as follows: Assume (as initial conditions) that gate A is outputting a high as a result of 0 volts input signal at point i and that gate B is outputting a zero. Let the signal voltage rise to the input threshold of gate A. Gate A switches to an output low, switching gate B to a high output. Because the two gates are inverting gates, the feedback through R_f to the input of gate A is positive and the action is regenerative, simulating the fast rise/fall time of the normal digital pulse.

Let the input signal start its negative-going transition toward 0 V. Because the feedback resistor has a high at the end connected to gate B, it pulls the input (point i) up toward V_{cc}. Therefore the input signal must drop considerably lower than the turn-on threshold before gate A can switch to a high, initiating the regenerative turn-off action. The voltage divider (R_2 – R_3) divides the signal setting the effective threshold voltage. R_1 isolates the signal source from point i, allowing it to float to prevent excessive loading of the feedback resistor R_f.

GATES WITH TRISTATE OUTPUTS

The tristate output gate was designed for systems where a large number of gates must tie to a common bus. The open collector devices have proved unsatisfactory for this application. Minicomputers, for example, are organized with a data-bus system in which many gate outputs are tied to a common line. Tristate logic uses the two standard logic levels and has an added third state that is not a logic level but an open-circuit condition that effectively disconnects the output of the gate from the bus. The gate does not interract with the bus in any way unless it is *enabled* by a signal on a special *enable* input. In the two normal logic states a totem-pole circuit provides active pull-up and pull-down with the speed of standard TTL gates (Figure 2-22a). When

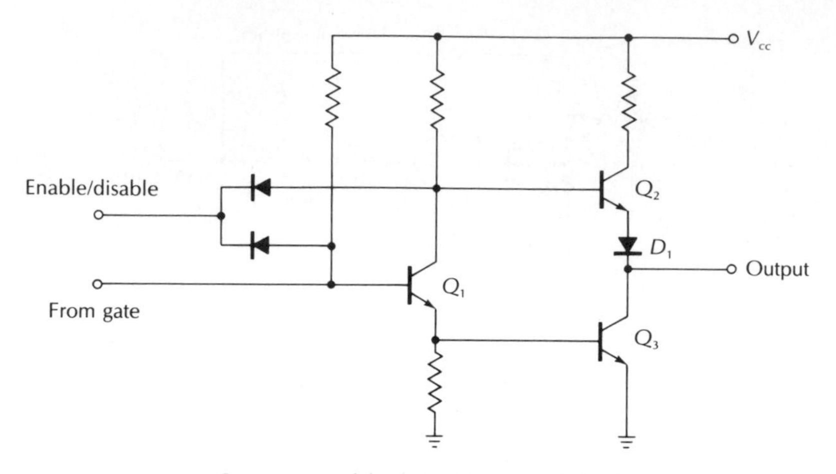

a. One version of the TTL tristate output circuit

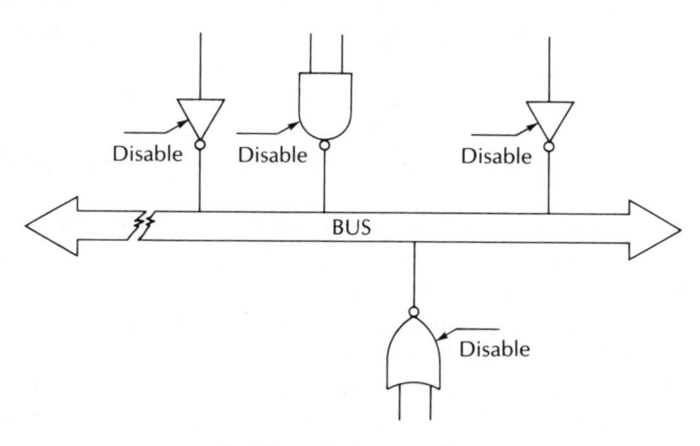

b. Tristate devices on a bus

Figure 2-22 Tristate Devices

tristate gates are used, two of them on the same bus must not be enabled and allowed to go to opposite logic levels at the same time. Since totem-pole output circuits cannot operate in this mode, this problem of "fighting for the bus" is one that must be carefully avoided. Figure 2-22b shows four tristate devices on a common bus.

Tristate devices are not used indiscriminately; but where they are necessary, there is no good substitute. Because of the extra enable inputs, a given package contains fewer tristate gates than would be usual for ordinary TTL.

-15 Power and Speed

FAN-OUT AND FAN-IN

Each family has a basic gate circuit (NAND, NOR, and so on) that is considered the standard gate for that family. The input current (at the appropriate input voltage) for an input on the standard gate is considered to be a unit load for that family. The fan-out number for a gate defines the number of standard gate inputs it can drive. A gate with a fan-out of 10 can drive 10 standard inputs.

A single unit load is typical of a simple gate input, but reset operations for an MSI counter or other circuit may require that several gate inputs (internally connected) be driven at the same time, counting as more than one unit load. If such an input requires three unit loads, it is said to have a fan-in of 3 and will require three fan-out units to drive it.

Fan-out numbers are valid only within a particular logic family. In the popular TTL family, there are several subfamilies, and fan-out or fan-in numbers are valid only within the subfamily. TTL subfamilies are only occasionally mixed within a particular logic system, but when they are, fan-out and fan-in numbers must be taken into account. For example, the low-power Schottky subfamily of TTL has a fan-out of 10 within the subfamily, but a low-power Schottky gate output can drive only five standard TTL gate inputs.

MOS logic gates have a typical fan-out of about 50, but they are low-power devices and can usually drive no more than one TTL input. It is very common to find a mixture of TTL and MOS devices in a given logic system. Each family and subfamily contains a few special gates and inverters with larger fan-out numbers than the rest of the family. These high fan-out devices can be used to drive an unusually large number of gate inputs, to couple a low-power TTL device or MOS device to a higher power TTL device, or to drive outside-world devices such as numerical displays.

POWER DISSIPATION AND SWITCHING SPEED

Power dissipation is a measure of circuit power dissipated in heat, plus the working power required by a gate or system of gates. It is often rated in milliwatts per gate.

Logic systems tend to become very complex. Because complex logic systems use large numbers of gates, the power consumption per gate must be kept low. Logic gates must also switch at very high speeds, and there is a definite trade-off between power consumption and speed. In general, the faster the switching speed the greater the power consumption. The cost of delivering regulated power is a factor, but a more important aspect of power is that it limits the circuit density on the chip. The more power a gate requires the more heat it produces. More heat requires more chip area to dissipate the heat. TTL technology is limited to less than 100 gates per chip because it is so power-hungry.

MOS gates, though slow compared to TTL, use very little power and thousands of MOS gates can be put on a chip.

2-16 MOS Gates

MOS gates are based on MOS field-effect transistors. MOS transistors exist in two complementary structures, *P*-channel and *N*-channel. *N*-channel devices are more common than *P*-channel devices because they are faster and more compatible with TTL power supply requirements. Both *P*-channel and *N*-channel devices are restricted to use in MSI and LSI circuits. Unlike TTL, they are not available as a family of general-purpose logic building blocks. These MOS devices are found in large-scale memory, complex interface devices, and complete microprocessor packages.

Figure 2-23 shows the structure and symbol for the *P*-channel MOS transistor. Notice that the gate is a metal or silicon film, a conductor. The substrate is also conducting silicon. The gate and substrate are separated by a very thin layer of silicon dioxide (glass). The gate-dielectric-substrate arrangement forms a tiny capacitor. The input (gate to source) resistance is extremely high. Appreciable current is required only while charging or discharging the input capacitor.

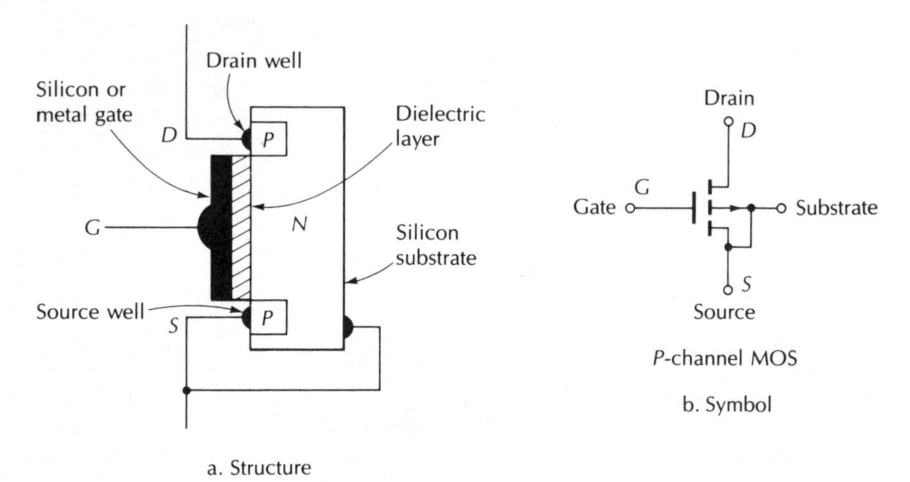

Figure 2-23 *P*-channel MOS Transistor

If you look again at Figure 2-23, you will see two wells of *P*-type silicon embedded in the *N*-type substrate. Each *P*-type well forms a *P-N* junction with the *N*-type substrate. The diode's junctions are intrinsically reverse-biased to provide the necessary dead-band. Because of the common *N*-region and because of certain voltage gradients, the amount of intrinsic reverse bias can be controlled to an extent. It can range from one to several volts. Figure 2-24 shows the structure of the *N*-channel transistor and its associated symbol. The only structural difference is that the *P* and *N* regions have been interchanged.

THE *N*-CHANNEL MOS INVERTER

Figure 2-25 shows an *N*-channel inverter. Assume that the input (gate) is low as shown in Figure 2-25a.

Input Low

The drain-well *N* region forms a diode junction with the *P*-type substrate. The junction is reverse-biased by $V+$. The *N*-type source well is at the same potential as the substrate, but it is still intrinsically reverse-biased. The path between source and drain consists of two reverse-biased diodes in series. There is no source-drain current. Because there is no current through R_L, there is no voltage drop across it, and the output is equal to $V+$. The gate output is high.

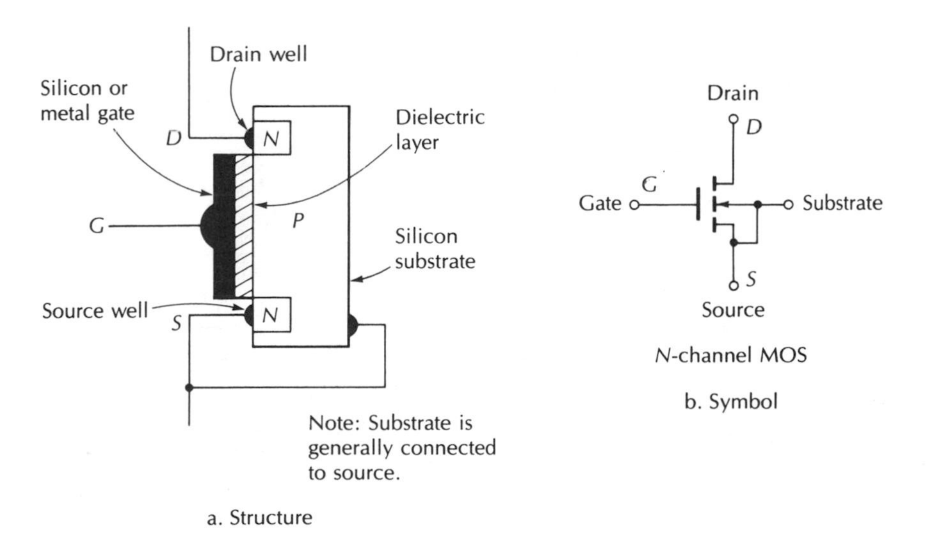

Figure 2-24 *N*-channel MOS Transistor

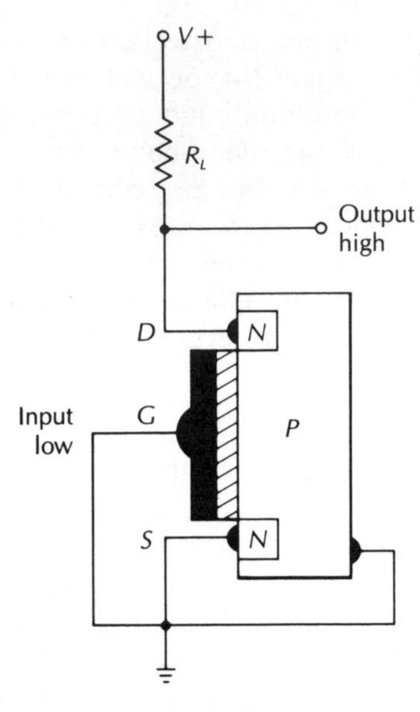

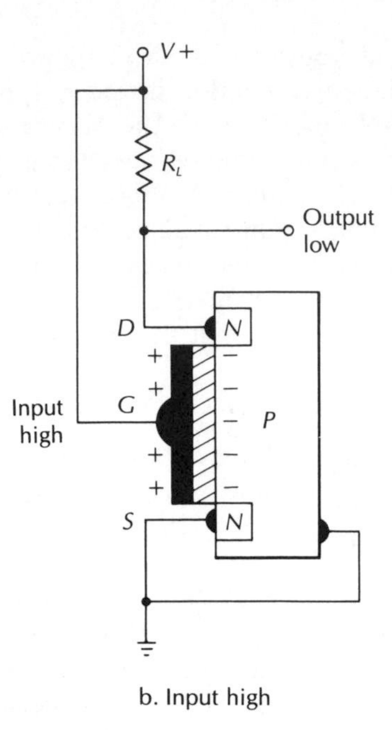

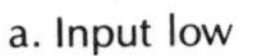
a. Input low

b. Input high

Figure 2-25 *N*-channel MOS Inverter

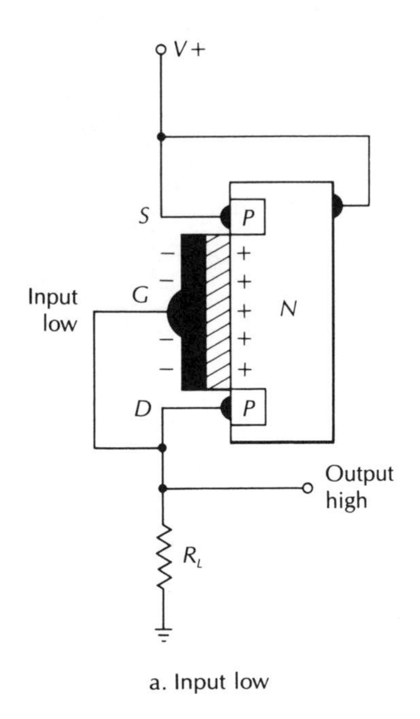

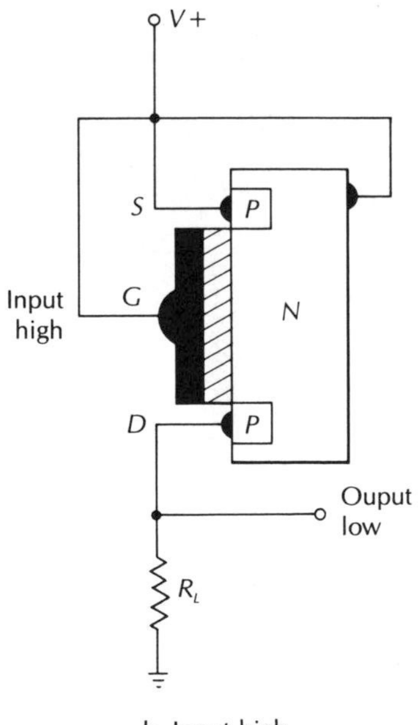

a. Input low

b. Input high

Figure 2-26 *P*-channel Inverter

Input High

Applying a positive potential to the gate (Fig. 2-25b) draws electrons up through the *P*-type substrate. Electrons enter the substrate from the negative (GND) terminal of the power supply. Electrons, attracted to the positive gate, congregate along the surface of the dielectric, combining with local holes in the *P*-substrate. Eventually all the holes along the surface of the dielectric will be filled and there will be some extra electrons in the area. The space between the drain and source wells will now act as a continuous channel of *N*-type silicon. This channel has a low resistance from drain to source and a large current flows through R_L. The output voltage drops to a few tenths of a volt and the output is low.

The *P*-channel inverter in Figure 2-26 operates exactly as the *N*-channel version except that all polarities are reversed and the channel (formed by a *negative* charge on the gate) is *P*-type silicon.

Because MOS devices tend to be quite complex, a general symbol that can represent either *P*-channel or *N*-channel devices is often used to make schematics easier to read. The symbol is shown in Figure 2-27a.

MOS GATES

In the MOS inverter in Figure 2-27b, the drain load resistor has been replaced by a MOS transistor (Q_2). An ordinary high-value resistor on an integrated circuit chip takes up far more chip space than a MOS transistor. Using the resistance of a MOS transistor channel instead of a separate resistor also simplifies the manufacturing process.

The circuit in figure 2-27b works the same way as the inverter in Figure 2-26a.

The MOS NAND gate in Figure 2-27c bears a close resemblance to its series mechanical switch equivalent. Transistors Q_1 and Q_2 must both be *on* to pull the output to ground (low).

The MOS NOR gate in Figure 2-27d is also very similar to the parallel mechanical switch equivalent. Either transistor, Q_1 or Q_2, or both will pull the output to ground (low) when it is turned on.

2-17 The C-MOS Family

Combining *N*-channel and *P*-channel transistors in the same device yields the highly versatile C-MOS or complementary MOS family. C-MOS is in direct competition with the TTL family and provides a complete array of individual gate packages, flip-flops, counters, and a number of other MSI devices. C-MOS devices are much slower than

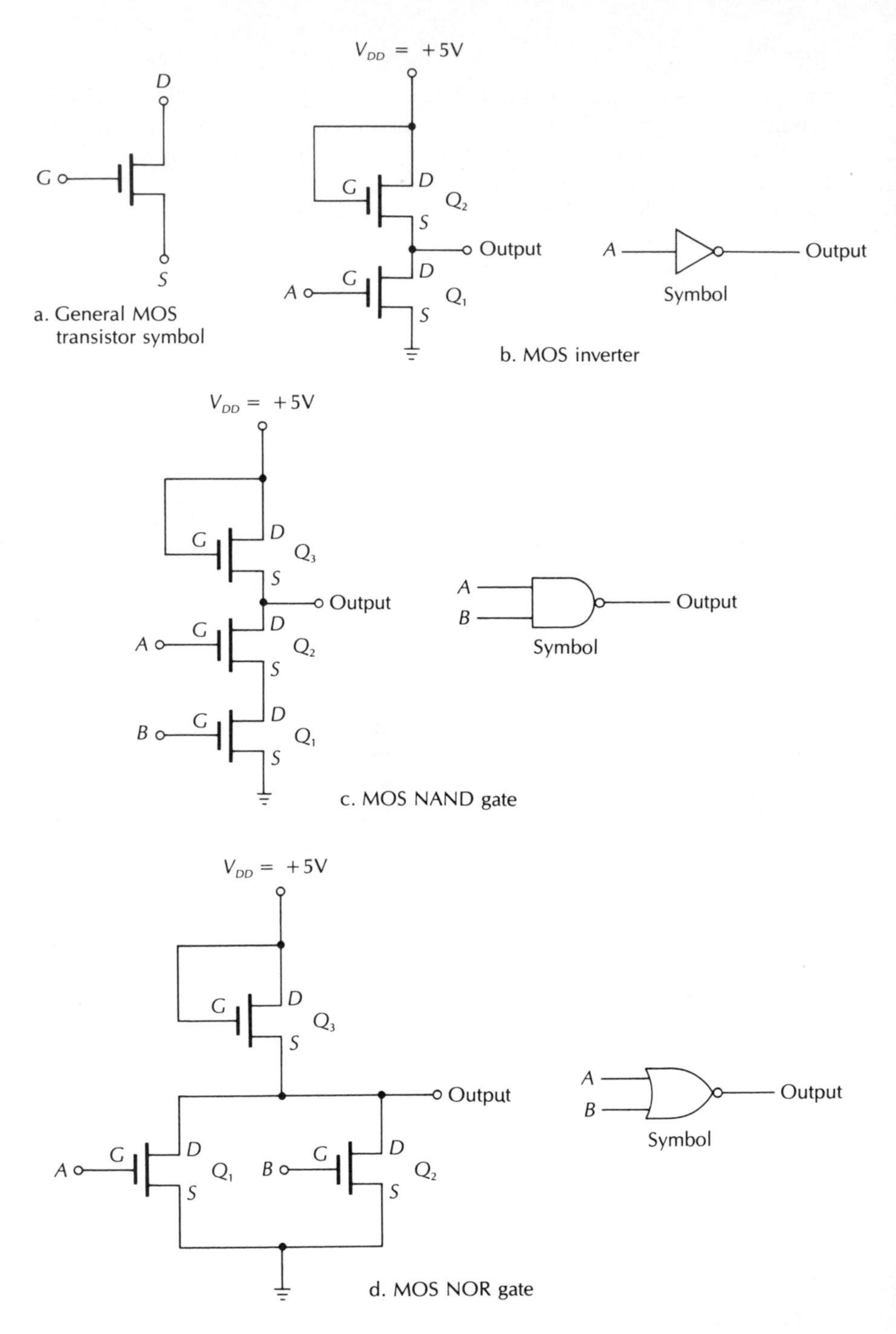

Figure 2-27 MOS Gates

TTL devices but have a number of offsetting advantages when TTL speed is not required.

SOME C-MOS FEATURES

1. Low power drain. C-MOS consumes from about 0.5 to 2 mW per gate as compared to 15 mW for a standard TTL gate.
2. Low input power. There is no input power required when the gate is in a stable on or off state. Input power is required only during the transition period from high to low or low to high. Input current is then required to charge or discharge the transistor gate capacitor. Because there are more changes of state per second at higher frequencies, the average power consumption of C-MOS devices goes up in direct proportion to the frequency of operation.
3. Good noise immunity. The logic state in C-MOS devices changes at halfway between ground and V_{cc}, yielding the best possible noise immunity. The noise immunity is approximately $\frac{1}{2}V_{cc}$.
4. No switching transients. In C-MOS there is some small overlap as one transistor turns off and another turns on. Unlike TTL, C-MOS does not generate current spikes.
5. No critical power supply voltage. C-MOS devices will work with power supply voltages from about +3 to +15 V. Power supply regulation is not very critical.
6. Large fan-out capability. Because the C-MOS input is simply a small capacitor, only a very small input current is required during switching and no input current at all is required in the steady-state condition. Theoretically, the output circuit of a C-MOS gate should be able to drive an almost infinite number of C-MOS gate inputs. In practice, a fan-out of 50 is about the maximum.

C-MOS TYPES

There are three types of C-MOS devices; the older A or unbuffered series, the improved B (buffered) series, and silicon-on-sapphire (SOS) series.

The A series devices are gradually being replaced by B series devices. The A series is now often called the UB (unbuffered) series.

The B series devices consist of the basic A series gate structure followed by two inverter stages (see Fig. 2-28.) The buffers provide an equal output drive in both high and low directions and a standardized output drive for all gate types. The buffers also improve the switching

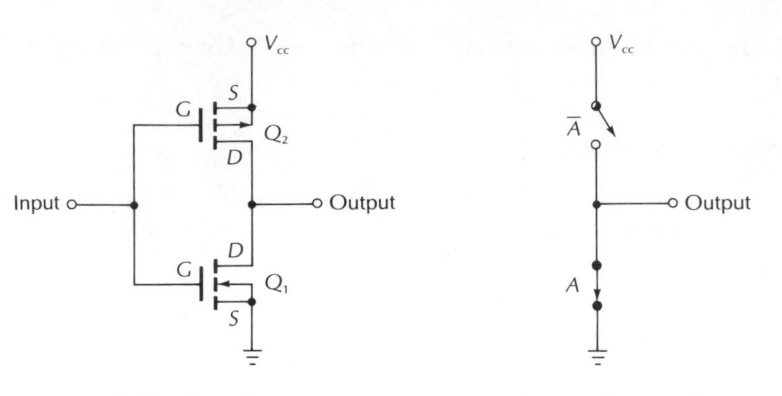

a. Unbuffered C-MOS inverter

b. Equivalent switching circuit

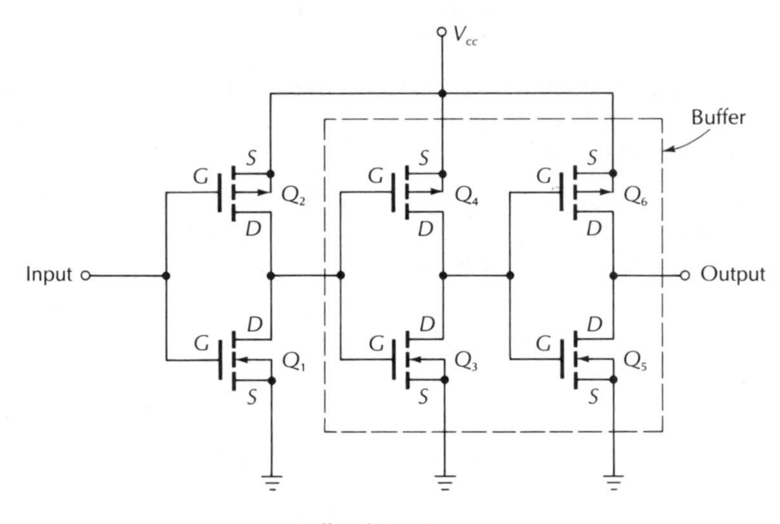

c. Buffered C-MOS inverter

Figure 2-28 C-MOS Inverters

characteristics, as shown in Figure 2-29. Improvements in fabrication technology have made up for the speed loss due to the buffer's added propagation delay. As a result, the buffered devices are often slightly faster than the unbuffered devices.

The silicon-on-sapphire (SOS) family is, at this writing, a comparatively expensive series that has a limited selection. SOS devices are about ten times as fast as B series devices and use less power but are not as easily available as the standard B series.

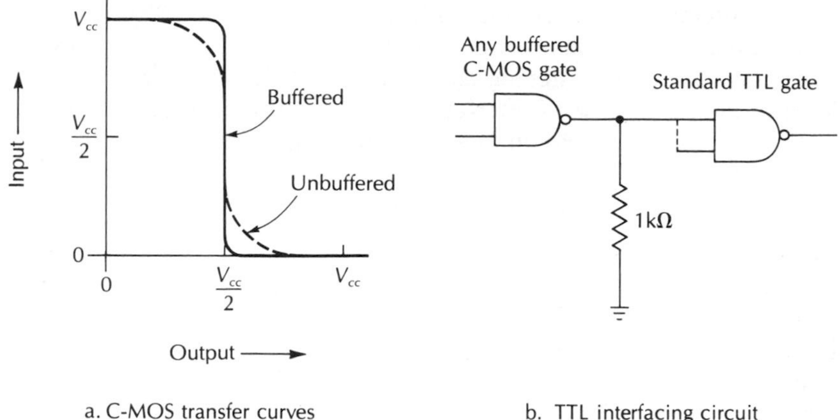

a. C-MOS transfer curves b. TTL interfacing circuit

Figure 2-29 C-MOS Transfer Curves and TTL Interfacing Circuit

2-18 Input Circuit Protection

The dielectric of the MOS gate capacitor is very thin and can be easily damaged by stray static charges. All C-MOS integrated circuits have built-in input circuit protection. Both Zener and ordinary diodes are used, and the protection arrangement varies slightly among manufacturers. In normal operation, protective circuits are inoperative and have no effect on circuit behavior. Occasionally, however, the protective circuit will play havoc with certain *R-C* time constant circuits. An extra series diode is sometimes required to get around the problem.

The input protection circuits provide good in-circuit protection. When the integrated circuits are handled off the circuit board, they should be kept in a conductive foam carrier. If C-MOS input leads leave the circuit board, a resistor to ground should be provided at the inputs.

2-19 C-MOS Gates

The C-MOS inverter circuit is shown in Figure 2-28a with its switching equivalent in Figure 2-28b. Notice that:

1. Q_1 is always on when Q_2 is off.
2. Q_1 is always off when Q_2 is on.

The transistor gate draws current only while the gate capacitor is being charged.

The buffered inverter circuit is shown in Figure 2-28c. The buffer for nearly all gates consists of two inverters. Figure 2-29a shows how the buffer improves the switching characteristics. The addition of the buffer also makes it easy to interface B series C-MOS to standard TTL inputs. The interface circuit is shown in Figure 2-29b.

C-MOS NAND AND NOR GATES

The C-MOS NAND gate schematic is shown in Figure 2-30 along with its switching circuit equivalent. Notice that all connections to V_{cc} must be open when the output is at ground. All connections to ground must be open when the output is connected to V_{cc}.

Figure 2-31 shows the schematic and equivalent switching diagrams for the C-MOS NOR gate. Here again, all switches (or transistors) connected to V_{cc} must be open when the output is grounded. And, conversely, all switches (or transistors) connected to ground must be open when the output is at V_{cc}.

By the time the buffer circuit is added to the transistors required for a basic C-MOS gate, the number of transistors is fairly high. Although MOS transistors are small (compared to bipolar), the large number of transistors per gate limits C-MOS to the lower edge of LSI circuit complexity.

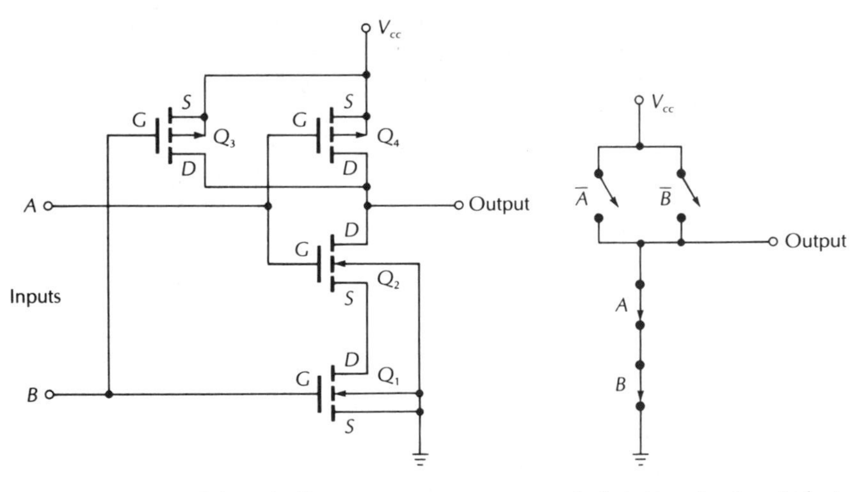

Figure 2-30 C-MOS NAND Gate

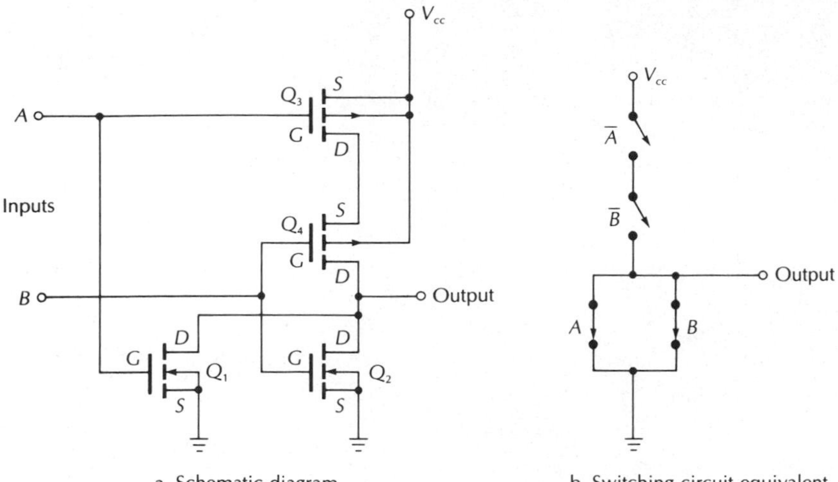

Figure 2-31 C-MOS NOR Gate

20 Transmission Gates

There is a C-MOS gate structure that has no bipolar (TTL) counterpart. It is called a *transmission gate* and its schematic diagram is shown in Figure 2-32. The transmission gate is bidirectional and can be used for analog as well as digital signals.

RESTRICTION

The signal voltage on either input/output connection must not be greater than V_{cc} nor less than ground. For AC signals, a negative voltage may be substituted for ground. This is often done when the transmission gate is used in analog circuits.

21 Dynamic MOS Logic

A form of logic circuit is possible in MOS circuits that has no counterpart in bipolar circuits. Dynamic logic is a sampling type of logic that works only because the input to an MOS transistor is almost completely capacitive. A logic circuit driving the input of an MOS transistor needs only to provide a current path for the short time required to charge the input capacitance. Once charged, the capacitor can hold the transistor in an *on* state until the charge leaks off. This dynamic form of

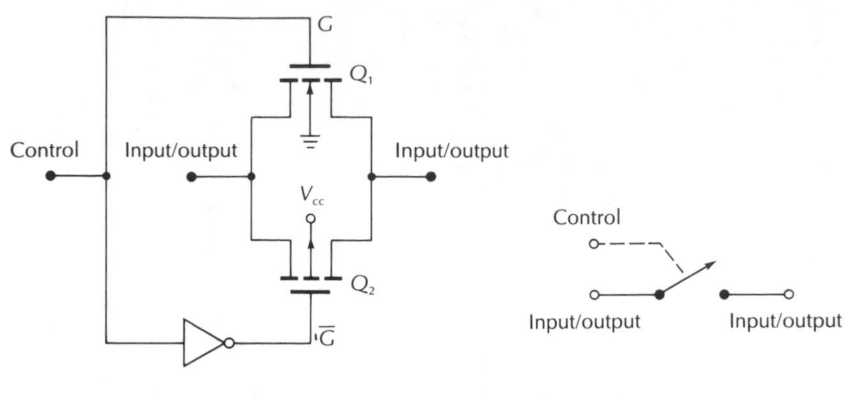

a. Schematic diagram

b. Switching circuit equivalent

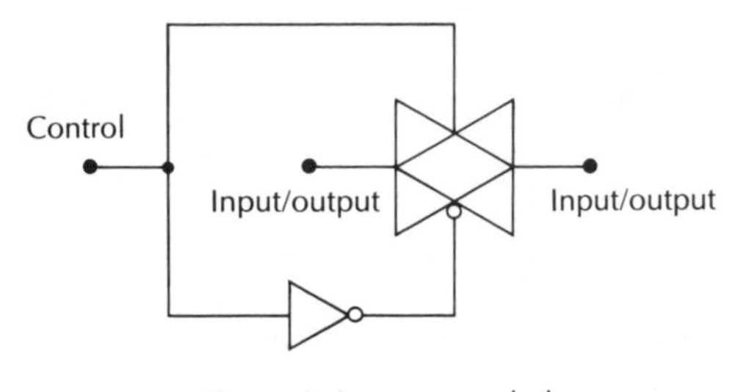

c. Transmission gate symbol with control circuit

Figure 2-32 C-MOS Transmission Gate

operation demands power only during switching and only enough to charge the very small capacitors involved.

The transistors provide the necessary rapid charge and discharge paths. Figure 2-33a shows a basic MOS dynamic inverter. Notice that two clock pulses are required to control the operation of the gate. The phase 1 and phase 2 clock pulses are 180° out of phase and not only serve to control the transfer of data but also control the energy so that they periodically refresh the capacitor charge that leaks off. The clocks must run continuously and the capacitors must be *refreshed* every few milliseconds. When the circuit is first started up (initialized), the first pair of clock pulses, Ø 1 and Ø 2, gate in *precharge* current for C_1 and C_2. In an IC with a large number of gates, the precharge operation demands the largest current of any phase of the operation.

The following is a step-by-step description of the dynamic MOS inverter in Figure 2-33.

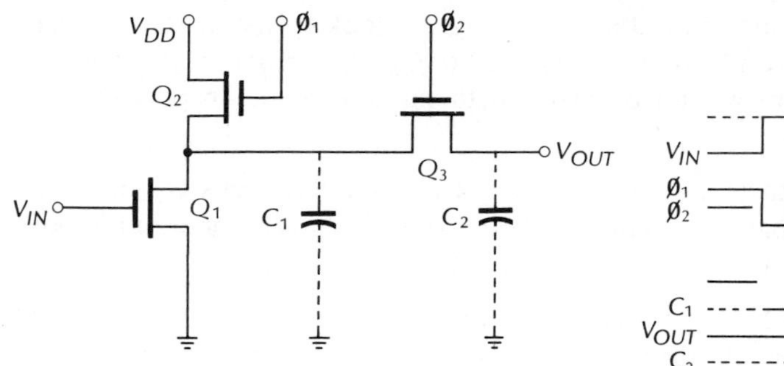

a. Dynamic two-phase inverter

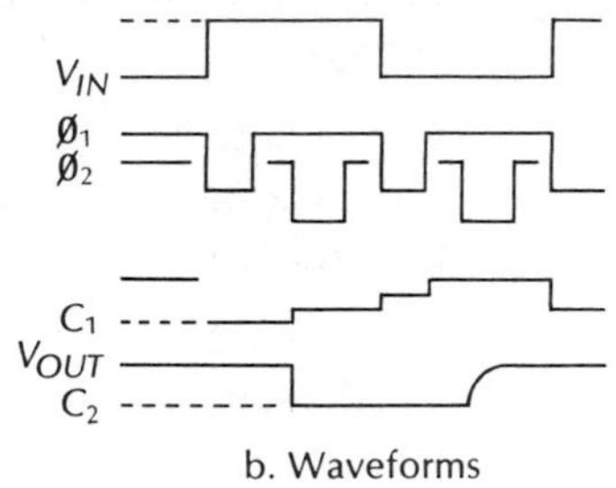

b. Waveforms

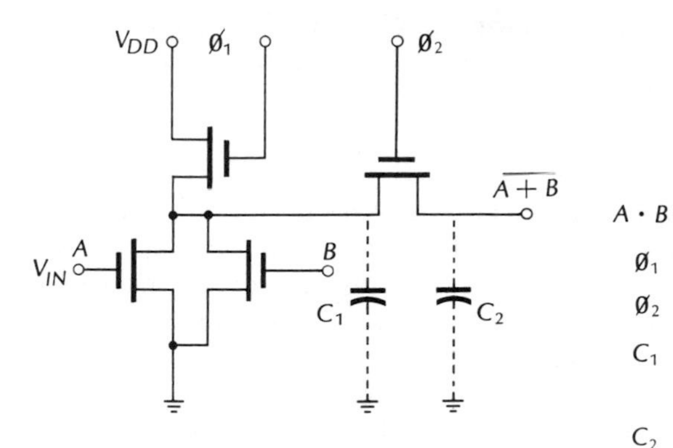

c. NOR gate

$A \cdot B$

$\emptyset_1$

$\emptyset_2$

C_1

C_2

d. Waveforms

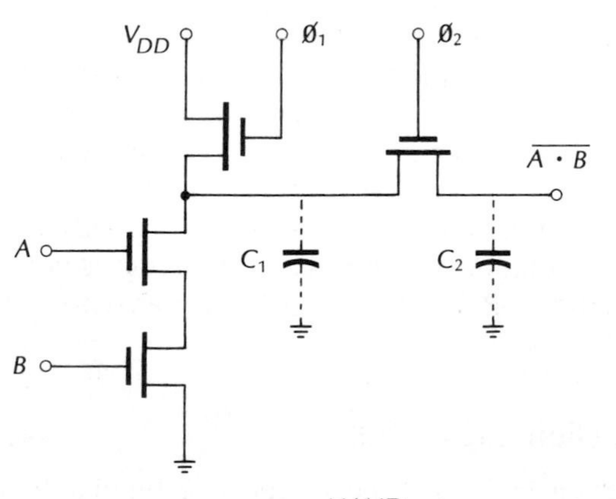

e. NAND gate

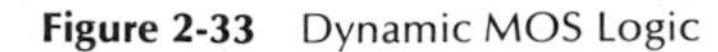

Figure 2-33 Dynamic MOS Logic

1. At t_1 the pulse from the phase 2 (∅ 2) clock turns on Q_3, causing the charge on C_1 to be transferred to C_2. C_1 is much larger than C_2 so the transfer can be made with no significant drop in the C_1 voltage.
2. At t_2 V_{IN} turns Q_1 on, and the ∅ 1 clock pulse turns Q_2 on. This causes a partial discharge of C_1. The discharge condition is not transmitted to C_2 because Q_3 is off.
3. At t_3 the ∅ 1 clock turns Q_2 off. A partial charge remains on C_1 and is subsequently discharged through Q_1, which remains turned on by V_{IN}.
4. At t_4 the ∅ 2 clock turns Q_3 on, and C_2 is also allowed to discharge through Q_1 to ground.
5. Drain current flows at t_1 (to charge up C_1) and during the short period that Q_1 and Q_2 are on simultaneously (from t_2 to t_3).

2-22 Emitter-Coupled Logic (ECL)

Emitter-coupled logic is a fast logic circuit used primarily in larger machines. ECL is available in several speeds, but the 1 and 2 nanosecond (propagation delay) versions are most common. The 1 ns ECL consumes 60 mW per gate while the 2 ns device consumes 25 mW per gate. The nearest TTL competitor is the standard Schottky with a 3 ns propagation delay and a power consumption of 20 mW per gate. Both ECL and Schottky TTL are non-saturating circuits, which contributes to their high speed.

ECL is based on a bipolar transistor current-switching circuit similar to the analog differential amplifier. The power supply voltage for ECL is a *negative* 5 V. Logic levels are both negative voltages; −0.8 V and −1.6 V. Figure 2-34 shows the two logic states. When the base of Q_1 is at the −1.6 V level, a fixed −1.2 V reference voltage on the base of Q_2 holds Q_2 in conduction. The collector of Q_2 is approximately −0.9 V and the collector of Q_1 is at ground. When the base of Q_1 goes to the −0.8 V level, Q_1 switches on and Q_2 turns off. The output is taken from the collectors. One output is inverted and the other is not. Figure 2-34c shows a combination OR/NOR ECL circuit. Transistors Q_4 and Q_5 are output buffers.

2-23 Integrated Injection Logic (I^2L)

Integrated injection logic is a recent form of bipolar logic that is as simple to fabricate and as inexpensive to produce as MOS. It is particu-

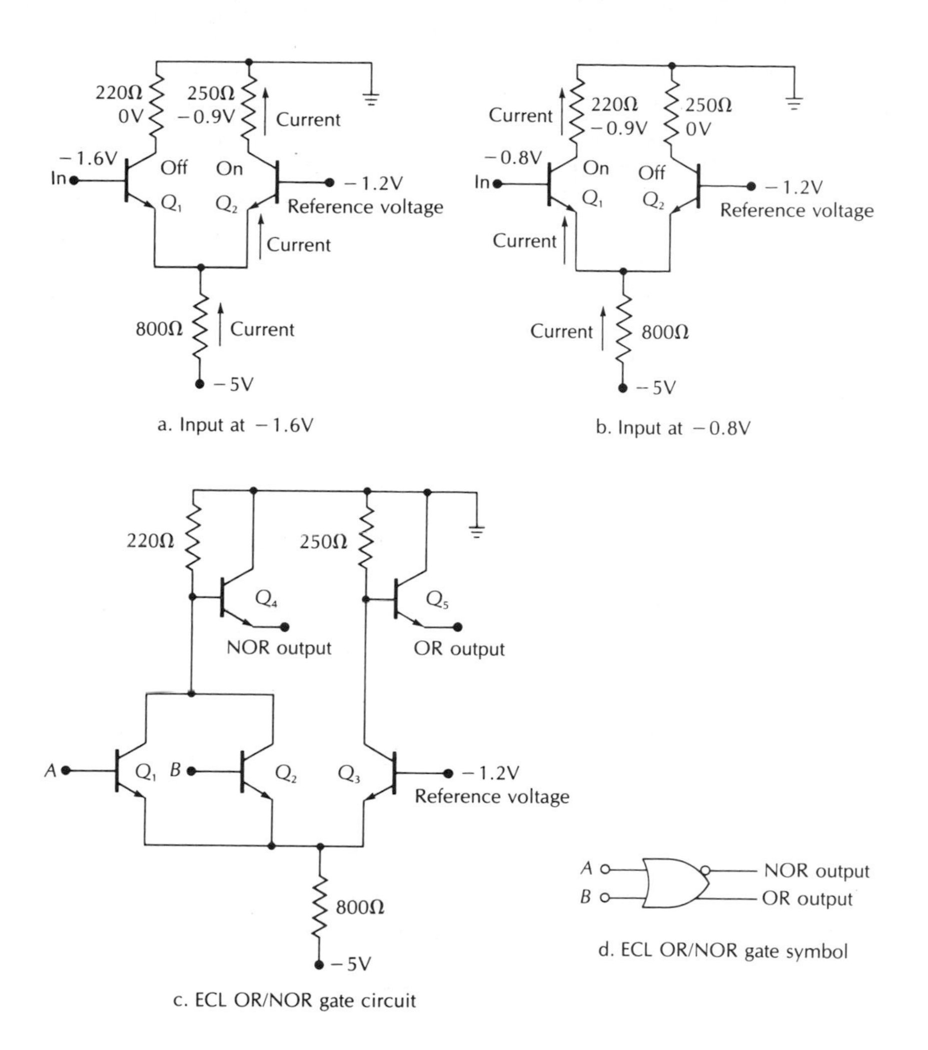

Figure 2-34 The Emitter-Coupled Logic Gate

larly well-suited to LSI processes, requires less chip space than MOS, and rivals the speed of TTL.

Bipolar logic faced some formidable problems in LSI packages. Circuits were too complex, occupied far too much chip space per gate, and required power dissipations that all too easily exceeded the modest practical limit of one-half watt per chip. The complexity of gate structures also made interconnection of individual gates difficult. Attempts to find ways to alter the geometry of existing bipolar logic circuits were doomed to failure from the beginning. The chief culprit was the passive load resistor required by each gate. The load resistors occupied the lion's share of the chip space and pushed the heat dissipation to the point of requiring exotic cooling methods for any significant number of gates.

Once the load resistor was replaced by a constant current source in the form of an active transistor, the injection logic gate became a reality. The injector transistor is a PNP transistor in the common base configuration—the same basic circuit frequently found as the active emitter resistor in differential amplifiers and other linear circuits.

Figure 2-35 illustrates the relative chip space occupied by a bipolar transistor and a passive load resistor. The illustration provides some insight into the magnitude of the space problem created by passive load resistors.

Next, the collector and emitter were exchanged on the chip resulting in a multicollector transistor operating in an inverse mode. With the upside-down transistor configuration, common emitter circuits can operate in a common N-type silicon bed. This eliminates the need for

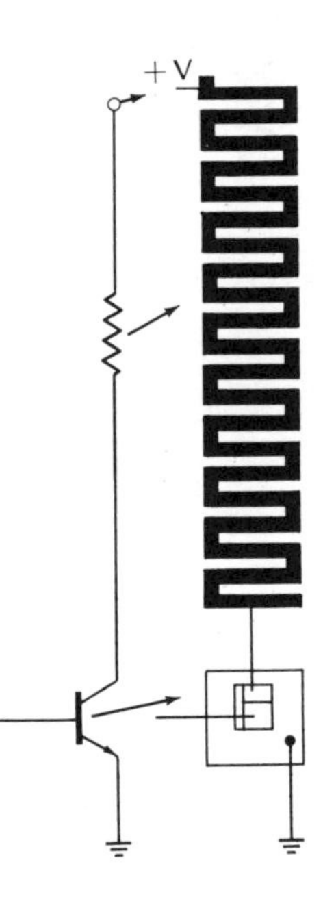

Figure 2-35 Relative Chip Space Occupied by Transistor and Passive Load Resistor

isolation space between transistors and allows one injector transistor to service several gate transistors.

The chip space required for isolation in right-side-up bipolar IC's is frequently twice as great as that required for a transistor. The inverted configuration allows the substrate to become a part of the interconnecting wiring rather than simply presenting an isolation problem. Those interconnections handled within the substrate minimize the complexity of the upper surface wiring pattern, reducing the cost still further. Figure 2-36b shows the basic inverter structure along with the PNP injector transistor. Figure 2-36a illustrates the simplicity of fabrication. Because of the way in which the transistors are "merged" within the substrate, the structure is often called *merged complementary bipolar* logic. The abbreviation MTL for merged transistor logic is used synonomously with I^2L.

By hard-wiring the collectors, more complex structures can be formed. Figure 2-37a shows how two inverters can be connected to form a NOR gate. Figure 2-37b is an inverter gate structure. The load is actually the driven gate, and the collector current for the driving gate

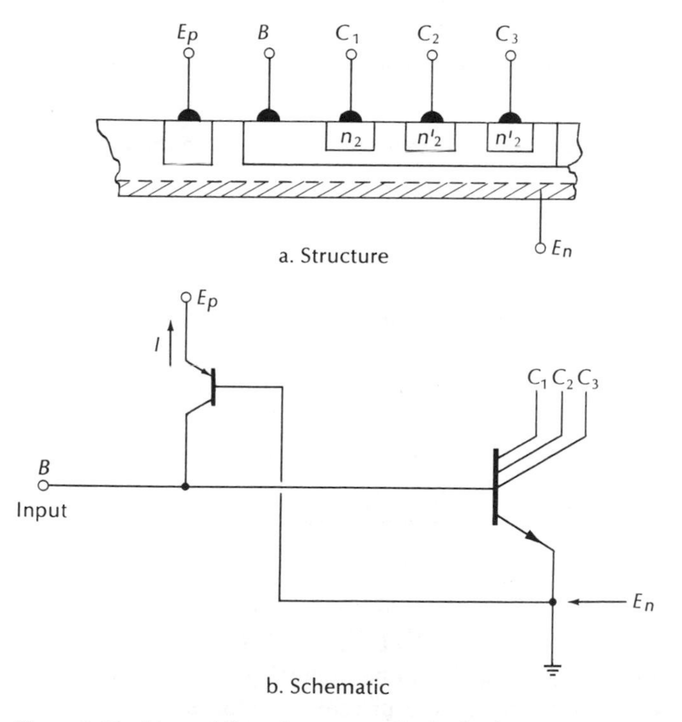

Figure 2-36 Merged Complementary Bipolar Logic

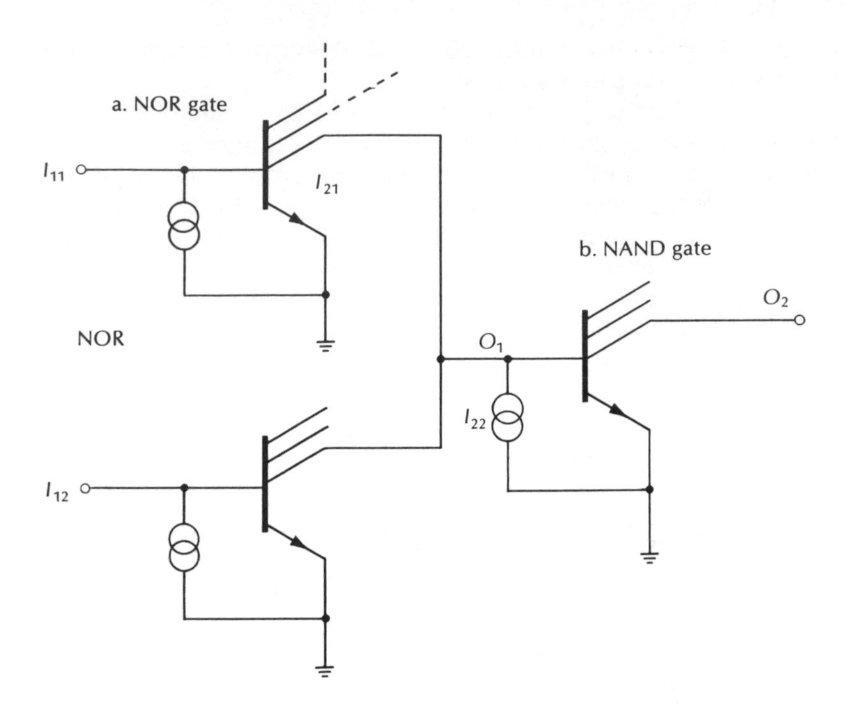

Figure 2-37 Inverters Combined to Form NOR and NAND Gates

is supplied by the injector transistor connected to the base of the driven transistor.

SUMMARY

Bipolar TTL and C-MOS dominate the small-scale and medium-scale integration (SSI and MSI) digital scene. Both families provide circuit packages ranging from a few simple independent gates to fairly complex counters, adders, and many-gate, general, and special-purpose logic structures. They have invaded the field of large-scale integration only to a relatively small extent. Neither form lends itself well to large-scale integration because of gate complexity and, in the case of TTL, excessive power dissipation.

TTL is considerably faster than C-MOS but consumes far more power. Where speed is the primary consideration, TTL has been the most popular choice. Both TTL and C-MOS can be operated from a single supply (+5 V for TTL and typically 5 to 18 V for C-MOS).

MOS devices are dedicated primarily to large-scale integration. The newer N-MOS is faster than the older P-MOS, uses a single +5 V power supply, and is generally more compatible with both I^2L and

C-MOS. P-MOS is an older process and requires a +5 V and −12 V power supply, making it more difficult to interface to TTL. P-MOS is also slower than N-MOS or C-MOS.

MOS devices are far more conservative in their power demands than TTL. C-MOS requires the least power of any presently available forms. Arithmetic systems, memories, and microprocessor units (the heart of a computer) are the special province of MOS devices. Individual gate packages, counters, and the like are not generally available in MOS integrated circuits.

Dynamic logic is a veritable miser when it comes to power consumption. It is more complex to operate because of the requirement of two or more clock signals. The timing required is fairly critical and speed is limited. This form of logic has been largely restricted to two kinds of large memory circuits and is not common in other applications.

Merged transistor logic (MTL or I^2L) is a recent bipolar logic that promises MOS simplicity and low cost with TTL speed

In any logic circuit there is always a trade-off between speed and power consumption. Each family has its own speed-power product, and process modifications can improve this product to some extent. However, for a given process an increase in speed is always obtained at the cost of increased power consumption.

Table 2-5 Comparison of MOS and Bipolar Logic Types

MOS lines / Gate Process	Power Dissipation per Gate	Supply Voltage (V_{DD}) (V_{GG}) (V_{CC})	Propagation Delay (ns/gate)	Freq. (MHz) (max.)	Noise Margin "1"	Noise Margin "0"
P-channel High threshold — Med. power	1.7mW	+12V, −12V	75ns	2	3	1.5
P-channel High threshold — Low power	0.45mW	+12V, −12V	300ns	500 kHz	3	1.5
Silicon gate	1.0mW	+5V, −12V	60ns	5	2	0.7
Ion implant, depletion loads	1.5mW	+5V, −12V	35ns	5	1.5	1
N-channel MOS						
Metal gate	1.0mW	+5V		10	1	1
Silicon gate	1.0mW	+5V		10	1	1
Complementary (C-MOS)						
Metal gate	50nW	+5V	40ns	20	$V_{DD}/2.2$	
Silicon gate	50nW	+5V	25ns	25	$V_{DD}/2.2$	
Bipolar lines						
TTL (standard)	15mW	5.0V +20% −10%	10ns	60	1.2	1.2
ECL	25-35mW	5.2V +20% −10%	1ns	400	0.4	0.4
Schottky TTL	19mW	5.0V +20% −10%	3ns	125	1.2	1.2

Problems

1. Define *LSI.*
2. Define *SSI.*
3. Define *MSI.*
4. Define *compatibility*.
5. Define *interface*.
6. Are most logic families completely *compatible* with most other logic families? Why?
7. What is propagation delay and what is the unit of measurement?
8. Define *noise immunity*.
9. What is meant by *speed* when applied to a given logic family?
10. Define *fan-out*.
11. Define *fan-in*.
12. List the 6 most popular logic families.
13. What two factors are most important in limiting the operating speed of a transistor switching circuit?
14. What is the purpose of adding a totem-pole output stage to the basic TTL gate?
15. What is the standard V_{cc} for the TTL family?
16. What is the maximum input voltage in TTL for a logic 0 output?
17. What is the minimum input voltage for a logic 1 in TTL?
18. What is maximum output voltage for a logic 0 in TTL?
19. What is the minimum output voltage for a logic 1 in TTL?
20. What is the guaranteed noise margin for a logic 0 in TTL? For a logic 1?
21. What are the differences between the 5400 and 7400 series TTL?
22. To which subfamily does each of the following belong?
    a. 7400 b. 74H00 c. 74S00 d. 74L00
23. When high speed is required, what must be traded for it? (See Table 2-4.)
24. Define the following:
    a. V_{cc} d. V_{OH} g. I_{IL}
    b. V_{IH} e. V_{OL} h. I_{OH}
    c. V_{IL} f. I_{IH} i. I_{OL}
    Find the value of each in the data manual for the 7400.
25. Why must unused gate inputs be returned to ground or V_{cc}?
26. Why is power supply line decoupling so important in TTL systems?
27. Wound mylar capacitors are not satisfactory for decoupling purposes. Why?
28. On a TTL power supply bus there is a 10 μF and a 0.1 μF disc capacitor. They are in different places along the line but effectively in parallel. Why can't the 0.1 μF be removed?

29. What is the difference between static and dynamic testing in troubleshooting digital systems?
30. What are the three levels in a TTL tristate output device?
31. Draw the schematic diagram of a two-input open collector TTL NAND gate.
32. Draw a logic diagram showing six open collector TTL gates in a wired AND configuration, with the node driving a single NAND gate.
33. What makes the Schottky-clamped gate faster than the standard TTL gate?
34. What do the Schottky gates have in common with ECL gates that improves their speed?
35. Which member of the ECL family is most popular?
36. Compare C-MOS and TTL with respect to the following parameters:
    a. speed
    b. propagation delay time
    c. V_{cc}
    d. noise immunity
    e. cost per gate
    f. fan-out
    g. chip space required per gate
    h. power consumption per gate
37. Match the following to items (1) through (4) below (more than one may be correct):
    a. MOS b. C-MOS
    (1) Uses both *N*- and *P*-channel devices on the same chip
    (2) The slowest member of the MOS group
    (3) Used primarily at MSI and SSI levels
    (4) Used primarily for LSI structures
38. What is the chief advantage of silicon-gate over metal-gate devices?
39. Is N-MOS compatible with TTL? Explain.
40. Is P-MOS compatible with TTL? Explain.
41. Is C-MOS compatible with TTL? Explain.
42. What is the basic gate structure in MOS?
    a. AND c. NOR e. Inverter
    b. OR d. NAND f. None of these
43. What kind of drain load resistor is used in MOS gates?
44. Which of the MOS devices consumes the least power?
45. In C-MOS the smallest noise voltage that will falsely turn on a gate is how many volts?
46. Explain the difference between static and dynamic MOS logic.

CHAPTER 3

ASYNCHRONOUS LOGIC

Learning Objectives. *Upon completing this chapter you should know:*

1. *The basic gates and their duals.*
2. *How to write equations from a truth table.*
3. *How to plot an equation on a truth table.*
4. *How to draw logic diagrams from equations or write a truth table from a logic diagram.*
5. *The terms minterm and maxterm.*
6. *How to simplify equations with up to four variables using a Karnaugh map.*
7. *How to plot a simplified equation on a truth table.*

A logic circuit is composed of one or more logic gates. The inputs to the gates consist of a high or low voltage, representing logic 1's and logic 0's. Circuit outputs are also either high or low, logic 1 or logic 0. There are two kinds of logic: combinatorial (also called asynchronous or direct logic) and sequential or synchronous logic. Synchronous logic (discussed in Chapter 5) responds to the input conditions only at specific times controlled by a master pulse generator called a *clock*. Circuit operation is synchronized to the clock. Asynchronous logic (the topic of this chapter) responds as the input conditions change. No clock input is provided and the circuit is not synchronized with the system clock.

We will examine logic operations starting with very basic ones and working up to fairly complex systems. At each level of complexity we will study the most appropriate methods of representing and manipulating equations and logic diagrams. At the lowest level of complexity we will examine the relationship among truth tables, equations, and logic diagrams. We will also investigate the dual nature of logic at this level.

At the next level of complexity we will work with Karnaugh mapping and simplification methods.

The various levels of complexity are dictated by the commercial availability of logic circuits, rather than purely by logic considerations. Certainly logic considerations have influenced manufacturing decisions, but their influence was tempered by practical considerations. Each method is best suited to a fairly specific complexity level and is either very difficult or completely impossible to use for greatly different levels. In Chapter 4 we will examine programmable logic on an MSI scale.

3-1 Single-Input Logic

There is only one valid logic function for a single input gate—inversion. Because there are two possible input values, 0 and 1, and two output values (also 0 and 1), we can make truth tables representing four different single-input gate functions as shown in Table 3-1. Truth tables 1 and 2 are useless because the output condition is not influenced by input conditions and thus performs no logic function. Table 3 is a logic "do-nothing" circuit because the output condition simply follows the input condition. Although the circuit represented by truth table 3 is a useful circuit element in the form of a non-inverting buffer amplifier, it cannot be considered as a logic element. The number 4 truth table is the table that describes the inverter function. Table number 4 is the only table which represents a valid logic function.

3-2 Two-Input Logic

For two-input devices it is possible to write sixteen different truth tables. Six of them are useless, six are common and most often used, and the remaining four are used only in very unusual situations.

The following are the useless truth table forms:

1. Where all outputs are always 0.
2. Where all outputs are always 1.
3. Where the outputs are identical to input A.
   Here input B does not affect the output under any condition. The circuit is, in effect, a single-input circuit with input A.
4. Where the outputs are identical to the inputs on B.
   In this case, we have a single-input circuit with input B.
5. Where the outputs are all inversions of input A.
   The circuit has degenerated into a single-input inverter for input A.

Table 3-1 All Possible Single-Input Logic Functions

1		2		3		4	
A	f	A	f	A	f	A	f
0	0	0	1	0	0	0	1
1	0	1	1	1	1	1	0

6. Where the outputs are the inversion of input B.
   Here we have a single-input inverter with input B.

These cases—in which the two-input circuit degenerates into a single-input non-inverting amplifier or an inverter—are useful in practice even though their logical significance is trivial.

Because actual IC gate packages come with several gates to a package, it is often practical, for example, to use an extra NAND gate in an existing package rather than adding an additional inverter package.

3-3 Logic State Definitions

There are two kinds of logic: positive and negative.

Positive logic definition

$$\begin{aligned} + &= \text{Logic } 1 \\ \text{Ground} &= \text{Logic } 0 \end{aligned}$$

Negative logic definition

$$\begin{aligned} \text{Ground} &= \text{Logic } 1 \\ + &= \text{Logic } 0 \end{aligned}$$

We stated earlier that TTL and most other modern logic circuits are positive-logic devices. This is true in the sense that manufacturers define their products' functions in terms of the positive-logic definition. If a manufacturer had been making 7400 NAND gates and arbitrarily decided to change the logic definition from positive to negative, the 7400 would have to be listed in the new data manual as a NOR gate. The same electronic circuit with a different logic definition performs a different logic function. Fortunately, manufacturers do not take such arbitrary liberties with the logic definition; however, it is often desirable for us to alter the logic definition for our own convenience.

3-4 Gates and Logic Duals

The concept of logic duals allows the selection of whichever logic definition best fits the problem at hand. The justification for logic duality is contained in the two laws of DeMorgan.

LOGIC DUAL DEFINED

From here on we will adopt the following definition for logic duals: *Two logic circuits are defined as logic duals when both have identical logic truth tables but opposite logic definitions.* One is defined as positive logic; the other is defined as negative logic.

With the preceding definition in mind, let us examine each of the most useful and common two-input logic circuits and their duals.

Figure 3-1 reviews bubble notation from Chapter 1. Figure 3-2 shows positive-logic gates and their negative-logic equivalents. Bubbles on the gate inputs define the gate as a negative-logic gate. The negative-logic symbols in Figure 3-2 (b, d, f, h) are sometimes found in logic diagrams even though negative-logic gates are rarely available in integrated circuit hardware. For example, you might find the negative-logic OR gate (Fig. 3-2b) in a logic diagram, but the hardware would actually be its positive-logic equivalent, the AND gate in Figure 3-2a. A logic input with a bubble is also called an *active-low input.*

3-5 Minterm and Maxterm Forms

Minterm and maxterm are two common forms for logic equations and the associated arrangement of logic gates, one or both of which is the

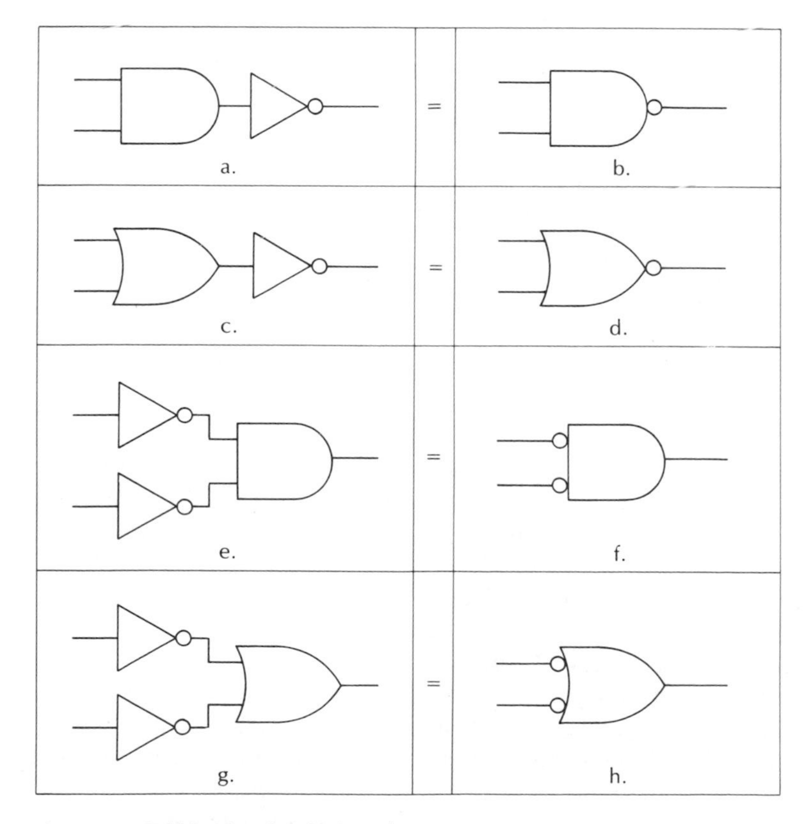

Figure 3-1 Bubble Notation Reviewed

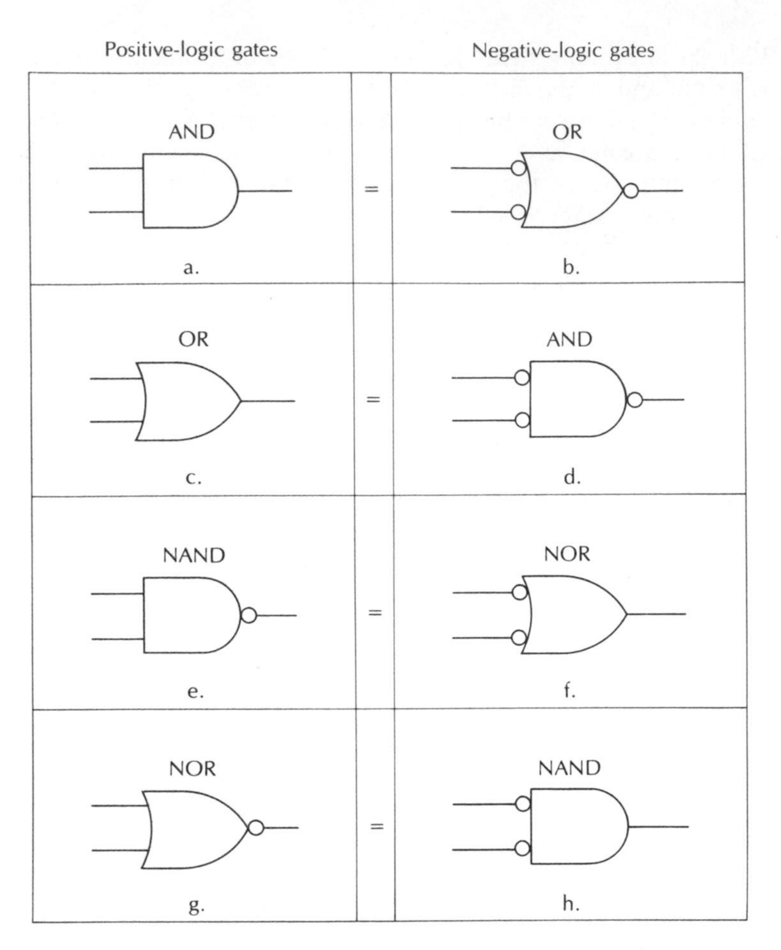

Figure 3-2 Positive-Logic Gates and their Negative-Logic Equivalents

basis for every asynchronous logic circuit. The names refer to an obsolete form of truth table and are abbreviations of the full names which are no longer used—minimum-area-term and maximum-area-term. Figure 3-3 shows the equation and logic diagram for the minterm and maxterm forms. In practical circuits, inverters (when called for) are generally found in series with selected input lines. Figure 3-4 shows the same circuits as Figure 3-3 but with inverters used to obtain the desired inverted input signals. In Figure 3-4a inputs are shown connected together—*A* on gate 1 to *A* on gate 2 and *B* on gate 1 to *B* on gate 2. In Figure 3-4b the connections are implied by the letter designations. Both are common methods of drawing the logic diagram. The minterm circuit in Figure 3-3a and the maxterm circuit shown in part b both perform the same logic function.

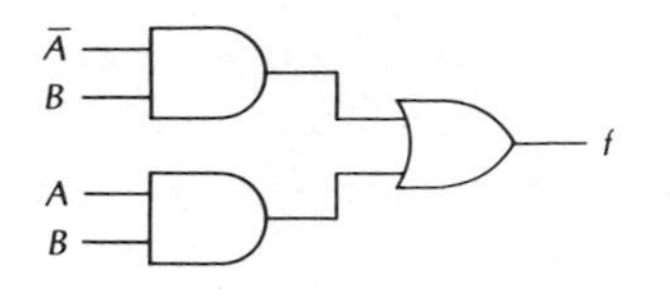

Equation: $f = (\overline{A} \bullet B) + (A \bullet B)$

a. Minterm-form logic diagram

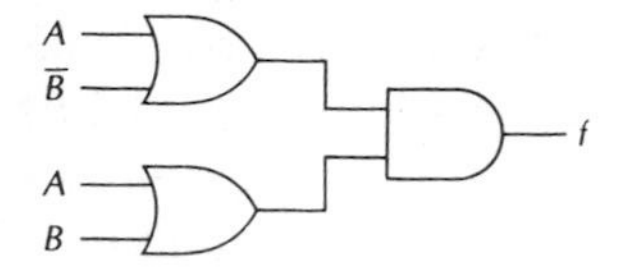

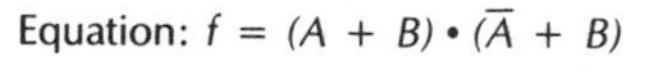

b. Maxterm form

Figure 3-3 Minterm and Maxterm Forms

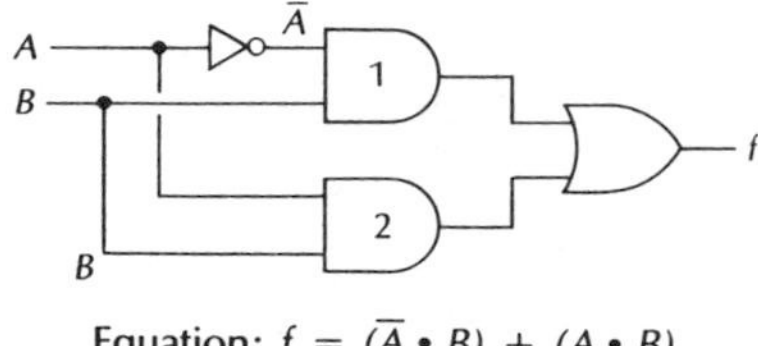

Equation: $f = (\overline{A} \bullet B) + (A \bullet B)$

a. Minterm-form logic diagram

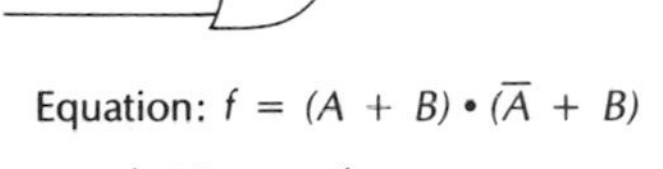

Equation: $f = (A + B) \bullet (\overline{A} + B)$

b. Maxterm form

Figure 3-4 Minterm and Maxterm Circuits with Inverters

EXAMPLES OF MINTERM AND MAXTERM EQUATIONS

The minterm form may be identified as a sum-of-products form.

Examples

1. $f = (\overline{A} \cdot \overline{B} \cdot C) + (A \cdot \overline{B} \cdot C) + (A \cdot B \cdot C)$
2. $f = (A \cdot \overline{C}) + (B \cdot D) + (\overline{A} \cdot \overline{C} \cdot \overline{D}) + (A \cdot B \cdot \overline{C} \cdot D)$
3. $f = (AB\overline{C}) + (B\overline{C}D) + (ABCD) + (A\overline{D})$
4. $f = (A \cdot \overline{B} \cdot \overline{C}) + (\overline{A} \cdot \overline{B} \cdot C) + (B \cdot C) + (A \cdot \overline{B})$

The maxterm form may be identified as a product-of-sums form.

Examples

1. $f = (A + B + \overline{C}) \cdot (\overline{A} + B + C) \cdot (\overline{A} + \overline{B} + C)$
2. $f = (\overline{A} + C) \cdot (\overline{B} + \overline{D}) \cdot (A + B + D) \cdot (\overline{A} + \overline{B} + C + \overline{D})$
3. $f = (\overline{A} + \overline{B} + C) \cdot (\overline{B} + C + \overline{D}) \cdot (\overline{A} + \overline{B} + \overline{C} + \overline{D}) \cdot (\overline{A} + D)$
4. $f = (\overline{A} + B + C) \cdot (A + B + \overline{C}) \cdot (\overline{B} + \overline{C}) \cdot (\overline{A} + B)$

As a direct result of DeMorgan's law:

1. *Any logic operation can be implemented in either a minterm or a maxterm form.*
2. *Any minterm-form logic circuit can be converted into a maxterm circuit.*
3. *Any maxterm-form logic circuit can be converted into a minterm circuit.*

The two forms are *complementary*. In order to understand what is meant by *complementary*, let us examine Tables 3-2 and 3-3.

In Table 3-2 we have made a truth table for some arbitrary function. Later we will deal with more realistic situations.

THE MINTERM FORM

The minterm type equation, often called the sum-of-products form, can be written directly from the truth table by interpreting a 0 under the A heading as $\overline{A}$ and a 1 in the A column as A. Entries under the B heading

Table 3-2 Minterm Truth Tables, Equations, and Logic Diagrams

	A	B	f	Minterms
m_0	0	0	1	→ $(\overline{A} \cdot \overline{B})$
m_1	0	1	0	
m_2	1	0	1	→ $(A \cdot \overline{B})$
m_3	1	1	0	

$$f = (\overline{A} \cdot \overline{B}) + (A \cdot \overline{B})$$

a. Truth table b. Minterms c. The minterm equation

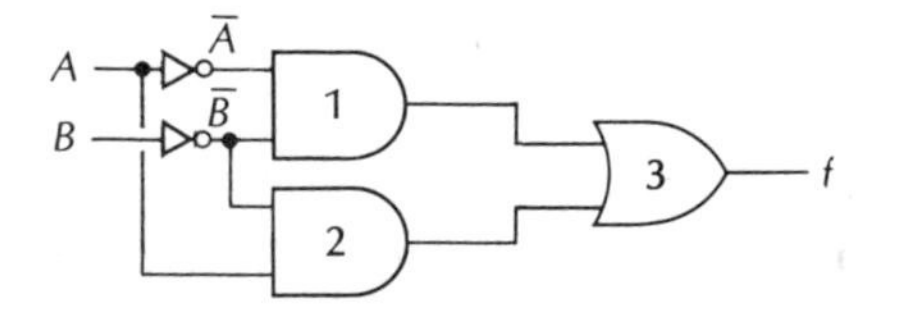

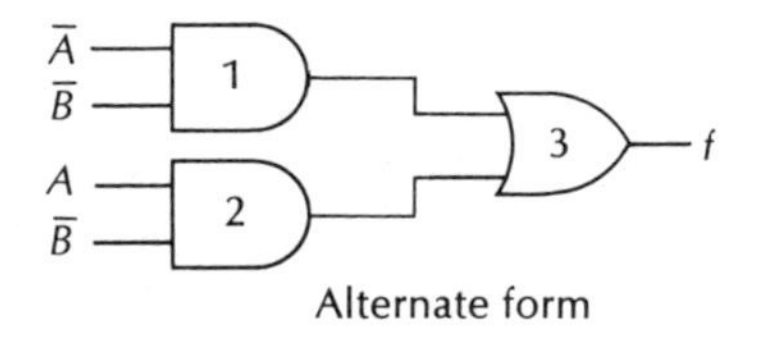

Alternate form

d. The logic diagrams

Table 3-3 Truth Table

m	A	B	C	f	Minterms
0	0	0	0	0	
1	0	0	1	1	$(\overline{A} \cdot \overline{B} \cdot C)$
2	0	1	0	0	
3	0	1	1	0	
4	1	0	0	0	
5	1	0	1	1	$(A \cdot \overline{B} \cdot C)$
6	1	1	0	1	$(A \cdot B \cdot \overline{C})$
7	1	1	1	0	

are treated in the same way. The conditions in the A, B, and other columns are interpreted and written only for rows in which there is a 1 in the f column. In Table 3-2a the m_0 row has a 1 in the f column. The 0, 0 under A, B is written as $(\bar{A} \cdot \bar{B})$. This is the first minterm. The next 1 entry in the f column is row m_2 and is written as $(A \cdot \bar{B})$, as shown under the heading *minterms*. The final equation is formed by joining the minterms (AND-terms) with OR (+) symbols as shown in part c of this table. This is the minterm, or sum-of-products form, equation.

The logic diagram for the equation in *d* is shown in two equally acceptable arrangements. The two input gates (1 and 2) are AND gates with the necessary inverters (shown or implied) to form the hardware equivalent of the two minterms. The OR gate joins the two products in the way indicated by the equation.

Example Write the equation and draw the logic diagram as described by the truth table 3-3. If we combine the terms to form an equation we get:

$$f = (\bar{A} \cdot \bar{B} \cdot C) + (A \cdot \bar{B} \cdot C) + (A \cdot B \cdot \bar{C})$$

Figure 3-5 shows the logic diagram.

Problems

Given the following equations, construct a truth table like the one in Table 3-3 for each problem (with a blank f column). Complete the f column to correspond to each equation.

1. $f = (\bar{A} \cdot \bar{B} \cdot \bar{C}) + (A \cdot \bar{B} \cdot C) + (A \cdot B \cdot C)$
2. $f = (\bar{A} \cdot B \cdot C) + (A \cdot \bar{B} \cdot \bar{C})$
3. Given the logic diagrams shown in Figure 3-6, write the equation for each.

THE MAXTERM FORM

Although the minterm form is the most common, the maxterm form is often used, so it is necessary to be able to translate from one form into the other. Truth tables can be assumed to be minterm truth tables

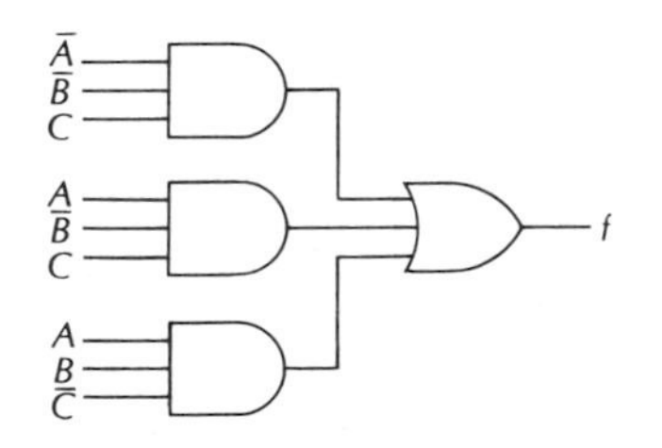

Figure 3-5 Logic Diagram for Truth Table 3-3

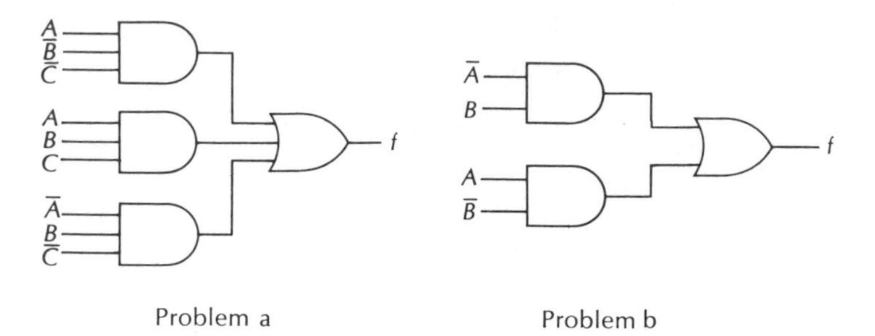

Figure 3-6 Logic Diagrams for Problem 3

unless otherwise specified. The maxterm truth table is primarily used only as an aid in translating from one form to another or for other special purposes.

In Table 3-4a we have repeated the minterm truth table from Table 3-2. Table 3-4b is the complementary (maxterm) truth table. Truth table b is formed by inverting all entries. All 0's are changed into 1's and all 1's become 0's. The maxterms in c are written from the maxterm truth table. The completed maxterm, or product-of-sums equation, is formed by joining the *sum* terms (maxterms) by an AND symbol as shown in Table 3-4d. The maxterm logic diagram is constructed as shown in Figure 3-7.

It is important to understand that the two logic circuits, the minterm and the maxterm forms, perform identical logic functions. From a functional logic standpoint, they are identical and completely interchangeable.

Table 3-4 Maxterm Truth Tables, Equations, and Logic Diagrams

	A	B	f						
m_0	0	0	1	M_0	1	1	0		
m_1	0	1	0	M_1	1	0	1	$(A+\overline{B})$	
m_2	1	0	1	M_2	0	1	0		
m_3	1	1	0	M_3	0	0	1	$(\overline{A}+\overline{B})$	$f=(A+\overline{B})\cdot(\overline{A}+\overline{B})$

a. Minterm truth table (repeated from Table 3-2) — b. Maxterm (complemented) truth table — c. Maxterms — d. Maxterm equation

Note: Lower-case m indicates minterm and upper-case M indicates maxterm

Table 3-5 Maxterm and Minterm Truth Tables

	A	B	C	f			A	B	C	f	
M_0	1	1	1	0		m_0	0	0	0	1	$(\bar{A} \cdot \bar{B} \cdot \bar{C})$
M_1	1	1	0	0		m_1	0	0	1	1	$(\bar{A} \cdot \bar{B} \cdot C)$
M_2	1	0	1	0		m_2	0	1	0	1	$(\bar{A} \cdot B \cdot \bar{C})$
M_3	1	0	0	1	$(A+\bar{B}+\bar{C})$	m_3	0	1	1	0	
M_4	0	1	1	0		m_4	1	0	0	1	$(A \cdot \bar{B} \cdot \bar{C})$
M_5	0	1	0	1	$(\bar{A}+B+\bar{C})$	m_5	1	0	1	0	
M_6	0	0	1	0		m_6	1	1	0	1	$(A \cdot B \cdot \bar{C})$
M_7	0	0	0	0		m_7	1	1	1	1	$(A \cdot B \cdot C)$

a. Maxterm truth table for $f = (\bar{A}+B+\bar{C}) \cdot (A+\bar{B}+\bar{C})$ b. Inverting to a minterm table c. Minterms

Equation: $f = (A+\bar{B}) \cdot (\bar{A}+\bar{B})$

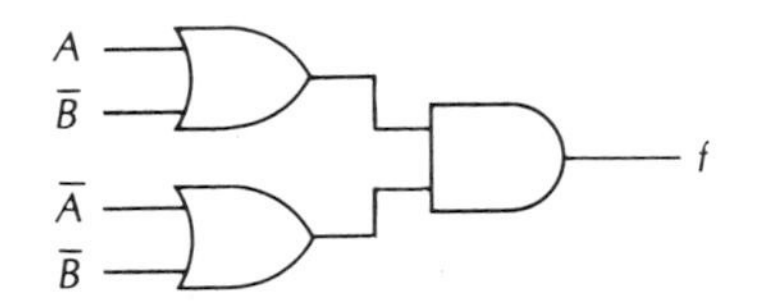

Figure 3-7 Maxterm Logic Diagram

Example Given the following maxterm equation, convert it into its minterm equivalent:

a. The equation:

$$f = (\bar{A} + B + \bar{C}) \cdot (A + \bar{B} + \bar{C})$$

b. The maxterm truth table is shown in Table 3-5a.

c. Inverting all entries in the maxterm table a yields table b, the minterm table. The minterms corresponding to the 1 entries in the minterm table f column are indicated in part c. Connecting the minterms by + (OR) symbols yields the equation:

$$f = (\bar{A} \cdot \bar{B} \cdot \bar{C}) + (\bar{A} \cdot \bar{B} \cdot C) + (\bar{A} \cdot B \cdot \bar{C}) + (A \cdot \bar{B} \cdot \bar{C}) + (A \cdot B \cdot \bar{C}) + (A \cdot B \cdot C)$$

On the surface it would appear that the minterm equation is far more complex than the maxterm version. What has actually happened is that a number of redundant terms are included that can be easily eliminated.

When a term accomplishes nothing that is not already accomplished by other terms, it is said to be *redundant* and can be dropped from the equation. We will find little or no difference between the complexity of the minterm and maxterm equation once redundant terms are eliminated. In the next section we will examine formal simplification (minimization) techniques that are used to eliminate the redundant terms. Redundancies can crop up in either or both forms of the equation.

Problems

Given the following maxterm equations, use an inverted truth table to convert them into minterm form. Draw the logic diagram for the given maxterm and the derived minterm equation.

4. $f = (\bar{A} + B + C) \cdot (\bar{A} + \bar{B} + C) \cdot (A + B + \bar{C}) \cdot (A + \bar{B} + \bar{C}) \cdot (\bar{A} + \bar{B} + \bar{C})$
5. $f = (\bar{A} + B + C) \cdot (\bar{A} + \bar{B} + C) \cdot (A + B + \bar{C}) \cdot (\bar{A} + \bar{B} + \bar{C})$
6. $f = (A + \bar{B} + C) \cdot (A + B + \bar{C})$
7. $f = (A + \bar{B} + \bar{C}) \cdot (\bar{A} + \bar{B} + C) \cdot (A + B + C)$
8. Design logic circuits to perform the above functions:
   a. Write a minterm truth table.
   b. Write the minterm equation.
   c. Draw the minterm logic diagram.
   d. Write the maxterm truth table.
   e. Write the maxterm equation.
   f. Draw the maxterm logic diagram.
9. There are 3 starting lanes (A, B, C) at a racecourse. A logic circuit is required that sounds an alarm whenever there happens to be an odd number of cars at the starting gates or if there are no cars at all at the gate. Let a 0 stand for no car at a gate and a 1 represent a car waiting in its lane at the gate.
10. In a particular building there are three doors. Fire regulations are such that there must never be one and only one door open at any given time. Design a logic circuit that produces a 1 output whenever the forbidden condition occurs.

3-6 Simplification

The process that we have been using to derive equations and logic diagrams from truth tables always yields valid results. But more often than not circuits derived from the truth table contain more hardware than is really necessary to get the job done. The problem is simply one of overlapping functions where part of a term will do the work of two or more complete terms. Suppose we have the equation: $f = (A \cdot B \cdot C) + (A \cdot B \cdot \overline{C})$. The equation says that we get the desired function when we have $(A \cdot B)$ combined with either C or $\overline{C}$. If either C or $\overline{C}$ will do, and every possibility includes one or the other, the term C is useless (complemented or uncomplemented). It can therefore be dropped from the equation. This leaves: $f = (A \cdot B) + (A \cdot B)$. Because the two terms are identical, one of them is a duplicate, or redundant, term. Therefore, we keep only one of them. The function can be performed by simply $f = (A \cdot B)$ just as well as by the more complex $f = (A \cdot B \cdot C) + (A \cdot B \cdot \overline{C})$.

AN ANALOGY

$f =$ the requirements for driving a nail:

$f =$ (a nail AND a hammer AND an apron) OR (a nail AND a hammer AND NO apron). The apron is obviously unimportant. Forgetting about the apron, we have:

$f =$ (a nail AND a hammer) OR (a nail AND a hammer). Obviously what we need is simply a nail and a hammer.

$f =$ (a nail AND a hammer)

The analogy is fine as far as it goes, but we need to build on a more rigorous premise. We find the foundation for simplification methods in the basic Boolean laws we examined in Chapter 1. The third law of complementation, $A + \overline{A} = 1$, forms the foundation for the discussion that follows.

THE RULES FOR COMBINING TERMS

Two terms may be combined whenever:

1. Each term contains exactly the same variables.
2. The terms to be combined are identical with the exception that one—and only one—variable appears in the complemented (barred) form in one term and in the uncomplemented form in the other.

Examples

1. $f = (A \cdot B \cdot C) + (\overline{A} \cdot B \cdot C)$

   Because A appears in both complemented and uncomplemented forms and because $A + \overline{A} = 1$, the A drops out, leaving $f = (B \cdot C) + (B \cdot C)$. Because the two new terms are identical, one of them is said to be redundant and is dropped. The final simplification of the equation is $f = (B \cdot C)$.

2. $f = (\bar{A} \cdot B \cdot \bar{C} \cdot D) + (\bar{A} \cdot B \cdot C \cdot D)$

   The C appears in the complemented form in the first term and in the uncomplemented form in the second term. The C drops out, leaving $f = (\bar{A} \cdot B \cdot D) + (\bar{A} \cdot B \cdot D)$. Because the two terms are identical, one of them is redundant and is dropped, leaving only $f = (\bar{A} \cdot B \cdot D)$.

3. Simplify the following equation by combining reducible pairs:

$$f = \underset{1}{(\bar{A} \cdot \bar{B} \cdot \bar{C})} + \underset{2}{(\bar{A} \cdot B \cdot \bar{C})} + \underset{3}{(\bar{A} \cdot \bar{B} \cdot C)} + \underset{4}{(\bar{A} \cdot B \cdot C)}$$

   a. If we combine terms 1 and 2:

$$\left.\begin{matrix}\bar{A} \cdot \bar{B} \cdot \bar{C} \\ \bar{A} \cdot B \cdot \bar{C}\end{matrix}\right\} \bar{A} \cdot \bar{C}$$

   b. If we combine terms 3 and 4:

$$\left.\begin{matrix}\bar{A} \cdot \bar{B} \cdot C \\ \bar{A} \cdot B \cdot C\end{matrix}\right\} \bar{A} \cdot C$$

   c. If we rewrite the equation using the simplified terms:

$$f = (\bar{A} \cdot \bar{C}) + (\bar{A} \cdot C)$$

   d. On inspection we can see that these new terms can be combined to further simplify the equation:

$$\left.\begin{matrix}\bar{A} \cdot \bar{C} \\ \bar{A} \cdot C\end{matrix}\right\} \bar{A}$$

   e. The final equation reduces to:

   $f = \bar{A}$, the equivalent of a single inverter for A.

4. An alternate pairing for example 3:

$$f = \underset{1}{(\bar{A} \cdot \bar{B} \cdot \bar{C})} + \underset{2}{(\bar{A} \cdot B \cdot \bar{C})} + \underset{3}{(\bar{A} \cdot \bar{B} \cdot C)} + \underset{4}{(\bar{A} \cdot B \cdot C)}$$

   a. Combining terms 1 and 3 we get:

$$\left.\begin{matrix}\bar{A} \cdot \bar{B} \cdot \bar{C} \\ \bar{A} \cdot \bar{B} \cdot C\end{matrix}\right\} \bar{A} \cdot \bar{B}$$

b. Combining terms 2 and 4:

$$\left.\begin{array}{l}\bar{A} \cdot B \cdot \bar{C} \\ \bar{A} \cdot B \cdot C\end{array}\right\} \bar{A} \cdot B$$

c. The simplified equation:

$$f = (\bar{A} \cdot \bar{B}) + (\bar{A} \cdot B)$$

d. Combining the two simplified terms:

$$\left.\begin{array}{l}\bar{A} \cdot \bar{B} \\ \bar{A} \cdot B\end{array}\right\} \bar{A}$$

e. The simplified equation:

$$f = \bar{A}$$

In this case the end result is the same whatever our selection of reducible pairs. This is not always true. Sometimes we can come up with two or more equally simple but different simplified equations.

So far the examples have been trivial and this pairing technique worked well, but as equations get more complex, the method becomes tedious and prone to error.

THE KARNAUGH MAP

This is a special truth table designed specifically for simplifying equations. There are several variations of this table (sometimes called a *map*), but they are all constructed so that all adjacent entries on the table represent reducible minterm pairs. Although there is more than one way that a Karnaugh map can be labeled and arranged, all versions are used and interpreted in exactly the same way.

Because the Karnaugh map is constructed so that all adjacent squares represent reducible pairs, it is really only an aid in using the pairing technique. The map makes the options obvious and clear because of their highly visual nature. It also allows us to perform multiple pairings in a single operation. Table 3-6 shows the Karnaugh maps for use with two, three, and four variable equations.

A careful examination of all the maps shows that each square represents one of the possible minterms on the standard truth table, and that any pair of adjacent squares represents a reducible pair of minterms. For the purpose of this discussion, the indicated minterm has been written for each square, but these are not normally included. Instead, 1's taken directly from the "f" column on the standard truth table are entered in the appropriate squares. The rest of the squares are left blank. The size

Table 3-6 Karnaugh Maps

	$\bar{A}$	A
$\bar{B}$	$\bar{A} \cdot \bar{B}$	$A \cdot \bar{B}$
B	$\bar{A} \cdot B$	$A \cdot B$

a. Two-variable Karnaugh map

	$\bar{A}\bar{B}$	$\bar{A}B$	AB	$A\bar{B}$
$\bar{C}$	$\bar{A}\bar{B}\bar{C}$	$\bar{A}B\bar{C}$	$AB\bar{C}$	$A\bar{B}\bar{C}$
C	$\bar{A}\bar{B}C$	$\bar{A}BC$	ABC	$A\bar{B}C$

b. Three-variable Karnaugh map

	$\bar{A}\bar{B}$	$\bar{A}B$	AB	$A\bar{B}$
$\bar{C}\bar{D}$	$\bar{A}\bar{B}\bar{C}\bar{D}$	$\bar{A}B\bar{C}\bar{D}$	$AB\bar{C}\bar{D}$	$A\bar{B}\bar{C}\bar{D}$
$\bar{C}D$	$\bar{A}\bar{B}\bar{C}D$	$\bar{A}B\bar{C}D$	$AB\bar{C}D$	$A\bar{B}\bar{C}D$
CD	$\bar{A}\bar{B}CD$	$\bar{A}BCD$	$ABCD$	$A\bar{B}CD$
$C\bar{D}$	$\bar{A}\bar{B}C\bar{D}$	$\bar{A}BC\bar{D}$	$ABC\bar{D}$	$A\bar{B}C\bar{D}$

c. Four-variable Karnaugh map

of the map is determined by the number of variables involved and will have as many squares as the standard truth table has rows: 2^n where n is the number of variables. In your examination of Table 3-6 you may have overlooked the fact that variables on the upper and lower edges and on the left and right edges are also reducible. In addition, the four corners are considered adjacent in all combinations except on the diagonal. No pairs of terms on a diagonal are reducible.

LOOPING TERMS AND READING OUT SIMPLIFIED EQUATIONS

Looping Rules

1. Each loop should be drawn around the largest group of 2, 4, 8, and so on—adjacent entries possible. The number of entries in a loop *must* be an integral power of 2.
2. An entry may be involved in any number of loops, but a new loop should not be added unless it includes at least one entry not included in any other loops. Typical looping patterns are shown in Table 3-7.
3. After looping, inspect the map for any loop that encloses entries where all of the 1's in the loop are involved in other loops. Remove any such loops.

Reading Out the Simplified Equation

1. Each loop represents a new simplified minterm for the simplified equation. All simplified minterms will be ORed together (connected by the + symbol).
2. Any variable in a given loop which appears in both the complemented and the uncomplemented form drops out.

Table 3-7 Looping Patterns

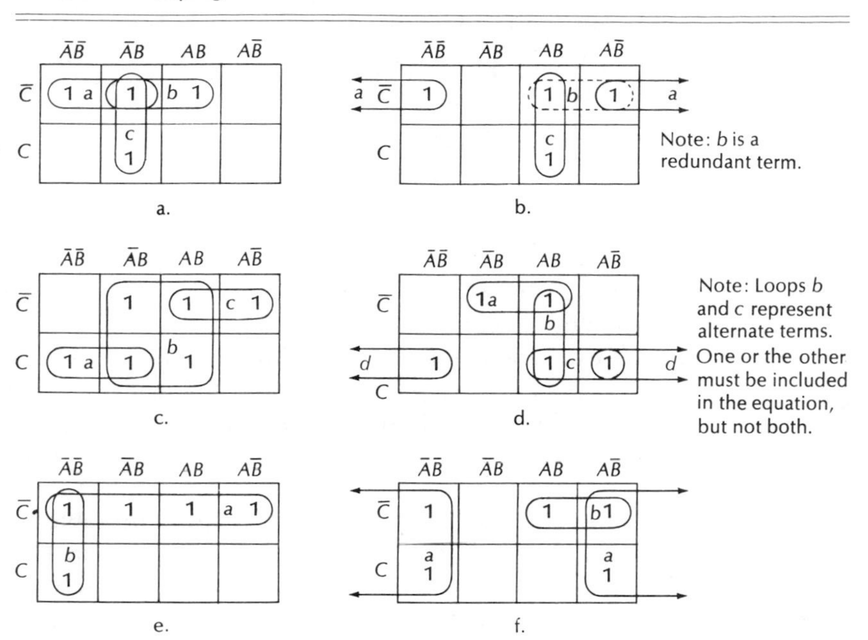

3. The variables left make up the simplified term for that loop. Individual variables are ANDed together (joined by the · symbol).

Example Write the simplified terms for each loop in Table 3-7a through f and combine the loops to get the simplified equation for each map.

Map	*Loop*	*Term*	*Equation*
a.	*a*	$\bar{A}\cdot\bar{C}$	
	b	$B\cdot\bar{C}$	$f = \bar{A}\cdot\bar{C} + B\cdot\bar{C} + \bar{A}\cdot B$
	c	$\bar{A}\cdot B$	
b.	*a*	$\bar{B}\cdot\bar{C}$	
	b	$A\cdot\bar{C}$	$f = \bar{B}\cdot\bar{C} + \cancel{A\cdot\bar{C}} + A\cdot B$
	c	$A\cdot B$	
c.	*a*	$\bar{A}\cdot C$	
	b	B	$f = \bar{A}\cdot C + B + A\cdot\bar{C}$
	c	$A\cdot\bar{C}$	
d.	*a*	$B\cdot\bar{C}$	
	b	$A\cdot B$	$f = B\cdot\bar{C} + A\cdot B + A\cdot C + \bar{B}\cdot C$
	c	$A\cdot C$	
	d	$\bar{B}\cdot C$	

Notice that term b is redundant on map d. Eliminating that term we get: $f = B \cdot \bar{C} + A \cdot C + \bar{B} \cdot C$.

$$\text{e.} \quad \begin{matrix} a & \bar{C} \\ b & \bar{A} \cdot \bar{B} \end{matrix} \Bigg\} f = \bar{C} + \bar{A} \cdot \bar{B}$$

$$\text{f.} \quad \begin{matrix} a & \bar{B} \\ b & A \cdot \bar{C} \end{matrix} \Bigg\} f = \bar{B} + A \cdot \bar{C}$$

Don't Cares

In some situations some of the combinations on a truth table are not defined. For example, a binary code for the decimal digits 0 through 9 requires that only ten of the combinations on a sixteen-row truth table be defined. The left over combinations are said to be *don't-care* terms. Because they will never occur as a part of the code, it is academic what the logic circuit would do if they did occur. We "don't care" whether the logic circuit produces a 1 or a 0 output for such will-never-happen combinations.

We can enter either a 0 or a 1 in the f column of the truth table for don't-care conditions. However, since the value we assign can make a difference in the simplification process, don't-care terms on the truth table are designated by an uncommitted symbol such as ∅ or X. When loops are being formed on the Karnaugh map, ∅ symbols are included in a loop if they serve to enlarge a loop—in this case the ∅ is treated as a 1 entry. If the ∅ entry cannot be used to enlarge a loop, it is treated as a 0 and ignored.

Problems

Given the Karnaugh maps in Table 3-8, loop each map and write the simplified equation for each problem.

Table 3-8 Karnaugh Maps for Problems 11 through 16

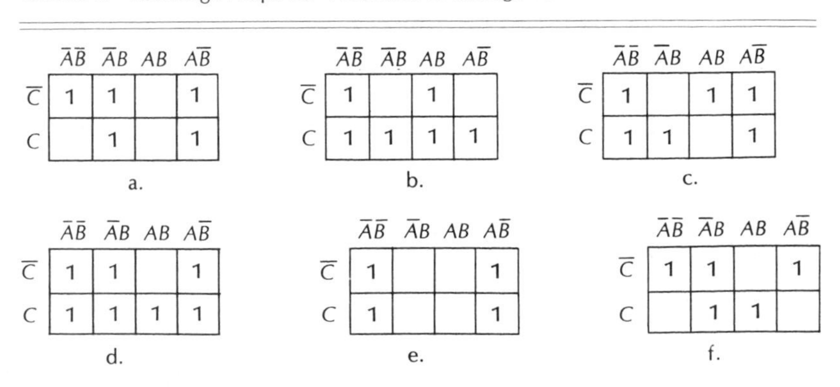

a.

	$\bar{A}\bar{B}$	$\bar{A}B$	AB	$A\bar{B}$
$\bar{C}$	1	1		1
C		1		1

b.

	$\bar{A}\bar{B}$	$\bar{A}B$	AB	$A\bar{B}$
$\bar{C}$	1		1	
C	1	1	1	1

c.

	$\bar{A}\bar{B}$	$\bar{A}B$	AB	$A\bar{B}$
$\bar{C}$	1		1	1
C	1	1		1

d.

	$\bar{A}\bar{B}$	$\bar{A}B$	AB	$A\bar{B}$
$\bar{C}$	1	1		1
C	1	1	1	1

e.

	$\bar{A}\bar{B}$	$\bar{A}B$	AB	$A\bar{B}$
$\bar{C}$	1			1
C	1			1

f.

	$\bar{A}\bar{B}$	$\bar{A}B$	AB	$A\bar{B}$
$\bar{C}$	1	1		1
C		1	1	

11. Map a. 12. Map b. 13. Map c.
    Hint: See Table 3-6b.
14. Map d. 15. Map e. 16. Map f.

Example Write the simplified terms for each loop in Table 3-9a and b. Combine the loops to get the simplified equation for each map. Study the table with care.

Map	*Loop*	*Term*	*Equation*
a.	a	$\bar{B} \cdot \bar{D}$	$f = \bar{B} \cdot \bar{D} + \bar{A} \cdot D + B \cdot \bar{C} \cdot D$
	b	$\bar{A} \cdot D$	
	c	$B \cdot \bar{C} \cdot D$	

Note: The loop in the dotted line is tempting but doesn't include any 1 entries that have not already been looped.

b.	a	$B \cdot D$	$f = B \cdot D + B \cdot C + A \cdot D$
	b	$B \cdot C$	
	c	$A \cdot D$	

Problem

17. Given the Karnaugh maps in Table 3-10, loop each map and write the simplified equation for each problem.

Example Given the truth table shown in Table 3-11a, simplify the circuit and draw the logic diagram for both the simplified and unsimplified circuits.

Table 3-9 Simplifying Equations in 4-Variable Karnaugh Maps

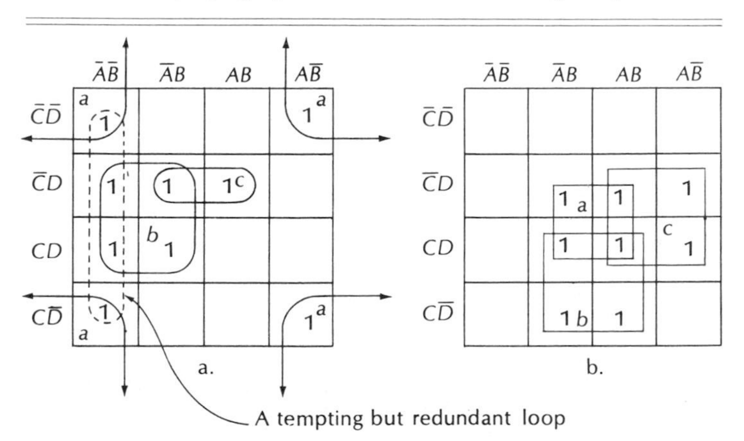

Table 3-10 Karnaugh Maps for Problem 17

a.

	$\bar{A}\bar{B}$	$\bar{A}B$	AB	$A\bar{B}$
$\bar{C}\bar{D}$	1			1
$\bar{C}D$				1
CD			1	1
$C\bar{D}$	1		1	1

b.

	$\bar{A}\bar{B}$	$\bar{A}B$	AB	$A\bar{B}$
$\bar{C}\bar{D}$	1		1	
$\bar{C}D$	1		1	1
CD	1		1	1
$C\bar{D}$			1	

Table 3-11 Table for the Example in Problem 17

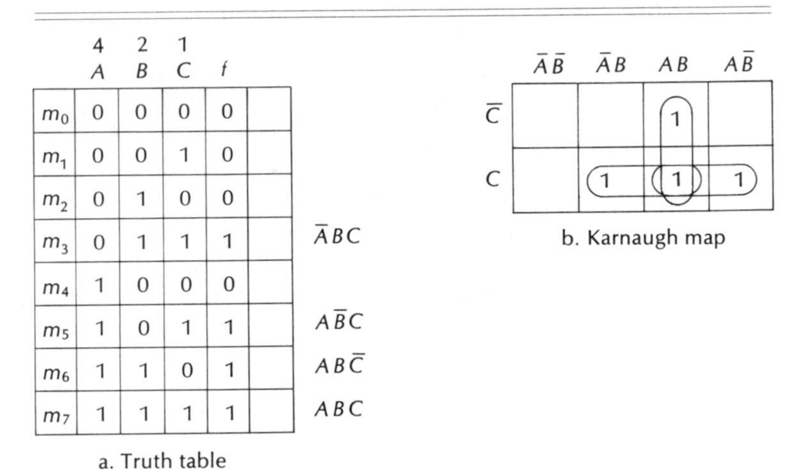

	4 A	2 B	1 C	f		
m_0	0	0	0	0		
m_1	0	0	1	0		
m_2	0	1	0	0		
m_3	0	1	1	1		$\bar{A}BC$
m_4	1	0	0	0		
m_5	1	0	1	1		$A\bar{B}C$
m_6	1	1	0	1		$AB\bar{C}$
m_7	1	1	1	1		ABC

a. Truth table

	$\bar{A}\bar{B}$	$\bar{A}B$	AB	$A\bar{B}$
$\bar{C}$			1	
C		1	1	1

b. Karnaugh map

Writing the minterm equation from the truth table:

$$f = (\bar{A} \cdot B \cdot C) + (A \cdot \bar{B} \cdot C) + (A \cdot B \cdot \bar{C}) + (A \cdot B \cdot C)$$

After simplifying with the Karnaugh map in Table 3-11b, we get:

$$f = (A \cdot B) + (A \cdot C) + (B \cdot C)$$

Figure 3-8 shows the unsimplified and simplified logic diagrams.

Problem

18. Given the simplified equation $f = (A \cdot B) + (A \cdot C) + (B \cdot C)$, plot it in the f_0 column of Table 3-12. Hint: The term $(A \cdot B)$ does not define the condition for variable C. Therefore, the term $(A \cdot B)$ represents two entries on the table: $(A \cdot B \cdot \bar{C})$ and $(A \cdot B \cdot C)$.

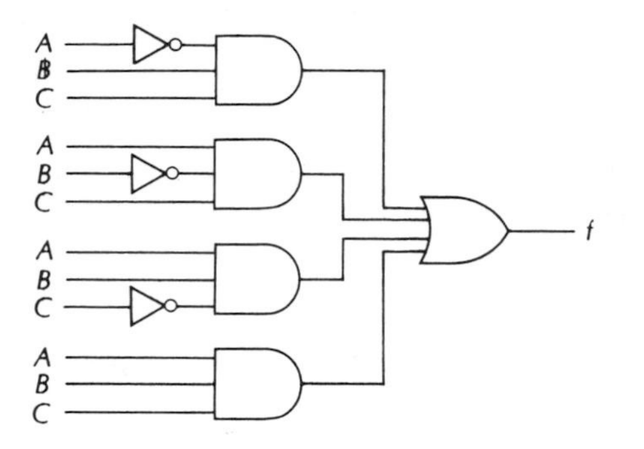

a. Unsimplified version

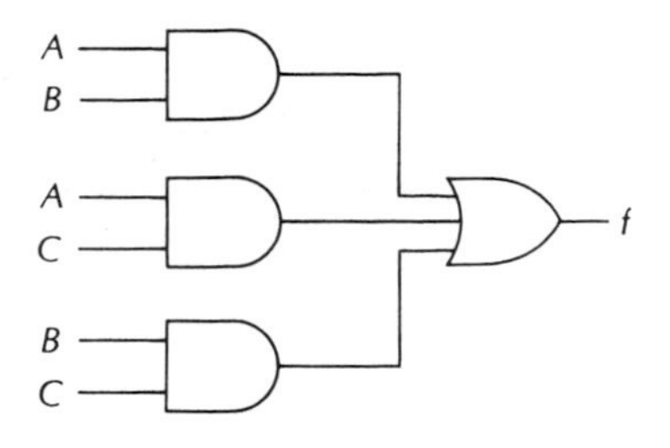

b. Simplified version

Figure 3-8 Logic Diagrams

Table 3-12 Table for Problem 18

	4 A	2 B	1 C	f
m_0	0	0	0	
m_1	0	0	1	
m_2	0	1	0	
m_3	0	1	1	
m_4	1	0	0	
m_5	1	0	1	
m_6	1	1	0	
m_7	1	1	1	

Truth table

3-7 A Case for Simplification

As a general rule, the more complex the specifications are (the more truth table entries involved), the more profitable simplification is likely to be. The following example illustrates the point.

Example Given the following truth table and Karnaugh map (Table 3-13), write the unsimplified and the simplified equations and draw both the simplified and unsimplified logic diagrams.

Table 3-13 A Case for Simplification

		8 *A*	4 *B*	2 *C*	1 *D*	*f*	Minterms
a	m_0	0	0	0	0	1	$\overline{A}\,\overline{B}\,\overline{C}\,\overline{D}$
e	m_1	0	0	0	1	1	$\overline{A}\,\overline{B}\,\overline{C}D$
i	m_2	0	0	1	0	1	$\overline{A}\,\overline{B}C\overline{D}$
	m_3	0	0	1	1	0	
b	m_4	0	1	0	0	1	$\overline{A}B\overline{C}\,\overline{D}$
f	m_5	0	1	0	1	1	$\overline{A}B\overline{C}D$
	m_6	0	1	1	0	0	
	m_7	0	1	1	1	0	
d	m_8	1	0	0	0	1	$A\overline{B}\,\overline{C}\,\overline{D}$
h	m_9	1	0	0	1	1	$A\overline{B}\,\overline{C}D$
j	m_{10}	1	0	1	0	1	$A\overline{B}C\overline{D}$
	m_{11}	1	0	1	1	0	
c	m_{12}	1	1	0	0	1	$AB\overline{C}\,\overline{D}$
g	m_{13}	1	1	0	1	1	$AB\overline{C}D$
	m_{14}	1	1	1	0	0	
	m_{15}	1	1	1	1	0	

a. Truth table

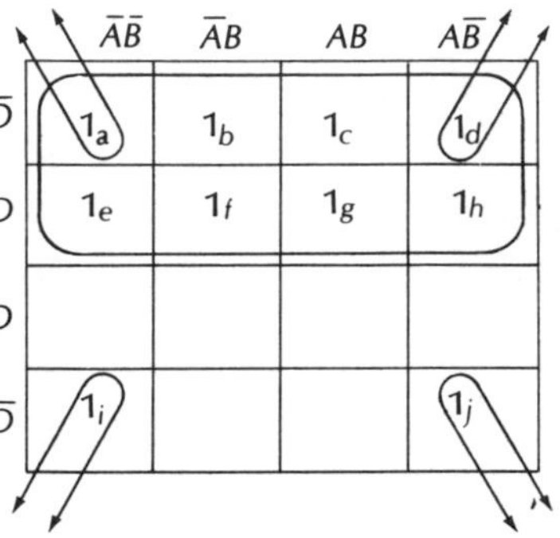

	$\overline{A}\,\overline{B}$	$\overline{A}B$	AB	$A\overline{B}$
$\overline{C}\,\overline{D}$	1_a	1_b	1_c	1_d
$\overline{C}D$	1_e	1_f	1_g	1_h
CD				
$C\overline{D}$	1_i			1_j

b. Karnaugh map for example 3-6

a. The unsimplified equation:

$$\begin{aligned} f = (\overline{A}\cdot\overline{B}\cdot\overline{C}\cdot\overline{D}) &+ (\overline{A}\cdot\overline{B}\cdot\overline{C}\cdot D) + \\ (\overline{A}\cdot\overline{B}\cdot C\cdot\overline{D}) &+ (\overline{A}\cdot B\cdot\overline{C}\cdot\overline{D}) + \\ (\overline{A}\cdot B\cdot\overline{C}\cdot D) &+ (A\cdot\overline{B}\cdot\overline{C}\cdot\overline{D}) + \\ (A\cdot\overline{B}\cdot\overline{C}\cdot D) &+ (A\cdot\overline{B}\cdot C\cdot\overline{D}) + \\ (A\cdot B\cdot\overline{C}\cdot\overline{D}) &+ (A\cdot B\cdot\overline{C}\cdot D) \end{aligned}$$

b. The simplified equation:

$$f = \overline{C} + (\overline{B}\cdot\overline{D})$$

The unsimplified logic diagram is shown in Figure 3-9a and the simplified version is shown in part b. Notice that the variable *A* has dropped out of the system entirely.

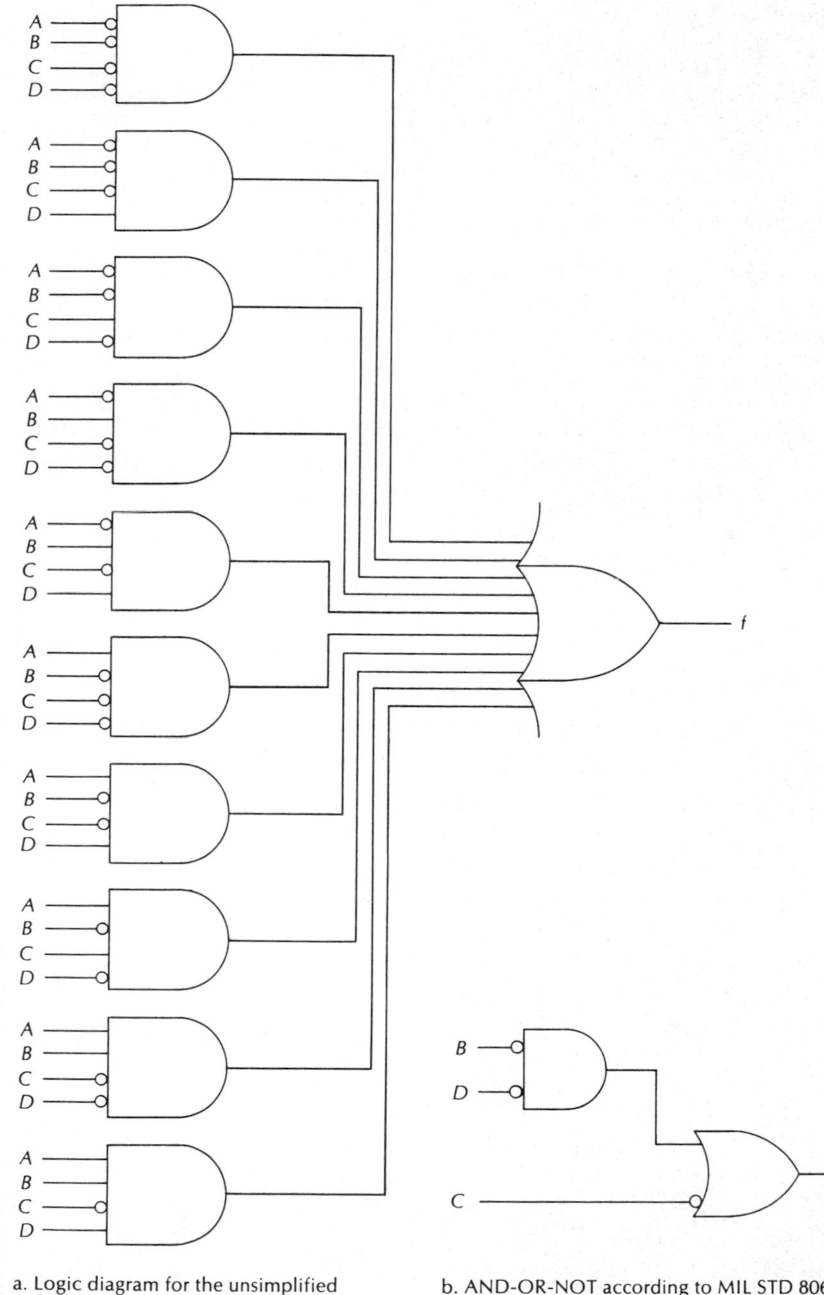

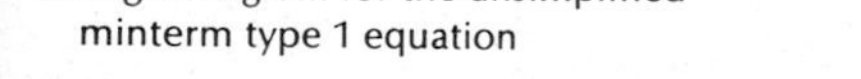
a. Logic diagram for the unsimplified minterm type 1 equation

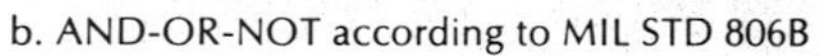
b. AND-OR-NOT according to MIL STD 806B

Figure 3-9 Logic Diagrams for the Minterm Type 1 Equation

Problem

19. Given the truth table (Table 3-14), plot the equation: $f = \overline{C} + (\overline{B} \cdot \overline{D})$ in the f column. (Notice that Table 3-14 is a 3 variable table because only the variables B, C, D appear in the equation.

3-8 Additional Methods of Minimizing Hardware

1. When more than one gate input in a logic structure requires a particular inverted variable, a single inverter can be used in place of multiple inverters as long as fan-out rules are observed. (See Figure 3-10.)
2. The distributive laws can sometimes be taken advantage of when differences in gate delay times for the different variables can be tolerated. An example of the first distributive law is shown in Figure 3-11 and the second distributive law in Figure 3-12.

Table 3-14 Table for Problem 19

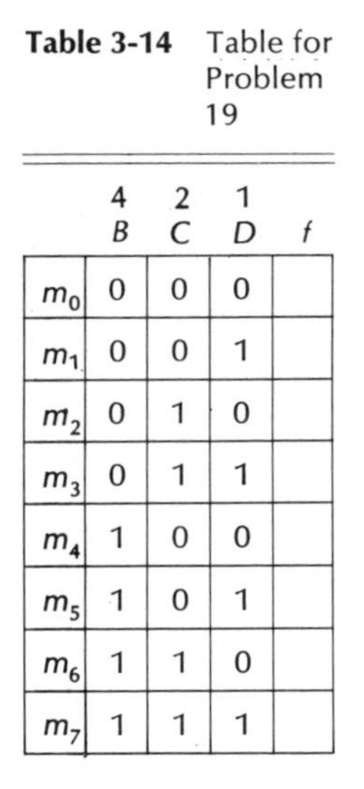

	4 B	2 C	1 D	f
m_0	0	0	0	
m_1	0	0	1	
m_2	0	1	0	
m_3	0	1	1	
m_4	1	0	0	
m_5	1	0	1	
m_6	1	1	0	
m_7	1	1	1	

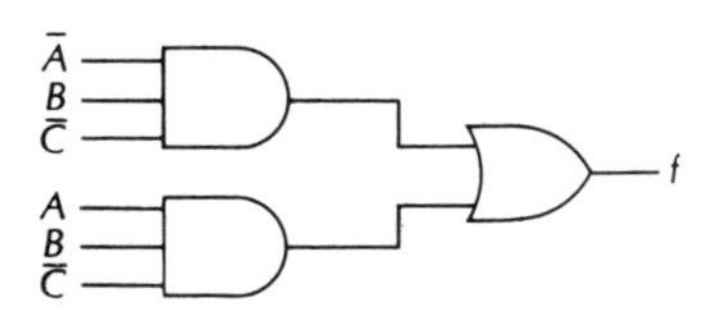

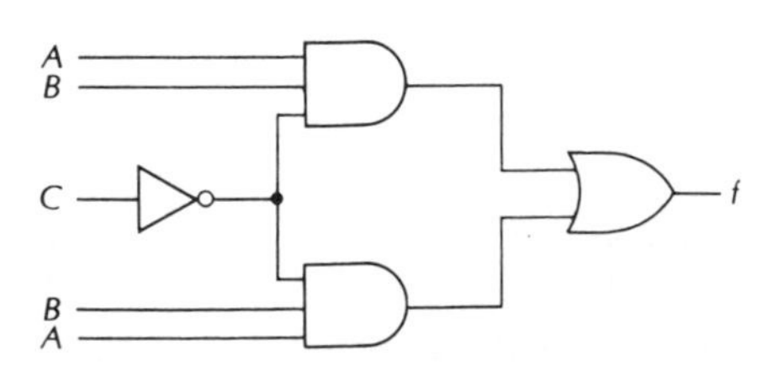

Figure 3-10 Combining Inverters

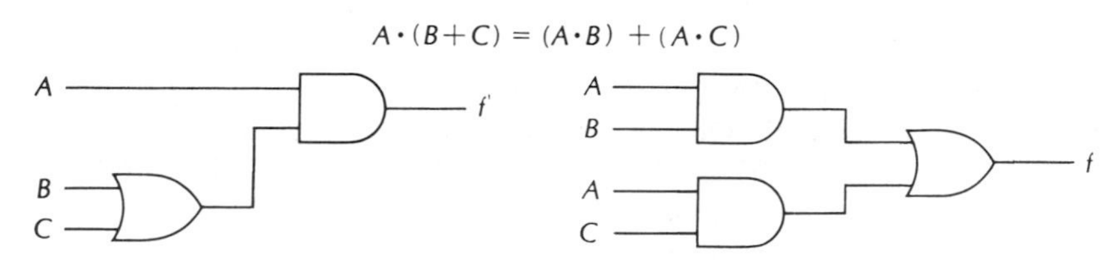

Figure 3-11 First Distributive Law

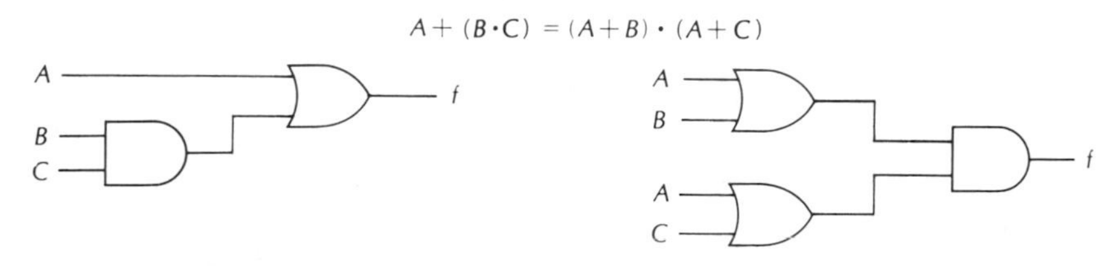

Figure 3-12 Second Distributive Law

3-9 Minterm and Maxterm Forms Using NAND and NOR Logic Gates

MINTERM CIRCUITS

In Figure 3-1 it was demonstrated that the truth table outputs (f) for a positive logic OR and an inverted negative logic AND (NAND) are identical. When NAND logic is used exclusively to form a minterm equation, a NAND gate can replace the OR gate.

In Figure 3-13a, a minterm logic circuit is shown using AND-OR-NOT logic. In part b the NAND gates (1 and 2) produce inverted

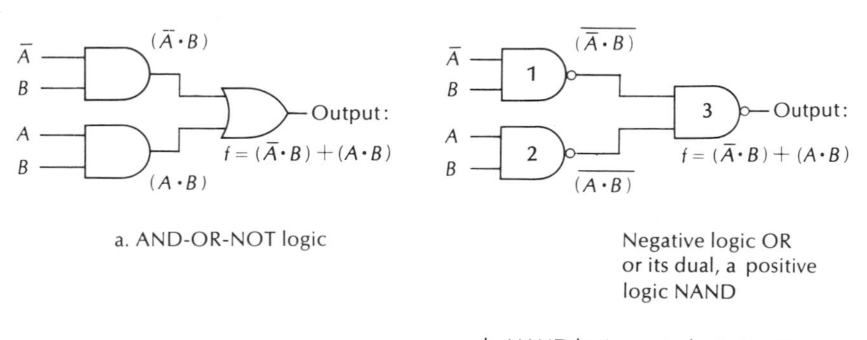

Figure 3-13 Minterm Circuits Using NAND Logic

outputs, so what we need is an inverted negative logic OR gate, which cannot be found in any of the modern logic families. However, the logic dual of a negative-logic OR is a positive-logic NAND (see Figure 3-1). If we substitute a positive-logic NAND for the negative-logic OR, we get the circuit shown in Figure 3-13b.

Because the NAND gate is the basic gate in TTL, most minterm circuits using TTL will be implemented as shown in Figure 3-13b instead of in the form shown in part a. In TTL AND and OR gates are more expensive and less available than NAND gates.

MAXTERM CIRCUITS

The maxterm form using NOR gates is less frequently used than the NAND minterm form, but it does occur. Figure 3-14a shows a maxterm circuit using AND-OR-NOT gates, and part b shows the NOR equivalent circuit.

Because the NOR gates (1 and 2) in Figure 3-14b invert the logic gate, 3 must be a negative-logic AND gate or its dual (inverted) a positive-logic NOR. (See Figure 3-18 for a summary of logic duals.)

Problems

Draw the NAND (minterm) logic diagram for each of the following equations:

20. $f = (\bar{A} \cdot B \cdot \bar{C}) + (A \cdot \bar{B} \cdot \bar{C}) + (A \cdot \bar{B} \cdot C)$
21. $f = (A \cdot \bar{B}) + (A \cdot B \cdot \bar{C}) + (B \cdot C) + (\bar{B} \cdot \bar{C})$
22. $f = (A\bar{B}CD) + (\bar{A}\bar{B}\bar{C}\bar{D}) + (ABC\bar{D})$
23. $f = (A \cdot B) + (C \cdot D) + (A \cdot \bar{C})$

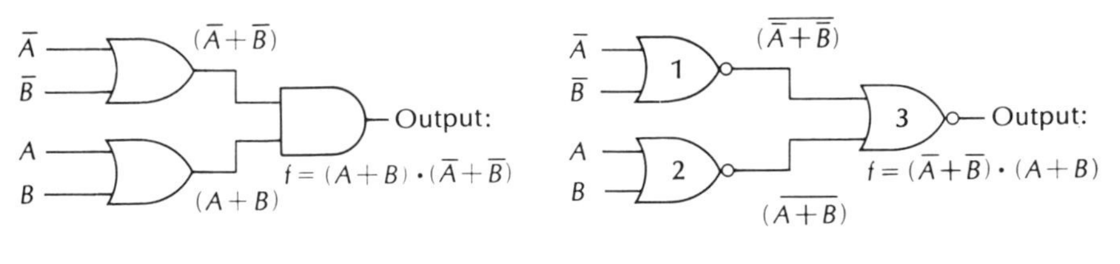

a. AND-OR-NOT logic

Negative logic AND or its dual, a positive logic NOR gate

b. NOR logic equivalent circuit

Figure 3-14 Maxterm Circuits Using NOR Gates

3-10 The Exclusive-OR Gate

The exclusive-OR is a two-input gate that finds a great many applications in digital systems. The truth table for the exclusive-OR is shown in Table 3-15, and Figure 3-15 is the X-OR (exclusive-OR) logic diagram and symbol.

The TTL 7486 package contains four exclusive-OR circuits that are functionally equivalent to Figure 3-15c. Each circuit is symbolized in Figure 3-15d.

Table 3-15 The X-OR Truth Table and Equation

A	B	f	
0	0	0	
0	1	1	$\overline{A} \cdot B$
1	0	1	$A \cdot \overline{B}$
1	1	0	

a. Truth table

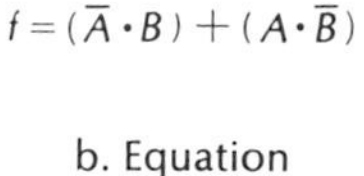

$$f = (\overline{A} \cdot B) + (A \cdot \overline{B})$$

b. Equation

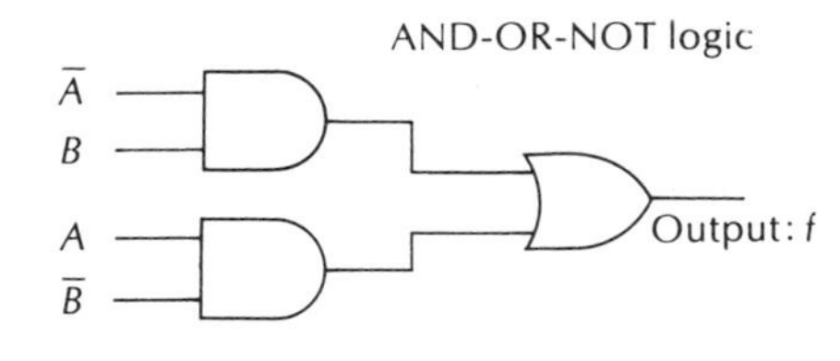

a. Logic circuit for $f = (\overline{A} \cdot B) + (A \cdot \overline{B})$

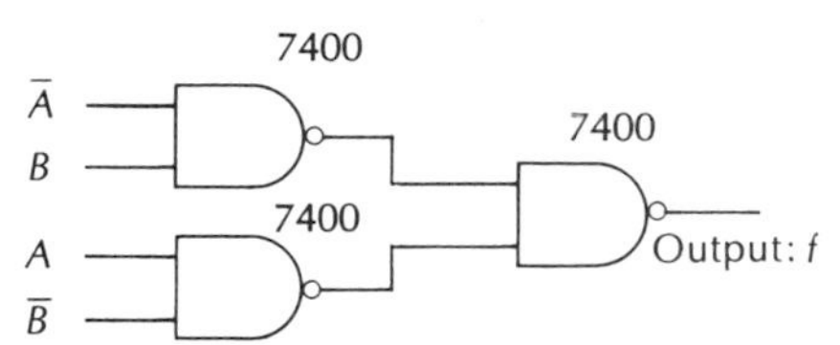

b. NAND logic with complements available

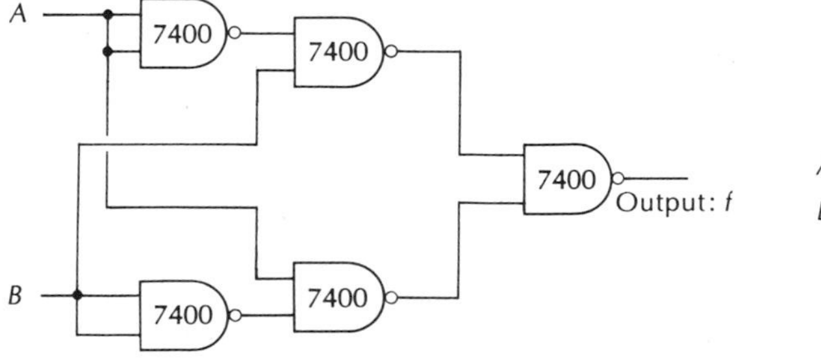

c. NAND logic when complements are not available

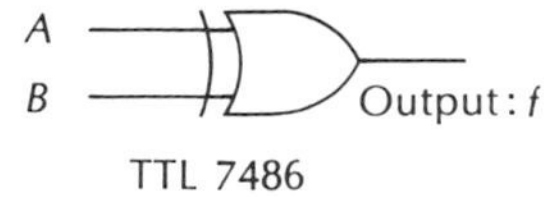

d. X-OR symbol

Figure 3-15 The X-OR Circuit and Symbol

Among the many applications for exclusive-OR gates, two important examples bear mentioning here.

1. Controlled Complementer

   The controllable complement circuit is sometimes called a *controlled inverter* or *true-invert gate*. In this application one input is used as a signal input that produces either a true (non-inverted) output signal or an inverted (complemented) signal, depending upon whether the second input is high or low. Look at truth table 3-15. Let input B be the signal input and put a fixed 0 level on input A. A zero on B produces zero out, and one on B produces a one out—the signal out is not inverted. Now, if input A is set to a fixed 1 level, a zero input on B produces a one output, and a one input produces a zero output—the signal is inverted.

   An examination of the truth table (3-15) indicates that control and signal inputs can be exchanged with the same results as before.

2. Binary Adder

   The truth table for binary addition and the exclusive-OR table are coincidentally identical, which makes the X-OR an ideal circuit element when binary addition is to be performed. We will examine these and other applications in some detail as we encounter them later in the text.

   The X-OR gate, like the inverter, is independent of logic definition. It performs the same logic function for both negative and positive logic. X-OR gates are not available with more than two inputs.

Problem

24. Draw the NOR (maxterm) logic diagram for an X-OR circuit.

3-11 Expanding the Number of Gate Inputs

It is frequently desirable to combine several gates with few inputs each to obtain a larger number of inputs, rather than to purchase a fairly expensive package with the required number of inputs. Figure 3-16 shows how this is accomplished.

3-12 AND-OR-Invert Gates

Minterm equations can often be implemented with fewer IC packages by using AND-OR-Invert gates like the one shown in Figure 3-17a. Some of the A-O-I gates can be expanded to obtain more inputs. Spe-

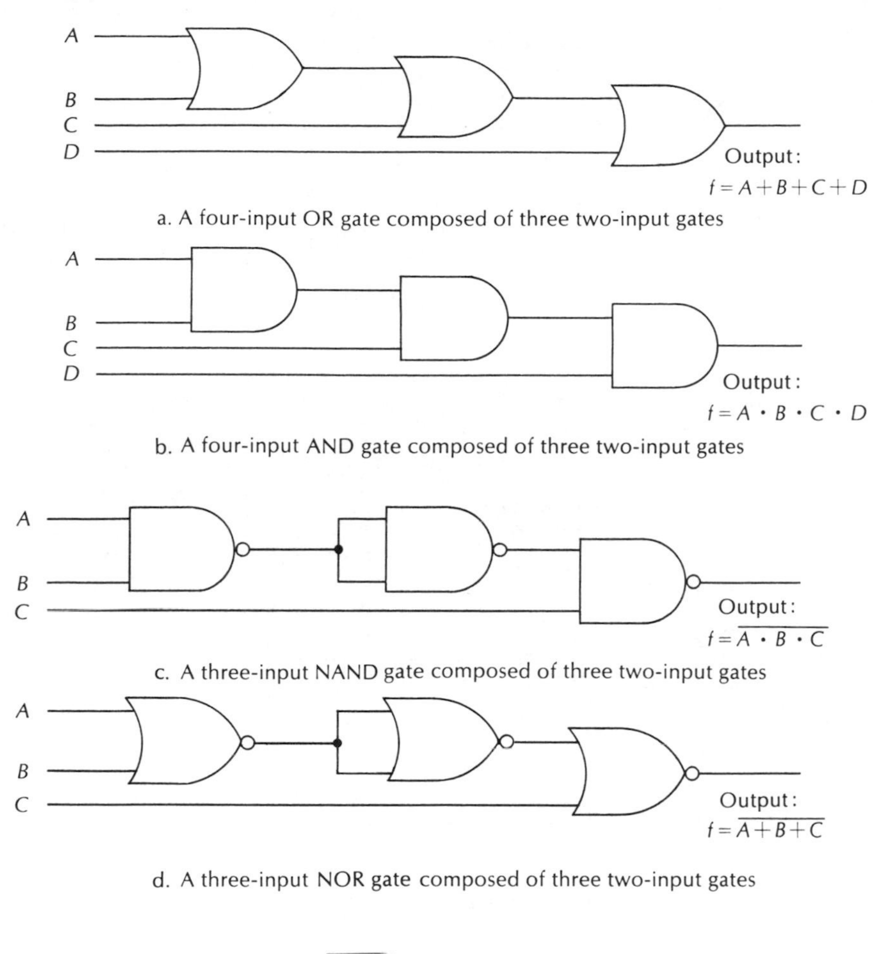

a. A four-input OR gate composed of three two-input gates

b. A four-input AND gate composed of three two-input gates

c. A three-input NAND gate composed of three two-input gates

d. A three-input NOR gate composed of three two-input gates

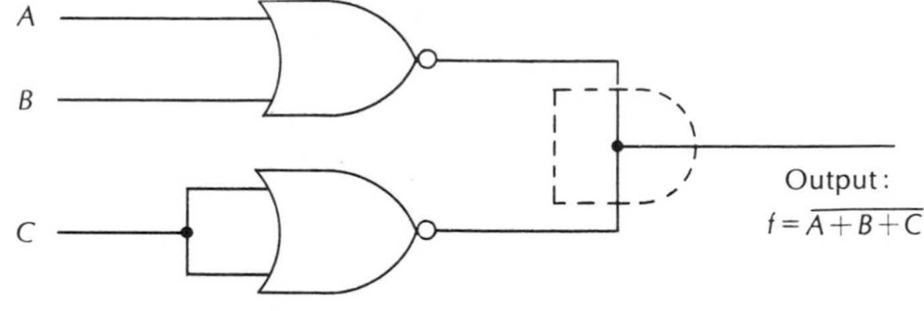

e. "Wired AND" three-input NOR gate (open collector)

Figure 3-16 Multiple Input Gates

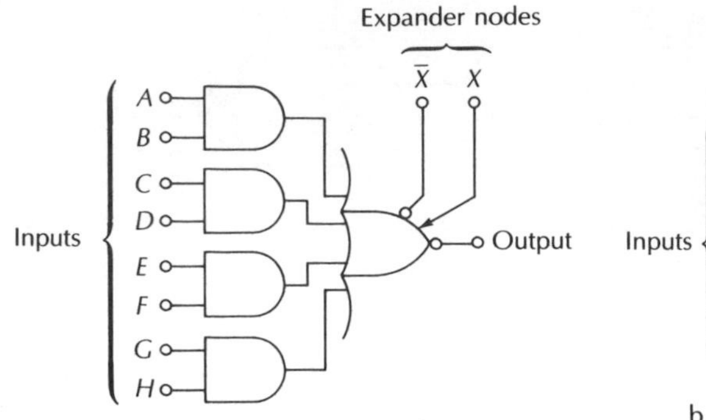

a. Four-wide, 2-input AND-OR-Invert gate

b. Frequently used symbol for AND-OR-Invert gates

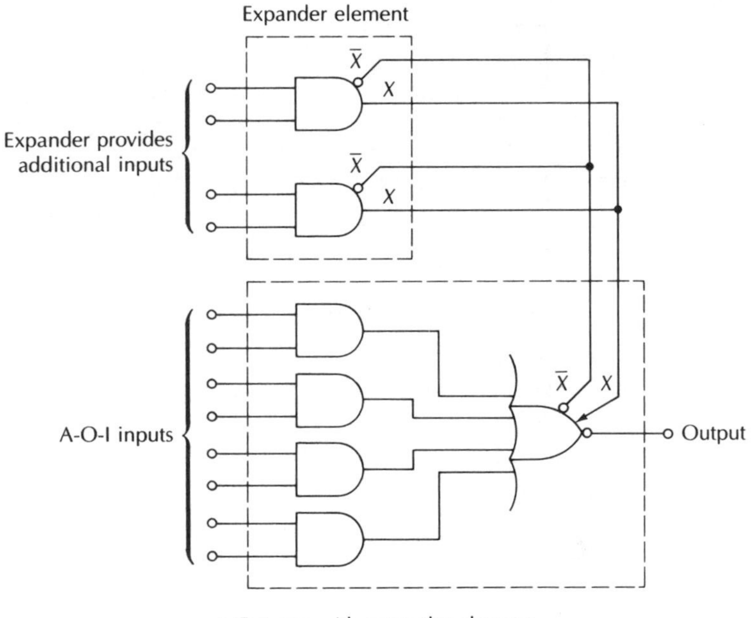

c. A-O-I gate with expander element

Figure 3-17 Expandable AND-OR-Invert Gates

cial expander elements are available for that purpose. The number of expanders that can be connected to an A-O-I gate is normally limited to four. When expandable A-O-I gates are used without an expander element, expander nodes must be left unconnected. Figure 3-17c shows how the A-O-I gate and expander element are interconnected.

Table 3-16 Common AND-OR-Invert Gates and Expander Elements

Type	Expandable	Description
7450	Yes	Dual 2-wide 2-input
74H51 74S51	No	Dual 2-wide 2-input
74L51 74LS51	No	a. 2-wide 2-input b. 2-wide 3-input
7452	Yes	4-wide inputs: 2,2,2,3 Note: This is an AND-OR gate. There is no Invert.
7453	Yes	4-wide 2-input
7454	No	4-wide 2-input
7455	Yes	2-wide 4-input

a. A-O-I gates

Type	Description	For use with gate numbers:
7460	2-wide 4-input	7423,7450,7453
7461	3-wide 3-input	7452 only
7462	4-wide inputs: 2,2,3,3	7450,7453,7455

b. A-O-I gate expanders

Table 3-16 lists common TTL A-O-I gates and associated expander elements.

SUMMARY

The five gate structures that are of importance in logic are the one-input inverter, two- (or more) input OR, AND, NAND, NOR gates, and the two-input exclusive-OR gate. All logic circuits are made up of combinations of these basic gates.

There are two standard arrangements for logic gates—the minterm form, called the sum-of-products form, and the maxterm form, called the product-of-sums. The minterm form is the more common because it is the natural form for NAND gate logic, the basic gate in TTL.

LOGIC DEFINITIONS

Positive Logic	*Negative Logic*
+ = logic 1	Ground = logic 1
Ground = logic 0	+ = logic 0

Simplification is required for gate-only logic circuits to minimize the number of packages. The Karnaugh map is the principal aid.

The logic duals in Figure 3-18 form the basis for NAND-only and NOR-only logic circuits, and can serve as a useful reference.

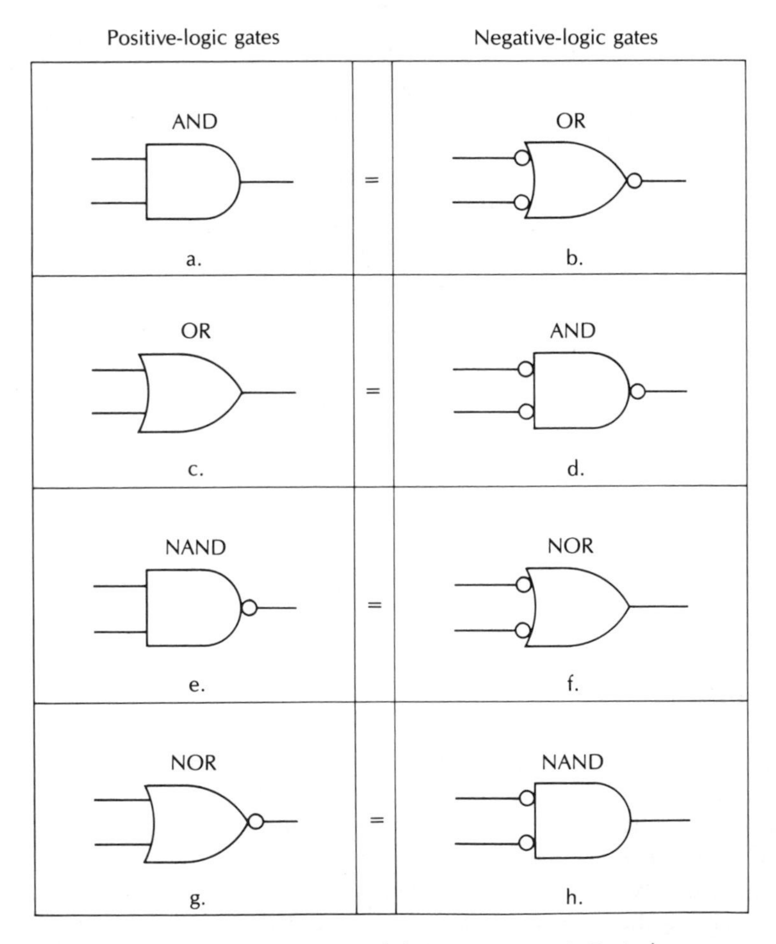

Figure 3-18 Positive-Logic Gates and their Negative-Logic Equivalents

Problems 25, 26, 27

Write the equations and draw the logic diagrams for the truth tables in Tables 3-17, 3-18, and 3-19.

28. Given the following equation, implement the logic diagram using AND-OR-Invert gates and expander elements:

$$f = \overline{\bar{A}\bar{C}D + \bar{A}BC + AB\bar{C} + ACD}$$

Table 3-17

A	B	C	f
0	0	0	1
0	0	1	0
0	1	0	0
0	1	1	1
1	0	0	0
1	0	1	0
1	1	0	0
1	1	1	0

Problem 25

Table 3-18

A	B	C	f
0	0	0	0
0	0	1	0
0	1	0	0
0	1	1	1
1	0	0	1
1	0	1	1
1	1	0	0
1	1	1	0

Problem 26

Table 3-19

A	B	C	f
0	0	0	0
0	0	1	1
0	1	0	1
0	1	1	0
1	0	0	1
1	0	1	0
1	1	0	1
1	1	1	0

Problem 27

CHAPTER 4

DATA CONTROL AND PROGRAMMABLE LOGIC CIRCUITS

Learning Objectives. *Upon completing this chapter you should know:*

1. *What an encoder is and where it is used.*
2. *The purpose of the priority function in commercial encoders.*
3. *The function of the decoder.*
4. *The function of special-purpose decoders, such as the BCD-to-7-segment decoder.*
5. *The properties of decoder-drivers and where they are used.*
6. *How he multiplexer/data selector works.*
7. *The relationship between the multiplexer and its mechanical switching equivalent.*
8. *How the multiplexer can be used to implement asynchronous logic functions.*
9. *How folded data selector logic is implemented.*
10. *How multiplexers can be expanded.*
11. *How the demultiplexer works.*
12. *How the C-MOS transmission gate multiplexer/demultiplexer works.*
13. *How the programmable logic array works and how it is programmed.*
14. *How the read-only memory works.*
15. *How to program a read-only memory for asynchronous logic functions.*

In this chapter we will study some medium-scale integration (MSI) asynchronous logic circuits. *Encoders* translate from decimal or some other single input signal into binary. *Decoders* translate from binary into decimal or some other one-at-a-time output. Special decoders are also available to drive seven-segment numerical displays. A *multiplexer* is the gate equivalent of a selector switch with many inputs and a single output. A *demultiplexer* is the gate equivalent of a selector switch with one input and many outputs. Bidirectional *transmission gates* in the C-MOS family form the basis for a single device capable of functioning as either a multiplexer or demultiplexer.

PROGRAMMABLE LOGIC

In the previous chapter we examined asynchronous logic circuits. The gate-only approach in Chapter 3 frequently leads to a great many IC packages as logic circuits become more complex. In this chapter we will see how the multiplexer can be used as a one- or two-package solution to many asynchronous logic problems.

THE PLA

We will also study the *programmable logic array* (PLA). The PLA is a package that contains the circuitry to produce a 1 output for all minterms in a given size truth table. The device is programmed to produce only the desired minterms. Programming is accomplished by blowing built-in fuses to eliminate unwanted minterm outputs.

ROM LOGIC

The *read-only memory* (ROM), originally intended as a permanent memory storage device for computers, can prove to be an economical solution to complex multiple-output asynchronous logic circuits. Some ROM's are programmed by blowing built-in fuses. Once programmed, the ROM becomes permanently dedicated to the task for which it was programmed.

4-1 Encoders

An *encoder* consists of a number of OR gates and is used to convert a single input into its binary equivalent output. An example is a decimal keyboard that must provide a binary output. The encoder converts each keypush signal into its binary equivalent.

An early encoder form but one that is still popular is constructed from diode OR gates. Diode encoders are often drawn and sometimes laid out on a circuit board in a rectangular matrix form. Figure 4-1 shows a typical *negative-logic* diode OR gate (a), how the gate is arranged in a matrix (b), and the shorthand method of drawing a matrix (c).

The matrix form consists of a rectangular grid with diagonal bars (see Fig. 4-1) representing inputs to gates. All wires in the grid are assumed to be insulated from each other, although no attempt is made to show this in the drawings. In actual circuits, the diagonal bars are a part of particular gates and may be diodes, resistors, or gate inputs. Either the vertical lines or the horizontal lines may denote the inputs; both conventions are used.

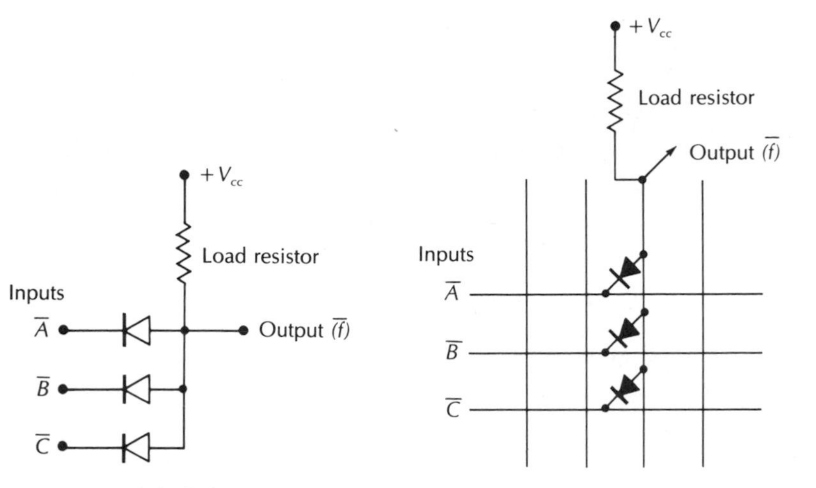

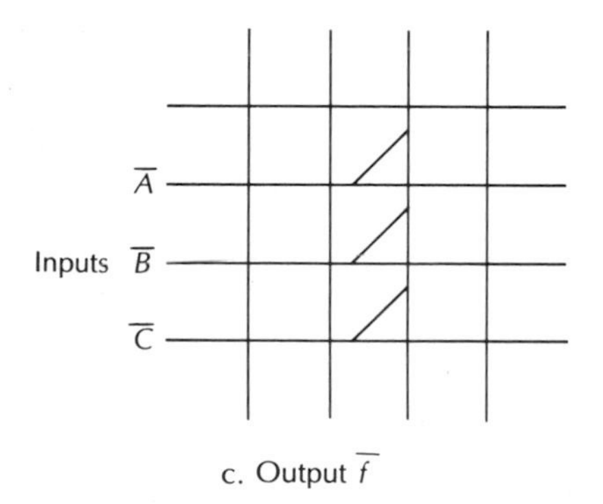

Figure 4-1 Diode OR Gate in Matrix Form

DECIMAL-TO-BINARY ENCODER

An *encoding matrix* consists of a number of OR gates and is often drawn on a rectangular grid, or matrix. It will become apparent that this layout for the gates is much easier to follow than the conventional one for logic diagrams in this particular application.

Let us examine an encoding matrix for translating from decimal into binary. The inputs to the circuit include the decimal digits 1 to 9, the zero requiring no translation. The circuit involves four OR gates as shown in Figure 4-2. Gate 1 must produce a 1 as a binary output for any odd decimal digit. For example, a decimal 7 will produce an output

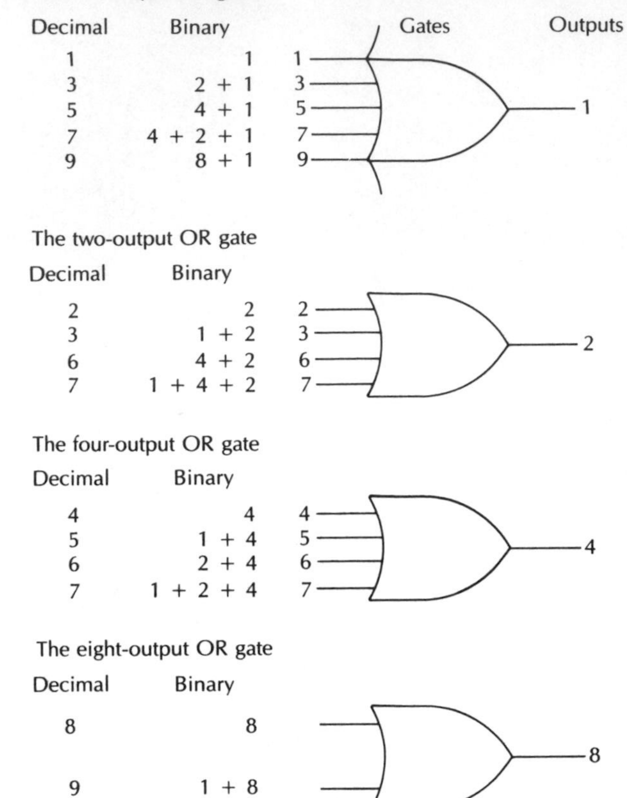

a. The four gates

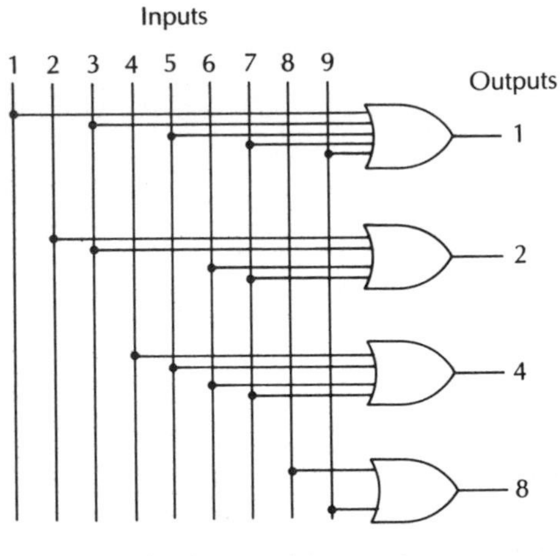

b. The complete encoder

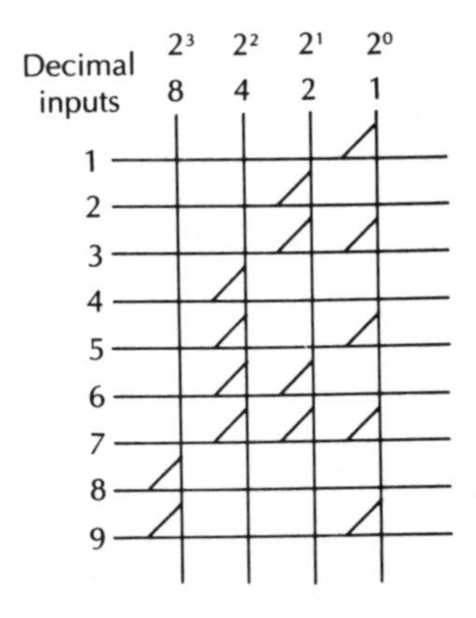

c. The encoder drawn in matrix form

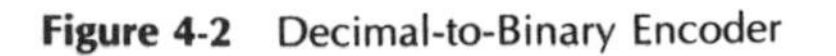

Figure 4-2 Decimal-to-Binary Encoder

from the 4 gate, the 2 gate, and the 1 gate. Figure 4-2b shows the complete decimal-to-binary encoder, and part c shows the complete encoder drawn in matrix form.

PRIORITY ENCODERS

There are two types of commercially available MSI encoders, an 8-line to 3-line and a 10-line to 4-line encoder. In the 7400 TTL series they are the 74147 (10-line to 4-line) and the 74148 (8-line to 3-line).

The 74147 converts 10 inputs to equivalent binary digits. It takes four binary digits (called bits for short) to represent the digits 0 through 9. There are six possible binary combinations that are never used. The 16 possible combinations are limited to only the 10 combinations corresponding to the decimal digits 0 through 9. When only the binary equivalent of decimal digits 0 through 9 are used it is called Binary Coded Decimal (BCD).

The 74148 converts eight inputs to their corresponding binary equivalents. The binary equivalents of decimal digits 0 through 7 requires three bits with no unused combinations.

Both of these encoders have additional circuitry that gives them the name *priority encoder*. If two or more inputs are activated simultaneously the output will correspond to the largest digit activated. For example, if the 7 and the 9 inputs were activated at the same time, the output would be the binary equivalent of 9(1001). The 7 would be ignored.

The priority feature is particularly useful with microcomputers. Assume that the president of Al Electronics, the chief engineer, and the stock clerk each have a terminal connected to the same microcomputer. The computer can perform only one task at a time, but it can be interrupted in the middle of the task to do another job and then take up where it left off in its original task. Each terminal could be assigned a priority number with the president having the highest number, the chief engineer the second highest, and the stock clerk the lowest. Should all request computer service at the same time, the president would be given priority. Even if someone else had a program working, the president's request would interrupt the program in progress while the computer dealt with him. It might appear that the stock clerk would become frustrated at all the interruptions, but in practice, he probably wouldn't be aware of them. The machine is so fast that it could effectively juggle the three sets of data using the priority numbers to keep track of what data goes with which terminal. A priority encoder is used to convert a *request for service key* or signal into its binary equivalent and to deliver only the highest priority number to the computer. Priority encoders can be cascaded for larger numbers.

There are specialized encoders specifically designed to interface typewriter-like keyboards and other larger keyboards to computers. We will look at those briefly in a later chapter.

4-2 Decoders

Decoders essentially reverse the encoding process by converting binary input data into a single output signal. Some special decoders have multiple outputs to drive seven-segment displays of the kind used for read-out in pocket calculators.

Most common decoding functions can be implemented with standard off-the-shelf IC's. The decoder samples a combination of bits and puts out a discrete single pulse for each desired combination. Decoders are used for translating from a binary code into decimal, for providing a discrete output pulse to interrogate a memory location or route data, or to perform some other discrete activity. Figure 4-3 shows a 2-line to 4-line decoder and truth table.

The truth table indicates that the two inputs provide four possible combinations for outputs. The circuit can have as many AND gates as there are possible combinations. Thus, three input lines could provide eight outputs (from eight AND gates), four lines in could produce 16 output lines, and so on. In general, n inputs can provide 2^n outputs. Figure 4-4 shows a 4-line to 16-line decoder.

Standard TTL decoders are often used to drive indicator lamps or other devices. TTL can sink fairly high currents in the low state but can source only a few hundred microamperes in the high state. For that reason, all common TTL 4-line to one-of-10 line and 4-line to one-of-16 line decoders have active-low outputs. For example, for the binary input 1001 (9), all outputs would be high except the 9 output. The 9

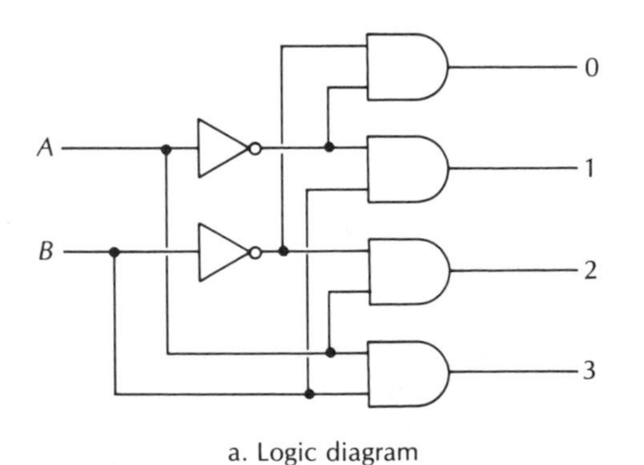

a. Logic diagram

m	A	B	f
0	0	0	0
1	0	1	1
2	1	0	2
3	1	1	3

b. Truth table

Figure 4-3 Two-Line to 4-Line Decoder

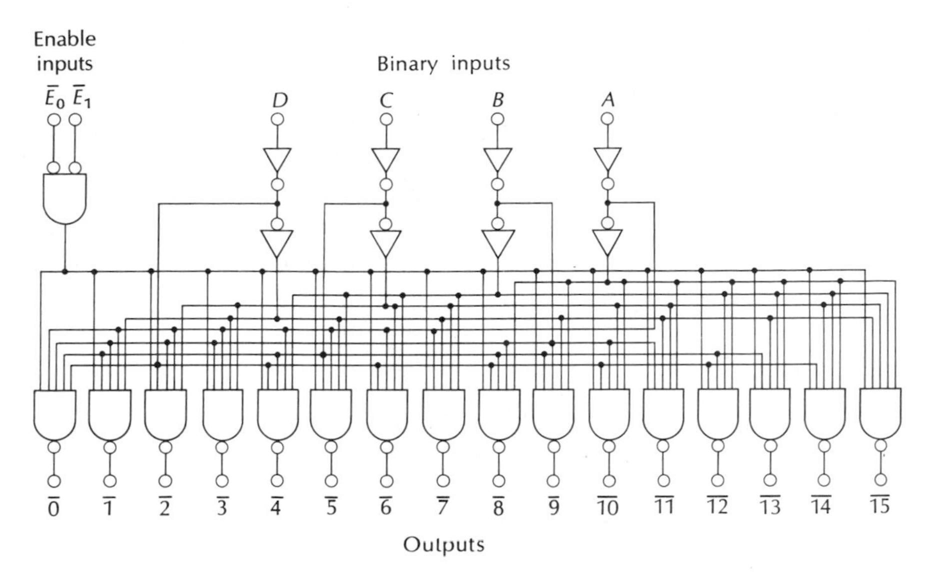

Figure 4-4 Logic Diagram for a 4-Line to 16-Line Decoder

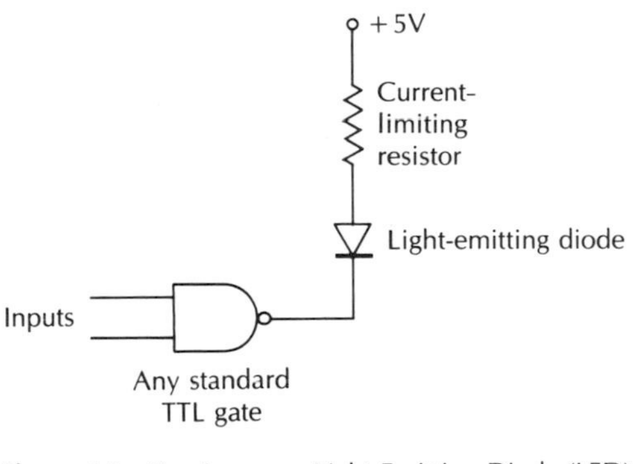

Figure 4-5 Turning on a Light-Emitting Diode (LED) in the Active-Low Mode

output would be low. Figure 4-5 shows how a light-emitting diode (LED) is connected for active-low operation.

In the C-MOS family, the 4-line to one-of-10 line decoder (4028) has active-high outputs. There are two common 4-line to one-of-16 line decoders in the C-MOS family; the 4514, which has active-high outputs, and the 4515, which has active-low outputs. The available output current from all three C-MOS decoders is 1mA at 5 V for each output, and 2 mA at 10 V. In contrast, TTL 4-line to one-of-16 line decoders can sink up to 20 mA in the low state. TTL 4-line to one-of-10 line decoders can sink up to 80 mA.

Four-line to one-of-16 line decoders are provided with two enable inputs (inputs to gates *G*1) in Figure 4-6. These inputs permit one to enable or disable the decoder and also allow for expansion. Figure 4-6 shows how a 4-line to one-of-16 line decoder can be expanded into a one-of 32 line decoder. Gates *G*1 in Figure 4-6 are part of the decoder package. The inverter is separate. Decoders can be further expanded by adding external gates.

4-3 Decoder-Drivers

There are several special decoders intended for driving numerical display devices. The decoders previously discussed have outputs that are TTL-compatible. Decoder-drivers have outputs compatible with whatever kind of display they are intended to drive.

Decoder-drivers designed to drive seven-segment display units of the

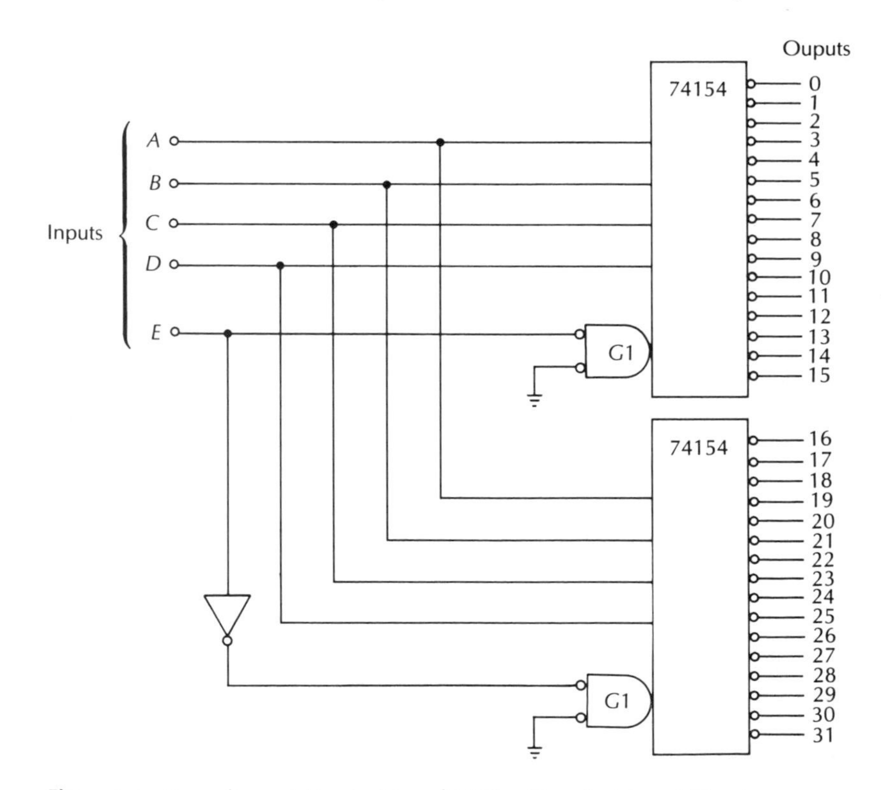

Figure 4-6 Expanding a 4-Line to One-of-16 Line Decoder into a 5-Line to One-of-32 Decoder

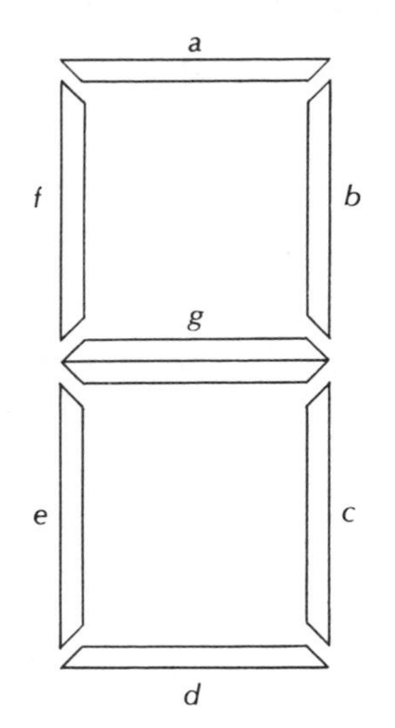

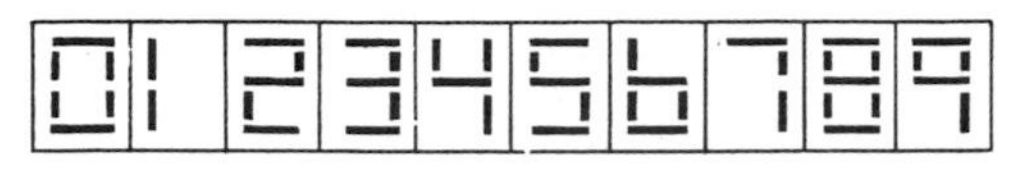

a. The display layout

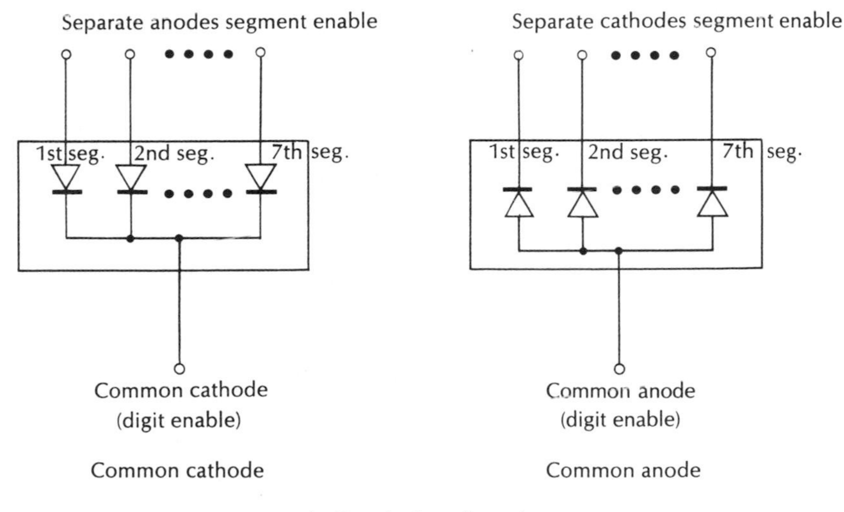

b. Electrical configurations

Figure 4-7 Seven-Segment Display

kind in Figure 4-7 actually consist of a combined decoder-encoder circuit. The four binary bits are decoded to one-of-10 (or one-of-16) outputs and then encoded into the appropriate segment patterns to form the individual digits. Figure 4-8 shows several common decoder-driver applications. Table 4-1 lists typical decoder-driver types.

Table 4-1 Types of Decoder-Drivers

Type Number	Type of Display	Display Voltage	Display Current
7445	10 individual incandescent lamps	5V	80mA
7447	7-segment common-anode LED	5V	10mA per segment typical
7447	7-segment RCA numitron incandescent	5V	10–15mA per segment
74141	Nixie gas-discharge display	60V	7mA

Current limiting resistors are required with light-emitting diodes and gas-filled Nixie® devices to prevent destructive currents. Each segment in seven-segment devices must have a separate resistor to prevent current hogging and unequal segment brightness. A few special seven-segment drivers have built-in current sources that make the resistor unnecessary.

4-4 Multiplexer/Data Selector

A *multiplexer/data selector* is the digital equivalent of a multiposition switch (see Table 4-2). It has from two to 16 data inputs and one output.

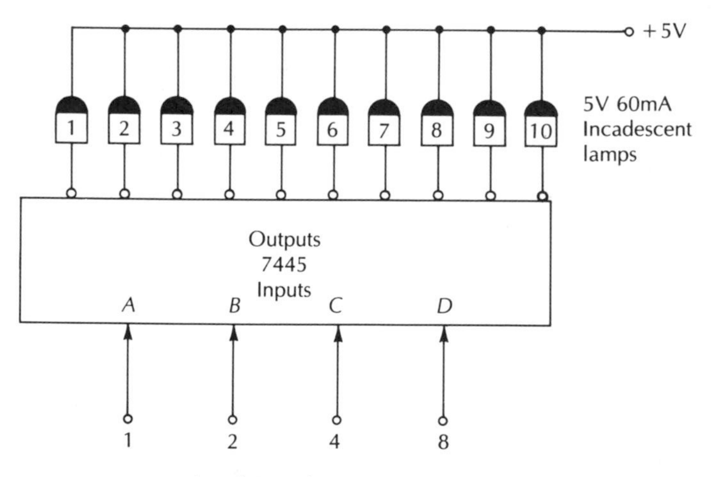

a. Incadescent lamp or printer hammers

Figure 4-8 Decoder-Driver Applications

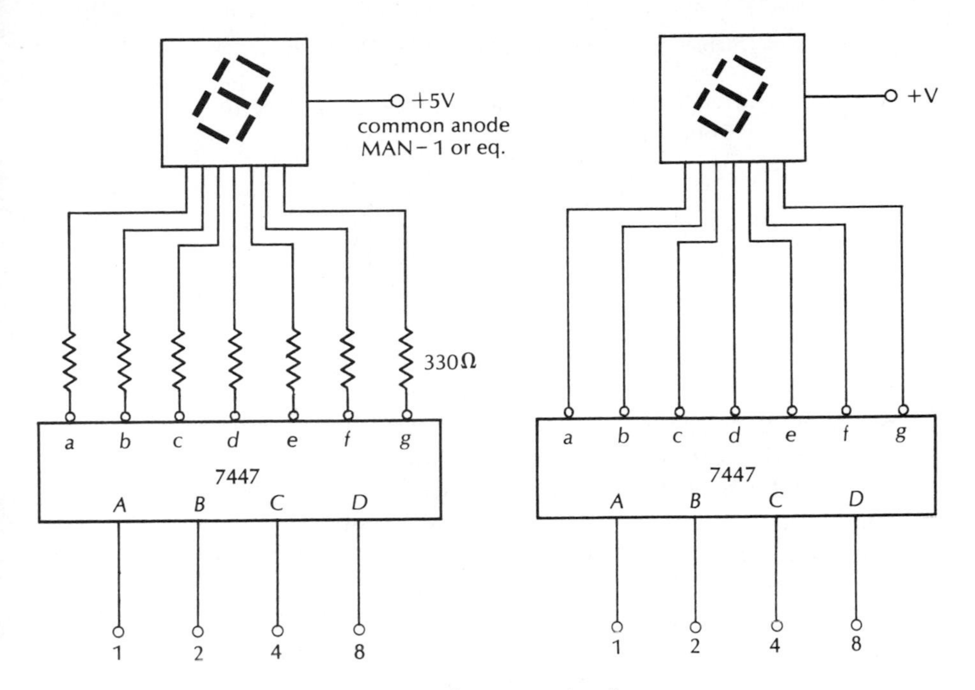

b. 7474 LED and 7-segment incadescent

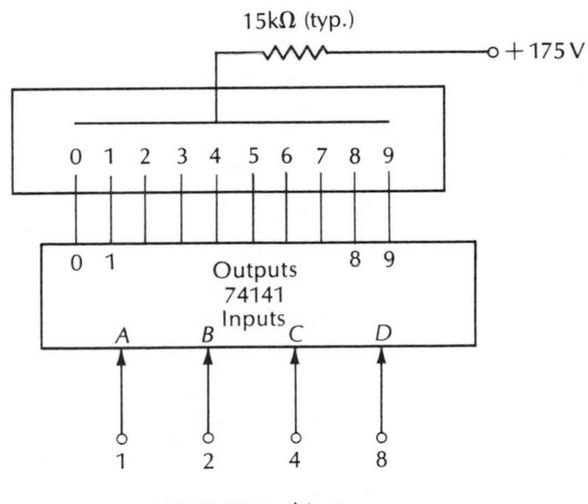

c. 74141 Nixie driver

Figure 4-8 continued

Table 4-2 Common TTL Multiplexers/Data Selectors

Switch Configuration	Multiplexer/Data Selector TTL Type
Quad SPDT	74157
Dual SP 4 position	74153
Single 8 position	74152
Single 16 position	74150*

* 24 pin DIP

Binary *select inputs* are provided to select one input (only) for each binary number applied to the select inputs. Figure 4-9 shows a four-input multiplexer circuit and its switching equivalent.

Figure 4-10a shows a NAND gate used as a controlled SPST switch and a logic circuit equivalent of an SPDT switch. Figure 4-10b is a simplified version of a typical data selector. Notice that the basic structure is a minterm logic form.

Figure 4-11 is the logic diagram of the 74151 multiplexer/data selector. The basic circuit again is a standard minterm-form logic circuit with inverters added to control the select inputs. An output inverter is provided so that both complemented and uncomplemented outputs are available. The strobe (enable) input behaves as a "master switch" to allow (enable) or inhibit the transmission of data to the output. A 0 (low) on this input allows data transmission; a 1 inhibits data transfer.

4-5 Implementing Asynchronous Logic Circuits with a Multiplexer

The multiplexer/data selector provides a one-package solution to many moderately complex logic circuits. An eight-input multiplexer can provide a one-package logic circuit to satisfy the requirements of any minterm-type logic circuit that can be defined by a three-variable truth table. No simplification is required because all possible minterms are available in the package. The multiplexer can be programmed by connecting appropriate data input lines to either +5 V or ground. Proper programming yields the desired logic function.

The design or analysis is an almost instantaneous process, involving no more than simple inspection. Let us see how it works.

Example Given the truth table (Table 4-3), implement the function with an eight-input multiplexer/data selector. Figure 4-12 shows how the programming works and Figure 4-13 shows the complete circuit along with its equivalent logic diagram. In this case the manufacturer has elected to defy the standard truth table arrangement for this device, so

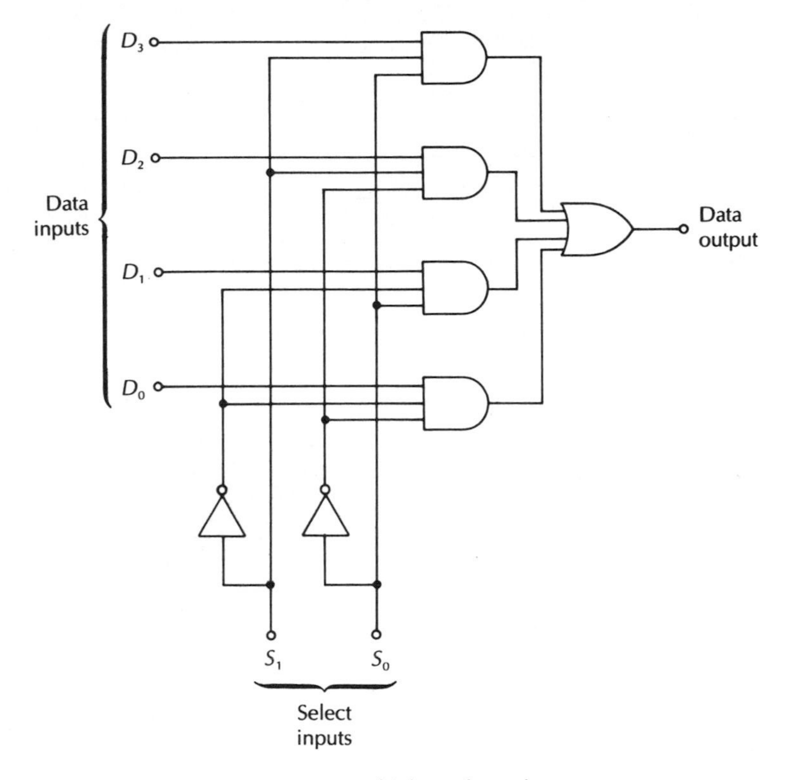

a. Four-input multiplexer/data selector

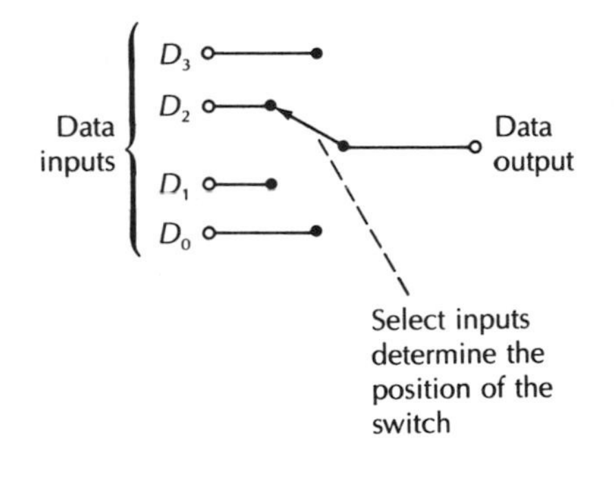

b. Switching equivalent

Figure 4-9 The Multiplexer/Data Selector

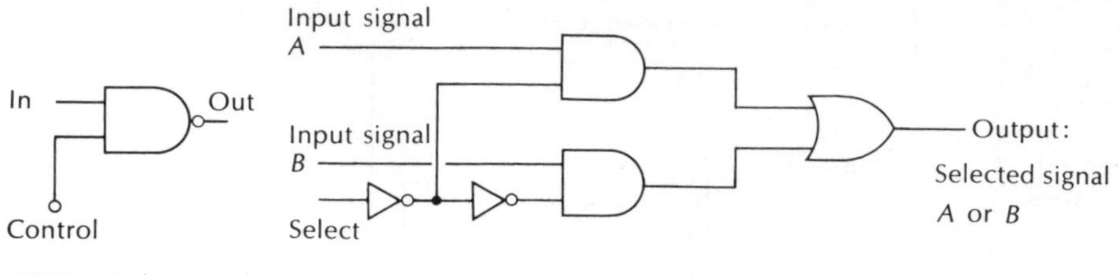

a. SPST switch equivalent

b. SPDT switch equivalent

Select	Signal out
0	A
1	B

Truth table

Figure 4-10 Logic Equivalents for SPST and SPDT Switches

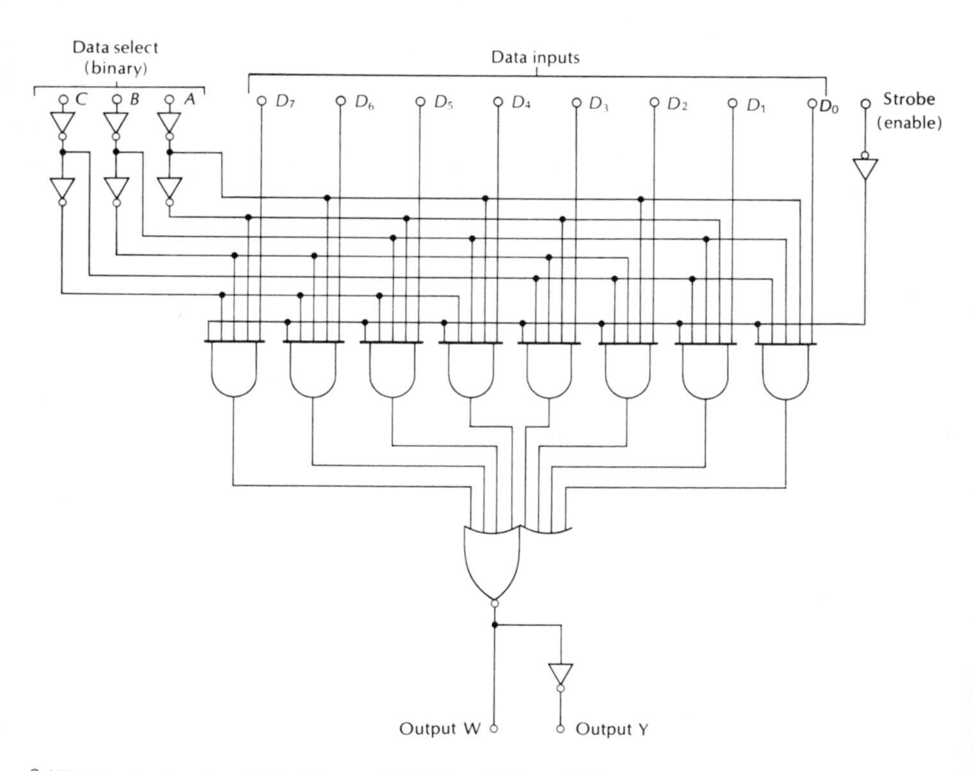

Figure 4-11 Logic Diagram for the 74151 Multiplexer/Data Selector

Table 4-3 The Truth Table

	C	B	A	f
m_0	0	0	0	1
m_1	0	0	1	0
m_2	0	1	0	1
m_3	0	1	1	1
m_4	1	0	0	1
m_5	1	0	1	0
m_6	1	1	0	0
m_7	1	1	1	1

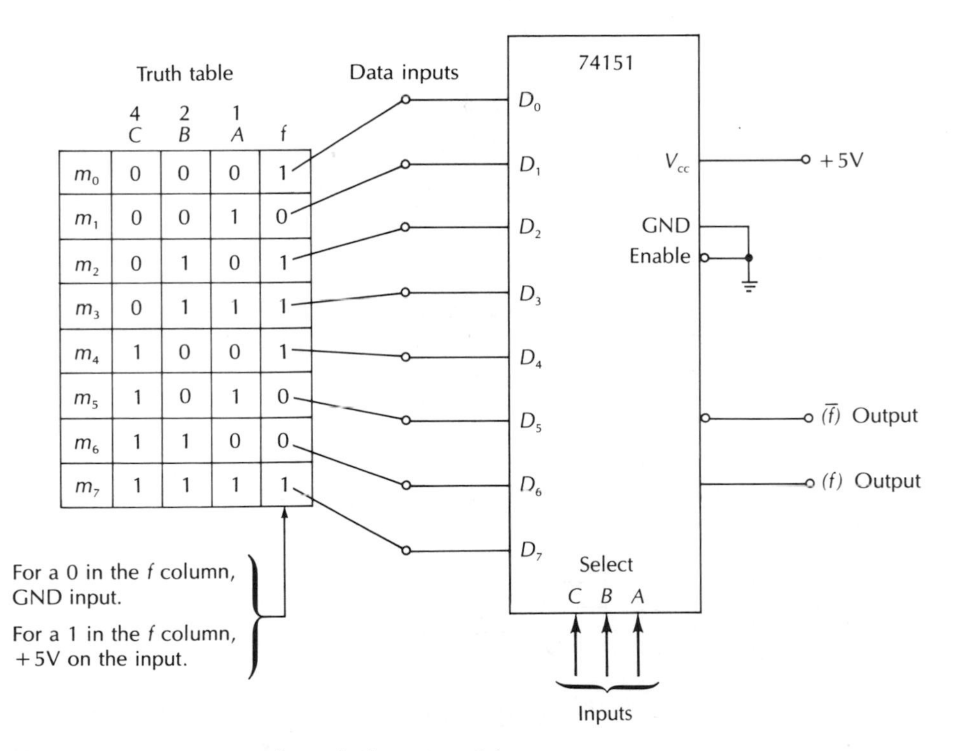

Figure 4-12 Programming the Multiplexer/Data Selector

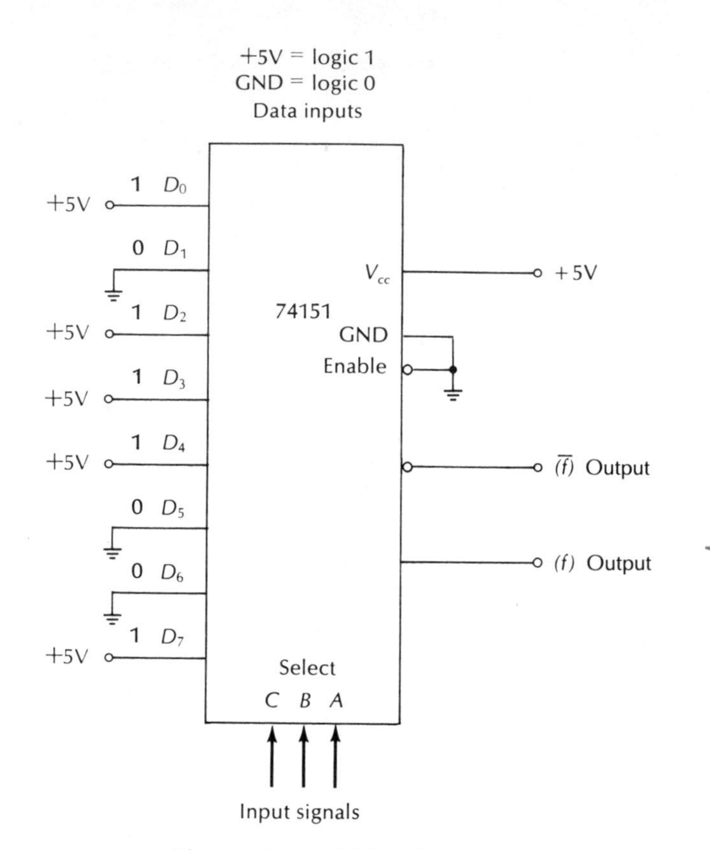

a. The complete multiplexer logic circuit

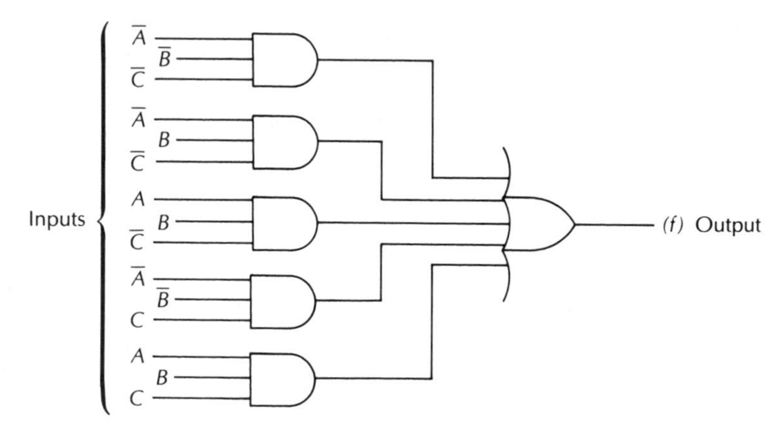

b. Equivalent logic diagram

Figure 4-13 The Finished Circuit

we must alter the truth table in our problem to conform to this change of convention.

FOLDED DATA SELECTOR LOGIC

With only a little extra effort a folded design can be used to reduce the size of the data selector to half the size required by the previous method. A four-variable truth table would normally require a 16-input data selector, but by using the folding technique we can use an eight-input data selector.

A four-variable truth table is shown in Table 4-4 (ignore the f column for now). An examination of the table reveals that there is a repeating pattern of zeros and ones under the A, B, and C headings. Ignore the entries under D for the time being. The entries under A, B, and C start with 000,001, . . . ending with 111 at m_7 and starting again

Table 4–4 Truth Table for Folded Data Selector Example

	D	C	B	A	f
m_0	0	0	0	0	0
m_1	0	0	0	1	1
m_2	0	0	1	0	1
m_3	0	0	1	1	0
m_4	0	1	0	0	0
m_5	0	1	0	1	1
m_6	0	1	1	0	1
m_7	0	1	1	1	0
m_8	1	0	0	0	1
m_9	1	0	0	1	0
m_{10}	1	0	1	0	1
m_{11}	1	0	1	1	0
m_{12}	1	1	0	0	1
m_{13}	1	1	0	1	1
m_{14}	1	1	1	0	0
m_{15}	1	1	1	1	0

at m_8 with 000. The entire pattern is repeated. Now examine the D column. You will notice that the D entries are all zeros through m_7 and all ones from m_8 through m_{15}. We can take advantage of this situation by considering pairs in which the A, B, and C entries are identical.

Example Given truth table 4-4, design a folded data selector logic circuit to implement the function.

The procedure is to examine each pair of identical pairs in A, B, and C (one for $D = 0$ and one for $D = 1$).

The first pair:

D	C	B	A	
0	0	0	0	(m_0)
1	0	0	0	(m_8)

We need an output of 1 when $D = 1$ and an output of 0 when $D = 0$. The required output is 0 when $D = 0$, and 1 when $D = 1$. We connect D to data input D_0 on the data selector in Figure 4-14. Truth table entries m_0 and m_8 are now accounted for.

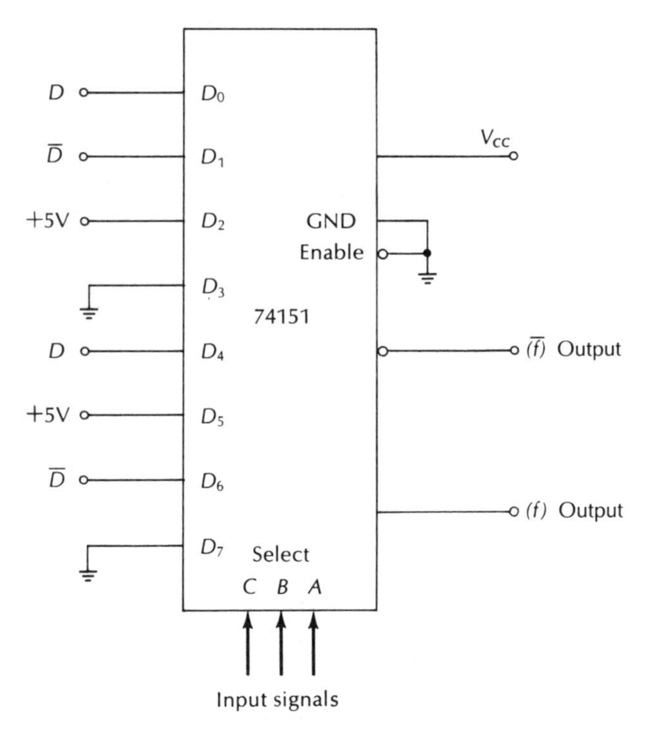

Figure 4-14 Folded Multiplexer/Data Selector Logic Circuit

The second pair:

D	C	B	A	
0	0	0	1	(m_1)
1	0	0	1	(m_9)

The truth table requires a 1 output for m_1 (where $D = 0$), and a 0 output for m_9, where $D = 1$. We need a 1 out when $D = 0$ and a 0 out when $D = 1$. These are complementary, so we need to have the complement of D $(\bar{D})$ connected to input D_1 of the data selector (see Fig. 4-14). The truth table entries m_1 and m_9 are now accounted for.

The third pair:

D	C	B	A	
0	0	1	0	(m_2)
1	0	1	0	(m_{10})

In this case both m_2 and m_{10} on the truth table produce a 1 output. The output is independent of D, so D_2 on the data selector is connected to a 1 (+5 volts), so truth table entries m_0, m_8, m_1, m_9, m_2, m_{10} have been satisfied.

The fourth pair:

D	C	B	A	
0	0	1	1	(m_3)
1	0	1	1	(m_{11})

The truth table calls for a 0 output for m_3 and m_{11}, and is independent of the state of D. Therefore, a 0 is placed on data selector input D_3. Entries m_3 and m_{11} are now satisfied.

If we follow this line of reasoning for the balance of the truth table we get the complete circuit as shown in Figure 4-14.

4-6 Expanding Multiplexer/Data Selectors

Data selectors can be connected to handle larger numbers of variables than can be accommodated by a single package. The select inputs are connected in parallel, and *enable* (sometimes called *inhibit*) *inputs* are used to select one or the other data selector, selecting a different package for each segment of the truth table. Figure 4-15 illustrates the method. Table 4-5 shows TTL package requirements for additional variables.

4-7 Demultiplexer/Data Distributor

The *demultiplexer/data distributor* routes data from one input to one of several outputs. Binary combinations applied to the select inputs de-

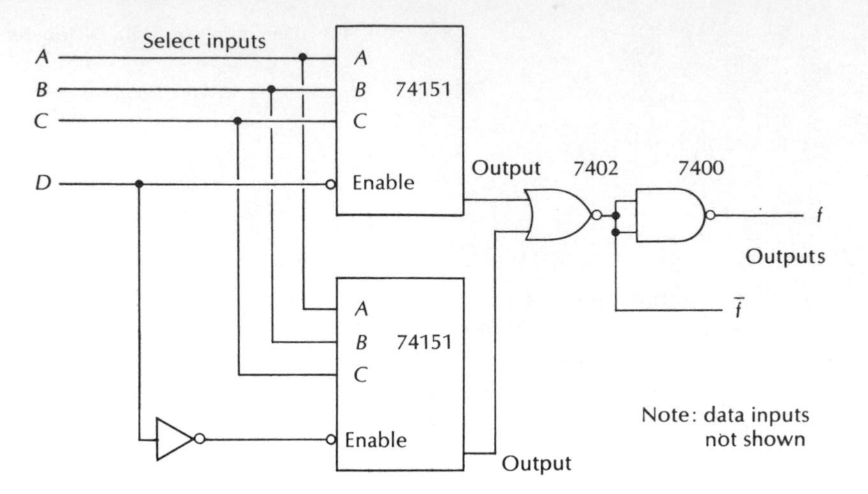

a. Two data selectors connected

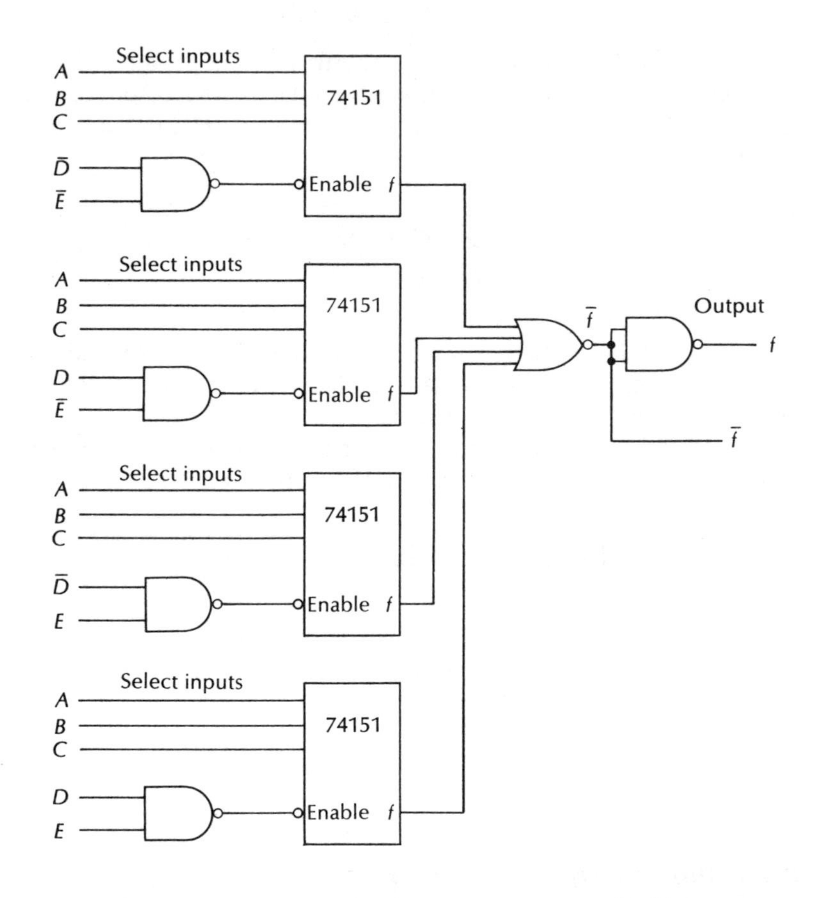

b. Using four data selectors for a thirty-two row truth table

Figure 4-15 Expanding Multiplexers

Table 4-5 Multiplexers for Additional Variables

Number of Input Variables	Selector Inputs	Packages Required
2	2	¼
3	4	½
4	8	1
5	16	1
6	32	2*
7	64	4*

* 24 pin packages

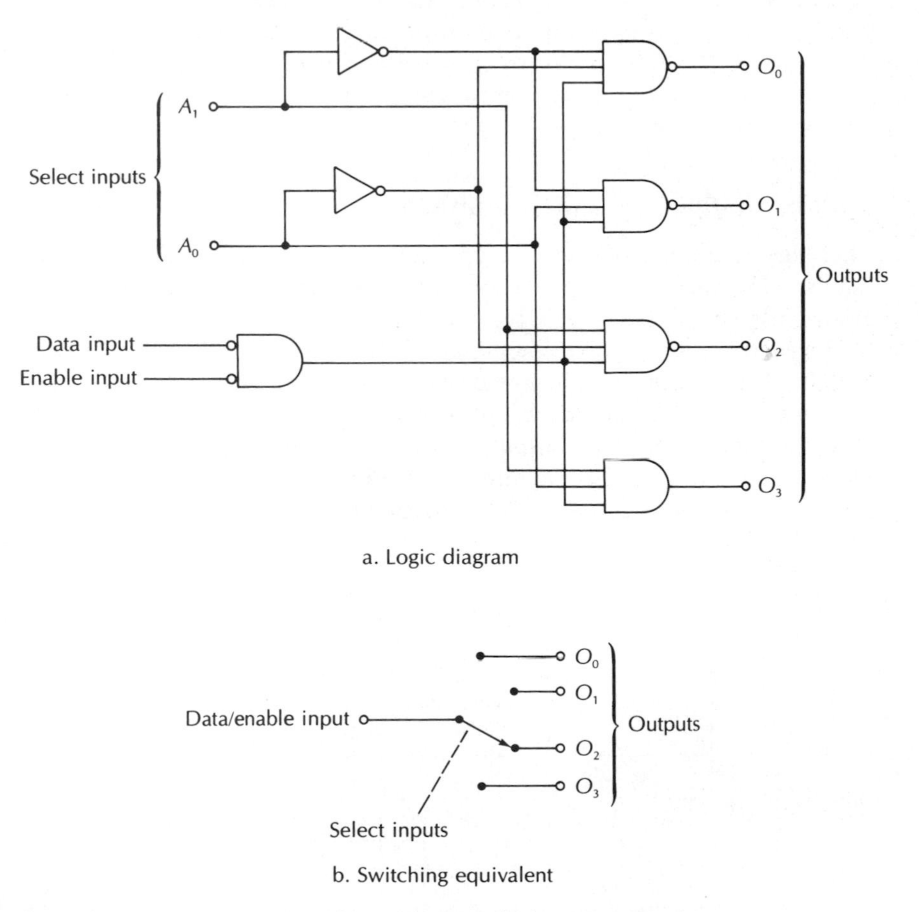

Figure 4-16 One-Input to One-of-4 Output Demultiplexer/Data Distributor

termine to which output the input data are applied. The demultiplexer/data distributor performs the reverse function of the multiplexer/data selector with its multiple selectable inputs and single output. Figure 4-16 shows the logic diagram and switching equivalent for a one-input to one-of-4 output demultiplexer/data distributor.

The logic diagram in Figure 4-16a is basically that of a decoder with an added input. If you compare Figure 4-16a with the decoder in Figure 4-3 you will see that an extra input has been added to each AND (or NAND) gate and these have all been connected together to form an additional input. A low-active AND gate has been connected to this new input. One input to the AND gate is labeled *data input* and the other *enable input*. The two inputs are interchangeable and each may be used for either function. Any decoder with one or more enable inputs can be used as a demultiplexer/data distributor. For example, the decoder in Figure 4-4 shows two enable inputs. One of these may be used as a data input for demultiplexer/data distributor service. Demultiplexer/data distributors are generally listed in the manufacturers' literature as decoder/demultiplexers. Table 4-6 lists popular TTL decoder/demultiplexer types.

4-8 C-MOS Multiplexers and Demultiplexers

There are two kinds of multiplexers and demultiplexers in the C-MOS family. One variety consists of ordinary C-MOS gates with logic diagrams very similar to their TTL counterparts.

The second variety is based on the C-MOS transmission gate (refer to Section 2-20). Transmission gate devices are bidirectional and the same device can serve as either multiplexer or demultiplexer.

A four-input or four-output multiplexer/data selector–demultiplexer/data distributor is diagrammed in Figure 4-17. Individual transmission gates are selected by applying binary data to the select input.

Table 4-6 TTL Decoder/Demultiplexers

Type Number	Number of Outputs	Number of Demultiplexers in the Package
74154	16	1
74138	8	1
74139 74155 74156	4	2

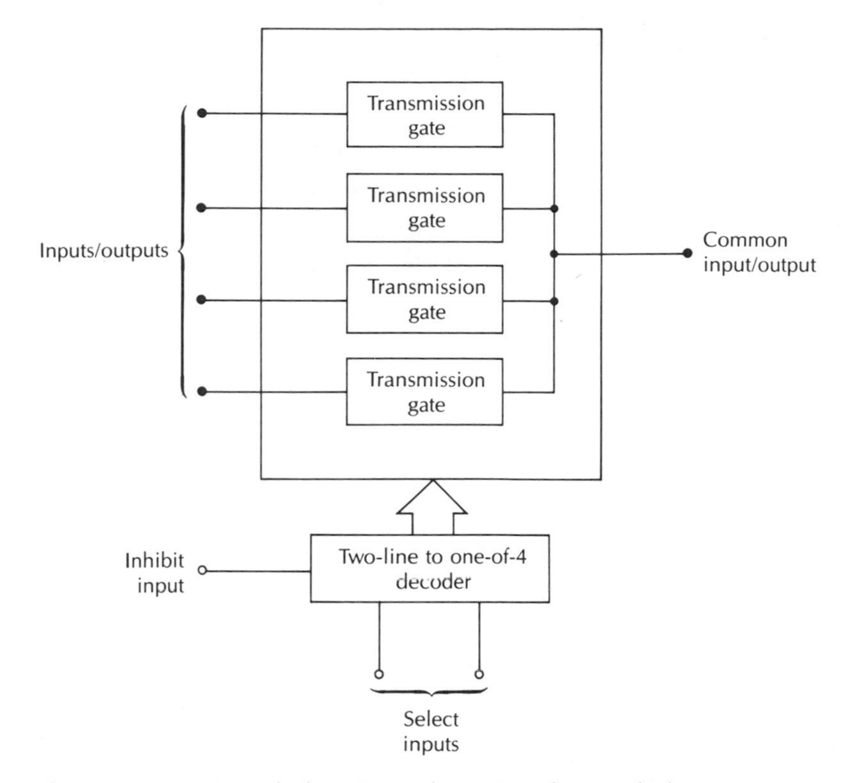

Figure 4-17 C-MOS Multiplexer/Data Selector–Decoder/Demultiplexer

Binary combinations on the select input are decoded to turn on one transmission gate for each of the four possible binary combinations. Only one transmission gate is turned on at a time. The decoder is a standard decoder circuit made up of C-MOS gates.

The transmission gate is essentially an open circuit in the off state and behaves like an ordinary 120 Ω resistor in the on state. Because of its purely resistive nature the transmission gate can be used with analog signals as well as with digital signals. If V_{EE} is supplied with a negative voltage (two power supplies) the transmission gate can pass zero crossing analog signals. V_{EE} is grounded for digital service. The maximum current through each transmission gate is 25 mA. Table 4-7 lists available C-MOS transmission gate multiplexers/demultiplexers.

4-9 Programmable Logic Arrays (PLA)

The programmable logic array is a medium-scale minterm-type circuit that must be programmed to perform the desired logic functions. The

Table 4-7 Transmission Gate Multiplexers/Demultiplexers

Type Number	Description
4051B	Single 8-channel
4052B	Differential 4-channel
4053B	Triple 2-channel

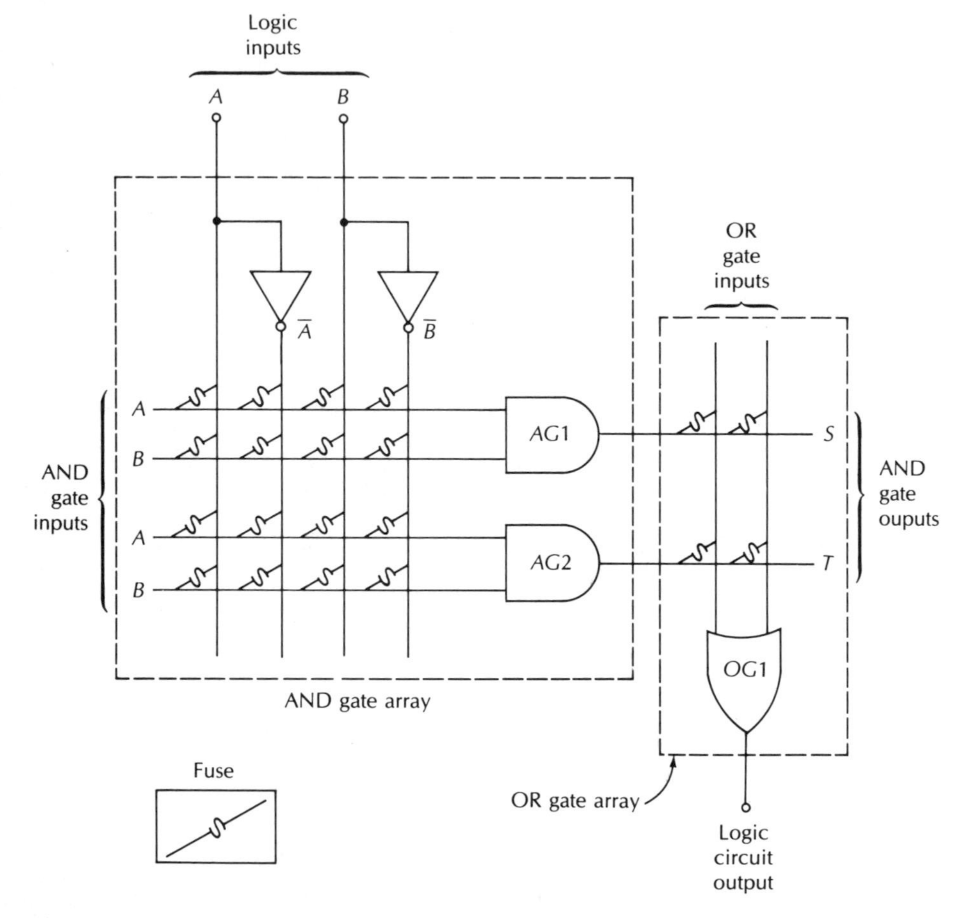

Figure 4-18 Programmable Logic Array (PLA)

basic device comes with everything connected to generate all the minterms possible for its size. The PLA is programmed by blowing internal fuses to disconnect unwanted circuitry. Figure 4-18 shows a simple unprogrammed logic array. In Figure 4-19 the PLA has been programmed to generate the logic function $f = (\bar{A} \cdot B) + (A \cdot \bar{B})$. Figure 4-20 illustrates a slightly more complex case where two separate output functions are generated. This is often called a *multiple-output circuit.* The dots at the grid intersections indicate fuses left intact (not blown). This is the common notation used to define the program for a PLA. Although typical PLA's have some nine inputs and eight outputs, Figure 4-20 illustrates a principal PLA limitation. Notice that the terms $A \cdot B \cdot C$ from AND gate $AG1$ and $\bar{A} \cdot \bar{B} \cdot C$ from $AG2$ are common to both output functions, $f = (ABC) + (\bar{A}\bar{B}C)$ and $f = (ABC) + (\bar{A}\bar{B}C) +$

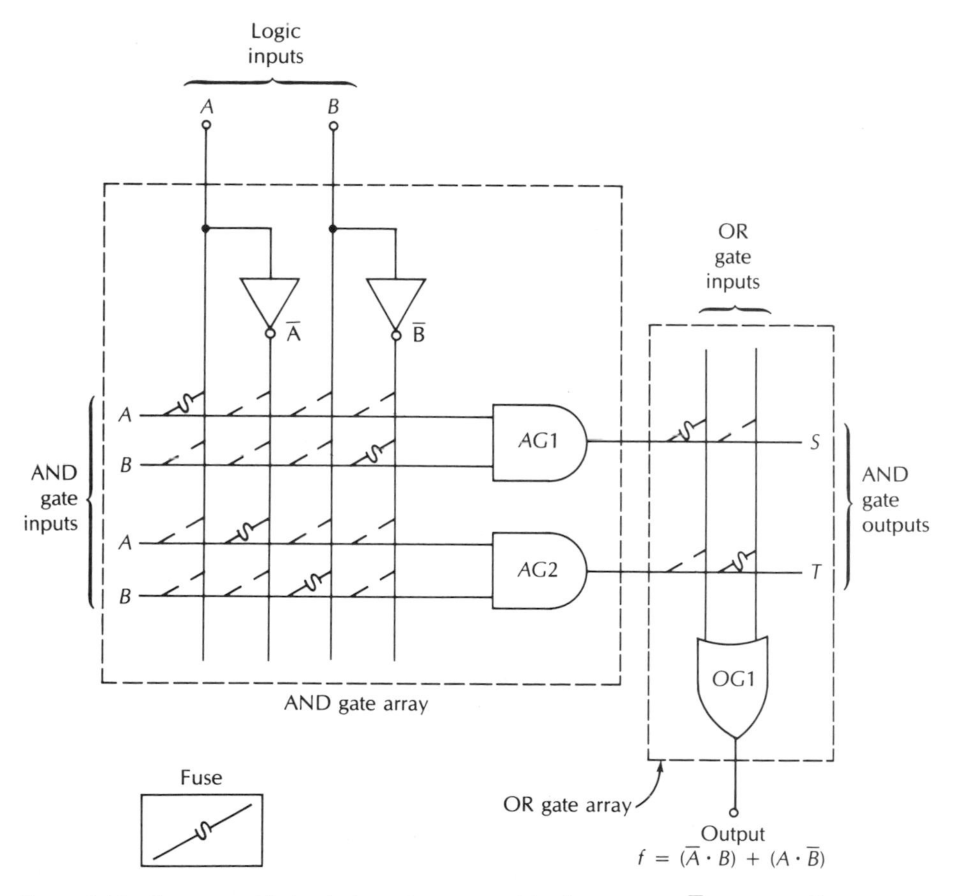

Figure 4-19 Programmable Logic Array Programmed for the Function $(\bar{A} \cdot B) + (A \cdot \bar{B})$

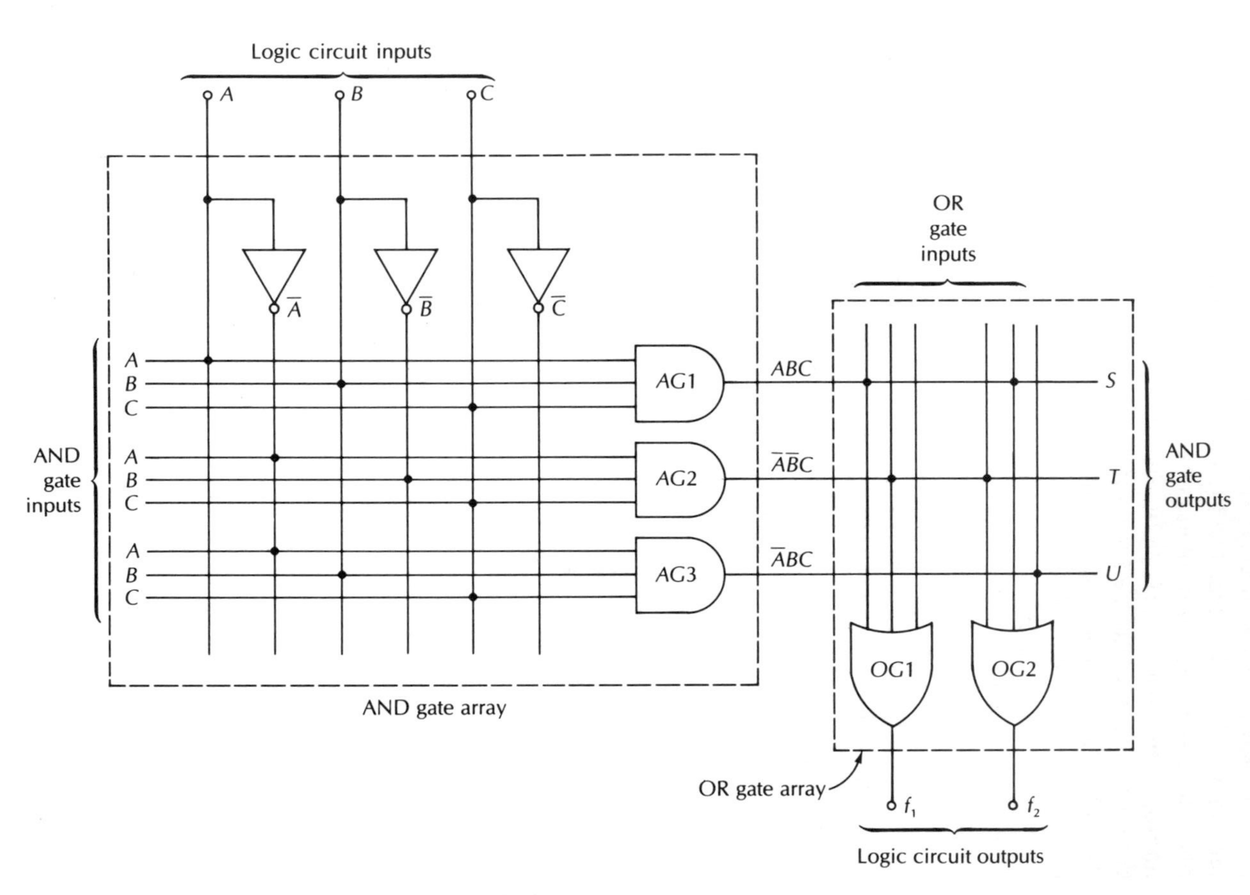

Figure 4-20 Programmable Logic Array with Standard Programming Notation

($\bar{A}BC$). Only the terms programmed into the AND gates $AG1$, $AG2$, and $AG3$ are available to the output OR gates. This restriction often results in fewer available functions from a given package than might initially be expected.

4-10 Read-Only Memory Logic

The read-only memory (ROM) is primarily intended for that section of computer memory where permanent data are stored. The computer can read the contents of the read-only memory, but it cannot add to, or alter its contents.

ROMs are available that can be programmed in the field. The PROM (programmable ROM) is programmed by blowing fuses in the same fashion as in the PLA. The EPROM (erasable programmable ROM) is programmed with an over-voltage that stores a trapped charge in the memory cell. EPROMs can be erased by exposing them to ultra-violet light. They can then be reprogrammed. Mask-programmable ROMs are programmed as the final manufacturing step, and are necessarily high-production-volume devices.

In addition to their permanent memory function, ROMs can also be used to implement asynchronous logic functions. Figure 4-21 illustrates a simple ROM consisting of a 3-line to one-of-8 line decoder and a diode

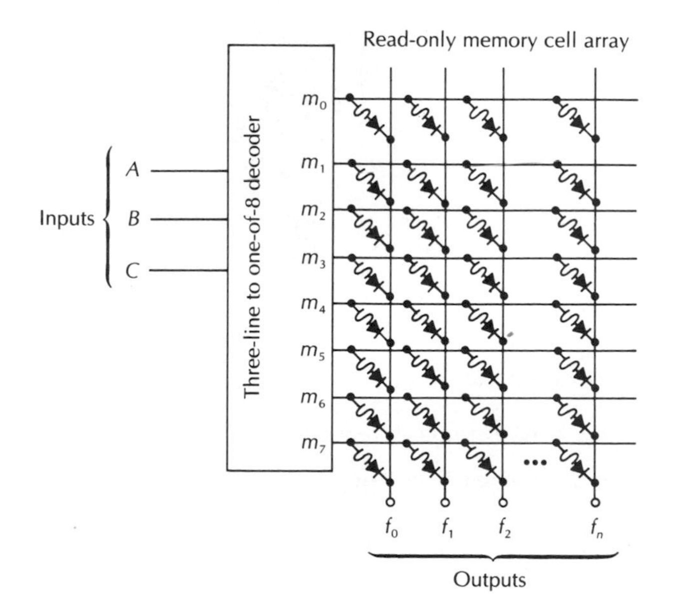

Figure 4-21 Simple Diode Read-Only Memory Circuit

Equation: $f_n = (\overline{A} \cdot \overline{B} \cdot \overline{C}) + (\overline{A} \cdot B \cdot \overline{C}) + (A \cdot \overline{B} \cdot \overline{C}) + (A \cdot B \cdot C)$

Truth table:

m	A	B	C	f_n	Minterms
m_0	0	0	0	1	→ $(\overline{A} \cdot \overline{B} \cdot \overline{C})$
m_1	0	0	1	1	
m_2	0	1	0	1	→ $(\overline{A} \cdot B \cdot \overline{C})$
m_3	0	1	1	0	
m_4	1	0	0	1	→ $(A \cdot \overline{B} \cdot \overline{C})$
m_5	1	0	1	0	
m_6	1	1	0	1	
m_7	1	1	1	1	→ $(A \cdot B \cdot C)$

a. Truth table and equation

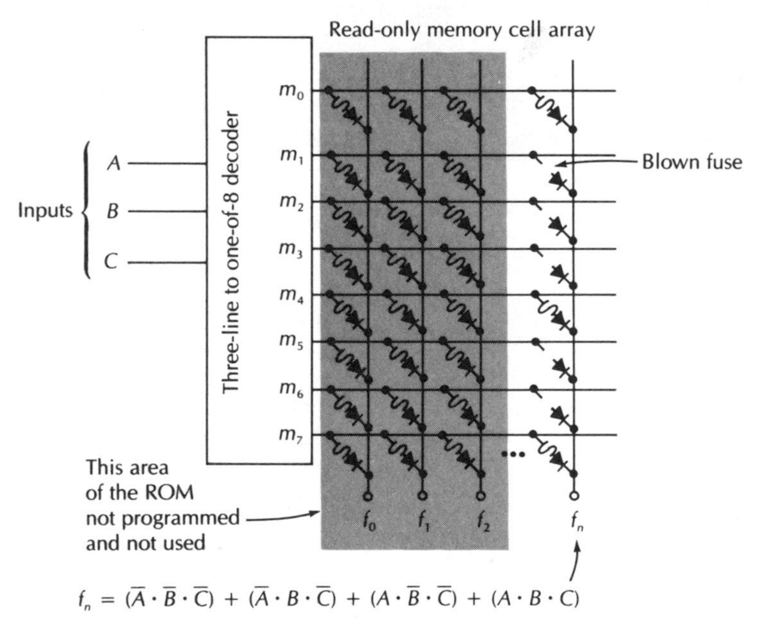

$f_n = (\overline{A} \cdot \overline{B} \cdot \overline{C}) + (\overline{A} \cdot B \cdot \overline{C}) + (A \cdot \overline{B} \cdot \overline{C}) + (A \cdot B \cdot C)$

b. ROM logic implementation of the truth table

Figure 4-22 Programmed Diode Read-Only Memory

read-only memory array. Each diode has a fuse in series with it to allow the ROM to be programmed. In most commercial ROM's, the necessary decoders are on the IC chip along with the memory cells. Figure 4-22 shows the diode PROM programmed to perform the function

$$f_n = (\bar{A} \cdot \bar{B} \cdot \bar{C}) + (\bar{A} \cdot B \cdot \bar{C}) + (A \cdot \bar{B} \cdot \bar{C}) + (A \cdot B \cdot C)$$

The outputs from the decoder are the decimal equivalents of the three-bit (A,B,C) inputs, and can be labeled m_0, m_1 and so on. The ROM outputs f_0, f_1, f_2 . . . (Fig. 4-22) are available for other logic functions.

The PROM in Figure 4-22 is not a commercially available item. Commercial PROM's and ROM's normally have the memory cells arranged in a rectangular coordinate form as illustrated in Figure 4-23, a

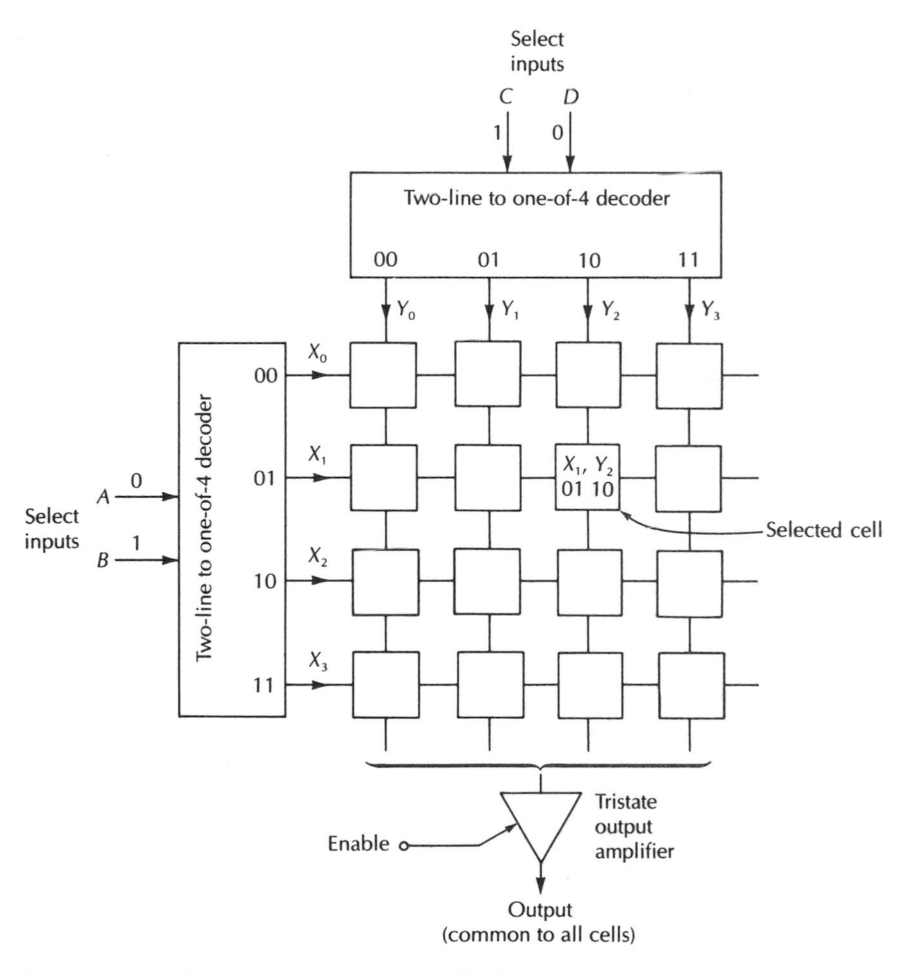

Figure 4-23 Read-Only Memory with $X-Y$ Cell Selection

technique that drastically reduces the decoder complexity and increases the flexibility of the device. Individual memory cells are normally bipolar or MOS transistors instead of diodes. The use of transistors reduces the drive requirements for the decoders and facilitates the selection of individual ROM cells.

Individual cells are selected by applying one of the 16 possible 4-bit binary numbers to the select inputs A, B, C, D. The row decoder decodes the input bits A and B into one-of-4 outputs X_0, X_1, X_2, X_3. In Figure 4-23, a 0 on input A and a 1 on input B causes the decoder to define the desired memory location as being somewhere in row X_1. A 1 on input C and a 0 on input D causes the column decoder to define the location of the desired cell location as being somewhere in column Y_2. The actual cell location is at the intersection of row X_1 and column Y_2.

The location of a selected cell, as defined by the binary values of the select inputs (A,B,C,D), is called the *address* of the cell. In Figure 4-23, the address of the selected cell is 0110.

The ROM cells are all connected to a common output (read) amplifier. The output will be a 0 or a 1, depending upon what has been permanently programmed into the selected cell. Output amplifiers are provided with an enable/disable input and are often tristate devices. The enable input is also called the *chip-select input* to describe its function in large memory systems where many ROM chips are required.

ROM chips are often organized so that a single address selects a group of four, eight, or 16 cells. The group of cells is called a *word* and a ROM so arranged is called a *word-organized memory chip*.

We will examine memory organization in detail in the chapter devoted to computer memories. In this chapter we will be concerned only with using ROM's instead of gate-only logic to implement standard asynchronous minterm logic functions.

EXAMPLE OF READ-ONLY MEMORY LOGIC

Figure 4-24 shows how a read-only memory is used to implement the logic function

$$f = (\bar{A} \cdot B \cdot C \cdot \bar{D}) + (A \cdot \bar{B} \cdot \bar{C} \cdot D) + (A \cdot B \cdot C \cdot \bar{D})$$

The desired logic function is programmed into the ROM by setting all cells to permanent zeros except for the cells required to implement the function. The select inputs to the ROM become the logic circuit inputs and the ROM output becomes the logic circuit output. In the example in Figure 4-24 the cells at the following addresses are programmed to a 1: 0110 ($\bar{A}BC\bar{D}$), 1001 ($A\bar{B}\bar{C}D$), and 1110 ($ABC\bar{D}$). All other cells are programmed to zeros. The output will produce a 1 for minterms (input conditions) $\bar{A}BC\bar{D}$, $A\bar{B}\bar{C}D$, and $ABC\bar{D}$. All other input conditions (minterms) will produce 0 outputs. The circuit will satisfy the appropriate truth table requirements.

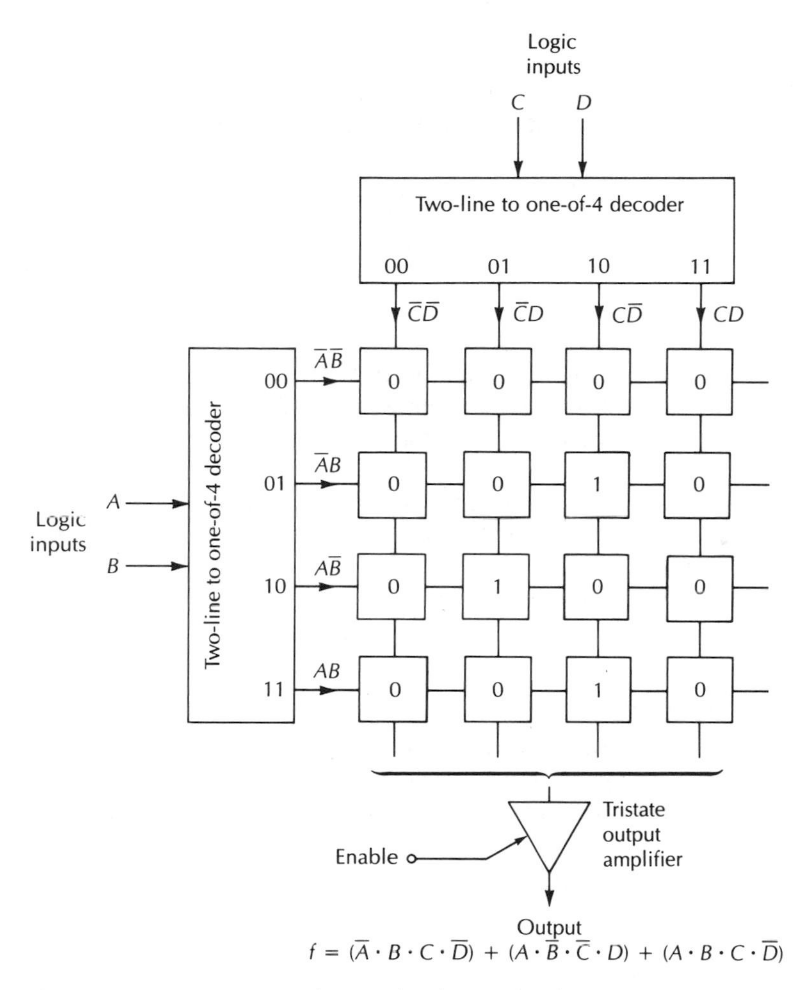

$$f = (\overline{A} \cdot B \cdot C \cdot \overline{D}) + (A \cdot \overline{B} \cdot \overline{C} \cdot D) + (A \cdot B \cdot C \cdot \overline{D})$$

Figure 4-24 Logic Circuit Implemented with a Read-Only Memory (ROM)

SUMMARY

ENCODERS

Encoders are made up of OR gates and convert data from a single input into its multiple output binary or binary coded equivalent. The encoder may use either diode logic gates or ordinary TTL gates. Commercial encoders are available in two types, 8-line to 3-line and 10-line to 4-line. Both commercial types contain a priority feature that outputs only the binary equivalent of the highest input value when two or more inputs are activated simultaneously. Priority encoders can be combined for larger numbers of inputs.

DECODERS

Decoders convert combinations of binary inputs into a single (one-at-a-time) output. The most common commercial decoders are the 4-line to one-of-10 line and the 4-line to one-of-16 line decoder. TTL decoders have active-low outputs. C-MOS decoders are available with either active-low or active-high outputs. Commercial decoders can be combined to expand for more inputs and outputs.

DECODER-DRIVERS

Special decoders with fairly high output drive capabilities are available to drive display units and other outside-world devices. The most common types are 4-line to one-of-10 and 4-line to seven-segment display decoder-drivers.

The seven-segment decoder-driver is the equivalent of a 4-line to one-of-16 line decoder with a built-in encoder to activate the proper segments in the display digits. Seven-segment decoder-drivers are commercially available for light-emitting diode, gas-discharge, and incandescent display devices.

MULTIPLEXER/DATA SELECTOR

The multiplexer/data selector is the logic gate equivalent of a multiposition selector switch. It has from two to 16 data inputs and one output. Select inputs are provided to control the equivalent position of the selector switch.

The internal multiplexer circuitry is of a modified minterm logic form, and a multiplexer, in addition to data routing chores, can be used to implement standard minterm logic circuits. Any minterm logic function, up to the size limit of the multiplexer, can be programmed into the device by connecting data input lines to +5 V or ground in the proper combination. Commercial multiplexers can be combined to expand the number of inputs.

DEMULTIPLEXER/DATA DISTRIBUTOR

The demultiplexer/data distributor is also the logic gate equivalent of a multiposition selector switch. However, unlike the multiplexer with its many inputs and single output, the demultiplexer has a single input and many outputs. The demultiplexer is a decoder circuit with one or more enable inputs. An enable input is interchangeable with a data input.

The C-MOS family contains two varieties of demultiplexer. One is based on ordinary C-MOS gates and the other is based on C-MOS transmission gates. Because transmission gates are bidirectional, the same device can be used for either multiplexer or demultiplexer service.

PROGRAMMABLE LOGIC ARRAY (PLA)

The programmable logic array (PLA) is a medium-scale integrated circuit containing an AND gate array and an OR gate array. The two arrays can be programmed by blowing internal fuses to implement any desired logic function of the standard minterm form. Several different logic circuits can be programmed into a single package. PLAs typically have nine inputs and eight outputs and can accommodate up to eight independent minterm-type asynchronous logic circuits.

READ-ONLY MEMORY (ROM) LOGIC

The read-only memory (ROM), originally designed for permanent data storage in computer systems, can also be used to implement standard minterm asynchronous logic functions.

Problems

1. What is the function of an encoder?
2. What is the priority function in a priority encoder?
3. Give an example (one not in the text) of how the priority function might be used.
4. Draw a diode matrix circuit for encoding the digits 0 through 7 into their three-bit binary equivalents.
5. What kind of logic gate is employed in the encoder circuit?
6. What is the function of a decoder?
7. How many output lines in a typical decoder are active at any given time?
8. Are the outputs from commercial 4-line to one-of-16 line TTL decoders active-high or active-low?
9. Are the outputs from commercial 4-line to one-of-16 line MOS decoders active-high or active-low?
10. How many outputs does a multiplexer/data selector have?
11. Draw the equivalent switching diagram for an eight-input multiplexer/data selector.
12. Given the truth table in Table 4-8, program the multiplexer/data selector in Figure 4-25 to implement the logic function indicated by the truth table.
13. Program the multiplexer/data selector in Figure 4-25 to implement the following logic equation:

$$f = (\bar{A} \cdot \bar{B} \cdot \bar{C}) + (\bar{A} \cdot \bar{B} \cdot C) + (\bar{A} \cdot B \cdot \bar{C}) + (A \cdot \bar{B} \cdot \bar{C}) + (A \cdot B \cdot \bar{C}) + (A \cdot B \cdot C)$$

Table 4-8 Truth Table for Problem 12

	A	B	C	f
m_0	0	0	0	1
m_1	0	0	1	1
m_2	0	1	0	1
m_3	0	1	1	0
m_4	1	0	0	1
m_5	1	0	1	0
m_6	1	1	0	1
m_7	1	1	1	1

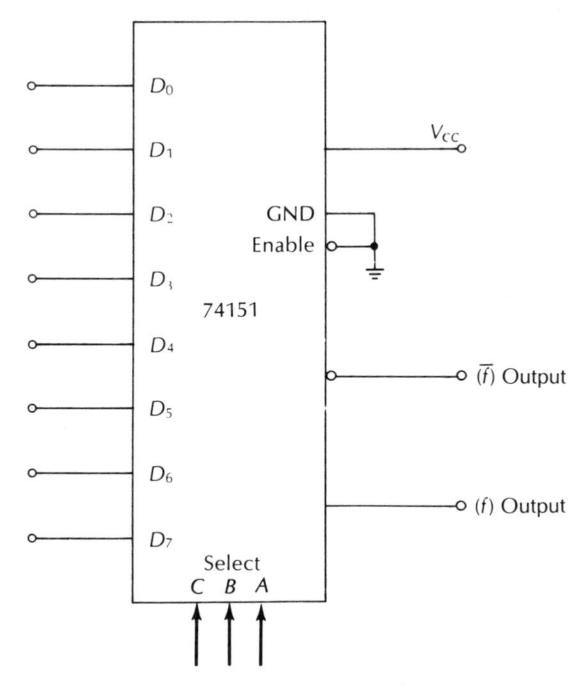

Figure 4-25 Multiplexer/Data Selector for Problems 12, 13, and 14

Table 4-9 Truth Table for Problem 14

	8 D	4 C	2 B	1 A	f
m_0	0	0	0	0	1
m_1	0	0	0	1	1
m_2	0	0	1	0	1
m_3	0	0	1	1	0
m_4	0	1	0	0	1
m_5	0	1	0	1	1
m_6	0	1	1	0	0
m_7	0	1	1	1	0
m_8	1	0	0	0	1
m_9	1	0	0	1	1
m_{10}	1	0	1	0	1
m_{11}	1	0	1	1	0
m_{12}	1	1	0	0	1
m_{13}	1	1	0	1	1
m_{14}	1	1	1	0	0
m_{15}	1	1	1	1	0

14. Given the truth table in Table 4-9, use the folded technique to program the multiplexer/data selector in Figure 4-25 to implement the function described by the truth table.
15. Write the equation and make a truth table for the multiplexer/data selector circuit in Figure 4-26.
16. Draw the switching equivalent of an eight-output demultiplexer/data distributor.
17. There is a particular device discussed in this chapter that can be used as a demultiplexer if it is provided with one or more enable inputs. What is the name of that device?
18. Under what heading would you look to find data distributors in the manufacturer's data manual?
19. What special property of a C-MOS transmission gate multiplexer/demultiplexer permits it to serve both multiplexing and demultiplexing functions?

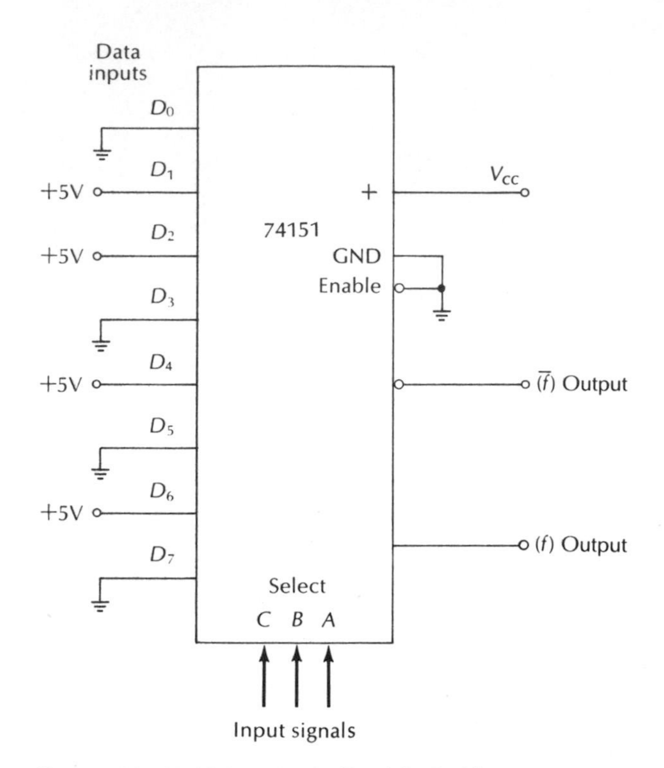

Figure 4-26 Multiplexer Logic Circuit for Problem 15

20. List the approximate on and off resistances for a C-MOS transmission gate.
21. Program the PLA in Figure 4-27 to implement the following functions:

$$f_1 = \bar{A} \cdot B \cdot C + A \cdot B \cdot C$$
$$f_2 = A \cdot B \cdot C + \bar{A} \cdot B \cdot C + A \cdot \bar{B} \cdot \bar{C}$$

Use dots to indicate fuses blown.

22. Given the PLA circuit in Figure 4-28, write the logic equations for f_1 and f_2.
23. (*Special challenge*) The gates are frequently left out of more complex PLA circuits. Figure 4-29 is an example. Program the PLA to activate the proper segments for the LED display when *A,B,C,D* are binary (BCD) inputs.

For the following problems, indicate which fuses in Figure 4-30 must be blown to implement the required function:

24. $f_0 = \bar{A} \cdot B \cdot \bar{C} + A \cdot \bar{B} \cdot C + A \cdot B \cdot C$

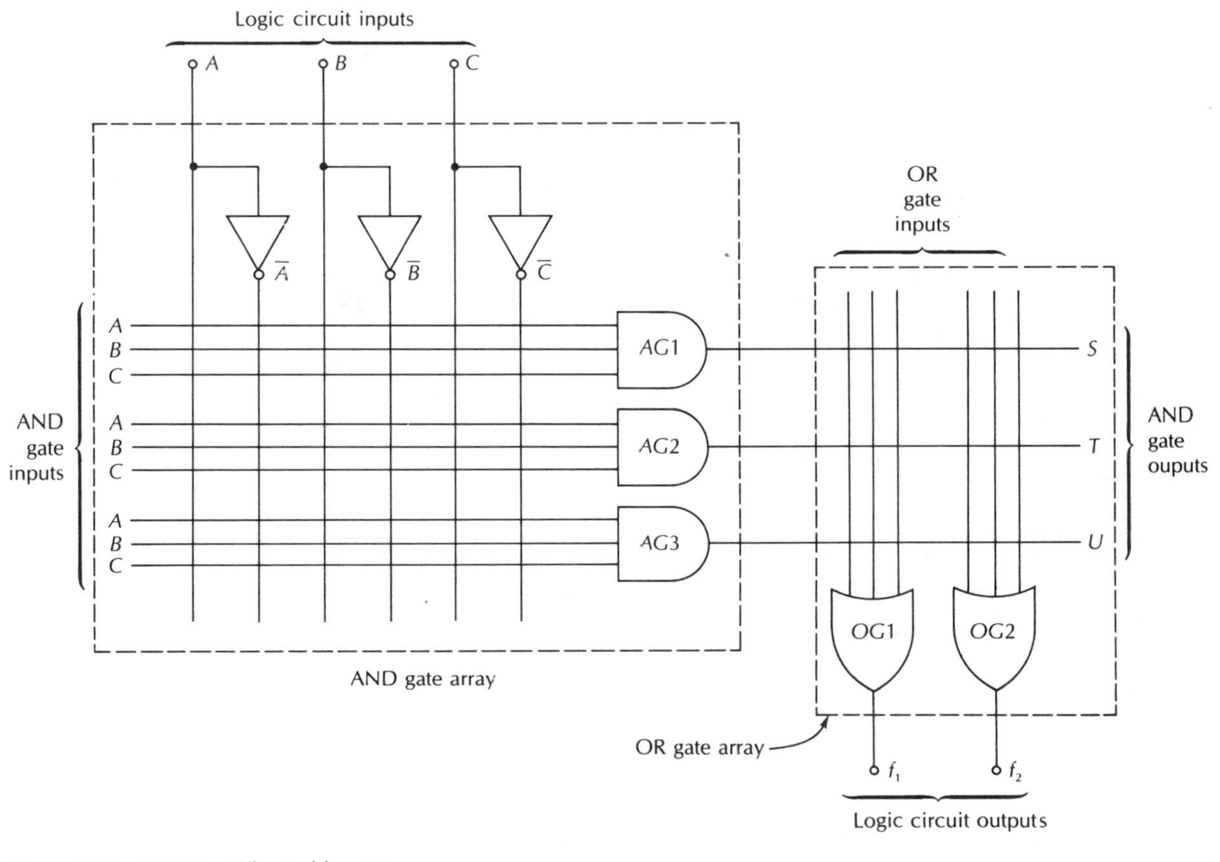

Figure 4-27 PLA Circuit for Problem 21

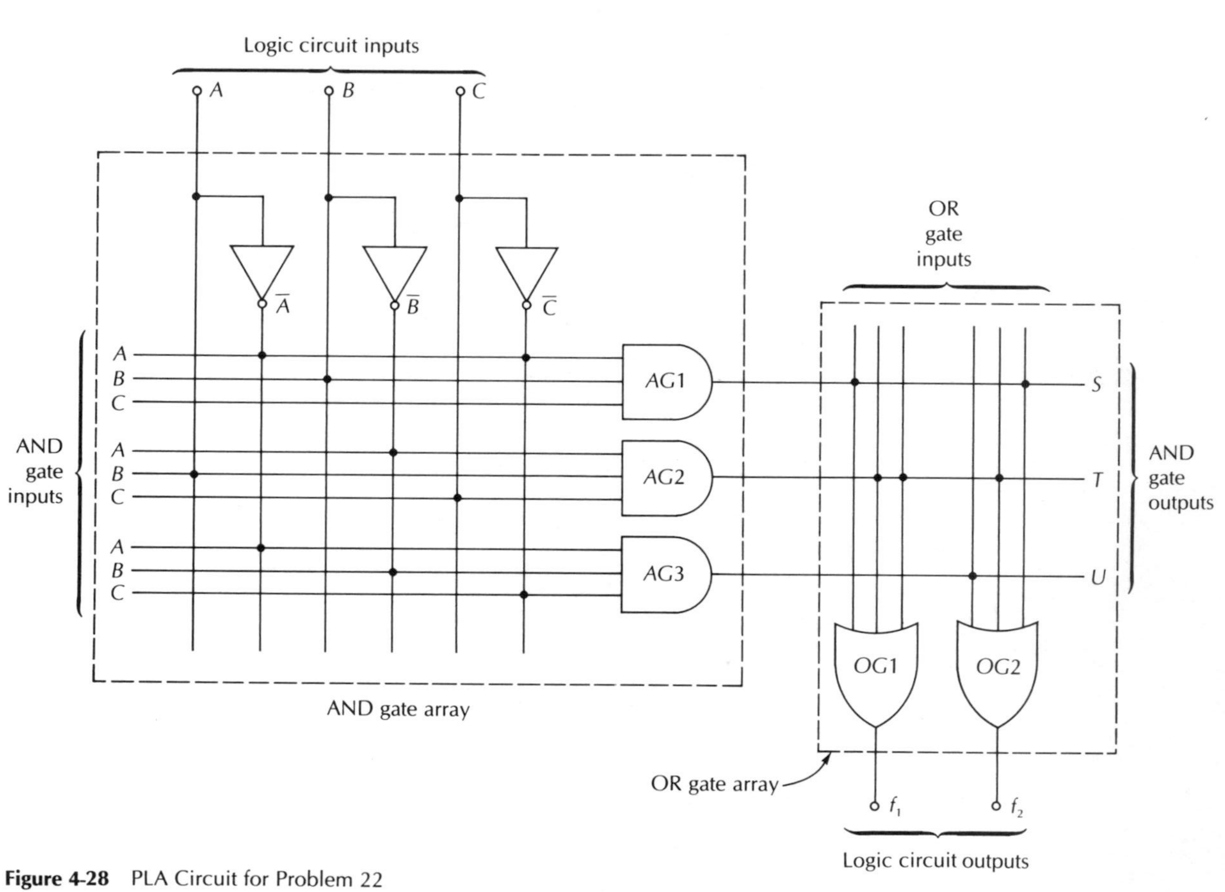

Figure 4-28 PLA Circuit for Problem 22

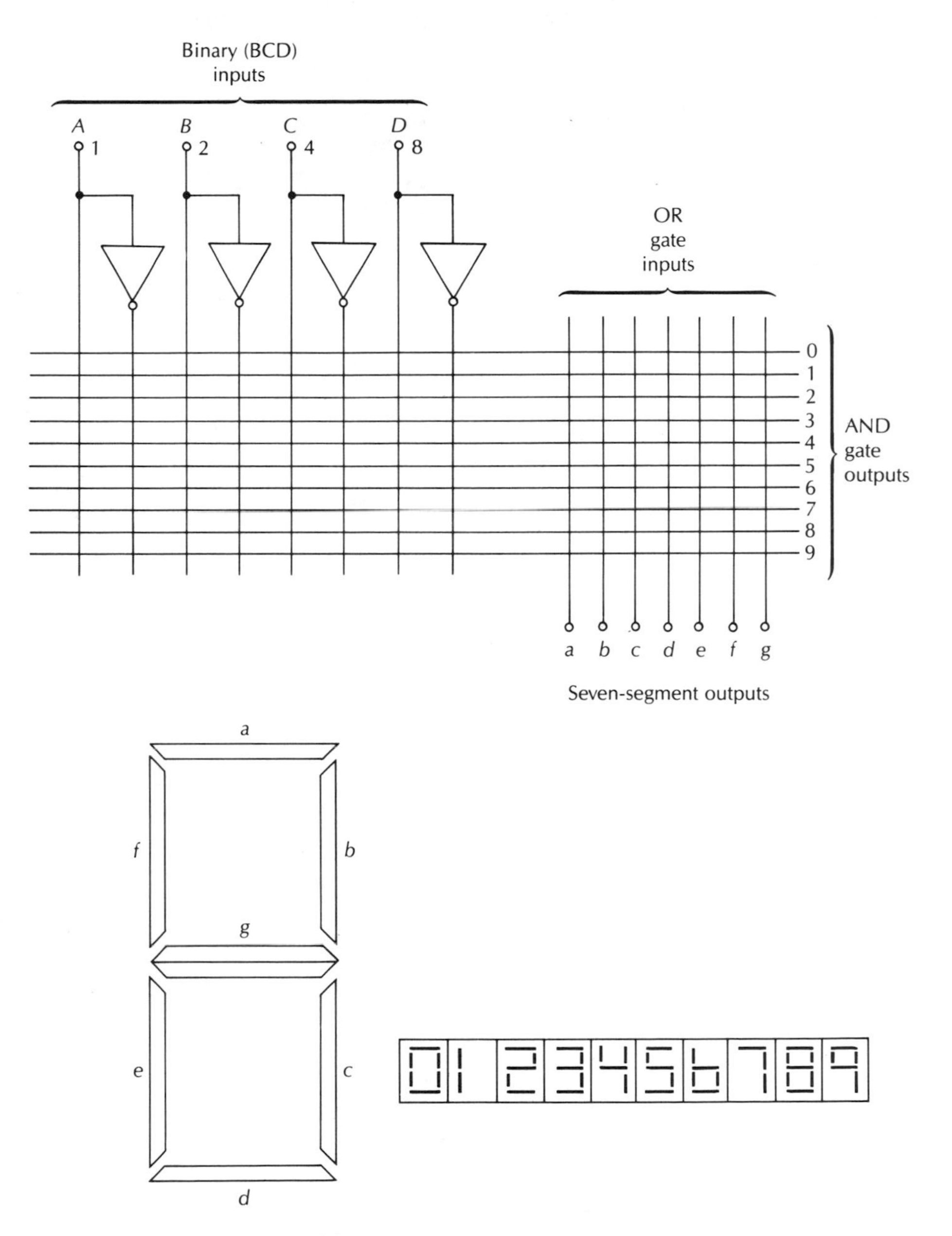

Figure 4-29 PLA Circuit for Problem 23

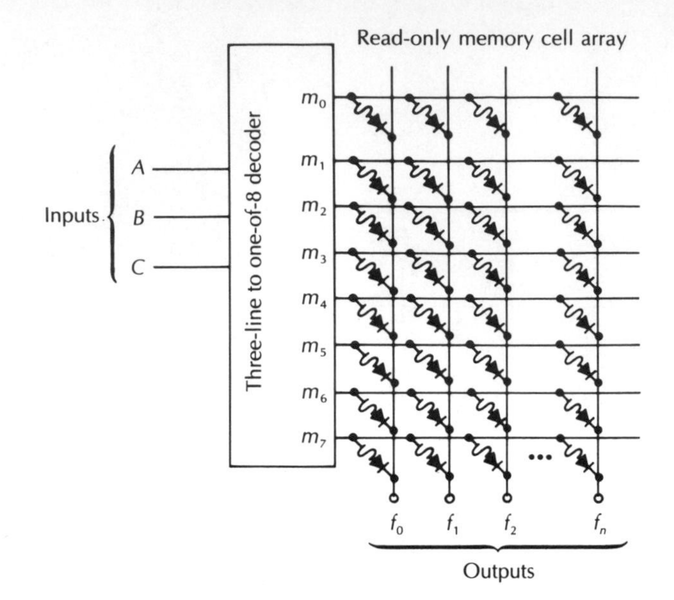

Figure 4-30 ROM Logic Circuit for Problems 24–27

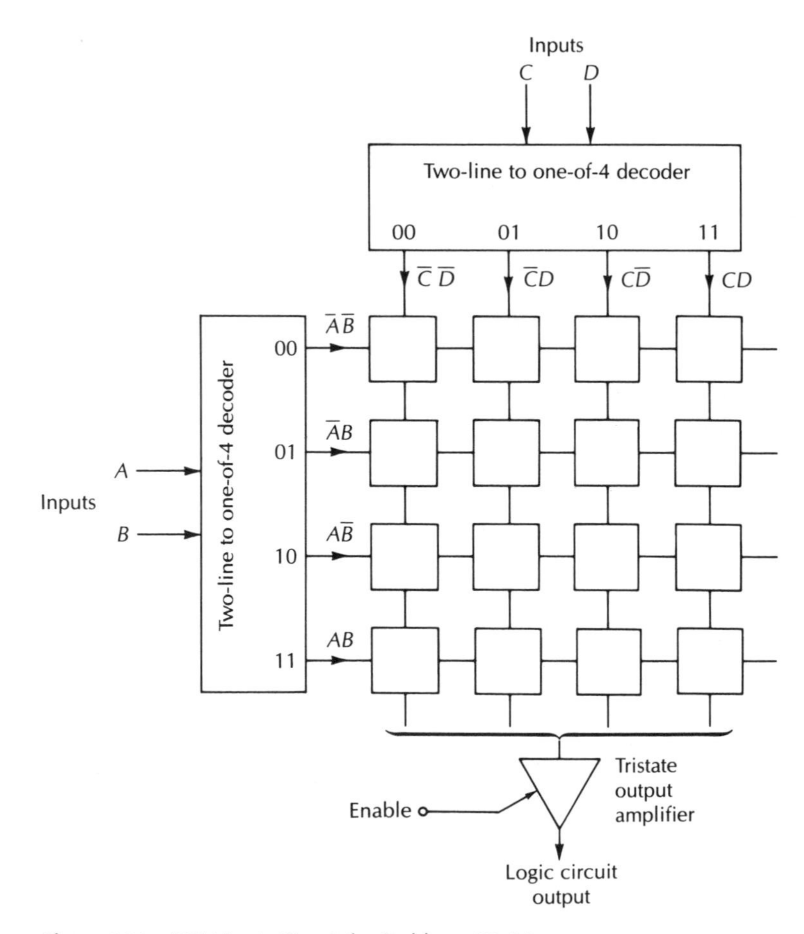

Figure 4-31 ROM Logic Circuit for Problems 28–30

25. $f_1 = \bar{A} \cdot \bar{B} \cdot \bar{C} + \bar{A} \cdot B \cdot \bar{C} + A \cdot \bar{B} \cdot C$
26. $f_2 = \bar{A} \cdot B \cdot C + A \cdot B \cdot \bar{C} + \bar{A} \cdot \bar{B} \cdot \bar{C} + \bar{A} \cdot \bar{B} \cdot C$
27. $f_n = \bar{A} \cdot \bar{B} \cdot \bar{C} + A \cdot B \cdot C + A \cdot \bar{B} \cdot C + \bar{A} \cdot B \cdot C + A \cdot B \cdot \bar{C}$

Given the ROM circuit in Figure 4-31, place zeros and ones in the proper boxes to implement the following logic functions:

28. $f = \bar{A} \cdot B \cdot C \cdot D + \bar{A} \cdot \bar{B} \cdot \bar{C} \cdot \bar{D} + A \cdot \bar{B} \cdot \bar{C} \cdot D + A \cdot B \cdot \bar{C} \cdot \bar{D}$
29. $f = A \cdot B \cdot C \cdot D + \bar{A} \cdot B \cdot C \cdot \bar{D} + A \cdot \bar{B} \cdot \bar{C} \cdot D$
30. $f = A \cdot B \cdot C + \bar{A} \cdot B \cdot C \cdot D + \bar{A} \cdot \bar{B} \cdot C \cdot D + \bar{A} \cdot \bar{B} \cdot \bar{C}$

CHAPTER 5

FLIP-FLOPS AND LATCHES

Learning Objectives. *Upon completing this chapter you should know:*

1. *How to identify:*
   - *a. an R-S flip-flop*
   - *b. a type T f-f*
   - *c. a type D f-f*
   - *d. a J-K f-f*
2. *Clock and input conditions for each type.*
3. *How a flip-flop works.*
4. *How edge triggering works and where it is used.*
5. *How the master-slave circuit works.*
6. *How to use a J-K as an R-S, a type T, or a type D.*
7. *The terms direct inputs, set, preset, reset, clear, and toggle.*
8. *How direct inputs are used.*
9. *The most important flip-flop IC packages.*
10. *How to follow the timing diagram for the edge-triggered type D flip-flop.*
11. *How to reproduce the sequence diagram for the J-K master-slave flip-flop.*
12. *The purpose of each f-f input configuration.*
13. *How to draw the logic diagram and explain the operation of:*
    - *a. NAND latch*
    - *b. NOR latch*
    - *c. clocked NAND latch*
    - *d. clocked NOR latch*
    - *e. basic type T*
    - *f. basic type D*
    - *g. master-slave J-K f-f*
    - *h. edge-triggered D f-f*

Basic memory elements come in two forms, static and dynamic. Static forms consist of two logic gates with the output of the first connected to the input of the second gate. The output of the second gate is in turn coupled back into the input of the first. The result is a regenerative circuit with two stable conditions (two stable states). Additional gating permits switching from one state to another. Once switched to a given state, the circuit—called a *flip-flop*—remains in that state as long as power is applied and no additional switching command pulse is provided.

There are two basic kinds of flip-flops: the type *D* and the *J-K*. Some flip-flops are operated in an asynchronous mode, without clock control, but most of them are operated under the control of a clock pulse in a synchronous system.

There are three distinct kinds of flip-flop memory arrays: rectangular coordinate memories, counting arrays, and shift register systems. These three kinds of memory arrays will be discussed in separate chapters. In this chapter we will concentrate on the operation and properties of the flip-flop elements themselves.

The dynamic memory element involves the control and transfer of a stored charge, and is based on the MOS dynamic logic form discussed in Chapter 2. There is no dynamic logic using bipolar technology. Dynamic memory cells are primarily used in rectangular coordinate memories and shift register memory systems. These will be discussed in a later chapter.

5-1 Basic Memory Latches

The term *latch* refers to a circuit's ability to remain at a particular logic level after having been driven to that state by some externally provided signal. The circuit must remain in that state even after the command or control pulse no longer exists. In a very real sense the latch* *remembers* that a command pulse once existed.

A very simple but limited memory latch circuit is shown in Figure 5-1. This circuit can be latched, but it is necessary to break the circuit or remove the power to reset it. Still, it is the basis of all of the more complex latches and flip-flops, and in some cases it can be a useful circuit in itself.

Figure 5-1a shows the initial conditions with a *low* output and a low on the latch command line, *B*. The feedback line, *A*, is connected to the output so that its condition will always be the same as the output level. By convention we will say that the *latch* is storing (remembering) a logic zero. The output *Q* is low. We will use the symbol *Q* instead of *f* as the output indicator for memory elements so that we can tell whether a Boolean expression represents simply asynchronous logic or memory elements. Now, if we want to store a 1, we change the *low* (0) latch command signal on leg *B* of the OR gate to a *high* (1) and leave it in that condition long enough for the gate to respond and drive the output *Q* to a *high* (1) state. This takes 10–50 nanoseconds. As soon as *Q* goes *high*, the feedback line feeds that *high* level back to the *A* input of the OR gate.

---

* *Latch* may be used as either a verb or a noun.

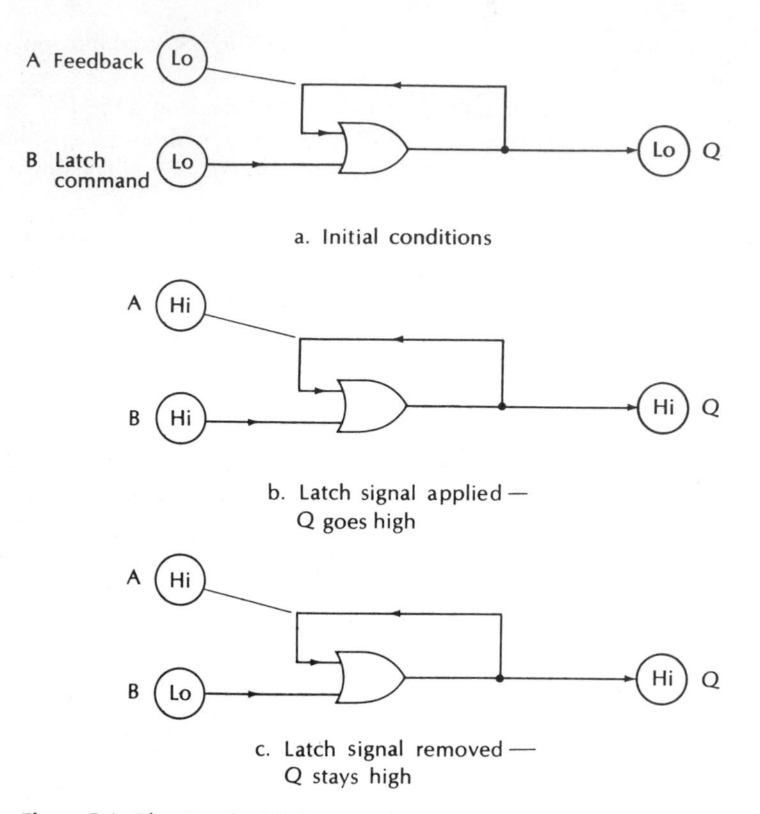

Figure 5-1 The Simple OR Gate Latch

The output *Q* stays *high* and holds the *A* input *high,* which latches the output at *high*. We can now allow the latch command input *B* to go *low* (Figure 5-1c) and the output *Q* remains *high*. The latch is now latched and the latch command input *B* has no further influence.

It is important to consider the time required to *set* the latch because this *set-up* time (delay), as short as it is, forces us to use some fairly complex latch and flip-flop circuits for most practical applications. Such increased complexity does not, however, significantly add to the cost of integrated circuits.

5-2 The NOR Latch

The OR latch in Figure 5-1 can be implemented using NOR gates as shown in Figure 5-2.

The rightmost NOR gate serves as an inverter which performs a double complement for the output *Q* and the feedback line to leg *A* of the

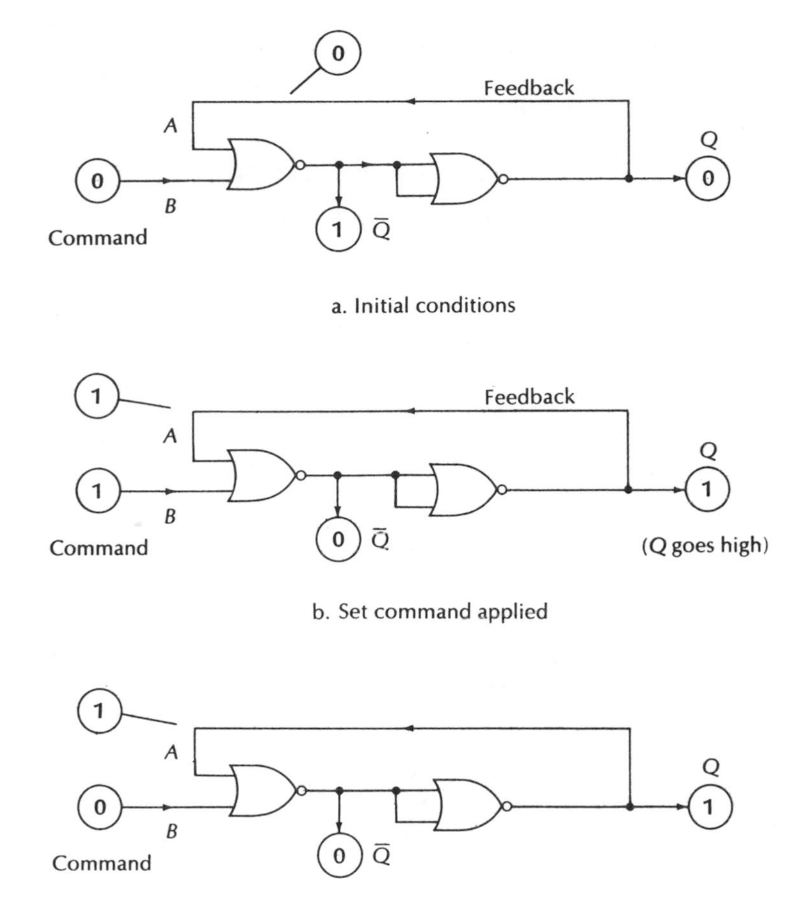

Figure 5-2 Simple Latch Using NOR Gates

lefthand (input) NOR gate. The result is the exact equivalent of the OR gate version. One significant difference between the two circuits is the fact that we now have a second output available which is the complement of the Q output. We will call that output $\overline{Q}$. If you will follow Figure 5-2 through, you will see that it operates in exactly the same fashion as the basic OR latch in Figure 5-1. The sequence of events is important because of the gate delay times. The process begins with the initial conditions of a *low* on the set command line B, a condition assumed to be stable and to have existed for as long as necessary for the circuit to *settle* down. The set command line is then raised to a *high* and held there until both gates have responded and a valid 1 *(high)* logic level rests on the feedback input A. The latch is now *set* and the command input line can go back to 0 (*low*) without affecting the *high* output (1) on Q. From this time on, the set command line may change states indefinitely without changing the output of the latch.

ADDING RESET CAPABILITY TO THE NOR LATCH

It is a simple matter to modify the NOR latch in Figure 5-2 to make it into a full-fledged *set-reset* latch. All that is necessary is to disconnect one of the legs on the NOR gate that was used as an inverter in the simple latch and to use that free gate leg as a *reset* input. Figure 5-3 shows the modified NOR latch with reset capability.

Follow the events in Figure 5-3 to see how the process works.

a. The initial condition requires a *low* (0) on the set command input. $\bar{Q}$ is *high* and Q is *low*. The reset input is *low*. The latch is *reset*.

b. The set line goes to *high* and the reset line remains *low*. The leftmost gate outputs a *low,* and $\bar{Q}$ goes *low*. The righthand gate outputs a 1 (*high*) which brings Q to a 1 and feeds a 1 (*high*) back to the lefthand gate.

c. After feedback input reaches a stable *high* logic level, the set line can return to *low* and the output Q will still remain *high*.

d. After the set line has settled to a logic *low,* the circuit is ready to receive a *reset* command pulse at the reset input. The reset line goes *high,* driving Q *low* and feeding a *low* back to the feedback input of the lefthand gate, which outputs a 1 and drives $\bar{Q}$ to *high*.

e. After $\bar{Q}$ has established a stable *high* logic level, the reset level may drop to *low*. The circuit is now reset and ready to accept a set command whenever we are ready to send it to the set line. Notice that there is a forbidden case where both the set and reset lines are *high* at the same time. If we hit both inputs (set and reset) with a 1 at the same time, both Q and $\bar{Q}$ will be *low*—a forbidden condition. Also you will notice that the only time we have a *low* on both set and reset inputs is after a *high* has remained on one of the lines long enough for the latch to settle. We can have a *low* sitting on both inputs at the same time but initially we cannot deliver them at the same time. If we do, there will be a race, with the final state unpredictable. Most of the refinements we will examine involve methods of avoiding indeterminate states and race problems.

The latch, or flip-flop, circuit is not generally drawn as it has been here. It was drawn this way to make the explanation easier to follow, but it will almost always be drawn as shown in Figure 5-4. A careful

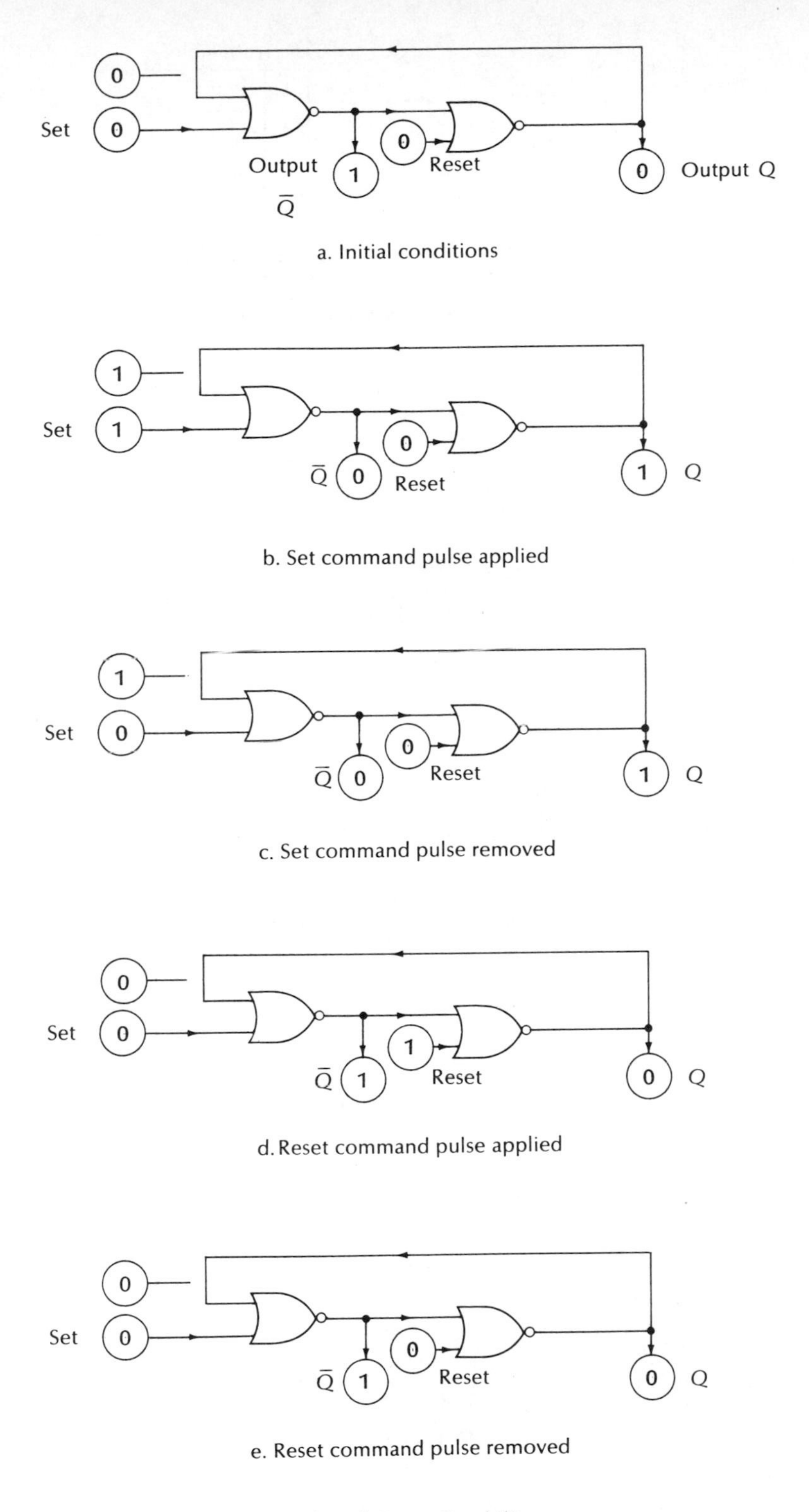

Figure 5-3 Basic NOR Latch with Reset Capability

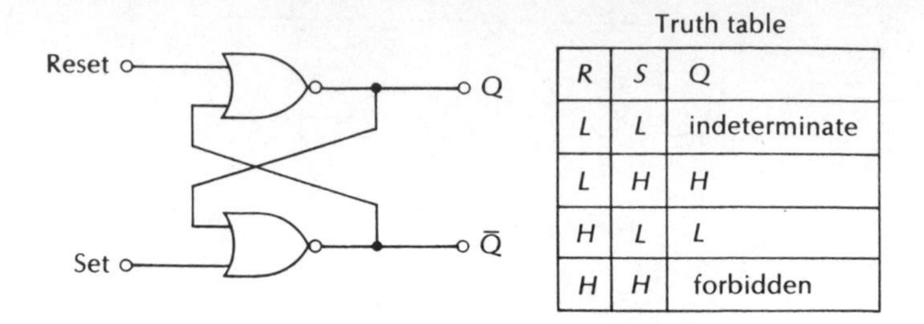

Truth table

R	S	Q
L	L	indeterminate
L	H	H
H	L	L
H	H	forbidden

Figure 5-4 *R-S* Latch Drawn in Conventional Fashion

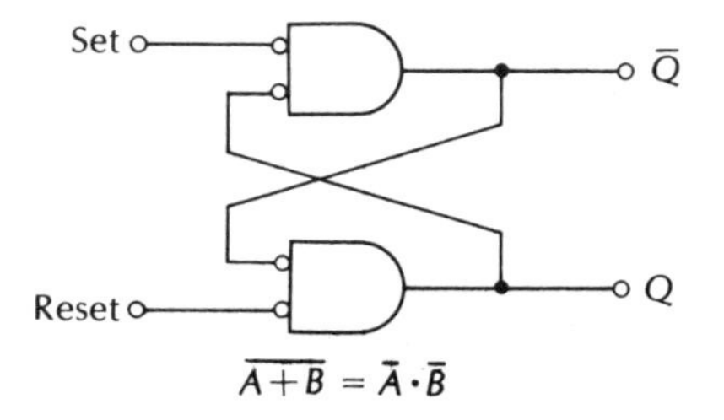

Figure 5-5 AND Gate Version of the NOR Latch

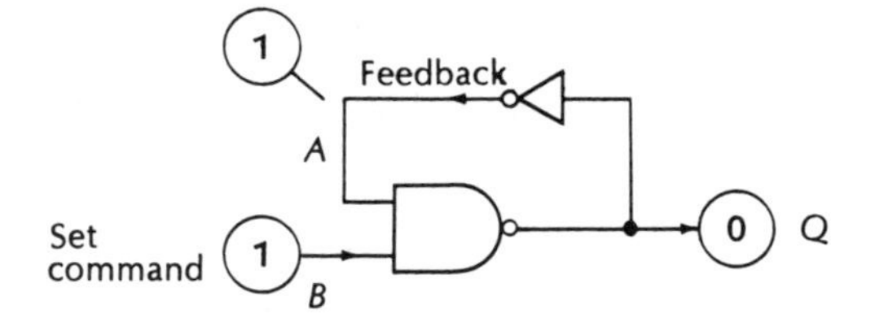

a. Initial conditions

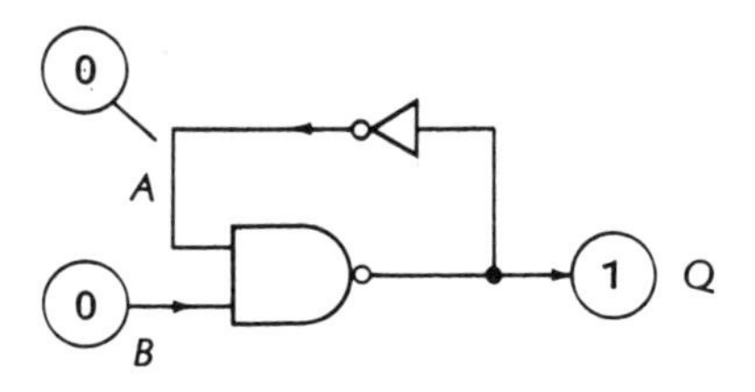

b. Set command goes low

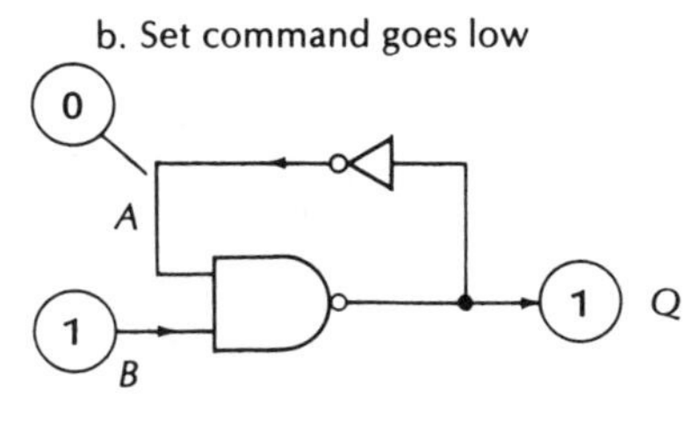

c. Set command returns to high

Figure 5-6 Simplest Form of the NAND Latch

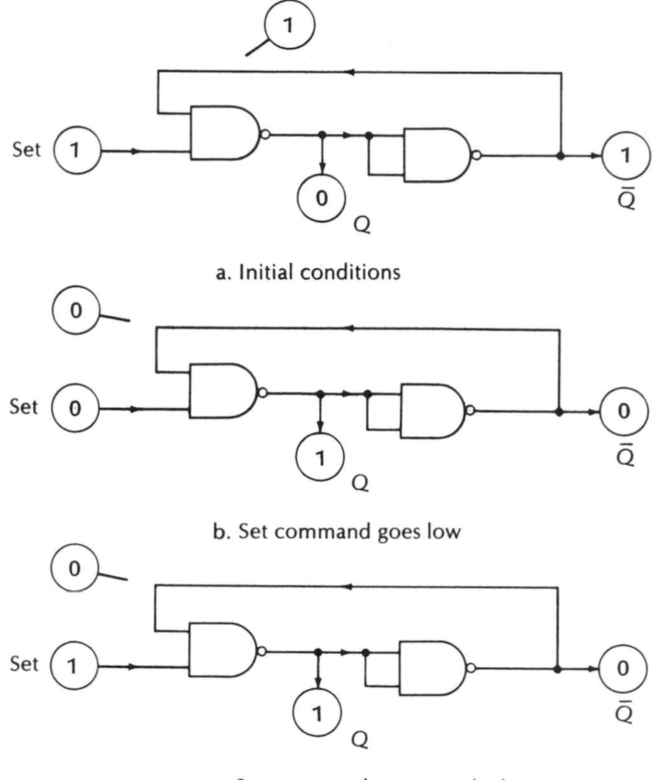

Figure 5-7 NAND Latch Circuit Using a NAND Gate to Replace the Inverter

examination will prove that Figures 5-3 and 5-4 are actually identical.
Sometimes the circuit is drawn as shown in Figure 5-5, which is based on DeMorgan's law: $\overline{A + B} = \overline{A} \cdot \overline{B}$. Both gates are converted to AND's having inverted inputs.

5-3 Basic NAND Latches

NAND latches may be fabricated using standard NAND gates. Figure 5-6 shows the circuit for a simple NAND latch without reset capability. Follow Figures 5-6 and 5-7 through. Figure 5-7 shows the same latch circuit with the inverter replaced by another NAND gate functioning as an inverter. This circuit provides two outputs $\overline{Q}$ and Q. Notice particularly that the NAND latches require a *low* for a set command. Also notice the Q output is taken from the lefthand gate and the $\overline{Q}$ is taken from the righthand gate. This is just the opposite of the NOR latch.

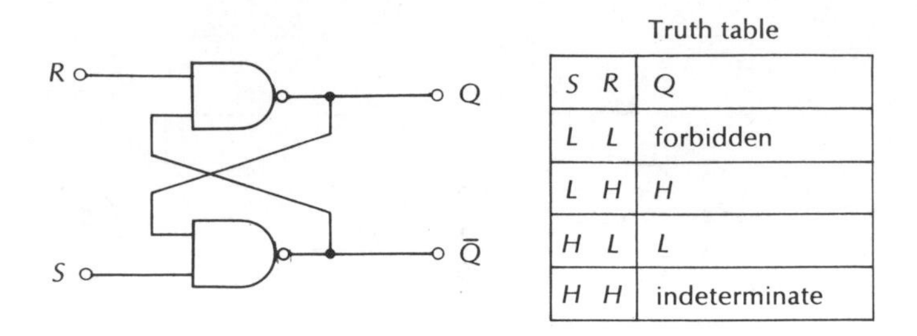

Truth table

S R	Q
L L	forbidden
L H	H
H L	L
H H	indeterminate

Figure 5-8 NAND Set-Reset Latch as Customarily Drawn

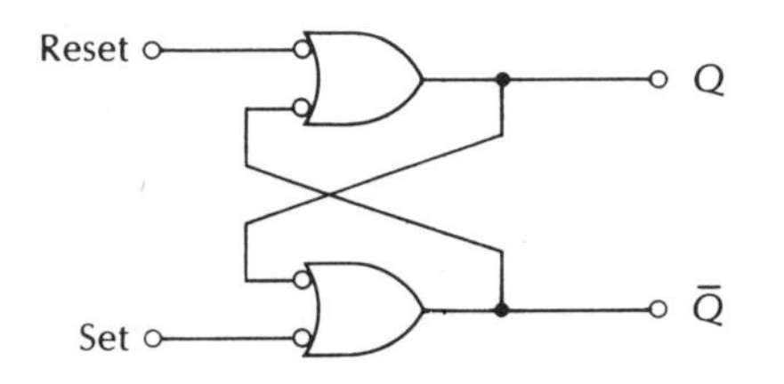

Figure 5-9 Alternate Representation of the NAND *R-S* Latch, Based on the Equality $\overline{A \cdot B} = \bar{A} + \bar{B}$

Figure 5-8 shows the NAND latch as it is customarily drawn. Figure 5-9 shows an alternate form based on DeMorgan's law: $\overline{A \cdot B} = \bar{A} + \bar{B}$.

Problem

1. Given the drawing (Figure 5-10), complete the entries (circles) to show how the circuit works.

CONTACT CONDITIONING

Mechanical contacts almost invariably bounce when contact is made. Ordinarily the bounce and multiple make-and-break action that results is no problem. However, when mechanical contacts are used with high-speed logic, the bounce is seen by the gate as multiple pulses. A simple NAND or NOR latch is often used as a switch debouncer (or bounceless pushbutton).

The set (or reset) action is initiated by the first contact closure pulse. Because of the regenerative behavior of the circuit, the switching action once initiated continues even with loss of switch closure.

The only pulse that counts is the first one—the rest are ineffective. NAND and NOR latches for contact conditioning are shown in Figure 5-11.

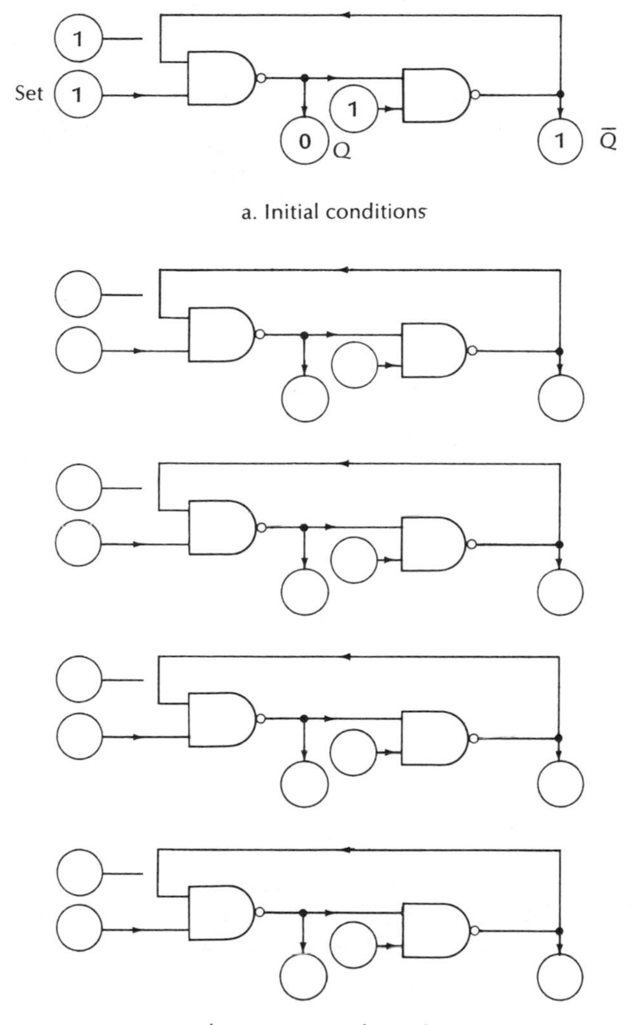

Figure 5-10 NAND Latch with Reset Capability

5-4 The D Latch

The simple NOR *R-S* (set-reset) latch we have examined has two potentially unpredictable conditions: *low's* on both set and reset lines at the same time and *high's* on both set and reset lines at the same time. A *low* state on both inputs at the same time is allowed, but only if the two inputs (*R* and *S*) did not start to go low at the same instant. The end

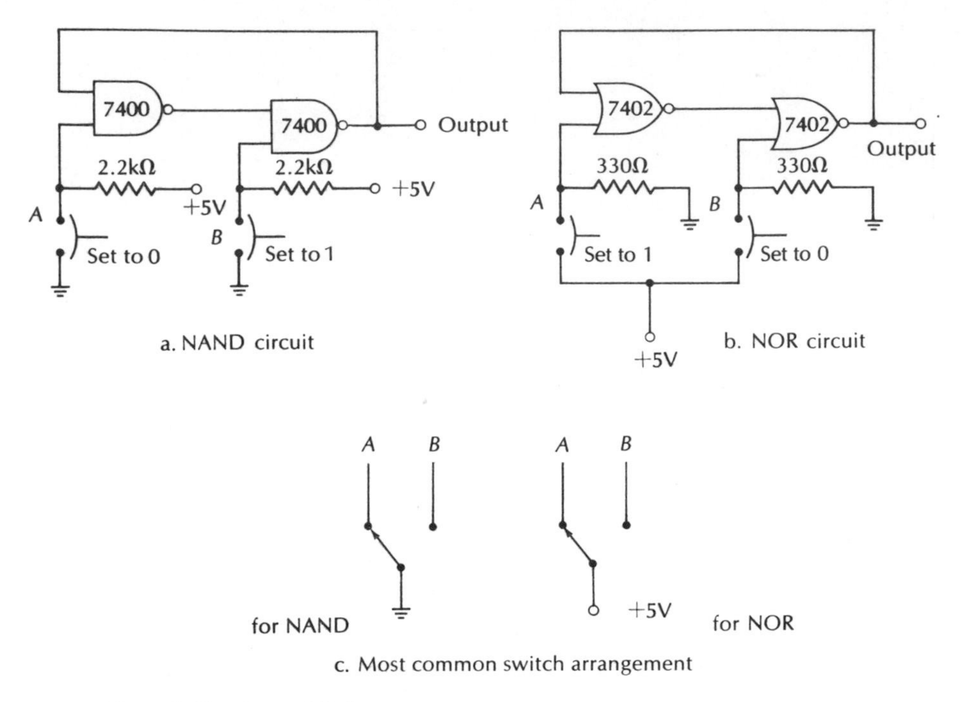

Figure 5-11 Contact Debouncers

result of *low*-going pulses arriving at both inputs simultaneously is indeterminate.

If both set and reset command pulses exist at the same time, both outputs of the latch will output a 1, a forbidden condition. When the input command pulses drop to 0 (theoretically at the same time), the result is again unpredictable.

One obvious way of preventing command pulses from arriving at both inputs at the same time (1, 1) or (0, 0) is simply to prevent these two conditions from ever coexisting at any time. This is easily accomplished by using a single inverter in an input arrangement called the D input configuration. Figure 5-12 shows a D input R-S latch circuit.

In Figure 5-12 the inverter makes it impossible for both inputs R and S to be at the same logic level at any time. The simple D latch can also be implemented in NAND form. Figures 5-13 and 5-14 show two versions of the NAND D latch, one with a *high-active* D input and one with a *low-active* D input. For a high-active input Q goes *high* when D goes *high*, and for a low-active input Q goes *high* when D goes *low*. Because the circuit is symmetrical, we could use either variation for either active-high or active-low, simply by changing the labels on the Q and $\overline{Q}$ lines. However, when the latch symbol (5-13b and 5-14b) is used, we

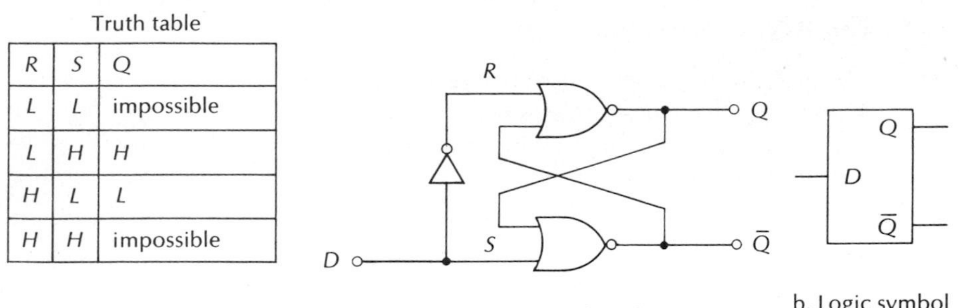

Truth table

R	S	Q
L	L	impossible
L	H	H
H	L	L
H	H	impossible

a. Logic equivalent diagram

b. Logic symbol for *D* latch

Figure 5-12 NOR *D* Input *R-S* Latch

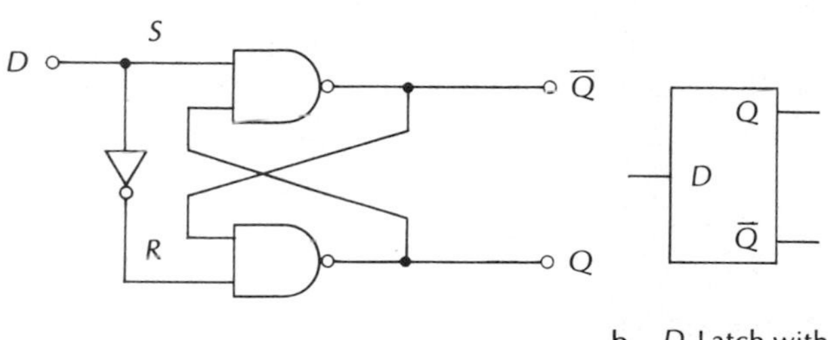

a. Logic diagram

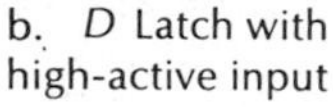
b. *D* Latch with high-active input

Figure 5-13 NAND Active-High *D* Latch

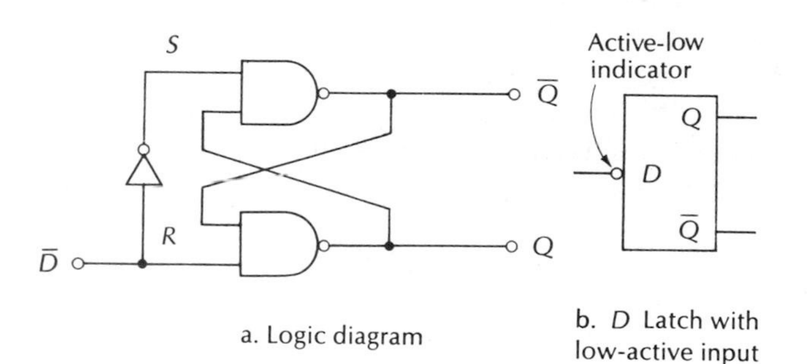

a. Logic diagram

b. *D* Latch with low-active input

Figure 5-14 NAND Active-Low *D* Latch

have no way of knowing how the circuit is configured inside the package without some additional notation. The bubble on the *D* line of Figure 5-14b is a state indicator that tells us that Q will go *high* when *D* goes *low*. Some flip-flops and latches have several inputs, some of which are active-high and some active-low. The bubble identifies the active-low inputs.

5-5 The R-S Flip-Flop Summary

The *R-S* flip-flop can be constructed with either NAND or NOR gates. The basic circuits are the same as the contact conditioning circuits except that the *R-S* f-f is switched (triggered) by pulses rather than by mechanical contacts. Figure 5-15 shows the NAND circuit, symbol, and truth table for the *R-S* flip-flop.

NAND R-S FLIP-FLOP OPERATION SUMMARY

1. Both inputs left positive: no change in state.
2. *Set* input momentarily grounded: Q goes positive, $\overline{Q}$ goes to ground.
3. *Reset* momentarily grounded: $\overline{Q}$ goes positive, Q goes to ground.
4. Both *set* and *reset* simultaneously grounded: Disallowed state with both Q and $\overline{Q}$ positive. If one input goes positive slightly before the other, the last input to go positive determines final state. This condition must be avoided.

Figure 5-16 shows the NOR gate *R-S* flip-flop, symbol, and its truth table.

NOR R-S FLIP-FLOP OPERATION SUMMARY

1. Both inputs grounded: no change in state.
2. *Set* input momentarily positive: Q output goes positive, $\overline{Q}$ goes to ground.
3. *Reset* momentarily made positive: $\overline{Q}$ goes positive, Q goes ground.
4. Both *set* and *reset:* Disallowed state with Q and $\overline{Q}$ grounded. In the event that one input goes to ground before the other, the final state is determined by the last input to go to ground. This condition is avoided.

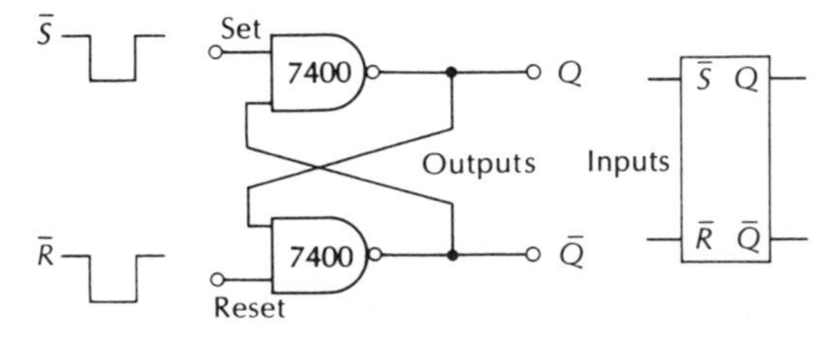

$\overline{S}$	$\overline{R}$	Q	$\overline{Q}$
1	1	stays the same	
1	0	0	1
0	1	1	0
0	0	disallowed	

a. Logic diagram b. Symbol c. Truth table

Figure 5-15 NAND *R-S* Flip-Flop

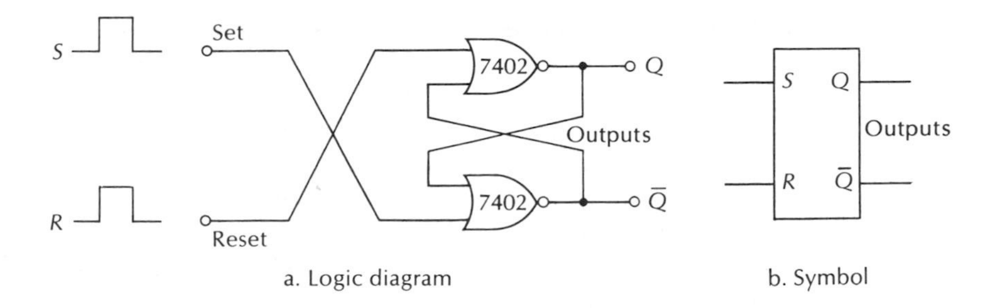

a. Logic diagram

b. Symbol

S	*R*	Q	$\bar{Q}$
0	0	stays the same	
0	1	0	1
1	0	1	0
1	1	disallowed	

c. Truth table

Figure 5-16 NOR *R-S* Flip-Flop

Problems

2. Can the NAND gate *R-S* flip-flop in Figure 5-15 be considered to operate as a negative logic element? Explain.
3. Is the output of the NAND *R-S* positive or negative logic? Can it be either? How?

5-6 Clocked Flip-Flops

In the majority of applications, flip-flops are required not only to store data (0 or 1) but also to pass stored data on to another flip-flop and to receive data from a previous f-f. Chains of flip-flops are connected in two special configurations called *counters* and *shift registers*.

A counter accepts a train of input pulses and stores a count that represents the total number of pulses entering the system. The action is similar to that of the odometer in a car that keeps track of total miles traveled. A transfer of carries from each flip-flop to the next highest order flip-flop is required, as the total is accumulated (as from the "units" odometer wheel to the "tens" wheel, and so on).

A register is a group of memory elements used for temporary storage. It differs from a *memory* array in that it usually has a much smaller storage capacity and generally is required to hold data for relatively short periods of time because of its location in the system's organization. The most common kind of register is also capable of shifting data on command from one element to the next within the register, a capability not usually found in other memory arrays. These shifting registers or *shift registers* do not keep a running total of incoming pulses as do counters. However, in both cases data transfer from flip-flop to flip-flop—involving from 2 to 4000 flip-flops in a chain—is required.

In these two flip-flop applications the simple latch we have previously examined is completely inadequate to the task. Because of system gate delays and flip-flop transition and settling times, timing becomes critical particularly at high data-shift or counting rates.

If the state of flip-flop A is to be transferred to flip-flop B, it is essential that both flip-flops have completed any previous state change and have had time to *settle in* before a transfer from A to B is made. Entering an input pulse commanding a change of state while a flip-flop is between states or in an unsettled condition invariably results in unreliable operation. There are two specific timing problems in counters and shift registers.

Every gate in the system has a delay time. These delays are variable from gate to gate—even of a given type number—and are cumulative. In addition, the time required for flip-flops to change states and settle in is quite variable. These delays give rise to the first timing problem.

The second timing problem is called the *race* problem. When a string of flip-flops is direct-coupled, a change of state in one flip-flop can race down the entire length of the string, resulting in a completely useless system.

Because of these two problems, flip-flops used in counters and shift registers cannot be of the simple *R-S* type. In order to make counters and shift registers possible, the basic *R-S* circuit is modified into clocked, or synchronous, circuits. In clocked circuits outputs do not change as soon as input conditions change but must wait for a *clock* command pulse before they can respond.

5-7 The Clocked R-S Flip-Flop

Figure 5-17 shows the necessary gates added to the *R-S* flip-flop to provide for clock control. The two additional NAND gates allow either the set or reset pulse to trigger the flip-flop only if the clock input is positive. The circuit in Figure 5-17 does not solve the basic timing problems, but it is the first step toward their solution.

5-8 The Type T Flip-Flop

The *T* modification in Figure 5-18 provides the *toggle* action required when flip-flops are connected in a counting arrangement. Cross-coupled

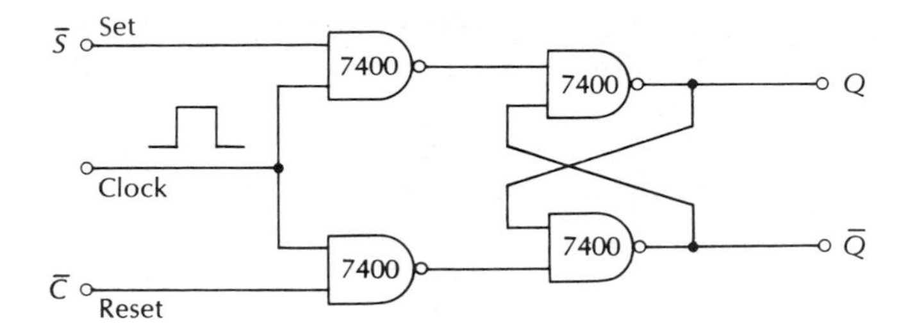

Figure 5-17 Clocked *R-S* Flip-Flop

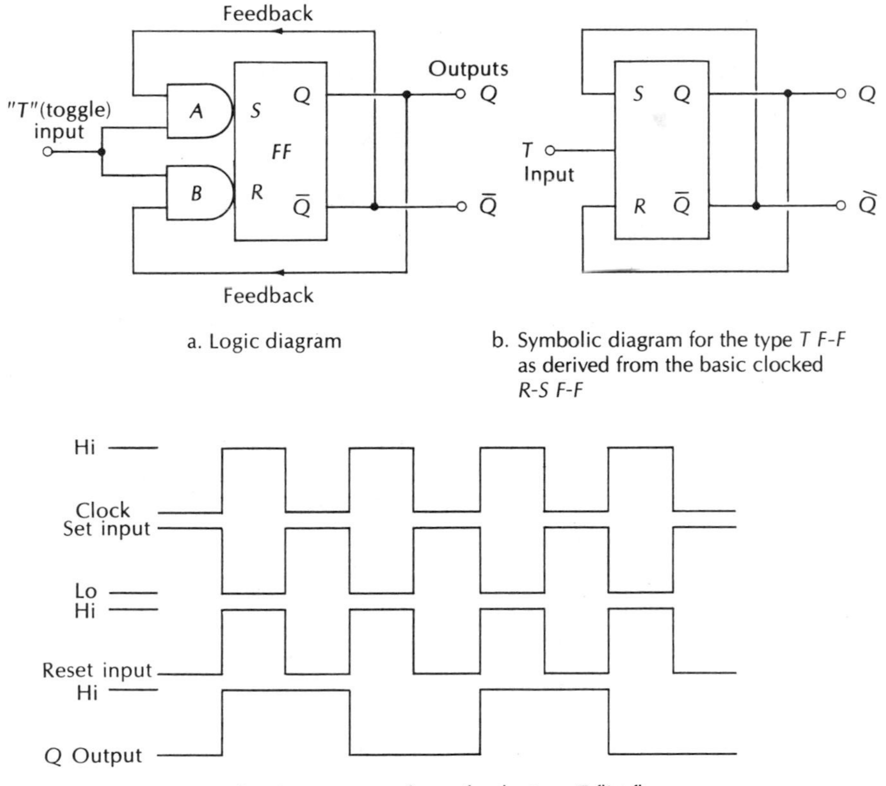

Figure 5-18 Type *T* Flip-Flop

feedback takes one input of gate A to $\overline{Q}$ while a second feedback path ties Q to one input of gate B. If the flip-flop is in the reset state $\overline{Q}$ *high*, there is a *high* on the upper input of steering gate A. This feedback signal constitutes a *set* command. Upon the arrival of a clock pulse the flip-flop *sets* to Q *high*. Now, the feedback places a *high* on the lower input of gate B and ground on the feedback input to gate A. The flip-flop is now conditioned to toggle to reset ($\overline{Q}$ *high*) with the arrival of the next clock pulse.

Notice that it takes two clock pulses to cause the flip-flop to make one complete transition from Q *low* to Q *high* and back to Q *low*. If a train of clock pulses is applied to the input at frequency f, the output of Q would be a pulse train with a frequency of $f/2$. The circuit is often called a divide-by-2 or binary. Any number of toggle flip-flops can be cascaded for division by successive powers of 2 as shown in Figure 5-19. The circuit is counting by twos.

Problems

4. Make a truth table for the type T flip-flop.
5. Given the type T flip-flop in Figure 5-20, draw the waveforms at the outputs Q and $\overline{Q}$.
6. Why does the disallowed state (Q and $\overline{Q}$ both *high*) not occur in the T flip-flop?
7. Does the toggle action occur when the clock goes positive or when it goes to ground?

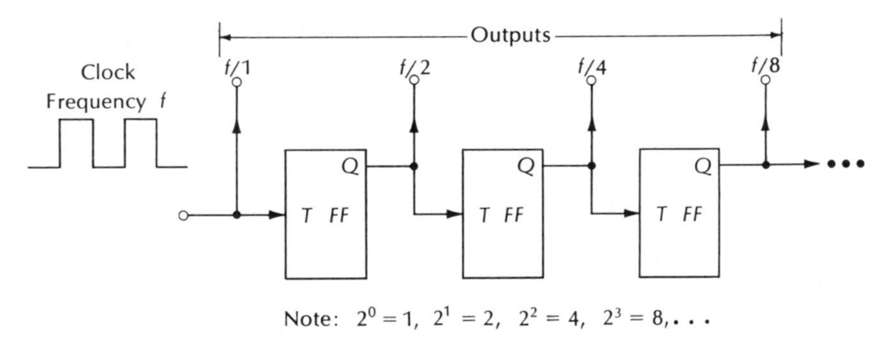

Figure 5-19 Type T Flip-Flops as a Frequency Divider by Powers of Two

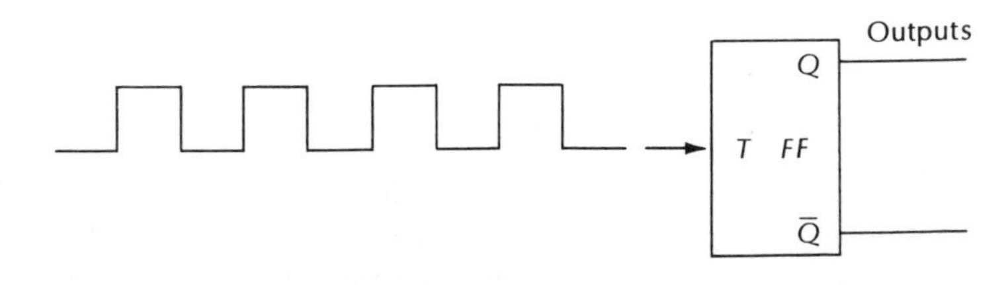

Figure 5-20 Figure for Problem 5

5-9 The J-K Flip-Flop

The circuit configuration in Figure 5-21 is the most versatile flip-flop circuit available. It can function as a clocked *R-S* flip-flop or a toggle flip-flop, as well as performing a number of more specialized functions. In addition to clock provisions and input flexibility, the *J-K* has the advantage that all four possibilities on its truth table are valid—none are forbidden conditions.

J-K OPERATING CHARACTERISTICS

1. *J* input grounded, *K* input grounded: when the clock goes *low*, nothing happens.
2. *J* input goes positive, *K* input grounded: when the clock goes *low*, *Q* goes (or stays) positive. $\overline{Q}$ is grounded. The 1 on the *J* input is "passed" directly to the *Q* output.
3. *K* input goes positive, *J* input grounded: when the clock goes *low*, $\overline{Q}$ goes *high* and *Q* goes to ground. The 0 on *J* input is transferred directly to the *Q* output.
4. *J* held positive, *K* held positive: the circuit toggles on each clock pulse. The circuit now behaves as a type *T* binary divider or counter stage.

These operating rules are summarized in Table 5-1. The *J-K* can be used as a type *D* flip-flop by adding an inverter, as shown in Figure 5-22. This is a usable circuit, although a packaged type *D* with special characteristics is more frequently used.

5-10 The Master-Slave J-K Flip-Flop

So far the flip-flops presented have represented stages in the evolution of the workhorse of flip-flops, the *master-slave J-K*. The master-slave

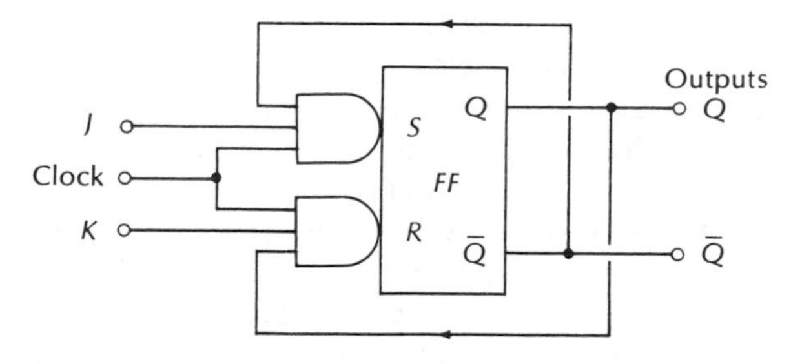

Figure 5-21 *J-K* Flip-Flop Reconfigured for Clock Input

Table 5-1 Truth Table for *J-K* Flip-Flop

Truth table

J	*K*	Q, after clock
0	0	no change
0	1	*F-F* resets
1	0	*F-F* sets
1	1	*F-F* toggles

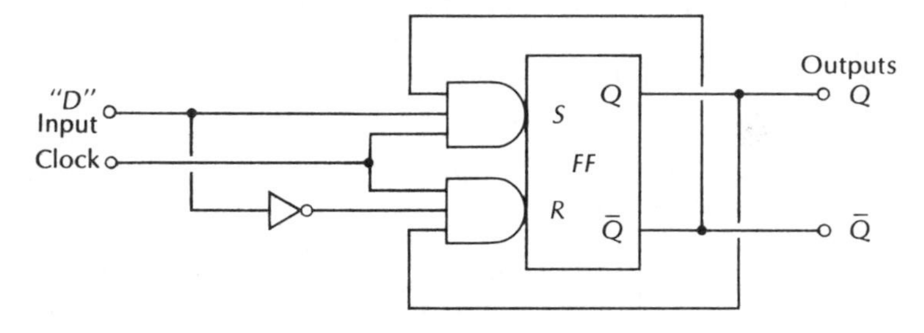

Figure 5-22 *J-K* Modified with a Type *D* Input

circuit shown in Figure 5-23 has the versatility of the simple *J-K* but effectively eliminates all timing problems including the race problem. The circuit consists of two flip-flops in series, with their clock inputs driven in complementary fashion. The slave stage serves as a kind of holding tank for data to be transferred to the next flip-flop in a counter or shift register chain. Data is buffered and held by the slave until all delays have been allowed for and all circuits have settled in.

HOW IT WORKS

On the leading edge of the clock pulse, data enters the *J* and *K* inputs on the master flip-flop. On the up-clock the inverter provides a down (*low*) clock for the slave flip-flop, locking out any data input. During the clock *on* time the master has plenty of time to set up (pulse widths are much greater than necessary for set-up). The slave waits during the clock *on* time and cannot change states because the inverter holds its clock input *low*. When the clock pulse first arrives at the master, only the *J* or the *K* input (but not both) is *primed* to cause the master to change state. This priming is accomplished by the feedback lines from the outputs of the slave. The master can only *set* if the slave is in the *reset* state—the reset

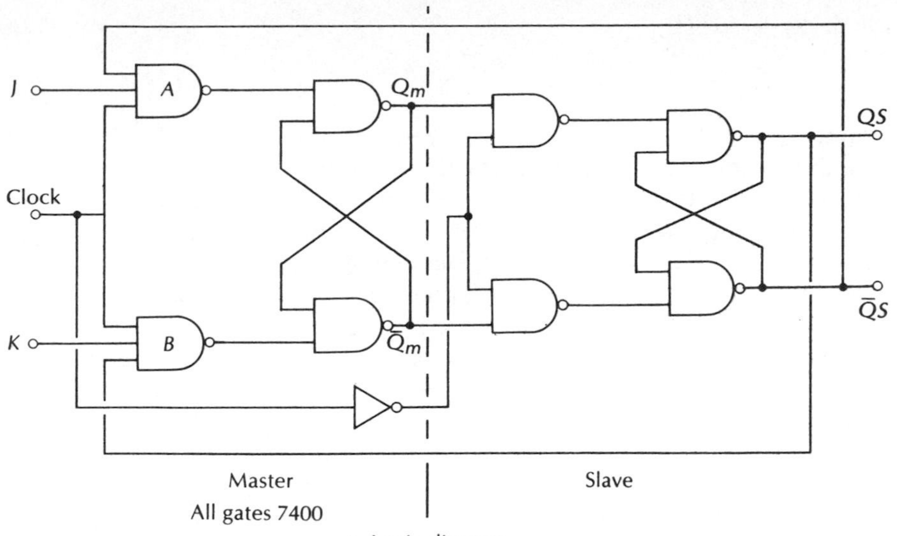

a. Logic diagram

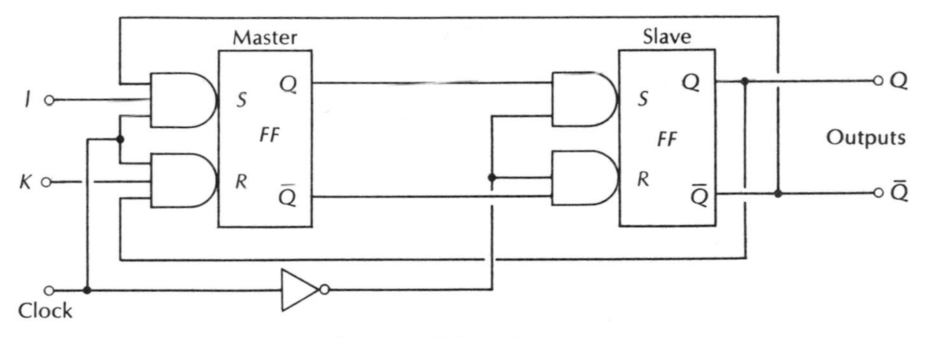

b. Functional block diagram

Clocked inputs

J	K	Clock ↓	Output, Q
0	0		no change
0	1		0
1	0		1
1	1		toggles to opposite state

c. *J–K* master–slave truth table

Clocking occurs when clock goes low. Data on J and K can only be changed immediately after clock goes low. Only one change of J and K conditions is allowed per clock cycle.

Figure 5-23 Master-Slave *J-K* Flip-Flop

input is disabled. Or, it can only *reset* if the slave is in the *set* condition—the master reset input is locked out.

Assume that the slave is in the *reset* condition. Before the up ↑ clock and during the clock *on* time, the master flip-flop can only respond to a set command. Even though the master changes state early in the clock *on* period, the slave stays as it was and the master can respond only to a *set* command during the clock *on* period. If a *reset* command should come down the reset line during the clock *on* time (the forbidden condition in the *R-S* f-f), that reset pulse is ignored, so the $Q = 1, \overline{Q} = 1$ condition doesn't exist for this circuit. At the end of the clock pulse *on* time, the clock pulse to the master falls. For the slave, because of the inverter, the clock is now rising and data from the master can be sampled. During the slave's clock *on* time the master's clock line is *off* and it cannot accept data at either of its inputs. This sequence of events is exactly what we assumed in the operational description of the *R-S* flip-flop. The sequence of events in the master-slave flip-flop can be graphically illustrated by a sequence drawing. (See Figure 5-24.) If we follow the clock waveform drawing in Figure 5-24 through, the sequence goes like this:

1. Primed input of master initiated change of state. Slave locked out by inverted (*down*) clock.
2. By 2 or a little later, master changes state and settles down. Slave still locked out by inverted clock pulse.
3. More than adequate set-up time has been allowed for the master. Slave clock pulse now rising and the slave begins to change states if the master has changed states between 1 and 3.
4. By 4 or a little later the slave has settled into new state. Master stays put during *off* time of clock and waits until the next *up* clock to sample any new data at its inputs.

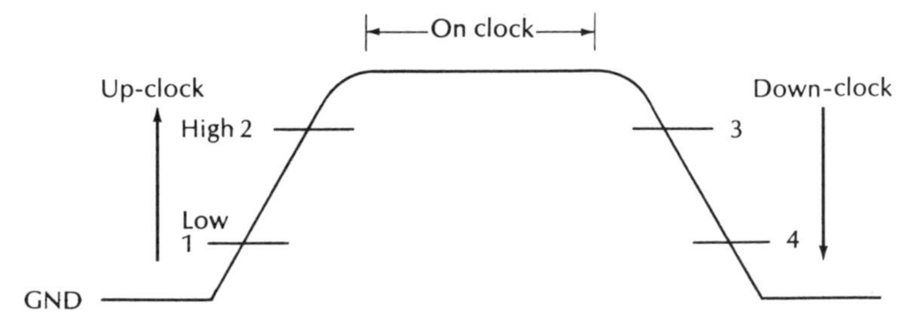

Figure 5-24 Clock Waveform for the *J-K* Master-Slave (Sequence Diagram)

The master slave concept can be applied to any flip-flop type, but the *J-K* is easily the most versatile. The master-slave circuit is the most nearly foolproof flip-flop configuration.

Problem

8. Draw a *J-K* master slave flip-flop based on NOR gates.

5-11 Direct Set and Clear

In Figure 5-25 direct *set* and *clear* lines have been added in the slave flip-flop stage. These inputs are override functions independent of the clock and other input signals. These inputs are used to clear a shift register or to reset a counter to 0 (or some predetermined count). Only one input may be used at a time. These direct inputs are normally used at the beginning or end of an operation and are not normally used as auxiliary data inputs.

The terms *set* and *preset* are often used interchangeably, as are the terms *reset, clear,* and *preclear.*

5-12 Commercial J-K Master-Slave IC's

The most common TTL *J-K* master-slave flip-flops are the 7473, 7476, and the 74107.

THE 7473

The 7473 is a dual *J-K* master-slave flip-flop with *clear* but no *set** direct input. The most popular and common *J-K* M-S has a nonstandard pin-out.

The clock goes *low* to toggle the flip-flop (level sensitive). The maximum toggle frequency is 20 MHz and the power supply current per package is 20 mA. The package is a 14 pin DIP.

THE 7476

This is a dual *J-K* master-slave with both *clear* and *set* direct inputs. The clock goes *low* to toggle the F-F (level sensitive). The maximum toggle frequency is 20 MHz and the supply current per package is 20 mA. The package is a 16 pin DIP.

THE 74107

This is identical to the 7473 but with standard pin-out and is preferred for new designs.

* Often called "preset"

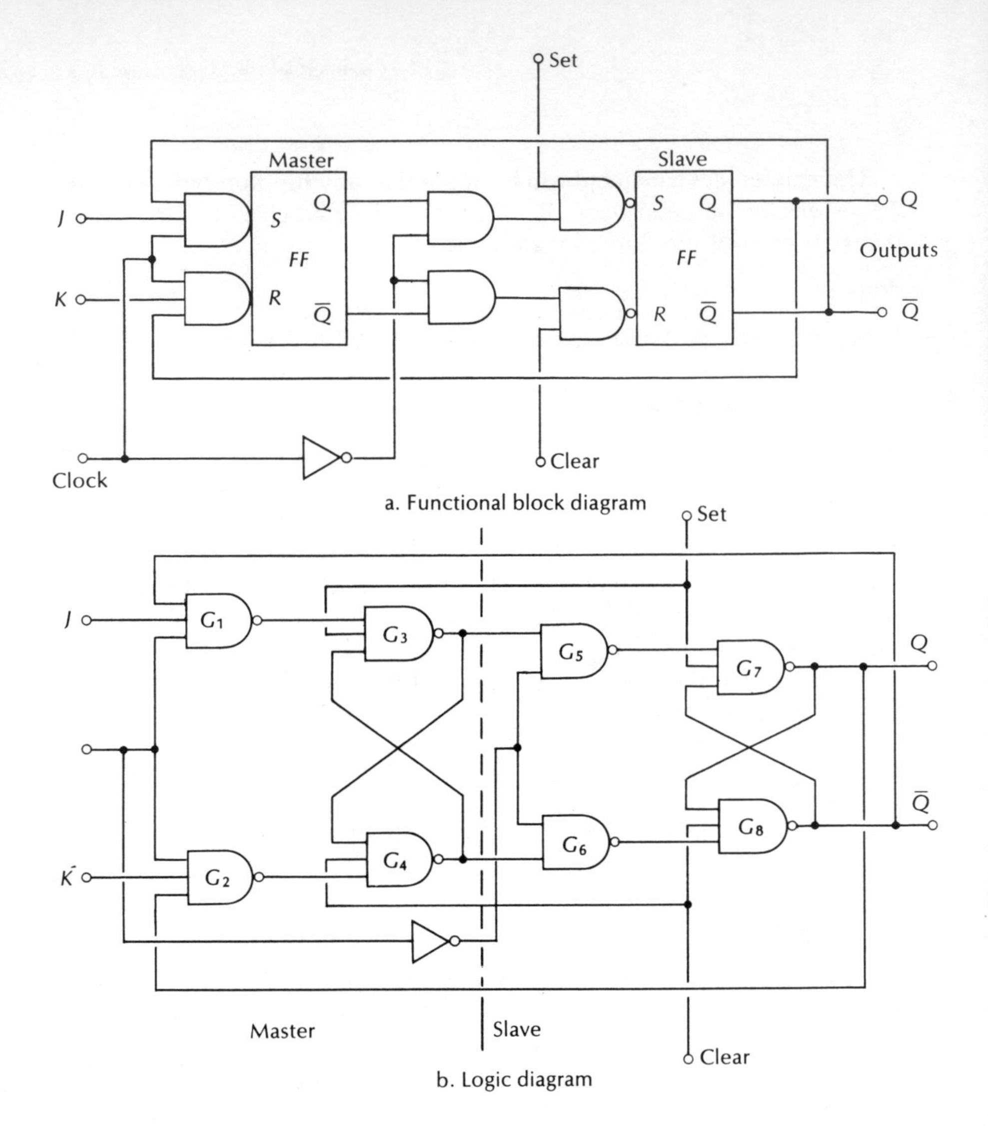

Set	Clear	Output, Q
0	0	disallowed
0	1	1
1	0	0
1	1	Normal clocked operation using J and K inputs

c. Truth table for direct set and clear inputs

Figure 5-25 Master-Slave J-K Flip-Flop with Direct Set and Clear

C-MOS: THE 4027 AD, AE, AK

The 4027 is a dual *J-K* M-S with both *clear* and *set* direct inputs. It has two *J* and two *K* inputs plus a clock input. The clock goes to *high* to toggle the flip-flop. Maximum toggle frequency is 8 MHz. Current per package (quiescent) is 60 μA, 16 pin DIP.

5-13 Level and Edge Triggering

The *J-K* flip-flops are designed to trigger when the clock pulse reaches a specific voltage level. Level triggering is suitable for synchronous operation but when incoming data is not synchronized to the clock, another form of triggering is required.

Edge triggering uses only the trailing or leading edge of the clock (or data) pulse. Triggering occurs only during the appropriate clock transition. Steady state conditions are ignored by edge-triggered circuits. The *D*-type input is required in the bulk of applications where input data are likely to be random or otherwise out of step with the system clock. As a result, standard TTL *D*-type flip-flops are edge triggered.

When the input data can enter the flip-flop while the clock pulse is changing from 0 to 1, the circuit is said to be *positive edge-triggered*. When the flip-flop can accept data during the clock pulse transition from 1 to 0, it is said to be *negative edge-triggered*.

5-14 The Type D Edge-Triggered Flip-Flop

The 7474 is a typical edge-triggered type *D* flip-flop. The logic diagram for one unit in the package is shown in Figure 5-26. The 7474 is a master-slave arrangement that achieves edge sensitivity without *R-C* differentiating circuits. *R-C* circuits are not satisfactory because the edge-triggered flip-flop must be capable of handling input signals from near 0 Hz to the maximum toggle frequency of the device. In some applications it must also handle completely random input data. The master circuit appears on the left and the slave on the right in Figure 5-26.

OPERATING RULES FOR THE 7474

1. *D* input positive: The *Q* output goes positive (or stays positive) when the clock makes the transition from ground to positive.
2. *D* input grounded: The *Q* output goes to ground (or stays at ground) when the clock makes the transition from ground to positive. Figure 5-27 is a timing diagram illustrating the edge-triggered behavior of the *D* edge-triggered flip-flop.

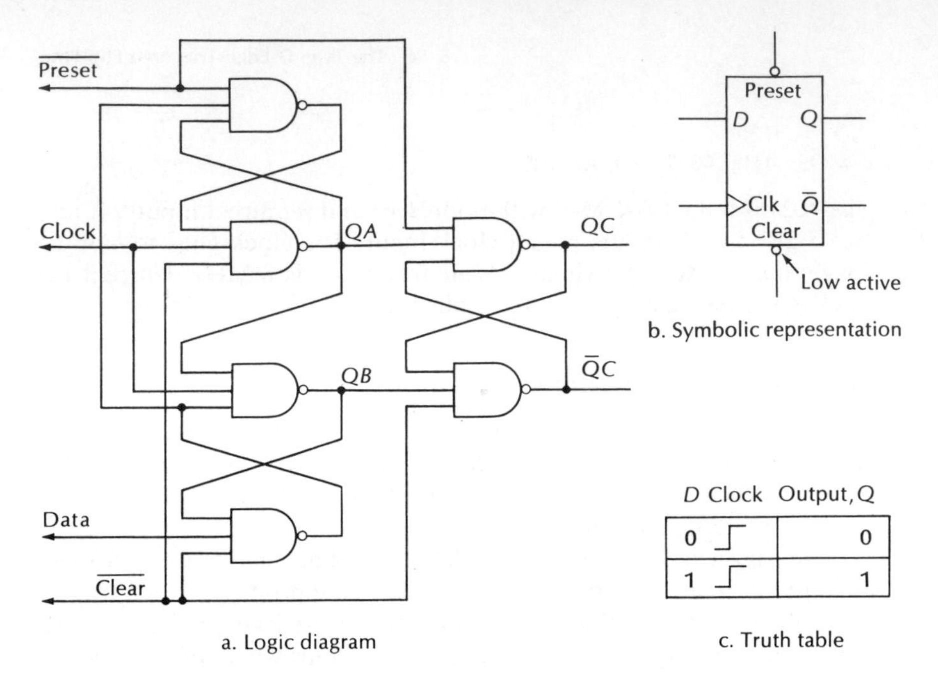

Figure 5-26 Type *D* Edge-Triggered Flip-Flop

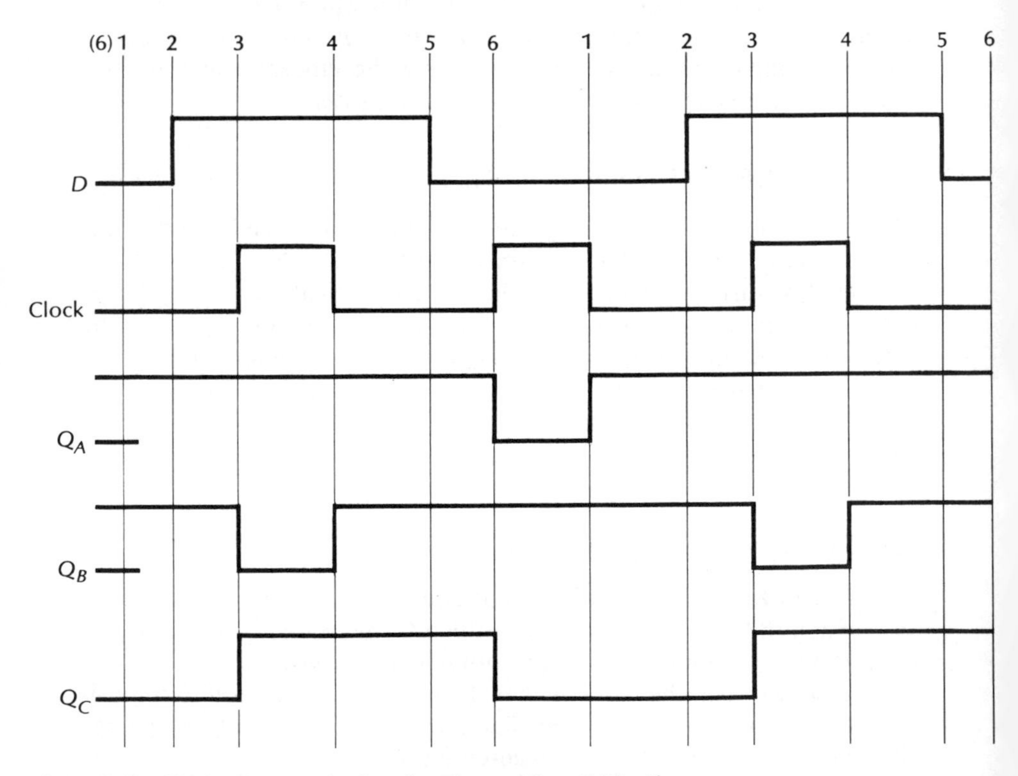

Figure 5-27 Timing Diagram for the Edge-Triggered Type *D* Flip-Flop

The following events occur in the process of setting and resetting the flip-flop. Note that events are numbered on the timing diagram to correspond with the following outline of events.

1. *Initial conditions (start to set)* (6) which occurred at the end of the previous reset part of the cycle.
   a. Data 0 *(low)*
   b. Clock 0
   c. Q_A 1
   d. Q_B 1
   e. Q_C 0
2. *Data line goes high* (1)—set command.
   a. Data 1
   b. Clock 0
   c. Q_A 1–no change
   d. Q_B 1–no change
   e. Q_C 0–no change
3. *Clock goes high* (1)—time to set Q_C.
   a. Data 1
   b. Clock 1
   c. Q_A 1–no change
   d. Q_B goes *low* (0)
   e. Q_C goes *high*—latch sets to $Q_C = 1$
4. *Clock goes low* (0).
   a. Data 1–no change
   b. Clock 0–goes *low*
   c. Q_A 1–no change
   d. Q_B 1–goes *high*
   e. Q_C 1–remains *high*—no change
5. *Data goes low*—start of reset phase–but no change in Q_C until after clock.
   a. Data 0 goes *low*—reset command
   b. Clock 0 still *low*
   c. Q_A 1–no change
   d. Q_B 1–still *high*
   e. Q_C 1–still *set*—must wait for next clock—*high*
6. *Clock goes high*—driving Q_A *low* and resetting Q_C to 0.
   a. Data 0–still *low*
   b. Clock 1–goes *high*
   c. Q_A 0–goes *low*
   d. Q_B 1–no change
   e. Q_C 0–goes to 0—reset
1. *Begin new set phase . . .*

Notice on the timing diagram that Q_C (and $\overline{Q}_C$) changes state only on the positive-going edge of the clock pulse. The circuit is edge-triggered.

PRESET AND CLEAR INPUTS

Preset and clear inputs for the edge-triggered *D* flip-flop are level sensitive, independent of clock and *D* inputs, and override any other existing input conditions. The direct inputs are not used for data but only to establish initial conditions for the system. Table 5-2 is the truth table for the direct inputs.

5-15 Commercial Type D Flip-Flops

TTL

The most popular SSI TTL type is the 7474, a dual (2 per package) edge-clocked *D* flip-flop. It is triggered on the *positive*-going edge of the *D* input (or clock) pulse. Maximum toggle frequency is 25 MHz and current per package is 17 mA, 14 pin package. Individual clock, set, and clear inputs are provided for each flip-flop.

The 74175 is a quad positive edge-triggered device. All four flip-flops have common clock and common clear inputs. It is intended for service as a memory buffer in counter systems and similar applications. The clear line is normally held *high* and dropped momentarily to *low* to clear (reset) the flip-flop. Because the quad *D* is generally used as a buffer memory, its frequency is called update frequency rather than toggle frequency. It is the maximum rate at which data can be changed. Maximum update frequency is 35 MHz and current per package is 30 mA, 14 pin package.

Table 5-2 Truth Table for the Edge-Triggered Type *D* Flip-Flop

Preset	Clear	Output Q
0	0	forbidden state
0	1	1
1	0	0
1	1	normal clocked operation ("D")

The 74174 is a hex positive edge-triggered type *D*. Except for the number of devices in the package and the current drain, it has the same specifications as the 74175. All six flip-flops share the same clock and clear inputs. Current per package is 30 mA, 16 pin package.

C-MOS

The CD 4013 is a dual C-MOS positive edge-triggered type *D* flip-flop. The two flip-flops feature separate *set* and *clear* and *clock* inputs. Direct inputs (set and clear) are held at ground for normal operation and momentarily taken *high* to set or reset the flip-flop. Maximum update (or toggle) frequency is 10 MHz and current per package is 60 to 120 μA, 14 pin package.

The CD 4042 A is a quad positive edge-triggered *D* latch (flip-flop). The four flip-flops share the same clock-reset lines. Maximum frequency is 2 MHz. Current per package is 60 to 120 μA, 16 pin package. Tristate output C-MOS *D* latches are also available.

Important note: Practically all C-MOS flip-flops are edge-triggered devices, including the *J-K* versions.

5-16 The Quad D Level-Triggered Flip-Flop

One commercial *D* level-triggered flip-flop type should be mentioned, the 7475. It is a special device intended for use as a buffer memory between counters and indicator units. There are two enable lines, one for each pair of flip-flops. These simple devices cannot be cascaded, so they cannot be used for shift register or counter service.

5-17 Other Flip-Flop Considerations

a. When edge-triggered *D* and level-triggered *J-K* devices are used in the same system, a clock line inverter must be provided. The edge-triggered *D* is triggered on the positive going (leading edge) part of the pulse, while the *J-K* transfers final data out on the negative going (trailing edge) part of the pulse. If the two clock signals are not opposites (180° difference), there will be a serious timing difference between their outputs.

b. Because of its ability to accept random (unequally spaced) data, the type *D* is often the choice for the first stage in counters and shift registers. It can be converted into a toggle (type *T*) for counter or frequency divider applications. Figure 5-28 shows the configuration.

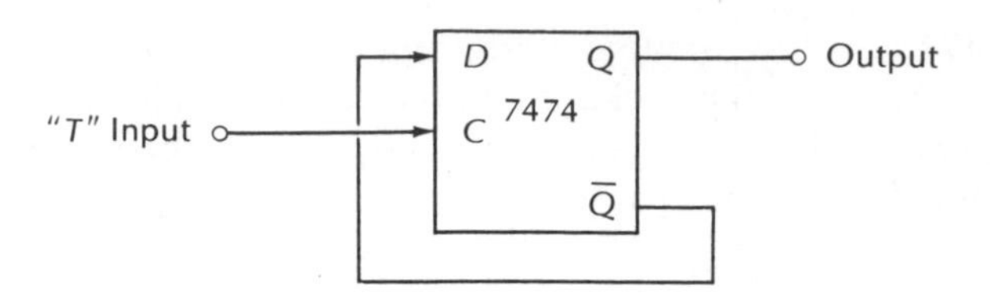

Figure 5-28 Edge-Triggered Type *D* Reconfigured as a Type *T*

c. Common reset (clear) or preset (set) lines should not leave a circuit board or go through a connector without being buffered by a gate or inverter.

d. Common *clear* and *set* lines should never be allowed to float. Noise pick-up can lead to erratic flip-flop operation. If these lines are not connected to the output of another gate, they should be tied +5 V through a pull-up resistor of 330 Ω or less (for TTL).

e. Reset pulses should last for at *least* 10 microseconds to insure the proper reset of all flip-flops on a common reset line.

SUMMARY

Flip-flops are memory elements that can be used to store data, pass it on, or to divide by 2 for counting operations. Unclocked flip-flops are rarely used in practice because they cannot be used to transfer data from one flip-flop to another. The clocked *R-S* flip-flop is used only occasionally as part of a more complex IC circuit.

The *J-K* level-triggered flip-flop and the edge-triggered *D* flip-flop are the two most important flip-flop circuits. Both circuits use the master-slave principle to eliminate timing problems when flip-flops are cascaded.

The TTL edge-triggered *D* is triggered on the positive going edge of the input (clock) pulse. The TTL level-triggered *J-K* is triggered (slave output) on the negative going transition of the input signal (clock). If both types are used in the same system, the two clock signals must generally be 180° out of phase with each other.

The *J-K* is more versatile than the type *D* but less suitable when input signals are unsynchronized or random. It can be converted into a *D* by adding an inverter. The *D* can be modified to toggle for counting applications by connecting the $\overline{Q}$ output back to the *D* input.

Problems

9. Match the following to items (1) through (7) below (more than one answer is possible):
   a. *J-K* master-slave
   b. Type *D*
   c. *R-S* (NAND)
   d. Clocked *R-S* (NAND)
   (1) Triggers on positive transition.
   (2) Triggers on negative transition.
   (3) Edge-triggered.
   (4) Can be used as an *R-S* or type *T* or modified for type *D* service.
   (5) The best choice for unequally spaced input pulses.
   (6) Infrequently used.
   (7) The most popular of the four types.
10. Draw the symbol for each of the following flip-flops:
    a. Type *T*
    b. *R-S*
    c. Clocked *R-S*
    d. Clocked *D*
    e. *J-K*
    f. Edge-triggered *D*
    g. Master-slave *J-K*
11. Examine the frequency divider in Figure 5-19. Now show how it can be constructed using *D* flip-flops.
12. Describe the sequence of events in a master-slave flip-flop.
13. What makes clocking and master-slave arrangements necessary?
14. Find two tri-state output flip-flops in the data manual. List the numbers and types.
15. Figure 5-19 shows a frequency divider using type *T* flip-flops. Draw a diagram of a similar one using *J-K* flip-flops.
16. In your own words, describe the operating sequence of a *J-K* master-slave flip-flop.
17. Under what conditions is the type *D* edge-triggered device preferred to the *J-K* master-slave flip-flop?
18. Using a manufacturer's TTL data manual, look under flip-flops to determine whether edge-triggered *J-K* flip-flops are available. Also, see if you can find any level-triggered *D* latches other than the 7475.
19. Are latches listed separately from flip-flops in the data manual? Read the spec sheets and see if you can find a reason why latches could be listed under a separate heading.
20. Look up IC numbers 7494, 74100, 74109, 7493, 7490 in the TTL data manual and answer the questions below.
    a. What type of circuit is it?
    b. What types of flip-flops are used?
    c. Is clear available?
    d. Is preset available?

e. Is the input positive or negative going for trigger?
f. Is the clock negative going or positive going for clocking?

21. In your own words describe each of the following input configurations:
a. *R-S* b. *J-K* c. *T* d. *D*

22. Draw a diagram showing how a *J-K* M-S can be configured into each of the following:
a. *R-S* b. *T* c. *D*

23. Draw the logic diagram for each of the following:
a. NAND latch
b. NOR latch
c. Basic type *T* flip-flop
d. Clocked NOR latch
e. Clocked NAND latch
f. Master-slave *J-K* flip-flop
g. Edge-triggered type *D*
h. Basic type *D* (clocked) level-triggered flip-flop

CHAPTER 6

NUMBER SYSTEMS AND CODES

Learning Objectives. *Upon completing this chapter you should know:*

1. *How to construct a positional number system using any integral radix.*
2. *How to write binary numbers in decimal form and decimal numbers in binary form.*
3. *How to translate numbers from one radix into another.*
4. *The octal and hexadecimal number systems.*
5. *How to write numbers in the BCD and XS-3 coded number systems.*
6. *How to define parity and explain its purpose.*
7. *The structure of the two most common communications codes.*
8. *The function and applications of code converter circuits.*

Digital machines operate in binary, binary-coded numbers, and binary-related number systems such as octal and hexadecimal.

Familiarity with structure and handling of these machine language number systems is essential to the understanding of computer arithmetic systems.

Translation among number systems and among numbers and codes is accomplished electronically by *encoders* and *decoders*. For human translations, there are several human-oriented methods.

This chapter is intended to provide the student with the necessary understanding of number systems and codes and to provide practice in writing computer language numbers and translating among number systems and number codes. Computer methods for translation will also be examined.

6-1 Number Systems

Although a positional number system can be constructed around any integral radix (base), we will examine only those most frequently encountered in digital systems—radix 10, radix 2, radix 8, and radix 16.

We will examine radix 10, which is so familiar to us that we rarely think about its structure. All other positional number systems have the same basic structure as radix 10. Table 6-1 summarizes the decimal system structure.

Table 6-1 Structure of the Decimal System (Radix 10)

3	2	Radix column 1	0	Position
Thousands	Hundreds	Tens	Units	Name of position
1000	100	10	1	Position value in decimal form
10 × 10 × 10	10 × 10	10	$\frac{10}{10}$	Use of radix to form value of each position
10^3	10^2	10^1	10^0	Position value in exponential form

Allowable weight digits: 0, 1, 2, 3, 4, 5, 6, 7, 8, 9

Note: The highest allowable weight digit is 9, which is the radix number 10 minus 1.

Base-10 weight digits

Position 1 (tens)	Position 0 (units)
0	0
1	1
2	2
3	3
4	4
5	5
6	6
7	7
8	8
9	9

Example

$$\begin{array}{llll} 10^2 & 10^1 & 10^0 & \\ 4 & 2 & 5 & = 4 \times 10^2 + 2 \times 10^1 + 5 \times 10^0 = 400 + 20 + 5 \end{array}$$

THE RADIX POINT

In the decimal system, the term *decimal point* is adequate to describe the dot separating the whole numbers from the fractions. In like manner, the dot used in the binary system to separate the group of whole numbers from the group of fractional numbers is a binary point. When it is necessary to discuss this dot without reference to any particular base, we will use the generic term *radix point*.

THE GENERAL POSITIONAL NUMBER SYSTEM

Positional values are as follows:

Whole numbers	*Units column*
$R^n \cdots R^3R^2R^1$	R^0
Fractions	
$R^{-1}R^{-2}R^{-3} \cdots R^{-n}$	R^0

The range of weight digits is zero through $R - 1$, where R is the radix of the system.

THE STRUCTURE OF THE BINARY SYSTEM

All digital computers on the market use the binary number system. The binary system has the advantage of being the simplest useful positional notation system. It has only two weight digits, 0 and 1. These two symbols may be further symbolized by on-off states in some device. Because of its importance, let us examine the binary system in light of the previous discussion of the positional notation scheme. Table 6-2 shows the structure of the binary system.

Example The subscript identifies the radix and is always written in decimal. Write 43_{10} in binary (radix 2). The position values are as follows:

2^6	2^5	2^4	2^3	2^2	2^1	2^0	Exponent value
64	32	16	8	4	2	1	Decimal value
0	1	0	1	0	1	1	Binary number

$2^6 = 64$. This is larger than 43. We enter a 0 in the 2^6 column.
$2^5 = 32$. This is less than 43. We enter a 1 in the 2^5 column.

Table 6-2 Structure of the Binary System (Radix 2)

4	3	2	1	0	Position
Sixteens	Eights	Fours	Twos	Units	Name of position
16	8	4	2	1	Position value in decimal form
2×2×2×2	2×2×2	2×2	2	$\frac{2}{2}$	Use of radix to form value of each position
2^4	2^3	2^2	2^1	2^0	Position value in exponential form

$2^4 = 16$ and $16 + 32 = 48$. This is larger than 43. We enter a 0 in the 2^4 column.
$2^3 = 8$ and $32 + 8 = 40$. This is less than 43. We enter a 1 in the 2^3 column.
$2^2 = 4$. $40 + 4 = 44$. This is more than 43. We enter a 0 in the 2^2 column.
$2^1 = 2$ and $40 + 2 = 42$. This is less than 43. We enter a 1 in the 2^1 column.
$2^0 = 1$; $42 + 1 = 43$. We enter a 1 in the 2^0 column, and the example is complete.
Thus $43_{10} = 101011$ (binary).

Example Write 101001 in decimal. Set up the following column headings for the binary system:

2^5	2^4	2^3	2^2	2^1	2^0	Exponent value
32	16	8	4	2	1	Decimal value
1	0	1	0	0	1	Binary number

Reading out, we have $32 + 8 + 1 = 41_{10}$.
Thus $101001_2 = 41_{10}$.

Problems

Write the following decimal numbers in binary (Hint: See Table 6-2 and expand as necessary):

1. 22_{10}
2. 69_{10}
3. 100_{10}
4. 37_{10}
5. 144_{10}
6. 41_{10}
7. 128_{10}
8. 36_{10}
9. 8_{10}

Write the following binary numbers in decimal:

10. 1000_2
11. 1001_2
12. 11111_2
13. 11001_2
14. 111010_2
15. 101011_2
16. 001101_2
17. 110101101_2

THE DECIMAL FRACTION

Table 6-3 shows the structure of decimal fractions.

THE STRUCTURE OF THE BINARY FRACTION

Table 6-4 shows the structure of the binary fraction. A comparison of the tables for the decimal system and the binary system shows that the structures are identical except for the radix. We often use radices 8 and 16 because each of these is a power of 2, enabling the machine to still operate in binary. More will be said later in this chapter about these related radices. Table 6-5 shows binary and decimal equivalents.

Table 6-3 Structure of the Decimal Fraction

Position	0	−1	−2	−3
Decimal fraction	1	0.1	0.01	0.001
Exponential form	10^0	10^{-1}	10^{-2}	10^{-3}

Table 6-4 Structure of Binary Fractions

Position	0	−1	−2	−3	−4
Binary fraction	1	0.1	0.01	0.001	0.0001
Use of radix to form value of each position	$\frac{2}{2}$ or $\frac{1}{1}$	$\frac{2}{4}$ or $\frac{1}{2}$	$\frac{2}{8}$ or $\frac{1}{4}$	$\frac{2}{16}$ or $\frac{1}{8}$	$\frac{2}{32}$ or $\frac{1}{16}$
Basic exponential form	$2^1/2^1$	$2^1/2^2$	$2^1/2^3$	$2^1/2^4$	$2^1/2^5$
Dividing through for preferred exponential form	2^0	2^{-1}	2^{-2}	2^{-3}	2^{-4}

Table 6-5 Equivalent Fractions

Binary	Decimal	Common
0.1	0.5	$\frac{1}{2}$
0.01	0.25	$\frac{1}{4}$
0.001	0.125	$\frac{1}{8}$
0.0001	0.0625	$\frac{1}{16}$
0.00001	0.03125	$\frac{1}{32}$

Example Write 0.11_2 as its decimal equivalent. Check by common fractions expressed in decimal digits:

Binary		*Common*		*Decimal*
0.1	=	½	=	0.5
+0.01	=	+¼	=	+0.25
0.11	=	¾	=	0.75

Example Write 0.10101_2 in radix 10.

Binary		*Common*		*Decimal*
0.1	=	½	=	0.5
0.001	=	⅛	=	0.125
+0.00001	=	+1/32	=	0.03125
0.10101	=	21/32	=	0.65625

Problems

Write the following binary fractions in decimal notation (Hint: Extend Table 6-4 as necessary):

18. 0.0001_2
19. 0.1101_2
20. 0.101011_2
21. 0.01110_2
22. 0.11011_2
23. 0.00001_2

THE OCTAL SYSTEM: RADIX 8

Another important number system in digital systems is the octal or radix 8 system. The octal system owes its importance to the fact that it is *related* to radix 2 (binary) in the sense that the radix number 8 can be formed by raising the radix number 2 to the third power ($2^3 = 8$). Table 6-6 shows the structure of the octal system.

Problems

Write the following octal numbers in radix 10:

24. 327_8
25. 71035_8
26. 6437_8
27. 46670_8
28. 001732_8
29. 7734_8

Table 6-6 The Structure of the Octal System (Radix 8)

	4	3	2	1	0	Position
	8^4	8^3	8^2	8^1	8^0	Exponent value
	4096	512	64	8	1	Decimal value
Example:	2	1	3	6	7	Number in radix 8

$8192 + 512 + 192 + 48 + 7 = 8951_{10}$

Weight digits: 0, 1, 2, 3, 4, 5, 6, 7

Example:

$$21367_8 = 2 \times 8^4 + 1 \times 8^3 + 3 \times 8^2 + 6 \times 8^1 + 7 \times 8^0$$
$$= 2 \times 4096 + 1 \times 512 + 3 \times 64 + 6 \times 8 + 7 \times 1$$
$$= 8192 + 512 + 192 + 48 + 7 = 8951_{10}$$

THE HEXADECIMAL SYSTEM (RADIX 16)

The hexadecimal system is also related to binary in the sense that the radix number 16 can be formed by raising the binary radix number (2) to the fourth power ($2^4 = 16$).

In the hexadecimal system 16 weight digits are required. We can take digit symbols 0-9 from the decimal system, but there are no *single-position* symbols for the equivalents of decimals 10, 11, 12, 13, 14, and 15. Since practical considerations dictate the use of symbols that are common in printer readouts and the like, the first six letters of the English alphabet have been almost universally adopted for the purpose. Table 6-7 shows the structure of the hexadecimal system and the extra weight digits.

Problems

Write the following hexadecimal numbers in decimal:

30. $A69_{16}$ 32. $429C_{16}$ 34. $AF32_{16}$
31. $BFA2_{16}$ 33. $2BC_{16}$ 35. 4679_{16}

6-2 Binary-Related Radices

In this section we will examine the translations among octal, hexadecimal, and binary. The procedure for translating from one radix into another related one is quite simple.

Table 6-7 Additional Symbols for Radix 16

Decimal equivalent	Symbol
10	A
11	B
12	C
13	D
14	E
15	F

Structure of the Hexadecimal (Radix-16) System

	3	2	1	0	Position
	16^3	16^2	16^1	16^0	Exponential form
	4096	256	16	1	Decimal value
Example:	0	B	A	3	Number in radix 16
	0	+ 2816 +	160 +	3	= 2979_{10}

TRANSLATION FROM BINARY INTO OCTAL

The procedure is to separate the binary number into groups of three digits and then simply translate each group into its octal equivalent, group by group. The resulting number is the octal equivalent of the original binary number.

The largest decimal digit which can be formed by one binary group (three binary digits) is 111, or decimal 7. The digits 0 through 7 in the units column have exactly the same values in both decimal and octal notation. Thus, if we treat each group of bits as a separate units column, we can translate each group into its decimal equivalent and still have an octal number.

Table 6-8 illustrates how direct translation between binary numbers and numbers written in related radices may be accomplished. The binary number 10110 (decimal 22) is used as an example. The first three rows illustrate a general rule for correct grouping. The last two provide a specific example. The grouping procedure starts at the radix point and proceeds to the left for whole numbers and to the right for fractions. An incomplete group should be completed by adding the appropriate number of zeros. Complete groups are a must in translating from a higher radix number to a lower radix number. It is good practice always to work with complete groups.

Example Translate 10110_2 into radix 8 (octal). Grouping the binary number and translating into the radix 10 equivalent of each group, we get:

010	110	Radix 2
2	6	Radix 8

Thus $10110_2 = 26_8$. Note: $26_8 = 22_{10}$.

Table 6-8 Binary-Related Radix Translation

	Radix 2	Radix 8	Radix 16
Power of 2	2^1	2^3	2^4
Number of digits per group	1	3	4
Example: 22_{10} =	010110_2	26_8	16_{16}
Binary equivalent (grouped)	010110	2 6 010 110	1 6 0001 0110

TRANSLATING FROM OCTAL INTO BINARY

Translation from octal into binary is accomplished by writing each octal digit as a three-bit binary number and combining the groups.

Example Translate 010011_2 into octal. Since $2^3 = 8$, the binary number is divided into groups of three. The octal equivalent of each group is

010	011	Radix 2
2	3	Radix 8

Thus $10011_2 = 23_8$.

Example Translate 26_8 into radix 2. The number to be translated is separated into the three-bit binary equivalent of each octal digit.

2	6	Radix 8
010	110	Radix 2

Combining the groups, we have 010110. Thus $26_8 = 010110_2$ or 22_{10}.

TRANSLATING FROM HEXADECIMAL INTO BINARY AND VICE VERSA

The following examples will illustrate the procedure:

Example Translate $24C_{16}$ into its equivalent in radix 2. $2^4 = 16$; therefore there will be four bits (binary digits) per group. Writing each hexadecimal digit as a four-bit binary number, we get:

2	4	C	Radix 16
0010	0100	1100	Radix 2

Combining groups yields $001\ 001\ 001\ 100_2$. Thus $24C_{16} = 001\ 001\ 001\ 100_2$.

Example Translate $00\ 010\ 011_2$ into hexadecimal. Since $2^4 = 16$, we divide the binary number into groups of four. Translating each group into hexadecimal, we get:

0001	0011	Radix 2
1	3	Radix 16

Thus $00010011_2 = 13_{16}$.

Example Translate $110\ 001_2$ into hexadecimal.

0011	0001	Radix 2
3	1	Radix 16

Thus $110\ 001_2 = 31_{16}$.

Problems

Translate the following numbers into binary.

36. 134_8
37. 136_8
38. $14C_{16}$
39. 4667_8
40. 1233210_8
41. $9B_{16}$
42. 9767610_{16}
43. 3214_8

Translate from binary into the indicated radix.

44. 001011101 into radix 8
45. 1011101 into radix 4
46. 111110110101 into radix 16
47. 101011010 into octal
48. 11101011111 into octal
49. 011010000 into hexadecimal

SHORTCUT TRANSLATION METHODS

Translation from Radices 16, 8, 2 into Radix 10

The procedure presented here is valid for translating from *any* radix into radix 10, but we will use only the most common radices (16, 8, 2) as examples. To translate a whole number in any radix into radix 10, we proceed as follows:

1. Multiply the most significant digit (MSD) for the given number by the given radix, and add the next digit.
2. Multiply this result by the given radix, and add the next digit.
3. Repeat the process until every digit in the given number has been used.

The final result will be the translated number in radix 10. The simplified form of this process is as follows:

1. MSD of number × radix + next digit = result
2. Previous result × radix + next digit = result
3. Repeat until all digits are used.

Example Translate 101010_2 into radix 10.

$$
\begin{aligned}
(0 \times 2) + 1 \ldots\ldots\ldots\ldots &= 1 \\
(1 \times 2) + .0 \ldots\ldots\ldots &= 2 \\
(2 \times 2) + ..1 \ldots\ldots\ldots &= 5 \\
(5 \times 2) + \ldots 0 \ldots\ldots &= 10 \\
(10 \times 2) + \ldots.1 \ldots\ldots &= 21 \\
(21 \times 2) + \ldots..0 \ldots.. &= 42_{10}
\end{aligned}
$$

(previous result) × (radix) + (next digit) = result

The same method used in the above example can be used for the following example. The tabular form used in the two previous examples was chosen because it best illustrates the method. A more practical form can be mastered with a little practice. The method is the same; only the form is changed.

Example Translate 101010_2 into radix 10. Note that for easy comparison the number used here is the same as that in the above example. From the simplified rule (for whole numbers only)

(MSD of number or previous result) × (radix) + (next digit) . . .
= result in radix 10

MSD					LSD	
1	0	1	0	1	0	Number in radix 2
1	2	5	10	21	42	Result in radix 10

If no confusion results, it is permissible to omit the zero at the beginning of the number in R_2.

Example

Translate 247_8 into radix 10.

2	4	7_8
2	20	167_{10}

Answer: $247_8 = 167_{10}$

Example Translate $5BF_{16}$ into radix 10.

5	B	F
5	91	1471

Answer: $5BF_{16} = 1471_{10}$

When the translation is from binary to decimal, it is often more convenient to translate by grouping (related radices), translating R_2 into R_8 or R_{16}, and then, using the method above, translate from R_8 or R_{16} into decimal.

Problems

Translate the following into radix 10.

50. 237_8	52. 370_8	54. 429_{16}
51. 665_8	53. $AB3_{16}$	55. $FF3_{16}$

TRANSLATION FROM RADIX 10 INTO RADICES 2, 8, 16

The following procedure is valid for translating numbers written in radix 10 into their numerical equivalents in *any* radix. Again, the examples will deal only with the most commonly encountered radices (2, 8, and 16).

Process

(*Note:* The following procedure applies to whole numbers only.)

1. Divide the given radix 10 number by the number of the radix into which you wish to translate.
2. If there is no remainder, record a 0 in the remainder row (see the examples). If there is a remainder, record that remainder in the remainder row.
3. Divide the quotient from the preceding step by the radix number and record any remainder.
4. Continue the process until the quotient becomes 0.

Example Translate 96_{10} into binary. (*Note:* The operations in the example proceed from right to left. If the process is carried out from left to right, the result will read out in reverse order.)

```
  0
2)1
          1
  0     2)3      3
  1       2    2)6       6
  .       1      6    2)12     12
  .       .      0      12   2)24     24
  .       .      .       0     24   2)48     48
  .       .      .       .      0     48   2)96   ←start
  .       .      .       .      .      0     96
  .       .      .       .      .      .      0
  .       .      .       .      .      .      .
  .       .      .       .      .      .      .
  .       .      .       .      .      .      .
  .       .      .       .      .      .      .
  .       .      .       .      .      .      .
  .       .      .       .      .      .      .    remainder
  1       1      0       0      0      0      0   ←row
 MSD                                         LSD
```

Answer: $96_{10} = 1100000_2$

Note: MSD = most significant digit
LSD = least significant digit

Example Translate 22_{10} into octal.

$$\begin{array}{ccl} \overset{0}{8\overline{)2}} & & \\ & \overset{2}{8\overline{)22}} & \text{start} \\ \cdot & \underline{16} & \\ \cdot & 6 & \\ \cdot & \cdot & \\ \cdot & \cdot & \\ \cdot & \cdot & \\ 2 & 6 & \text{remainder row} \\ \text{MSD} & \text{LSD} & \end{array}$$

Answer: $22_{10} = 26_8$

Example Translate 150_{10} into hexadecimal.

$$\begin{array}{ccl} \overset{0}{16\overline{)9}} & \overset{9}{16\overline{)150}} & \text{start} \\ \cdot & \underline{144} & \\ \cdot & 6 & \\ \cdot & \cdot & \\ \cdot & \cdot & \\ \cdot & \cdot & \\ 9 & 6 & \text{remainder row} \end{array}$$

Answer: $150_{10} = 96_{16}$

Problems

Translate the following radix 10 numbers into binary.

56. 75	58. 76	60. 33
57. 197	59. 225	61. 44

Translate the following radix 10 numbers into octal.

62. 75	64. 76	66. 27
63. 197	65. 225	67. 100

Translate the following radix 10 numbers into hexadecimal.

68. 75	70. 76	72. 4390
69. 197	71. 225	73. 1975

6-3 Coded Number Systems

There are a number of numeric and several alphanumeric codes. Alphabet symbols are encoded into bits, as are special symbols and instructions such as carriage return (for an input-output typewriter, for example). First let us examine numeric codes and their properties.

THE BINARY-CODED DECIMAL SYSTEM

In a BCD number system the original positional decimal structure is retained, but each decimal digit is represented by a four-bit number:

9	5	4	Decimal number
1001	0101	0100	BCD equivalent

Four binary digits is the minimum number by which all decimal digits 0 to 9 can be represented. Table 6-9 shows the structure of the BCD system.

Example Write 234_{10} in BCD.

2	3	4	Decimal
0010	0011	0100	BCD

Further examples are shown in Table 6-10.

Problems

Write the following decimal numbers in BCD:
74. 1234 75. 8765 76. 98910 77. 5809

The BCD code we have just examined is the *natural* BCD (NBCD) code and is only one of several BCD codes in use. The reason for having other codes lies in certain properties, such as ease in checking errors or in complementing the code for subtraction, a process we will

Table 6-9 Structure of the BCD System

3	2	1	0	Position
Thousands	Hundreds	Tens	Units	Position value
$2^3\ 2^2\ 2^1\ 2^0$	$2^3\ 2^2\ 2^1\ 2^0$	$2^3\ 2^2\ 2^1\ 2^0$	$2^3\ 2^2\ 2^1\ 2^0$	BCD value in exponential form

Table 6-10 Decimal Numbers and Their BCD Equivalents

Decimal numbers	BCD equivalent
001	0000 0000 0001
123	0001 0010 0011
546	0101 0100 0110
879	1000 0111 1001

examine in the chapter on computer arithmetic circuits. Each of the codes has advantages and disadvantages. The selection of a code involves choosing the one with the properties most desired for a particular application.

All basic BCD codes use only 10 of the 16 possible combinations of 4 bits since they must represent only digits 0–9. The unused combinations can be treated as error combinations because they are not a part of the code.

ERROR-DETECTOR CIRCUIT FOR NBCD

Here we will develop a circuit to detect those combinations that are not part of the NBCD code. This same circuit is used in BCD adders to detect when a correction must be made to the sum. We will look at that application in the chapter on arithmetic circuits.

The error-detector circuit must detect those combinations not used in the NBCD code. These are the combinations of four bits in binary that represent the binary equivalents of decimal digits 10, 11, 12, 13, 14, and 15. Table 6-11 is the truth table for the NBCD error detector.

Plotting the problem on the truth table (see Table 6-11) and writing the type-one minterm equation, we get:

$$f = (A \cdot \overline{B} \cdot C \cdot \overline{D}) + (A \cdot \overline{B} \cdot C \cdot D) + (A \cdot B \cdot \overline{C} \cdot \overline{D}) + (A \cdot B \cdot \overline{C} \cdot D) + (A \cdot B \cdot C \cdot \overline{D}) + (A \cdot B \cdot C \cdot D)$$

Plotting the equation on the Karnaugh map (Table 6-12) and reading out the simplified equation from loop a, we get $A \cdot C$ and from loop b we get $A \cdot B$. Combining the terms we get $f = (A \cdot B) + (A \cdot C)$ as the final simplified equation. The logic diagram is shown in Figure 6-1.

THE EXCESS-3 CODE (XS-3)

The excess-3 code is another form of BCD code. It differs from natural BCD in that the 10 combinations required to encode the decimal digits (0–9) are taken from the middle of the table starting with the binary equivalent of decimal 3. Table 6-13 shows the decimal digits 0–9 and their excess-3 equivalents.

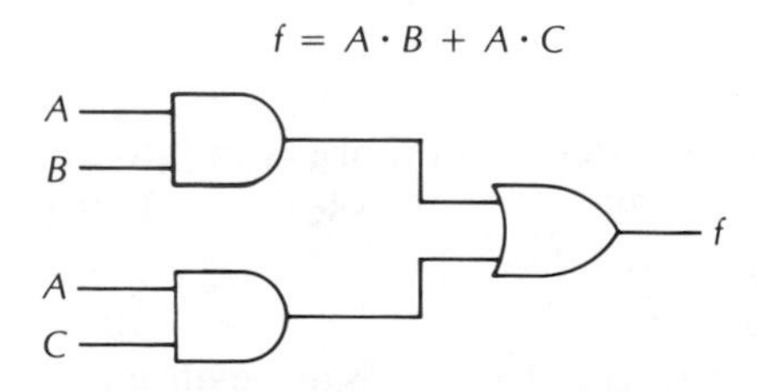

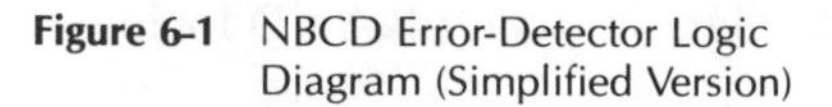

Figure 6-1 NBCD Error-Detector Logic Diagram (Simplified Version)

Table 6-11 Truth Table for the NBCD Error Detector

	2^3 A	2^2 B	2^1 C	2^0 D	f
0	0	0	0	0	0
1	0	0	0	1	0
2	0	0	1	0	0
3	0	0	1	1	0
4	0	1	0	0	0
5	0	1	0	1	0
6	0	1	1	0	0
7	0	1	1	1	0
8	1	0	0	0	0
9	1	0	0	1	0
10	1	0	1	0	1
11	1	0	1	1	1
12	1	1	0	0	1
13	1	1	0	1	1
14	1	1	1	0	1
15	1	1	1	1	1

Table 6-12 Karnaugh Map for the NBCD Error Detector

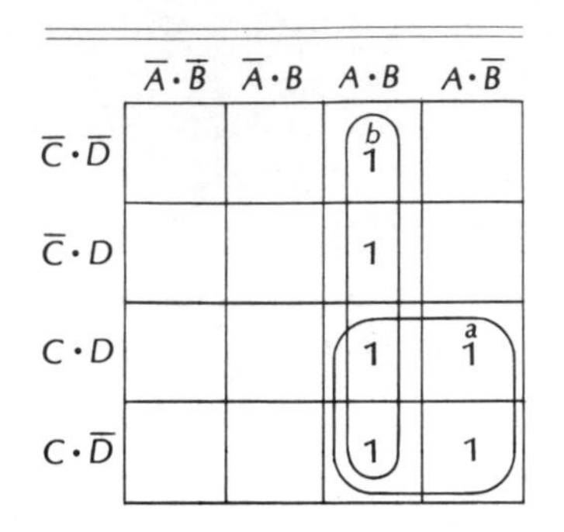

	$\overline{A}\cdot\overline{B}$	$\overline{A}\cdot B$	$A\cdot B$	$A\cdot\overline{B}$
$\overline{C}\cdot\overline{D}$			1 (b)	
$\overline{C}\cdot D$			1	
$C\cdot D$			1	1 (a)
$C\cdot\overline{D}$			1	1

Examples

1. 425_{10} = 0111 0101 1000_{XS-3}
2. 379_{10} = 0110 1010 1100_{XS-3}
3. 146_{10} = 0100 0111 1001_{XS-3}
4. 80_{10} = 1011 0011_{XS-3}

Problems

78. Write the following decimal numbers in XS-3 code.
    a. 6421 b. 5790 c. 386 d. 9700
79. Design an XS-3 error detector.
    a. Make a truth table.
    b. Simplify if possible, using a Karnaugh map.
    c. Draw a logic diagram using NAND logic.

Table 6-13 Decimal Equivalents in the XS-3 Code

Decimal number	Excess−3 equivalent
0	0011
1	0100
2	0101
3	0110
4	0111
5	1000
6	1001
7	1010
8	1011
9	1100

THE BINARY-CODED OCTAL SYSTEM

Because radix 2 and radix 8 are related, we can form a coded structure similar to BCD but composed of binary-coded octal (BCO) groups.

Table 6-14 shows the BCO structure and equivalent weight digits.

Examples

$$147_{10} = 001 \quad 100 \quad 111_{BCO}$$
$$760_{10} = 111 \quad 110 \quad 000_{BCO}$$
$$235_{10} = 010 \quad 011 \quad 101_{BCO}$$

Problem

80. Write the following BCO numbers in decimal.
    a. 001 010 111 110_{BCO}
    b. 110 101 100 000_{BCO}

SUMMARY OF NATURAL CODES AND VARIATIONS

Two of the most important natural codes are binary-coded octal and binary-coded decimal (and the variations on these two). These are called *natural* codes because the column headings are the natural radix 2 headings. The excess-3 code (sometimes abbreviated XS-3) is a BCD scheme that has been modified to make it possible to generate the complement for subtraction by simple inversion of 0's and 1's.

Table 6-14 The BCO Structure and Equivalent Weight Digits

	3	2	1	0	Position
	512	64	8	1	Position value
	8^3 $2^2 2^1 2^0$	8^2 $2^2 2^1 2^0$	8^1 $2^2 2^1 2^0$	8^0 $2^2 2^1 2^0$	BCO value in exponential form
Example	000	001	011	101	Number in BCO
	0	1	3	5	Octal equivalent

a. Structure of the BCO system

Octal digit	BCO Equivalents
0	000
1	001
2	010
3	011
4	100
5	101
6	110
7	111

b. Octal weight digits and their BCO equivalents

The hexadecimal (radix 16) system is also popular in modern computers. Hexadecimal numbers are more compatible with human communications than straight binary numbers and, since this system is a *relative* of radix 2, it is also compatible with the computer's basic binary language. The computer operates in straight binary, but the hexadecimal grouping is more convenient for humans to deal with.

Because arithmetic is performed in straight binary without the need for correction circuits, hexadecimal numbers can be complemented for subtraction by simply changing all 0's to 1's and all 1's to 0's. The translation from decimal into hexadecimal is fairly easy to mechanize.

Many computers are organized into dual hexadecimal groups. Each hexadecimal group consists of four binary bits and each pair of groups (eight bits) is called a *byte*. (A group of four bits is called a *nybble*.) Such an organization allows considerable flexibility in that the programmer has the option of treating a byte as either a single eight-bit character or as two BCD digits. Within the range of eight bits there is adequate room to encode alphabetic information, special symbols, a sign bit, and extra bits for error-detecting purposes.

WEIGHTED CODES

Weighted codes are not generally used as a computer internal language. Internal computation is performed in straight binary, octal, or hexadecimal. Some pocket calculators operate in BCD as an internal language, but they are an exception to the general rule. Weighted codes are used primarily in conjunction with input-output devices and numerical displays. These codes are similar to the natural codes in that they follow a sort of positional notation structure, but they differ in that the position values are not ascending powers of some base. In the weighted codes the position values are quite arbitrary; they may be anything the code designer desires.

An example of a weighted code is the four-bit binary code called the 2, 4, 2, 1 code. Table 6-15 illustrates the similarities and differences between it and a natural four-bit BCD system and shows the binary code group 1001 translated from each code into its corresponding decimal value.

Table 6-15 Comparison of Natural BCD and Weighted 2,4,2,1 Codes

Natural BCD

	2^3 2^2 2^1 2^0	Exponent values
	8 4 2 1	Decimal position value
	1 0 0 1	Binary number (code group)
$(1 \times 8) + (0 \times 4) + (0 \times 2) + (1 \times 1) = 9_{10}$		

Weighted 2, 4, 2, 1

	None	Exponent values
	2 4 2 1	Decimal position value
	1 0 0 1	Binary number (code group)
$(1 \times 2) + (0 \times 4) + (0 \times 2) + (1 \times 1) = 3_{10}$		

Another popular weighted code is the 6, 4, 2, 1 code. Any four-bit BCD system, weighted or natural, has some built-in self-checking properties because there are always six unused combinations. The occurrence of one of these forbidden combinations indicates an error. Unfortunately, these forbidden combinations do not provide an optimum error-detecting situation. To provide more nearly optimum error-detecting facility, five-bit codes of two distinct types have been adopted. In one case a fifth bit is added to an existing four-bit code so that each five-bit group always contains either an odd number (odd parity) or an even number (even parity) of 1's. The added bit is called a *parity bit*. Both odd and even parity codes are used. In the second type of self-checking code a constant number of 1's is built into the structure of the code.

Table 6-16 shows an odd-parity natural BCD code, an even-parity natural BCD code, and a structured self-checking code. This structured code is a weighted code with position values 7, 4, 2, 1, 0. It is commonly called the *two-out-of-five code* because all valid combinations contain two 1's and three 0's.

THE BI-QUINARY SYSTEM

The bi-quinary system illustrates an important class of number codes in which two digits are required to indicate the value in each position. In the quinary portion of the two-digit group, the weight digits 0, 1, 2, 3, 4 may represent either the corresponding decimal digits or decimal digits 5 to 9, depending on the digit in the *bi* portion. A 0 in the bi portion means that the quinary digit represents decimals 0 to 4, and a 1 in the bi portion means

Table 6-16 Examples of Error-Detecting Codes

Decimal number	Odd-parity natural BCD	Even-parity natural BCD	Structured weighted code
	P8421	P8421	74210
0	10000	00000	11000
1	00001	10001	00011
2	00010	10010	00101
3	10011	00011	00110
4	00100	10100	01001
5	10101	00101	01010
6	10110	00110	01100
7	00111	10111	10001
8	01000	11000	10010
9	11001	01001	10100

that the quinary digit represents decimals 5 to 9. Thus decimal 75 written in bi-quinary is

Bi	*Quinary*	*Bi*	*Quinary*	
1	2	1	0	Bi-quinary number
	7		5	Decimal equivalent

Table 6-17 shows the bi-quinary structure.

A weighted seven-bit code can be derived from the bi-quinary system. This code, with position values 5, 0 and 4, 3, 2, 1, 0, is illustrated in Table 6-18.

The bi-quinary code is another example of a code with an intrinsic error-detection capability. It is a two-out-of-seven code in which the representation of each decimal digit contains a single 1 in the bi group and a single 1 in the quinary group. There is a similarly constructed code called the qui-binary code, which has position values 8, 6, 4, 2, 0 and 1, 0.

Table 6-19 shows three additional codes that have distinctive *patterns* that make error checking relatively easy and certain.

6-4 Parity Checkers and Generators

Exclusive-OR gates can be used to check for parity. The circuit is a modulo 2 adder, which is the same as a binary adder except that no carries are generated. An even number of *ones* added together always yields a sum function of zero while an odd number of *ones* added together produces a sum function of 1. Because the exclusive-OR truth

Table 6-17 Decimal and Bi-Quinary Equivalents

Decimal	Bi	Quinary
0	0	0
1	0	1
2	0	2
3	0	3
4	0	4
5	1	0
6	1	1
7	1	2
8	1	3
9	1	4

Table 6-18 Bi-Quinary Code

Decimal number	Bi	Quinary
	50	43210
0	01	00001
1	01	00010
2	01	00100
3	01	01000
4	01	10000
5	10	00001
6	10	00010
7	10	00100
8	10	01000
9	10	10000

Table 6-19 Some Additional Codes

Decimal	51111	Shift-counter	Ring-counter 9876543210
0	00000	00000	0000000001
1	00001	00001	0000000010
2	00011	00011	0000000100
3	00111	00111	0000001000
4	01111	01111	0000010000
5	10000	11111	0000100000
6	11000	11110	0001000000
7.	11100	11100	0010000000
8	11110	11000	0100000000
9	11111	10000	1000000000

table is the same as the truth table for binary addition, it makes an ideal adder element for the parity detector. The parity detector can also be used as a parity generator by arranging it so that the existing parity is determined and the appropriate output from the detector is used to add a parity bit as needed.

Figure 6-2 shows a four-bit parity detector. Figure 6-3 shows a nine-bit (eight bits plus a parity bit) parity detector generator. The appropriate output is applied to the ninth bit position to achieve parity.

THE GRAY CODE

In machine and process control it is often necessary to translate the angular position of a shaft into digital information. The most common method for accomplishing this task is the use of an optically read binary code disc mounted on the shaft in question. Figure 6-4 shows the arrangement and the code disc layout for Gray code representation.

The Gray code is used because only one bit in the group changes at a

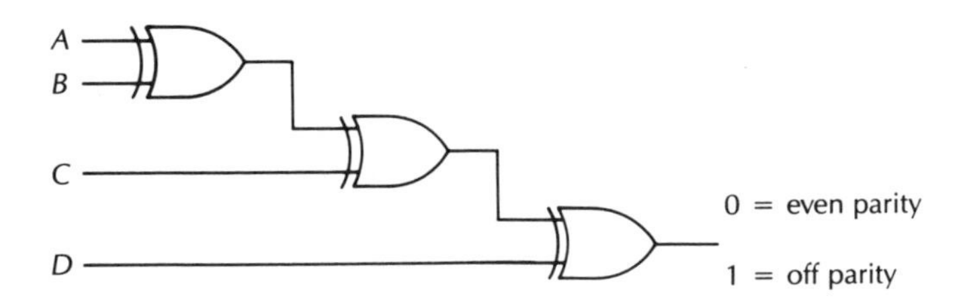

Figure 6-2 Four-Bit Parity Detector

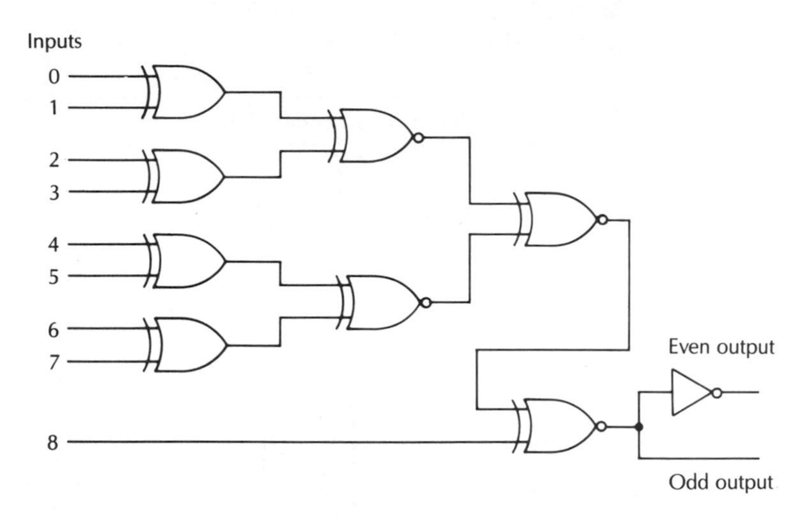

Figure 6-3 Nine-Bit Parity Detector-Generator

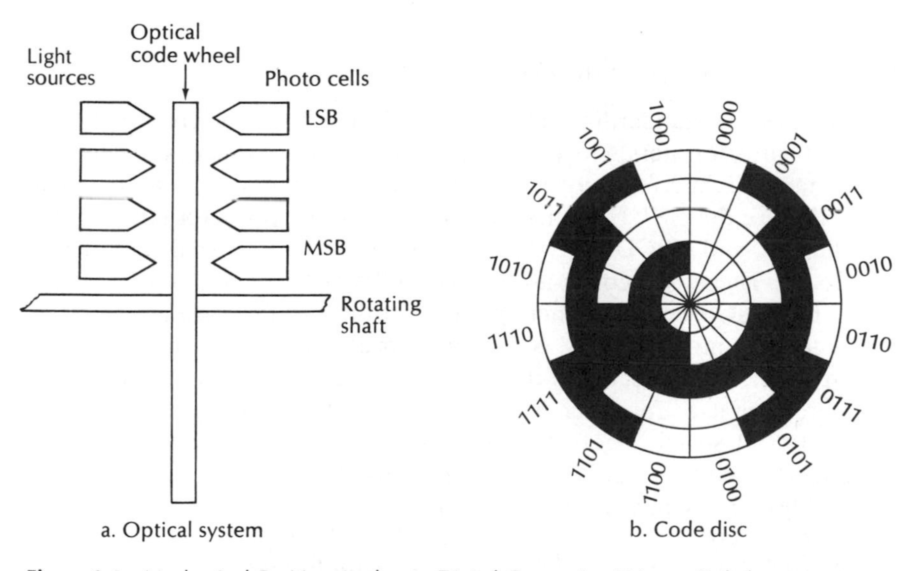

Figure 6-4 Mechanical Position Analog-to-Digital Conversion Using a Coded Optical Disc

time, which improves resolution and minimizes conversion error. The Gray code is not suitable for processing purposes, and data are normally translated into normal binary as an integral part of the conversion process. Table 6-20 details the Gray code.

Table 6-20 Gray Code

Decimal	Gray code
0	0000
1	0001
2	0011
3	0010
4	0110
5	0111
6	0101
7	0100
8	1100
9	1101
10	1111
11	1110
12	1010
13	1011
14	1001
15	1000

6-5 Computer Information Codes

In order to standardize interface hardware between computers and keyboards, printers, video displays, and the like, it is necessary to have a standard interface code. The two most common communications codes are the ASCII (American Standard Code for Information Interchange), and EBCDIC (Extended Binary-Coded-Decimal Interchange Code). These two codes are shown in Tables 6-21 and 6-22.

The ASCII code, the more popular of the two for most applications, is divided into word sets. Sixty-four words are used for the upper case alphabet, numbers, often-used punctuation, and a blank. Thirty-two words are used for machine commands such as carriage return, stop, start, and so on. These "words" do not appear in printouts or displays. Thirty-two additional words are used for the lowercase alphabet and less frequently used punctuation.

In this code, seven bits are required for the 128 ASCII code words. An eighth bit is provided for use as a parity bit, which may be used or left blank. The code may be transmitted one bit at a time, in serial, or all bits simultaneously in parallel. The EBCDIC code is similarly structured but not directly interchangeable with ASCII.

Although the IBM Selectric® data typewriter code is based upon the EBCDIC code, a number of small variations from the standard code can be found in commercial equipment. These small variations can sometimes be of minor importance, but in other cases they create such serious problems that the slightly modified code must be treated as an

Table 6-21 American Standard Code for Information Interchange (ASCII)

	Column	0	1	2	3	4	5	6	7
Row	Bits → 765 ↓ 4321	000	001	010	011	100	101	110	111
0	0000	NUL	DLE	SP	0	@	P	\	p
1	0001	SOH	DC1	!	1	A	Q	a	q
2	0010	STX	DC2	"	2	B	R	b	r
3	0011	ETX	DC3	#	3	C	S	c	s
4	0100	EOT	DC4	$	4	D	T	d	t
5	0101	ENQ	NAK	%	5	E	U	e	u
6	0110	ACK	SYN	&	6	F	V	f	v
7	0111	BEL	ETB	'	7	G	W	g	w
8	1000	BS	CAN	(	8	H	X	h	x
9	1001	HT	EM	)	9	I	Y	i	y
10	1010	LF	SUB	*	:	J	Z	j	z
11	1011	VT	ESC	+	;	K	[	k	{
12	1100	FF	FS	,	<	L	\	l	¦
13	1101	CR	GS	−	=	M	]	m	}
14	1110	SO	RS	.	>	N	⌢	n	~
15	1111	SI	US	/	?	O	—	o	DEL

Example: { bits : 7 6 5 4 3 2 1
Code A = 1 0 0 0 0 0 1 }

a. Table of codes

NUL	Null	DLE	Data link escape
SOH	Start of heading	DC1	Device control 1
STX	Start of text	DC2	Device control 2
ETX	End of text	DC3	Device control 3
EOT	End of transmission	DC4	Device control 4
ENQ	Enquiry	NAK	Negative acknowledge
ACK	Acknowledge	SYN	Synchronous idle
BEL	Bell (audible signal)	ETB	End of transmission block
BS	Backspace	CAN	Cancel
HT	Horizontal tabulation (punched card skip)	EM	End of medium
LF	Line feed	SUB	Substitute
VT	Vertical tabulation	ESC	Escape
FF	Form feed	FS	File separator
CR	Carriage return	GS	Group separator
SO	Shift out	RS	Record separator
SI	Shift in	US	Unit separator

b. Legend for control codes in columns 0 and 1

Table 6-22 Extended Binary-Coded-Decimal Interchange

Code (EBCDIC)

*Decimal	Row (HEX)	Bits 0123 → / 4567 ↓	0 0000	1 0001	2 0010	3 0011	4 0100	5 0101	6 0110	7 0111	8 1000	9 1001	A 1010 (*10)	B 1011 (11)	C 1100 (12)	D 1101 (13)	E 1110 (14)	F 1111 (15)
	0	0000	NUL	DLE	DS		SP	&	–									0
	1	0001	SOH	DC1	SOS						a	j			A	J		1
	2	0010	STX	DC2	FS	SYN					b	k	s		B	K	S	2
	3	0011	ETX	DC3							c	l	t		C	L	T	3
	4	0100	PF	RES	BYP	PN					d	m	u		D	M	U	4
	5	0101	HT	NL	LF	RS					e	n	v		E	N	V	5
	6	0110	LC	BS	EOB ETB	UC					f	o	w		F	O	W	6
	7	0111	DEL	IL	PRE ESC	EOT					g	p	x		G	P	X	7
	8	1000		CAN							h	q	y		H	Q	Y	8
	9	1001	RLF	EM							i	r	z		I	R	Z	9
*10	A	1010	SMM	CC	SM			!		:								
11	B	1011	VT				.	$		#								
12	C	1100	FF	IFS		DC4		*	%	@								
13	D	1101	CR	IGS	ENQ	NAK	(	)	—	'								
14	E	1110	SO	IRS	ACK		+	;		=								
15	F	1111	SI	IUS	BEL	SUB			?	"								

Example: - - - - - - - - bits: 0 1 2 3 4 5 6 7
Code for letter D = 1 1 0 0 0 1 0 0
column row

a. Table of codes

Code	Meaning
NUL	Null
SOH	Start of heading
STX	Start of text
ETX	End of text
PF	Punch off
HT	Horizontal tab
LC	Lower case
DEL	Delete
RLF	Reverse line feed
SMM	Start of manual message
VT	Vertical tabulation
FF	Form feed
CR	Carriage return
SO	Shift out
SI	Shift in
DLE	Data link escape
DC1	Device control 1
DC2	Device control 2
DC3	Device control 3
RES	Restore
NL	New line
BS	Backspace
IL	Idle
CAN	Cancel
EM	End of medium

Code	Meaning
CC	Cursor control
IFS	Interchange file separator
IGS	Interchange group separator
IRS	Interchange record separator
IUS	Interchange unit separator
DS	Digit select
SOS	Start of significance
FS	Field separator
BYP	Bypass
LF	Line feed
EOB/ETB	End of block/End of transmission block
PRE/ESC	Prefix/escape
SM	Set mode
ENQ	Enquiry
ACK	Acknowledge
BEL	Bell
SYN	Synchronous idle
PN	Punch on
RS	Reader stop
UC	Upper case
EOT	End of transmission
DC4	Device control 4
NAK	Negative acknowledge
SUB	Substitute
SP	Space

b. Machine control codes for columns 0–4

entirely new code. Integrated circuits for translation from ASCII into EBCDIC and from EBCDIC to ASCII are available. The most available code conversion IC's normally translate only the capital letters, numbers, and symbols that fit within those code groups. Most of the code variations in EBCDIC occur in the lowercase and machine command sections of the code.

THE BAUDOT CODE

The Baudot code is a five-bit code used in older model teletype machines. Because so many of these machines are still in operation, code translation chips are available for translating between Baudot and either ASCII or EBCDIC. Table 6-23 shows the Baudot code.

6-6 Code Converter Circuits

It is often necessary to convert from one code into another within a digital system. For example, the Gray code is ideal for providing a

Table 6-23 Five-Bit Baudot Teletype Code

Code	Letters	Figures	Code	Letters	Figures
00000	blank	blank	10000	T	5
00001	E	3	10001	Z	+
00010	linefeed	linefeed	10010	L	)
00011	A		10011	W	2
00100	space	space	10100	H	#
00101	S	'	10101	Y	6
00110	I	8	10110	P	Ø
00111	U	7	10111	Q	1
01000	car ret.	car ret.	11000	O	9
01001	D	acknowledge	11001	B	?
01010	R	4	11010	G	&
01011	J	bell	11011	Figs.	Figs.
01100	N	,	11100	M	.
01001	F	!	11101	X	/
01110	C	:	11110	V	=
01111	K	(	11111	Letters	Letters

digital representation of mechanical shaft position, but it is not a convenient code when calculations or counting operations must be performed by common electronic digital circuits. The most reasonable approach in this case and many others is to convert the code into ordinary binary or some related binary code.

Code conversions can sometimes be accomplished with fairly simple hardware. A case in point is the Gray code-to-BCD converters in Figure 6-5. Conversion from BCD to excess-3 is only slightly more complex than the BCD-to-Gray converter in Figure 6-5. The basic BCD/excess-3 converters are shown in Figure 6-6. Most practical code converters involve several decades or a fair number of binary digits. The result is that the basic converter circuit must be repeated a number of times, which makes the total circuit fairly complex.

Most real code conversion circuits are complex enough to demand

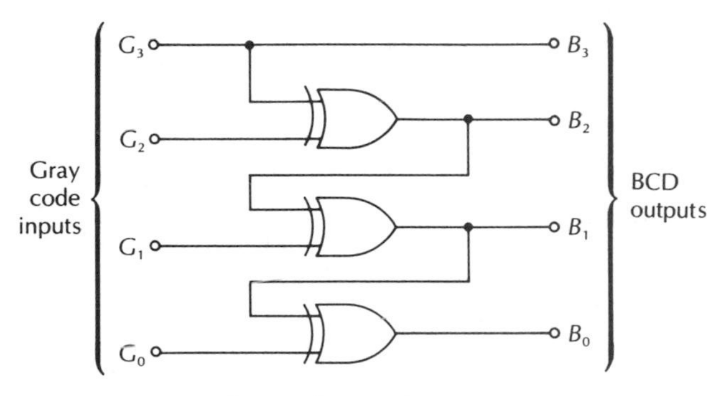

a. Gray code-to-BCD converter

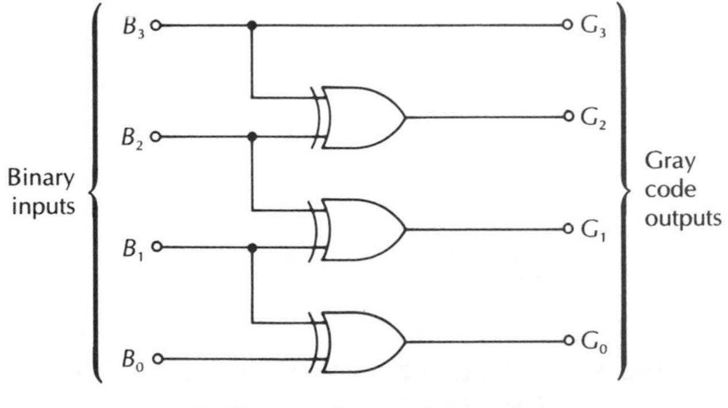

b. Binary-to-Gray code converter

Figure 6-5 Gray Code-to-BCD and BCD-to-Gray Code Converters

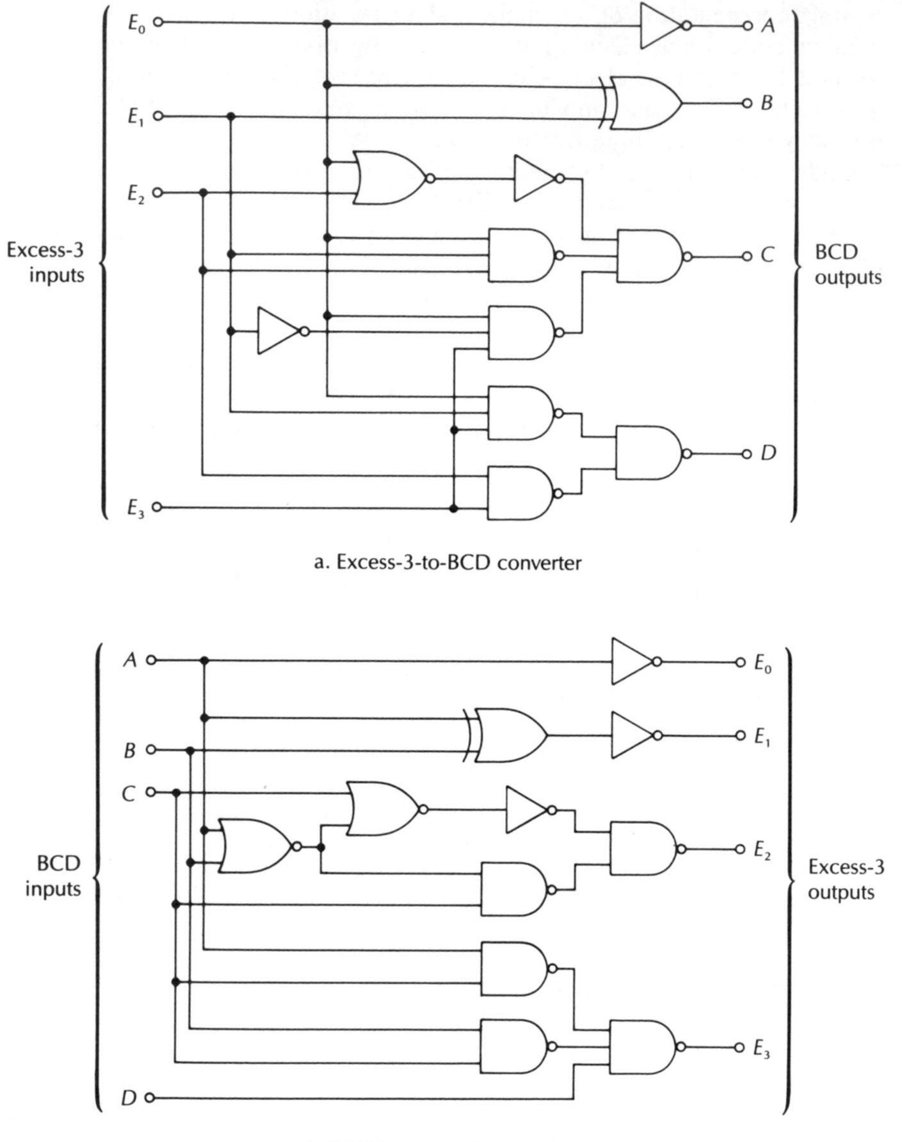

a. Excess-3-to-BCD converter

b. BCD-to-excess-3 converter

Figure 6-6 Excess-3-to-BCD and BCD-to-Excess-3 Converters

read-only-memory (or PLA) logic for an economical solution to the code conversion problem. The ROM logic circuit in Figure 6-7 illustrates, in a simplified form, how ROM logic can be used to convert the BCD code into excess-3 code equivalents. The X symbols in minterms

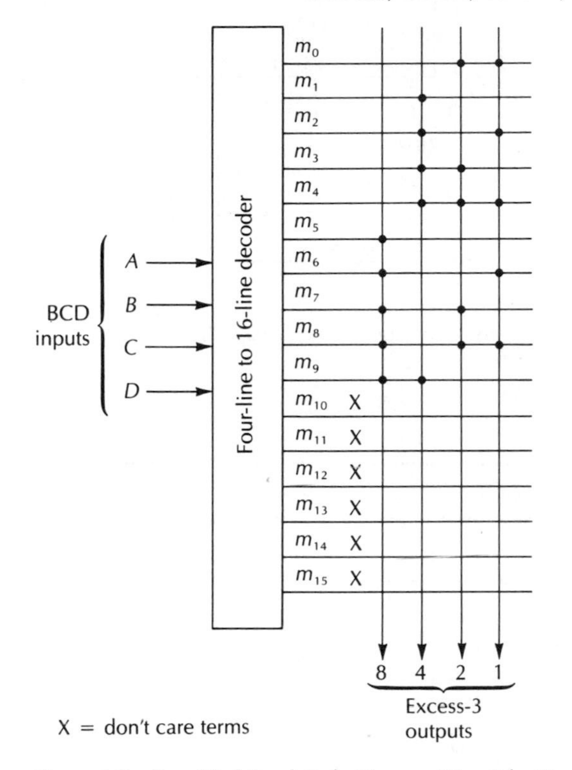

Figure 6-7 Simplified Read-Only Memory Circuit for Converting BCD to Excess-3

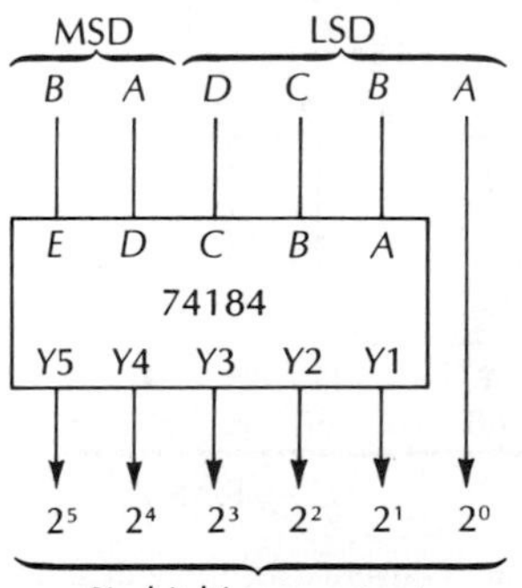

Figure 6-8 74184 BCD-to-Binary Converter

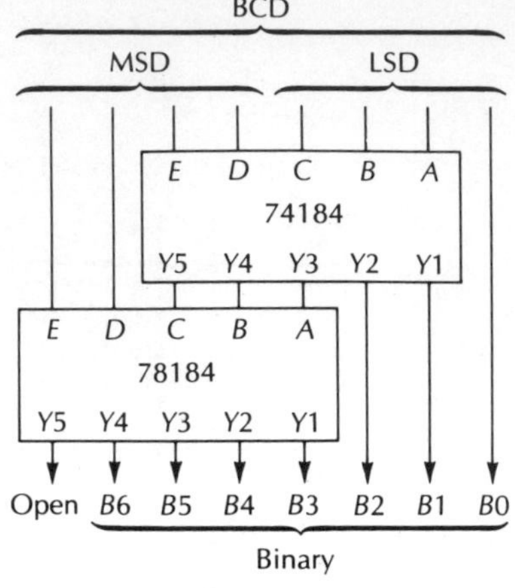

a. BCD-to-binary converter for two BCD decades

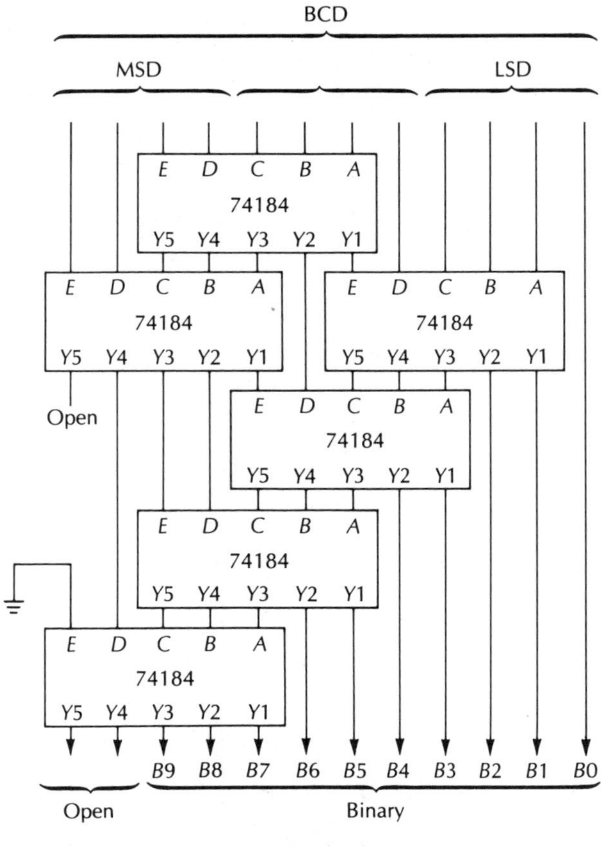

b. BCD-to-binary converter for three BCD decades

Figure 6-9 Expanding the 74184 BCD-to Binary

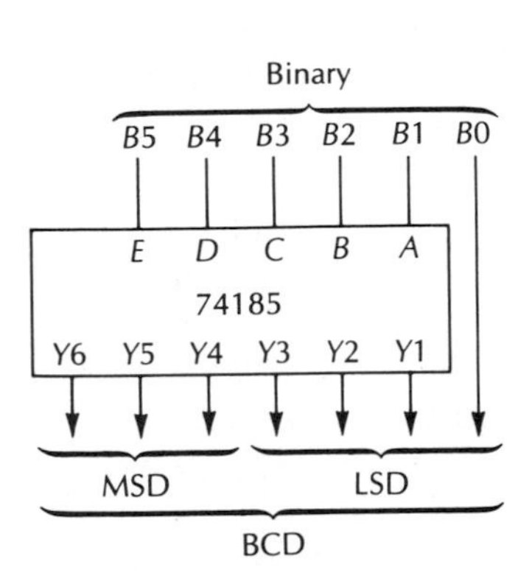

a. 74185A 6-bit binary-to-BCD converter

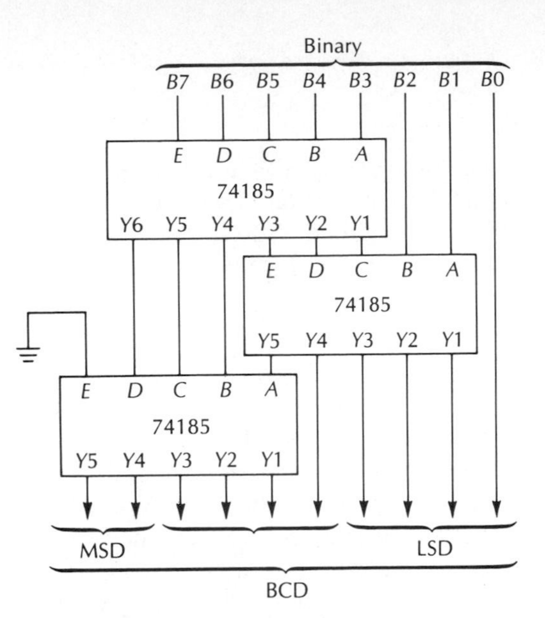

b. Expanding the 74185 binary-to-BCD converter to 8 bits

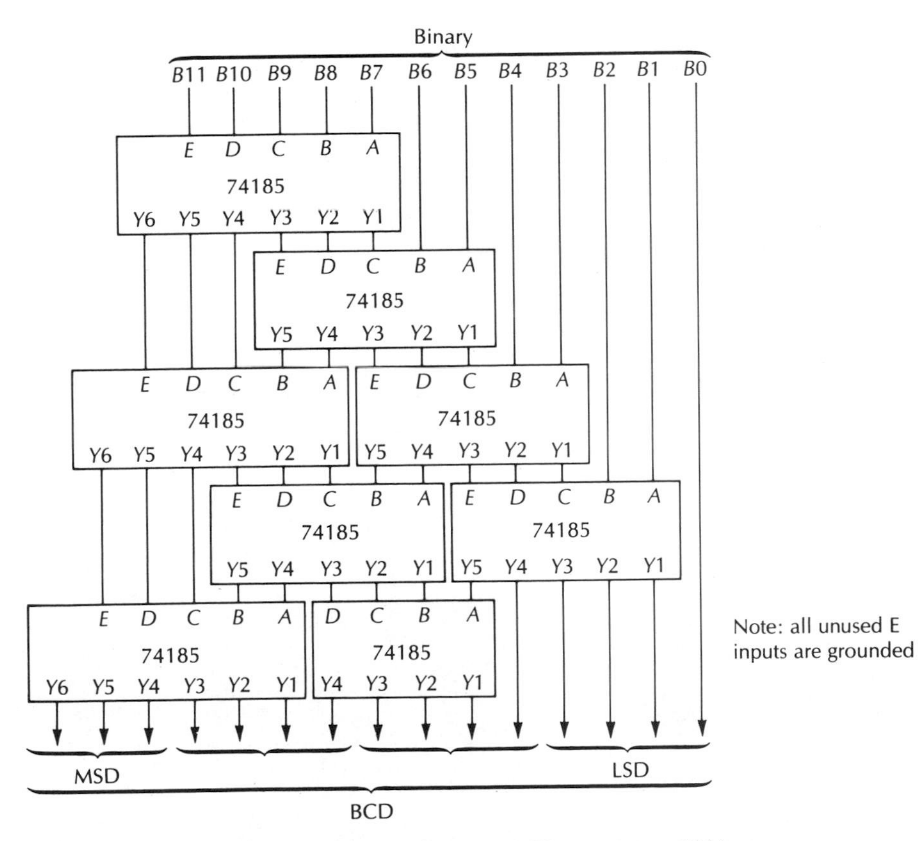

c. Expanding the 74185 binary-to-BCD converter to 12 bits

Figure 6-10 The 74185A Binary-to-BCD Converter

m_{10} through m_{15} in Figure 6-7 represent don't-care terms. It is assumed that these particular binary combinations will never occur because they are not legitimate BCD code groups. Legitimate BCD code groups include only the binary equivalents of the decimal digits 0 through 9.

BCD-TO-BINARY AND BINARY-TO-DECIMAL CONVERTERS

BCD-to-binary and binary-to-BCD converters are available as a mask-programmed 256-bit ROM (TTL). The 74184 mask-programmed ROM 6-bit BCD-to-binary and the 74185 mask-programmed ROM binary-to-BCD converters are off-the-shelf items. Figure 6-8 shows the basing diagram for the 74184 6-bit BCD-to-binary ROM converter. The 74184 converter can be expanded for any number of decades, as illustrated in Figure 6-9.

Figure 6-10 shows the basing diagram of the 74185 binary-to-BCD converter and how it can be expanded for extra decades.

CHAPTER 7

COUNTERS

Learning Objectives. *Upon completing this chapter you should know:*

1. How to draw logic diagrams for the following counters:
  - *a. Binary ripple up-counter*
  - *b. Binary ripple down-counter*
  - *c. Binary ripple up/down counter*
  - *d. Synchronous up-counter*
  - *e. Synchronous down-counter*
  - *f. 5 × 2 decimal counter*
  - *g. 6, 2, 3 mod 12 counter*

2. When to use a D flip-flop and when a J-K is more appropriate.

3. How to explain the operation of:
  - *a. The binary ripple up-counter*
  - *b. The binary ripple down-counter*
  - *c. The binary ripple up/down counter*
  - *d. The synchronous up/down counter*
  - *e. Presettable counters*

4. How to compute the maximum operating frequency for a ripple counter.

5. What modulus counters are and the methods for obtaining a modulus that is not an integral power of 2.

6. How to explain the operation of programmable counters.

7. How to program a programmable counter.

8. Some typical decoder circuits.

9. Some common C-MOS counter types.

Counters are among the most common digital circuits. They are used to count people, items, pulses, and events. They are also used to count program steps in computers and to advance the program to the next step. Frequency can be measured by counting cycles referenced to a precise time period. There is an almost unlimited variety of counter applications. In this chapter we will examine the basic counter circuits and configurations.

7-1 Ripple Counters

The ripple counter is the simplest and most basic counter. It is easily implemented with *J-K* flip-flops. The term *ripple* is derived from the fact that the output of each flip-flop is connected to the input of the following flip-flop, so that the count must propagate down the line, activating each flip-flop in sequential order. The count seems to *ripple* down the chain. Figure 7-1 shows a four-stage binary ripple counter.

Initially all flip-flops are reset to 0 (*Q-low*). The clock input to F-F *A* sets the F-F to 1 on the negative-going transition of the clock pulse. F-F *B* does not change states (assuming master-slave F-F's) until its clock input goes to 0, which happens on the second clock pulse when Q_A goes to zero. Q_A changes states at every clock pulse, Q_B changes states every second clock pulse, F-F *C* changes states every fourth clock pulse, and so on (see Figure 7-1b). At the end of the fifteenth clock pulse, all F-F's are at $Q = 1$. On the sixteenth pulse Q_A is reset to zero, and this command ripples down the string to bring all of the F-F's back to the $Q = 0$ state. The circuit is said to be asynchronous because only one of the F-F's is under the direct control of the clock. The rest of them are clocked by the output of the preceding flip-flop.

The maximum operating frequency of a simple ripple counter of the form shown in Figure 7-1 is determined at the fifteenth to sixteenth transition where the counter must go from 1111 to 0000. In this case all of the counters must change states and the clock pulse must ripple through all four F-F's. This particular transition is selected because it requires the longest time of any count in the sequence and involves the cumulative delays of *all* of the F-F's in the chain. In longer chains the condition in which the counter changes from all ones to all zeros is used to determine the maximum speed.

Maximum counting frequency is given by:

$$\frac{1}{f}(\text{max}) = N(T_p) + T_s$$

where

$$f = \text{frequency}\left(\frac{1}{f} = \text{period}\right)$$

N = number of F-F's in the string
T_p = propagation delay of one F-F
T_s = strobe time – pulse width of decoded output

Decoders are often used with counters, and when decoder outputs set flip-flops or drive logic circuits a problem arises. The counter-output lines produce some spurious spikes while the flip-flops are changing states. These spurious signals are called *glitches* and may last long enough to set a flip-flop or make other logic circuits produce false data. The most common method of solving the glitch problem is to disable the decoder until the flip-flops have had time to settle into a stable condition. The enable/disable decoder input is controlled by a

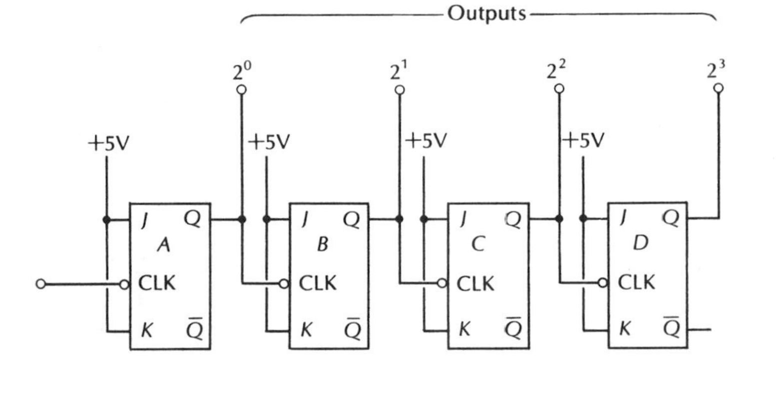

a. Logic diagram

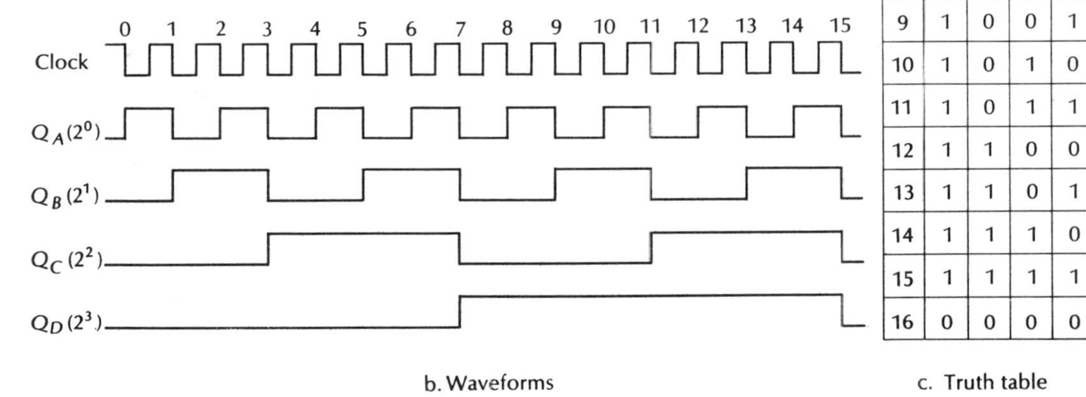

b. Waveforms

Count	D	C	B	A
0	0	0	0	0
1	0	0	0	1
2	0	0	1	0
3	0	0	1	1
4	0	1	0	0
5	0	1	0	1
6	0	1	1	0
7	0	1	1	1
8	1	0	0	0
9	1	0	0	1
10	1	0	1	0
11	1	0	1	1
12	1	1	0	0
13	1	1	0	1
14	1	1	1	0
15	1	1	1	1
16	0	0	0	0

c. Truth table

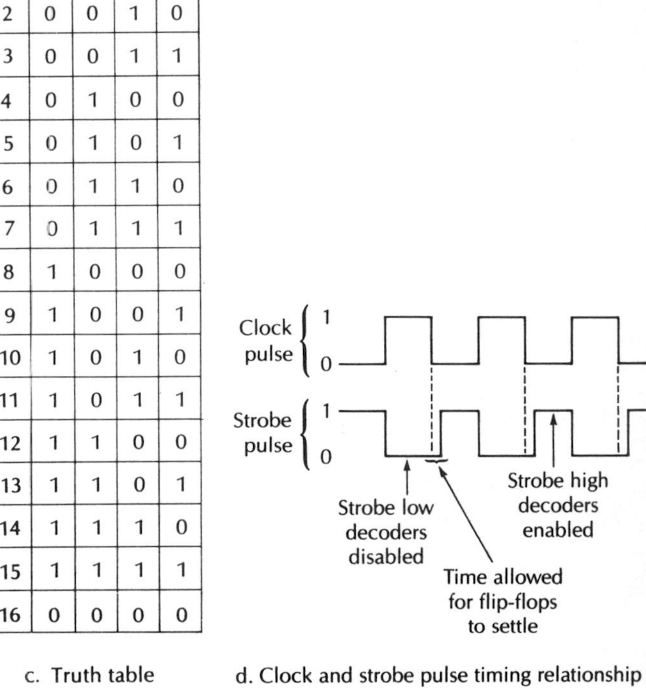

d. Clock and strobe pulse timing relationship

Figure 7-1 Binary Ripple Counter (Up-Counter, Asynchronous)

delayed clock pulse called a *strobe* pulse. Figure 7-1d shows the timing relationship between clock and strobe pulses.

The glitch problem is primarily a ripple counter problem but can occur in synchronous counters as well. Glitches are generally no problem when the decoder is used directly to drive a display device.

Example If four F-F's are arranged as in Figure 7-1, assume a propagation delay of 50 ns per F-F and a decoding time (strobe time) of 50 ns.

$$\frac{1}{f}(\max) = 4(50) + 50 = 250 \text{ ns}$$

$$f \max = 4 \text{ MHz}$$

The binary ripple counter in Figure 7-1 is also a frequency divider because F-F A requires two master-clock pulses for one complete ($\overline{Q}$-Q-$\overline{Q}$) output transition, thus producing a subordinate clock pulse with a frequency of half the master-clock frequency. F-F B produces a second subordinate clock frequency (one-fourth the master clock frequency and one-half the first subordinate clock frequency; see the timing diagram in Figure 7-1.) Two F-F's is a divide-by-4 circuit, three F-F's is a divide-by-8 circuit, and so on. The frequency division is given by 2^n, where n is the number of cascaded F-F's. Keep in mind that the longer the chain, the lower the maximum master-clock frequency must be. The primary deficiency of the simple ripple counter is the low speed due to cumulative propagation delays and the frequency limitation imposed by the number of flip-flops in the chain. Later in this chapter we will examine the faster synchronous counter that requires only one F-F propagation delay regardless of how many F-F's are in the chain.

7-2 Ripple Counters Using the *D* Flip-Flop

The type D flip-flop can be configured as a toggle flip-flop by connecting $\overline{Q}$ back to the D input as shown in Figure 7-2a. Figure 7-2b shows D flip-flops connected as a ripple *up*-counter and part c shows a *down*-counter using D flip-flops.

7-3 Down-Counters

Down-counters count backward from some predetermined value toward zero. They are subtracting counters, reducing the count by one unit per input pulse. An ordinary up-counter can be converted to a down-counter simply by taking the outputs from the $\overline{Q}$ outputs of the flip-flops. If the $\overline{Q}$ outputs are not available, the Q outputs can be inverted with the same results. Assume that the circuit in Figure 7-3 has been

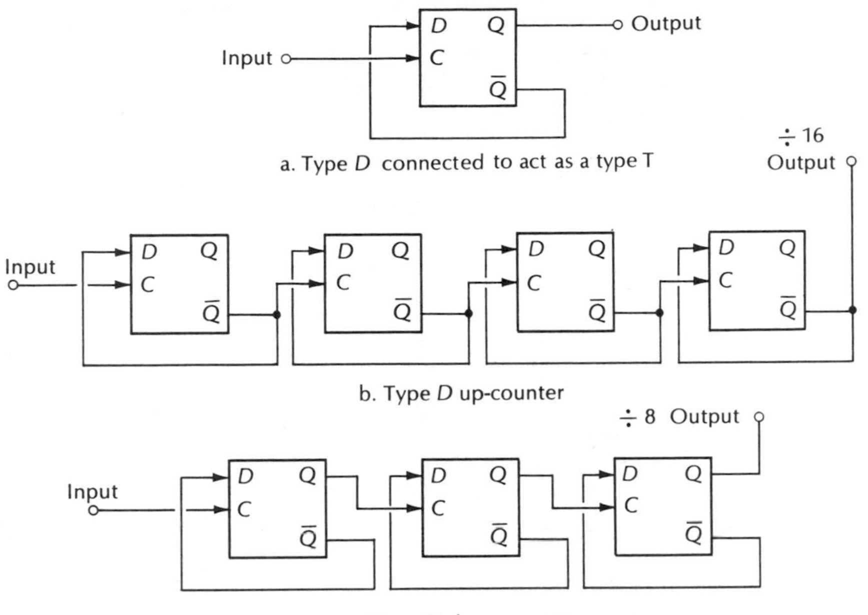

Figure 7-2 Type *D* Flip-Flop as Counter/Divider

cleared to all Q's = 0. Compare the truth tables in parts b and c. Notice that, while Q outputs represent an upward count, the $\overline{Q}$ outputs are counting down.

The outputs for a down-counter can be taken from the Q outputs if the circuit in Figure 7-3 is modified as shown in Figure 7-4.

If gate circuits are added that can effectively route the circuit paths as shown in either Figure 7-3 or 7-4 on command, the result is a reversible counter called an *up/down counter.* Figure 7-5a shows how this can be done. Figure 7-5b shows a complete up/down counter circuit. Up/down counters are more expensive than up-counters because of small demand rather than increased complexity.

7-4 Synchronous Counters

Fully synchronous counters eliminate the problem of cumulative F-F delays because all data transfers are clocked at the same time. The propagation delay is only one flip-flop delay regardless of the number of flip-flops in the counter chain. In addition to one flip-flop delay, there is also a gate delay and generally a decoder delay, but these are fairly constant and not a function of counter length.

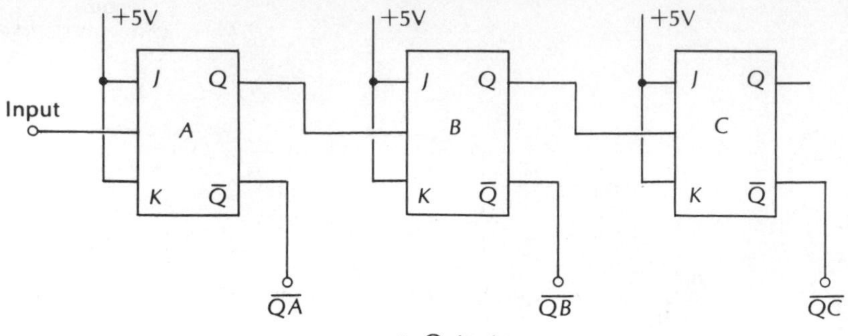

a. Outputs

Count	QA	QB	QC
0	0	0	0
1	0	0	1
2	0	1	0
3	0	1	1
4	1	0	0
5	1	0	1
6	1	1	0
7	1	1	1

b. Q outputs

Count	$\overline{QA}$	$\overline{QB}$	$\overline{QC}$	Decimal equivalent
0	1	1	1	7
1	1	1	0	6
2	1	0	1	5
3	1	0	0	4
4	0	1	1	3
5	0	1	0	2
6	0	0	1	1
7	0	0	0	0

c. $\overline{Q}$ outputs

Figure 7-3 Binary Ripple Down-Counter (Asynchronous) with Outputs Taken from $\overline{Q}$

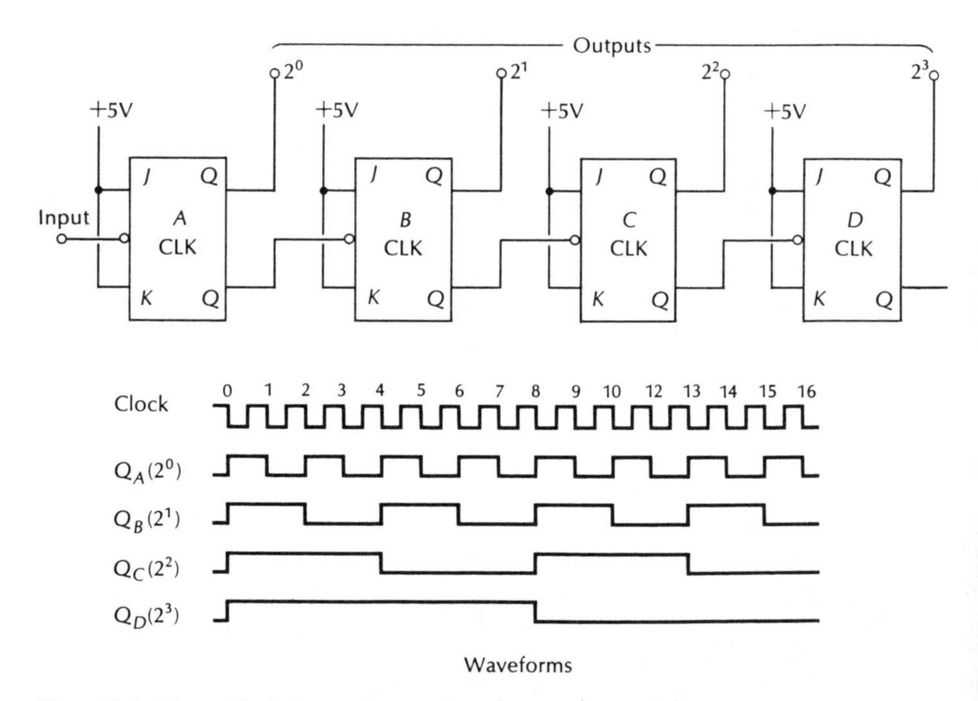

Figure 7-4 Binary Ripple Down-Counter (Asynchronous) Using Q Outputs

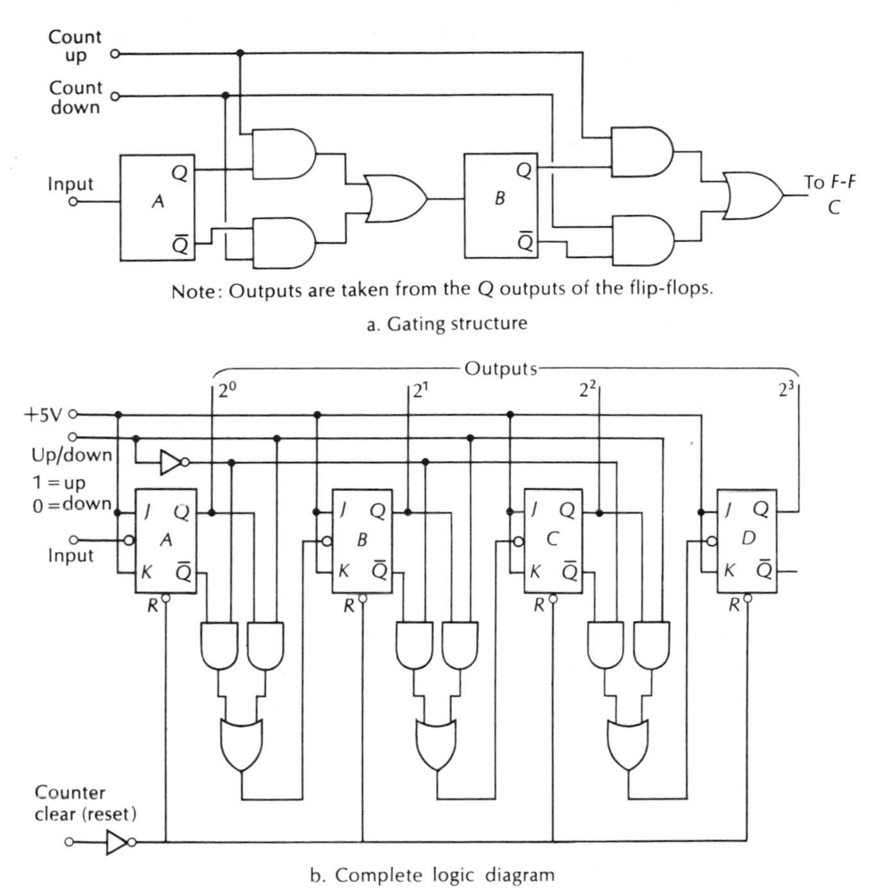

Figure 7-5 Binary Up/Down Counter (Ripple Asynchronous)

The maximum clock frequency is given by:

$$\frac{1}{f} = T_p + T_g$$

where

$$T_p = 1 \text{ flip-flop delay}$$
$$T_g = 1 \text{ gate delay}$$

Example Assume a flip-flop delay of 50 ns and a gate delay of 30 ns.

$$\frac{1}{f} = 50 + 30 = 80 \text{ ns}$$

$$f = \frac{1}{80} \text{ ns} = 12.5 \text{ MHz}$$

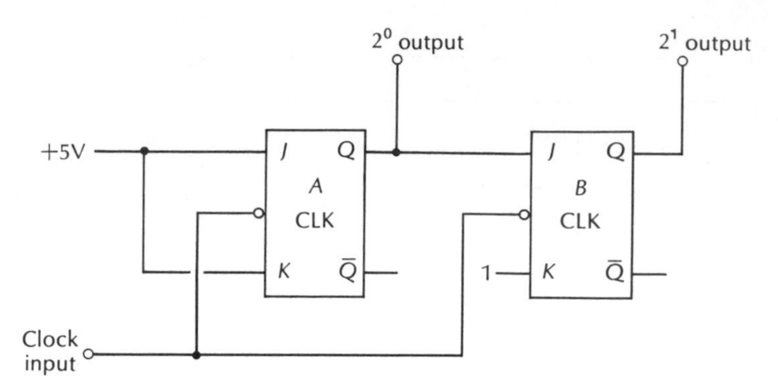

Figure 7-6 Parallel Clocked Synchronous Counter

Figure 7-6 shows the simplest synchronous counter. Notice that both flip-flops are clocked at exactly the same time since the second flip-flop does not have to wait for the first to pass along the clock pulse.

In order to extend the length of the counter and maintain synchronous clocking, additional gates are required which make the counter more complex. A four-stage synchronous counter is shown in Figure 7-7.

7-5 Synchronous Up/Down Counters

A synchronous down-counter can be constructed by using the $\overline{Q}$ outputs in the same fashion as in the ripple counter and using the same gating structure as in the up-counter in Figure 7-7. The synchronous down-counter circuit is shown in Figure 7-8. By adding gates to transfer data out of the Q or the $\overline{Q}$ outputs on command, we arrive at the reversible (up/down) synchronous counter shown in Figure 7-9.

RIPPLE-CARRY SYNCHRONOUS COUNTERS

An examination of the logic circuit for the fully synchronous counter reveals that each additional F-F requires an additional gate and that each successive gate requires one more input than the gate for the preceding stage. A compromise configuration that retains the synchronous operation of the flip-flops with a simpler gate structure is called the *ripple-carry synchronous counter*.

Though flip-flop delay is not cumulative, ripple-carry gate delays are. Thus, the circuit is faster than the simple ripple counter but slower than the fully synchronous parallel-carry counter. The logic diagram for a synchronous ripple-carry up/down counter is shown in Figure 7-10.

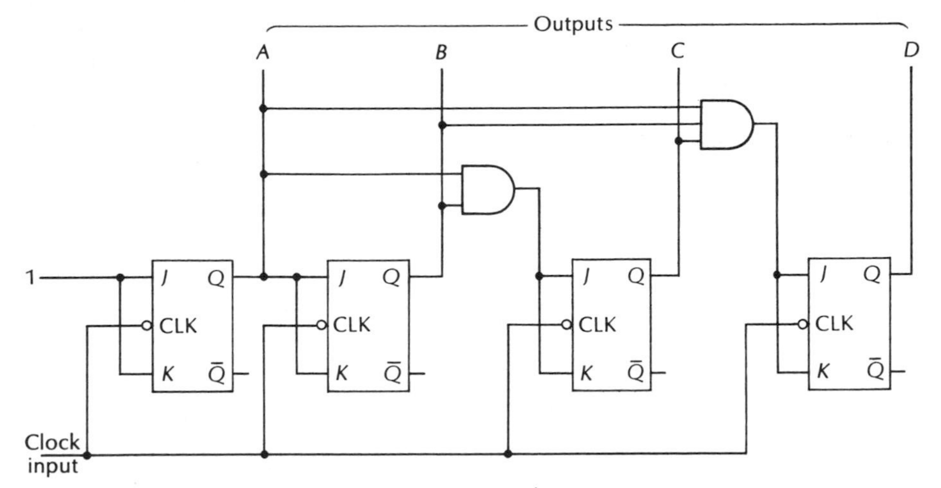

a. Logic diagram

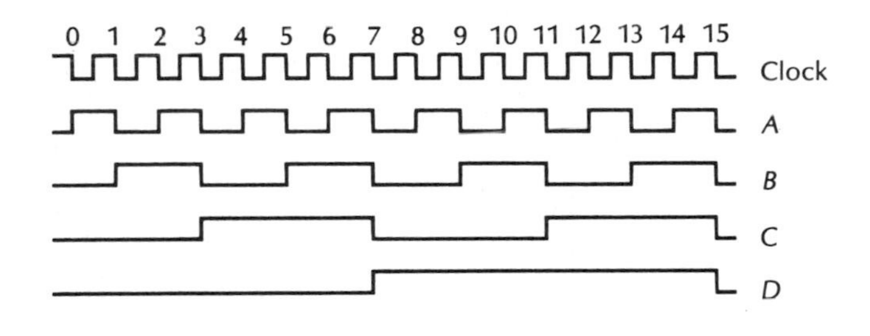

b. Waveforms

Figure 7-7 Fully Synchronous Up-Counter

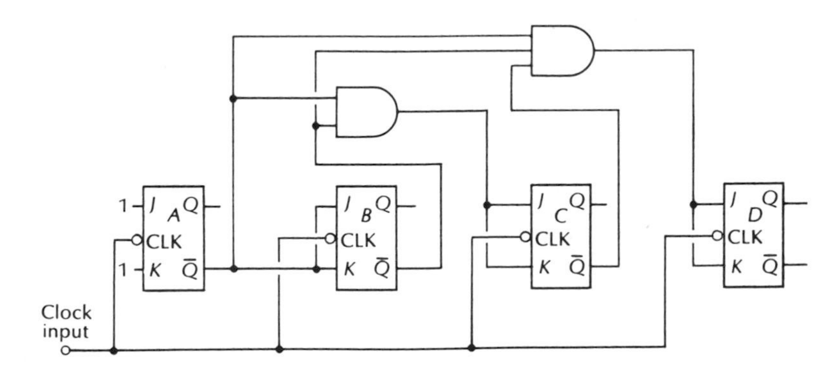

Figure 7-8 Synchronous Down-Counter

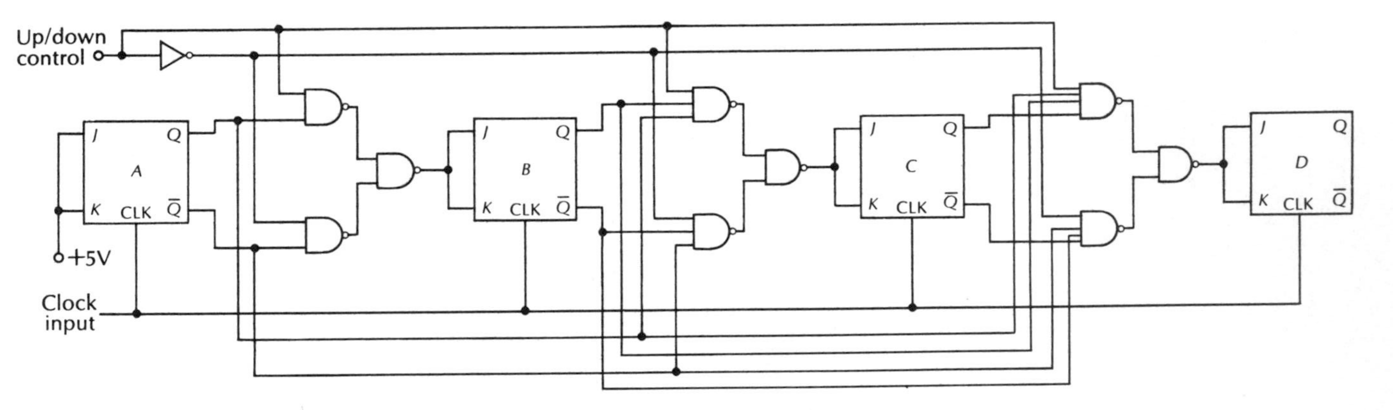

Figure 7-9 Synchronous Up/Down Counter

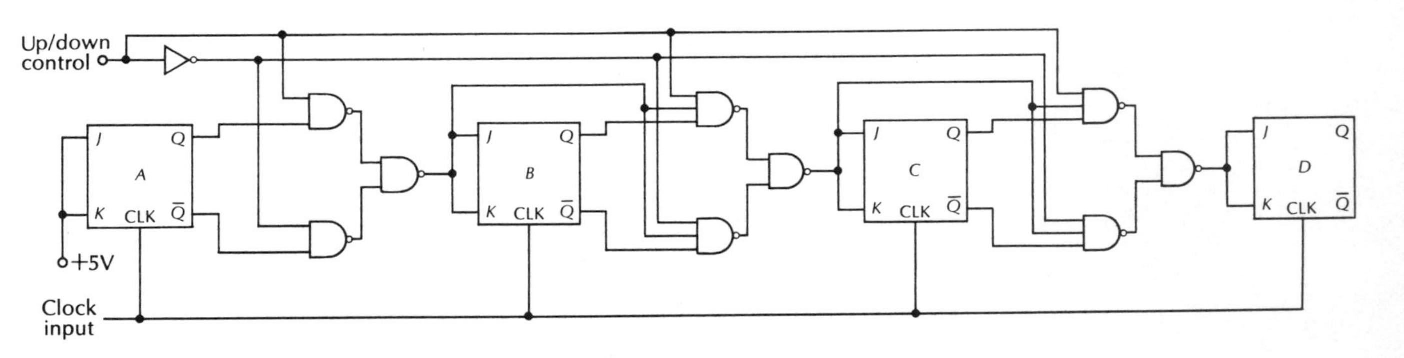

Figure 7-10 Ripple Carry Synchronous Up/Down Counter

7-6 Modulus Counters

The *modulus* (mod) of a counter is simply the number of counting states before it begins to repeat itself. A four-stage counter has a natural modulus of 16, a three-stage counter has a modulus of 8, and so on.

So far the counters we have examined have all operated at their natural modulus, using their full-count length. A counter can, however, be made to count to any desired modulus by selecting a counter with a natural modulus that is the next step higher than the desired modulus and then causing it to skip the proper number of steps. Counts can be skipped anywhere in the natural sequence.

Usually some form of feedback is used to force the counter one or more extra counts ahead or to clear the counter before the end of the natural count sequence.

When we want to have a ripple counter operate in some modulus that is not a power of 2, the following procedure can be used (for preset mode counters):

1. Find N (the number of flip-flops required); that is, find the nearest power of 2 that yields a higher count than the modulus number, N.

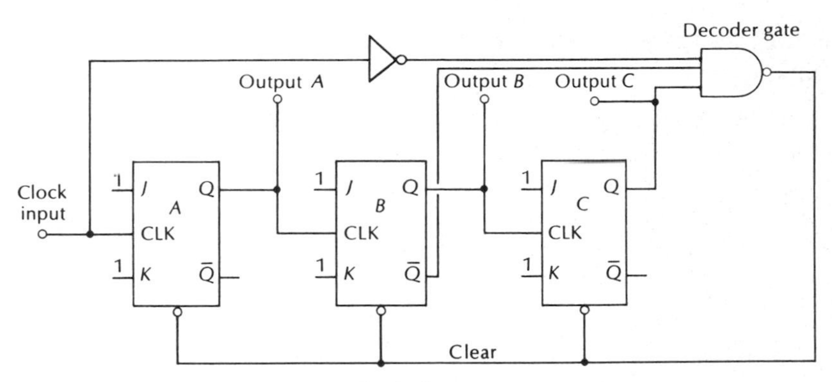

a. Logic diagram

Count	C	B	A
0	0	0	0
1	0	0	1
2	0	1	0
3	0	1	1
4	1	0	0
5	0	0	0

— Count $N-1$ (row 4)

b. Truth table

Figure 7-11 Logic Diagram for a Mod 5 Counter Using F-F's with Preset Capability

If we assume a modulus (mod) of 5, $2^2 = 4$ is not enough, but $2^3 = 8$ is sufficient. The exponent is the number of flip-flops, so three flip-flops are required. Note that three F-F's can count up to 111, or 7_{10}, before resetting naturally to 000. Because 5 (the modulus) is less than 7, three F-F's are required.

2. Connect all flip-flops as a ripple counter, as shown in Figure 7-11.
3. Find the binary number $N - 1$.
4. Connect all flip-flop outputs (Q) that are 1 at count $N - 1$ to one of the inputs on a NAND gate. Use one NAND gate input for the clock.
5. Connect the NAND gate output to the *preset* input of all flip-flops that are at $Q = 0$ at count $N - 1$.

Example Draw the logic diagram of a modulus 5 ripple counter using presets.

1. The number of F-F's required is 3.
2. $N - 1 = 5 - 1 = 4$

   CBA
3. 4_{10} in binary = 1 0 0

(Figure 7-11 shows the logic diagram and truth table for this device.)

MODULUS COUNTERS USING FLIP-FLOPS WITHOUT PRESET CAPABILITY

Many IC flip-flops do not have preset inputs. When such flip-flops are used, they can be configured as follows:

1. Find the number of flip-flops required. Find the nearest power of 2 larger than the mod number. Assuming a modulus of 5, $2^2 = 4$, $2^3 = 8$; three flip-flops are required.
2. Connect all flip-flops as a ripple counter.
3. Find the binary number N (the modulus number).
4. Find the flip-flops that have 1's on Q at the Nth count. In this example it would be flip-flops A and C.
5. Connect a NAND gate with inputs connected to Q of the flip-flops that have $Q = 1$ at count N (the highest count in the sequence). Connect the output of the NAND gate to the clear (reset) inputs on *all* flip-flops.

The circuit will count up to $N - 1$ and on the Nth count will go to the number N momentarily—long enough for the NAND gate to detect the presence of the number N and reset the counter to all zeros. Figure 7-12 shows the logic diagram.

One problem with this scheme is that a fairly broad range of reset

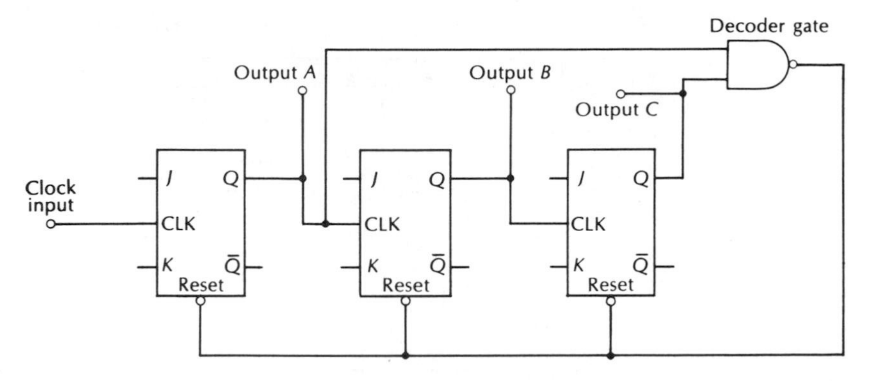

Figure 7-12 Mod 5 Ripple Counter Using the Reset Method

times exists among the various flip-flops in a string. Because of different loading conditions from one flip-flop to another and because of normal tolerances in circuitry, the reset time may vary. The pulse may not always be long enough to reset all counters.

There are two basic methods of overcoming this difficulty. One involves the use of a latch with clock reset and the other uses some form of delay (pulse stretching). Figure 7-13 shows a mod 5 counter with

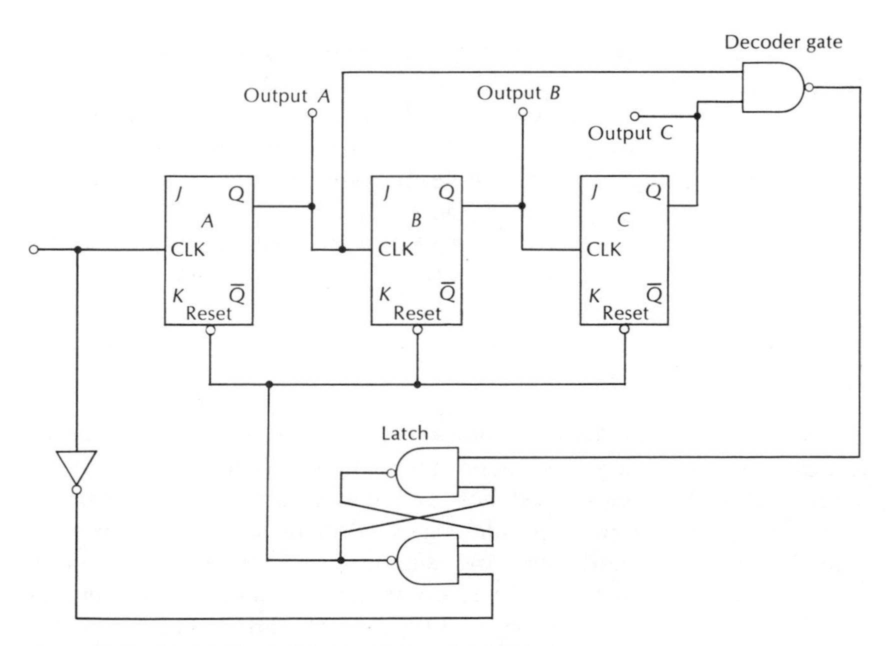

Figure 7-13 Mod 5 Ripple Counter Using a Latch Reset

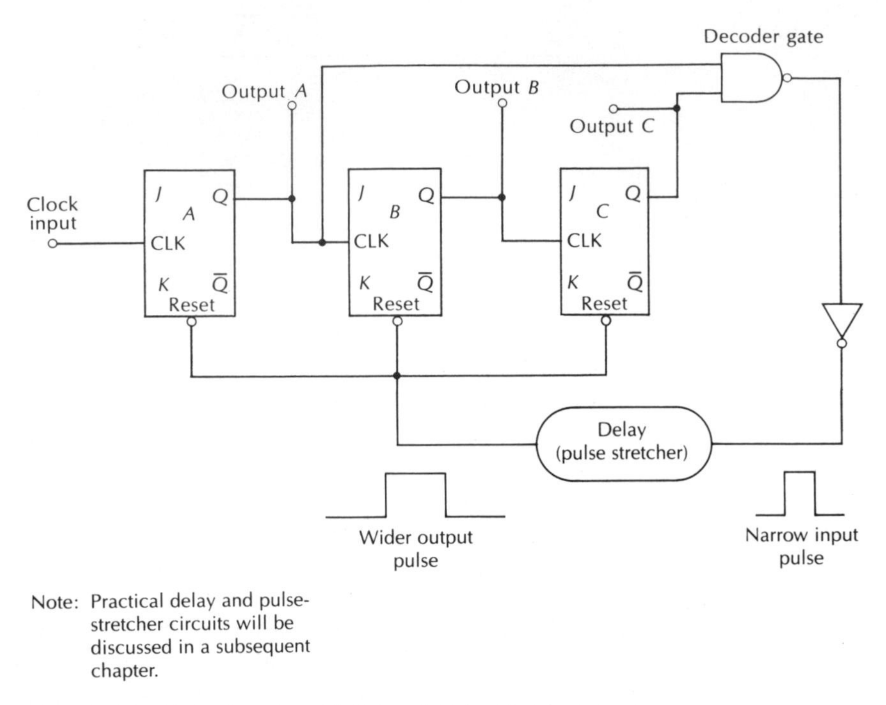

Figure 7-14 Mod 5 Counter with Delayed (Pulse Stretcher) Reset

latch reset, and Figure 7-14 shows a mod 5 counter with delayed reset. Notice that *all* F-F's are reset at count N, including those that were naturally in a $Q = 0$ state at count N. This is done to insure that the new cycle begins at zero and to prevent any miscount from the previous cycle from carrying over into the next. Some counter circuits include the 0 states of the flip-flops not in the 1 state at count N as inputs to the NAND gate. In that case, they are taken from the $\overline{Q}$ outputs of the flip-flops involved. The NAND gate used in preset and clear mode counters is a single condition decoder.

7-7 IC Counter Techniques

Modern IC counters often combine several counter techniques in an IC package. Combinations of synchronous and asynchronous techniques combined with gateless feedback and inhibit circuits are common. Many of these counters provide a great deal of flexibility as well as simplicity and standardization by using combinations of two or more different modulus counters. For example, a mod 2 and a mod 3 counter can be combined as a divide-by-3-by-2 combination, providing two versions of a mod 6 counter.

A divide-by-3, divide-by-2 pair can be combined with another divide-by-2 counter to get a mod 12 counter. Two versions of decimal counters can be formed using a divide-by-2, divide-by-5 or a divide-by-5, divide-by-2 configuration.

In addition, each modulus counter in a given package can be arranged for external connections that allow the individual modulus counters to be used separately or configured for two or more different counting modes.

DIVIDE-BY-4 HYBRID PARALLEL-CLOCKED GATELESS COUNTER

Figure 7-15 shows the logic diagram of a mod 4 (binary two-stage) counter using a parallel (synchronous) clock and ripple count propagation. This configuration provides the flexibility necessary for most of the hybrid counter forms.

Problem

1. Using what you know about *J-K* flip-flops, analyze the operation of the circuit in Figure 7-15. (Hint: see the operation of the divide-by-3 counter.)

DIVIDE-BY-3 COUNTER

Figure 7-16 shows a divide-by-3 (mod 3) counter. The following shows how it operates (see Table 7-1):

1. Initial conditions (Count 0)
   a. F-F *A* is reset to 0.
   b. F-F *B* is reset to 0.
   c. $\overline{Q}$ of F-F *B* is at a logical 1 and is connected to the *J* input of F-F *A*. F-F *A* will operate in a toggle mode as long as $\overline{Q}$ of F-F *B* is at 1.

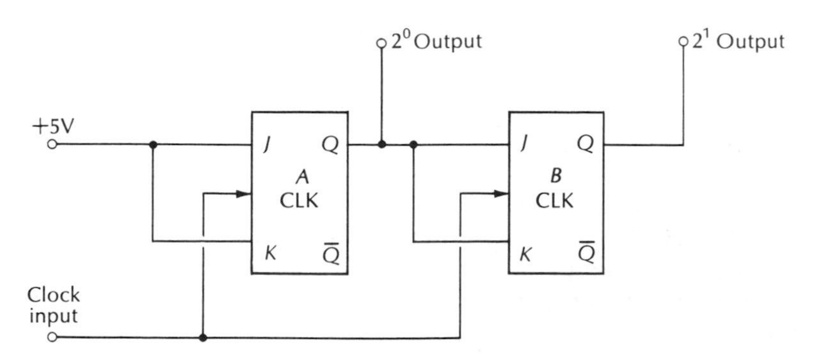

Figure 7-15 Parallel-Clocked Gateless Counter

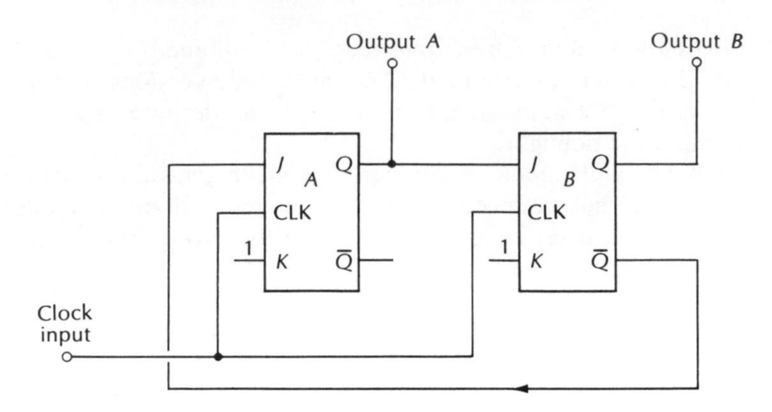

Figure 7-16 Mod 3 Counter

Table 7-1 Mod 3 Hybrid Counter Truth Table

Count Number	Q_b	Q_a
0	0	0
1	0	1
2	1	0
3	0	0

2. Count 1
   a. F-F A toggles to $Q = 1$, F-F B no change
3. Count 2
   a. F-F A toggles to $Q = 0$, F-F B toggles to $Q = 1$
   b. $\overline{Q}$ of F-F B goes to logical zero, preventing F-F A from setting to $Q = 1$.
4. Count 3
   a. QB goes to 0
   b. QA locked out—stays at 0

DIVIDE-BY-5 COUNTER

Figure 7-17 shows the logic diagram of a mod 5 counter with two feedback control paths. Table 7-2 shows the truth table for the mod 5 hybrid counter. Notice particularly the two feedback control paths: from Q_c to K_c and $\overline{Q}_c$ back to J-K_a. The counting sequence goes as follows:

1. Count 0
   Initial conditions: F-F's A, B, and C reset to $Q = 0$

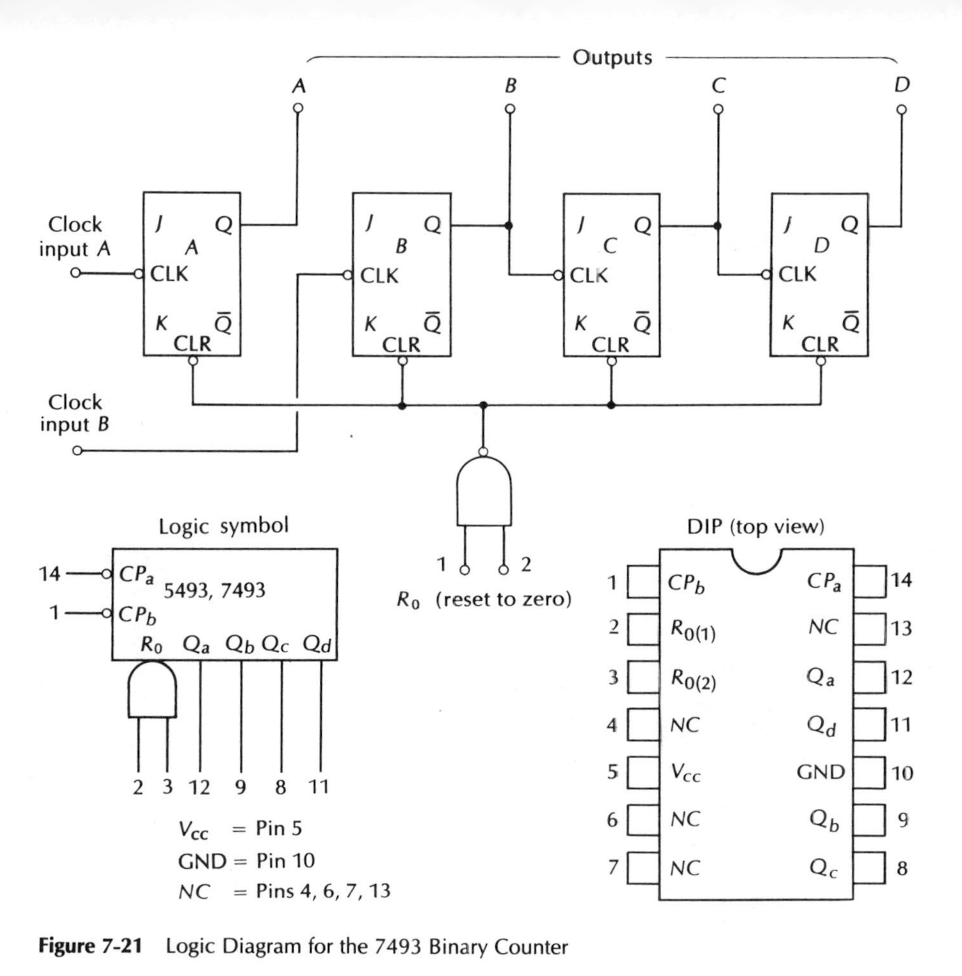

Mode 1 (divide-by-16)

D	C	B	A
0	0	0	0
0	0	0	1
0	0	1	0
0	0	1	1
0	1	0	0
0	1	0	1
0	1	1	0
0	1	1	1
1	0	0	0
1	0	0	1
1	0	1	0
1	0	1	1
1	1	0	0
1	1	0	1
1	1	1	0
1	1	1	1

Mode 2 (divide-by-8)

B	C	D
0	0	0
1	0	0
0	1	0
1	1	0
0	0	1
1	0	1
0	1	1
1	1	1

Truth table

Figure 7-21 Logic Diagram for the 7493 Binary Counter

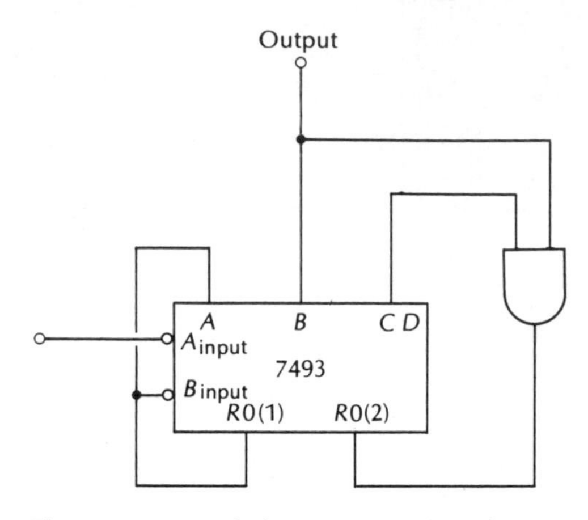

Figure 7-22 Divide-by-7 Counter Using the 7493

frequency of 20 MHz. To make it a mod 16 counter, the clock is connected to input *A* and an external jumper is connected from Q_A (pin 12) to clock *B* (pin 1). The 7493 counts up only. Figures 7-22 through 7-26 show the 7493 used in several common configurations.

7493 CONFIGURATIONS

1. Divide-by-7 (7493)

   Operation: Decodes 6 and resets to zero at momentary 111.

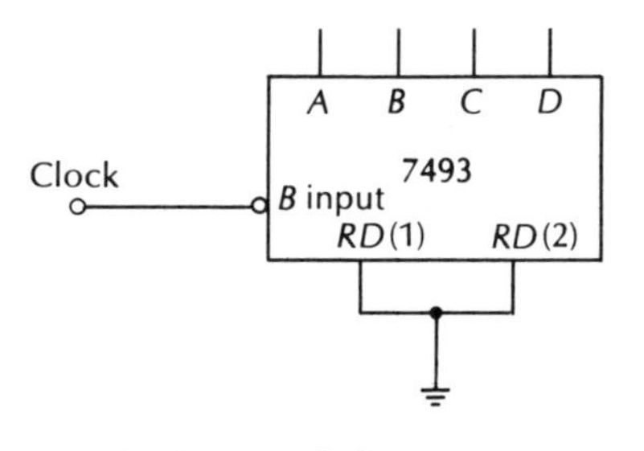

Figure 7-23 Divide-by-8 Counter Using the 7493

2. Divide-by-8 (7493)

   Operation: Input to divide-by-8 section only.

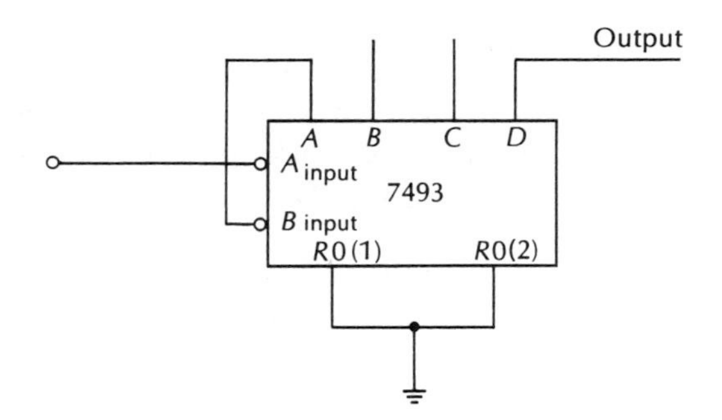

Figure 7-24 Divide-by-16 Ripple Counter Using the 54/7493

3. Divide-by-16 (7493)

   Operation: Jumper from the divide-by-2 output to the divide-by-8 input.

Problems

2. Explain the operation of the divide-by-9 circuit in Figure 7-25.
3. Explain the operation of the divide-by-10 counter in Figure 7-26. Use a count truth table.

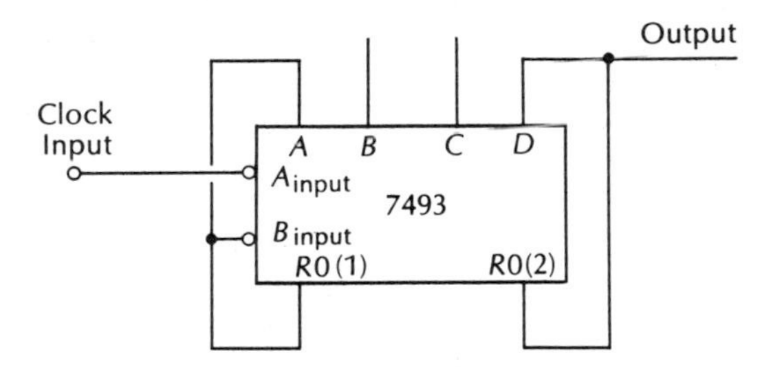

Figure 7-25 Divide-by-9 Ripple Counter Using the 54/7493

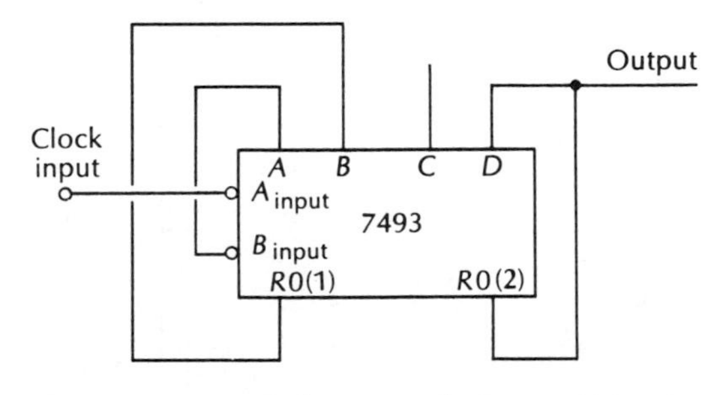

Figure 7-26 Divide-by-10 Ripple Counter Using the 54/7493

7-9 The 7490 Decade Counter

The 7490 decade counter contains separate divide-by-2 and divide-by-5 counters composed of a total of four master-slave flip-flops. Gated inputs are provided to reset all outputs to 0 or to preset to BCD 9 (1001_{BCD}).

The 7490 is a ripple counter, with a maximum clock rate of 18 MHz. The count is advanced on the negative-going clock edge. The clock signal must be TTL compatible. The counter requires normal TTL rise

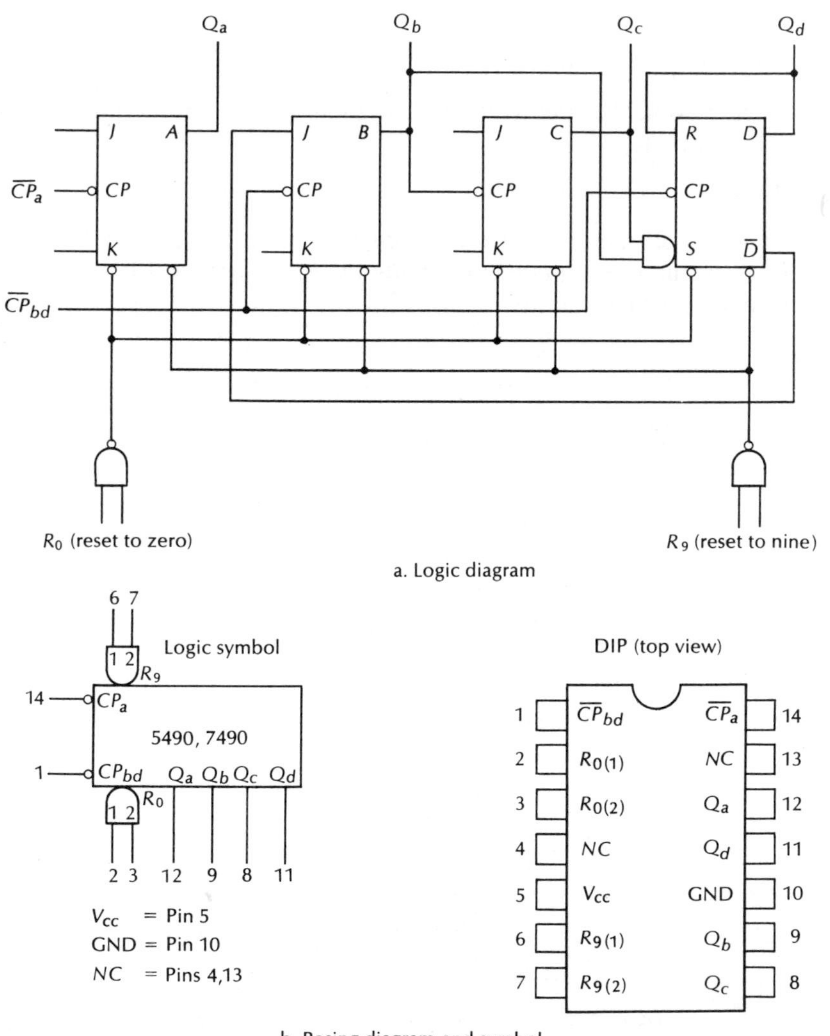

Figure 7-27 Logic Diagram for the 54/7490 Decade Counter

and fall times for proper operation. For normal counting operation the two set-to-0 and the two set-to-9 inputs must be grounded.

To set to 0, both of the set-to-0 inputs must go positive (*high*). To set to 9, both of the set-to-9 inputs must go positive. Reset/set pulses should be as wide as possible—10 microseconds or wider.

There are three basic counting modes: mod 5 and mod 2 separately, a divide-by-2, a divide-by-5, and a divide-by-2/divide-by-5 decade counter. It may also be easily configured for divide-by-6 and divide-by-7 counting modes. Figure 7-27 shows the complete logic diagram for the 54/7490 decade counter. Figure 7-28 shows external configurations and truth tables for the three primary counting modes.

The BCD mode is natural BCD, the most commonly used mode for decimal counting, and can be decoded by several off-the-shelf decoders. The symmetrical divide-by-10 mode provides a symmetrical output waveform and must be divided by a power of 10.

Problems

4. Draw a timing diagram for each of the two decimal counting modes.
5. Describe how the count progresses starting with a set-to-9 condition followed by grounding the set-to-9 input and starting the count.

7-10 The 54/7492

The 7492 is a divide-by-2, divide-by-6 pair of ripple up-counters. Figure 7-29 shows the block diagram. The count advances on the negative transition of the clock, and the maximum clock rate is 18 MHz. One or both set-to-0 lines are grounded for normal counting operation. Both set lines are taken *high* to set-to-0 (reset). The preferred mod 12 count requires a jumper from Q_A to $\overline{CP}_{BC}$. The clock input is on $\overline{CP}_A$. The other possibility, a jumper from Q_D to $\overline{CP}_A$ with input to $\overline{CP}_{BC}$ yields a different sequence that is not (normal) binary related.

Problem

6. Draw a timing diagram for the preferred counting mode and another for the alternate connection for the divide-by-12 7492.

7-11 The 7492 and 7493 as Various Modulus Counters

Modulus counters using 7492 and 7493 configured for modulus 7, 9, 11, 12, 13, 14, and 15 are shown in Figure 7-30.

7-12 Programmable Counters

The most popular programmable counters are typified by the 74161 mod 16 and the 74160 mod 10 presettable counters. These units can be

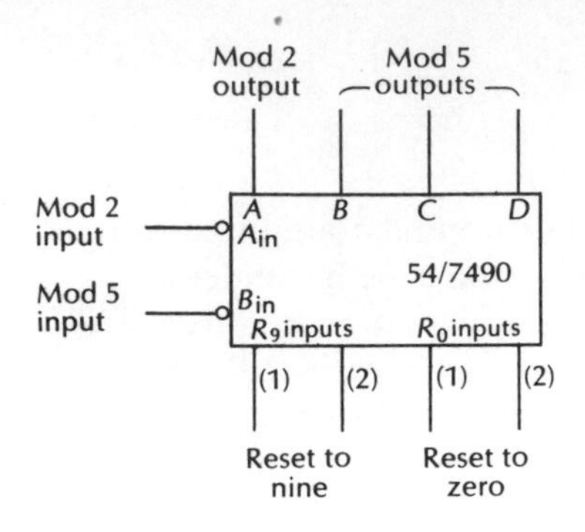

a. Mod 2 and Mod 5 counter

Count	D	C	B
0	0	0	0
1	0	0	1
2	0	1	0
3	0	1	1
4	1	0	0
5	0	0	0

b. Divide-by-5 truth table

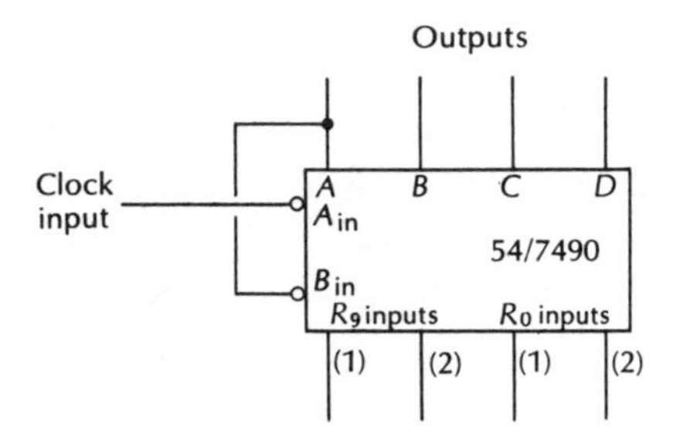

c. Binary coded decimal count

Count	D	C	B	A
0	0	0	0	0
1	0	0	0	1
2	0	0	1	0
3	0	0	1	1
4	0	1	0	0
5	0	1	0	1
6	0	1	1	0
7	0	1	1	1
8	1	0	0	0
9	1	0	0	1
10	0	0	0	0

d. Truth table for BCD mode

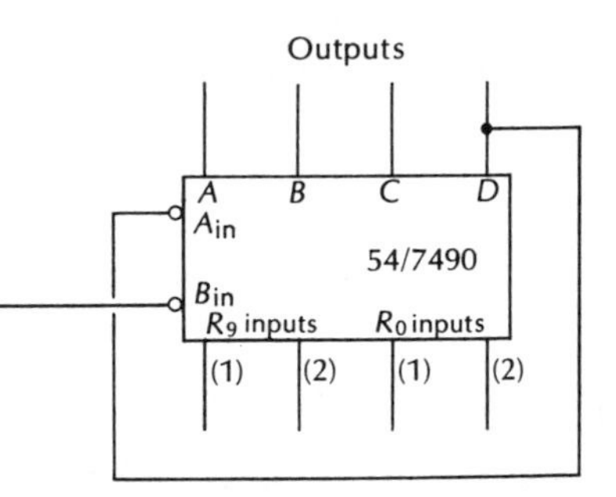

e. Symmetrical divide-by-10 mode using the 54/7490

Count	D	C	B	A
0	0	0	0	0
1	0	0	1	0
2	0	1	0	0
3	0	1	1	0
4	1	0	0	0
5	0	0	0	1
6	0	0	1	1
7	0	1	0	1
8	0	1	1	1
9	1	0	0	1
10	0	0	0	0

f. Truth table for the symmetrical divide-by-10 mode

Figure 7-28 Configurations and Truth Tables for the Three Primary Counting Modes of the 54/7490

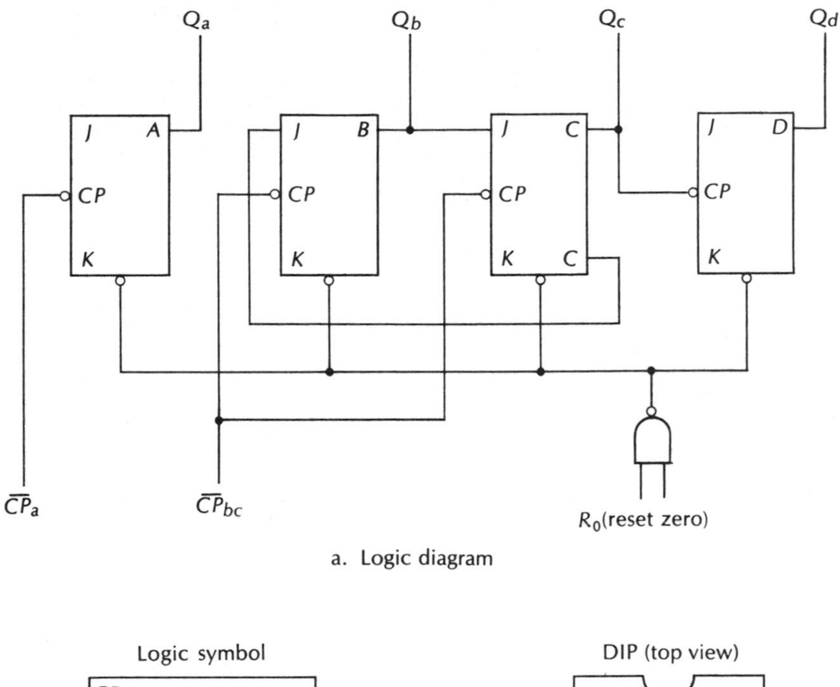

a. Logic diagram

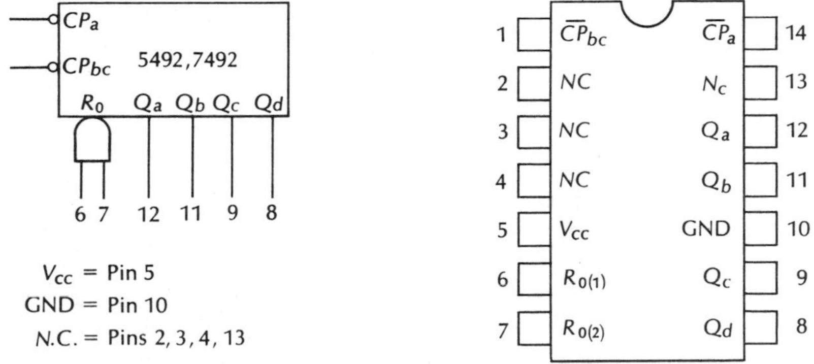

b. Basing diagram and symbol

Figure 7-29 The 54/7492 Divide-by-Twelve Counter (Divide-by-2, Divide-by-6)

programmed to operate in any modulus by using enough packages and programming. Programming consists of connecting program inputs to 5 V or ground as required by the desired count. This can be accomplished by using BCD switches to vary the count at will. Figure 7-31 shows the input, output, and control terminals for the 74161.

Mode 1 (Divide-by-12)

A	B	C	D
0	0	0	0
1	0	0	0
0	1	0	0
1	1	0	0
0	0	1	0
1	0	1	0
0	0	0	1
1	0	0	1
0	1	0	1
1	1	0	1
0	0	1	1
1	0	1	1

Mode 2 (Divide-by-6)

B	C	D
0	0	0
1	0	0
0	1	0
0	0	1
1	0	1
0	1	1

Mode 3 (Divide-by-12)

A	B	C	D
0	0	0	0
0	1	0	0
0	0	1	0
0	0	0	1
0	1	0	1
0	0	1	1
1	0	0	0
1	1	0	0
1	0	1	0
1	0	0	1
1	1	0	1
1	0	1	1

c. Truth tables

Figure 7-29 continued

Functions:

1. Q_A, Q_B, Q_C, Q_D: outputs
2. P_1, P_2, P_4, P_8: presetting inputs
3. Load held at +5 V for normal counting. Taken to ground to enter preset values.
4. Clear, set-to-0: Normally held +5 V for counting. Taken to ground to clear-to-0.
5. P connected to the carry-out of the previous stage for cascaded synchronous operation.
6. Carry-out: Goes *high* at the maximum count: $Q_A = 1$, $Q_B = 1$, $Q_C = 1$, $Q_D = 1$. Produces *low* output for all other counts.
7. T: Enable; held positive for normal counting.
8. Clock advances count on *up* clock.

PROGRAMMING THE COUNTER

The counter is programmed by setting the counter to an initial count using the preset inputs P_1, P_2, P_4, P_8. When the load control goes to +5 V, the counter begins to advance on the positive edge of each clock pulse. The counter starts its counting at the preset number and con-

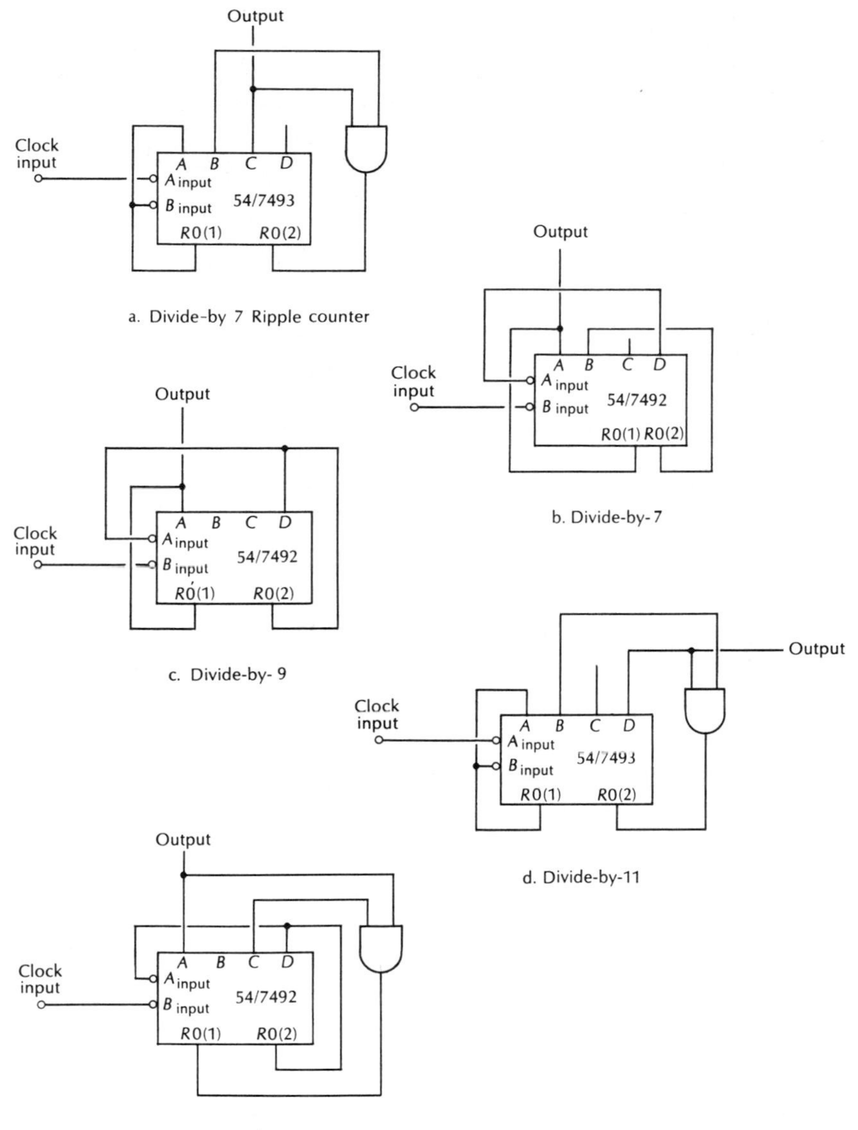

a. Divide-by 7 Ripple counter

b. Divide-by-7

c. Divide-by-9

d. Divide-by-11

e. Divide-by-11

Figure 7-30 Various Mod Ripple Counters Using the 54/7493 and the 54/7492

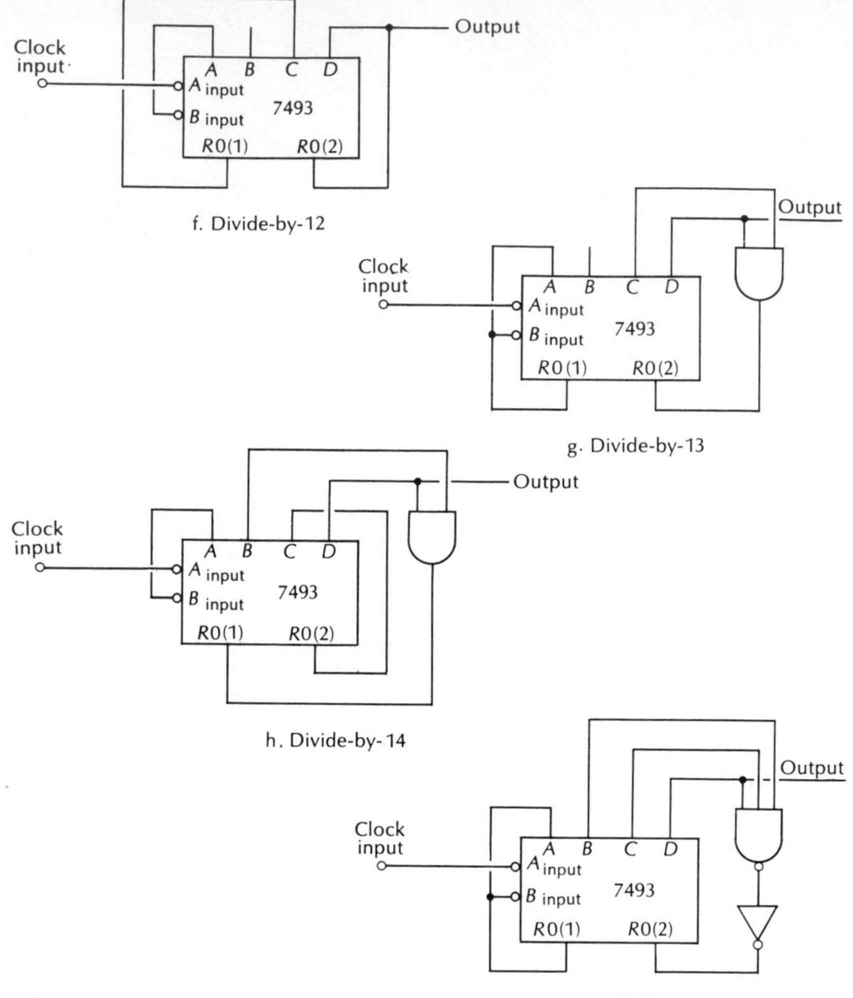

Figure 7-30 continued

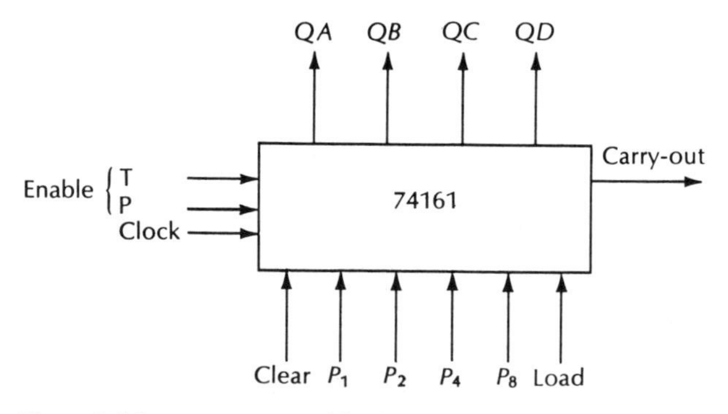

Figure 7-31 74161 Presettable Counter

tinues counting until it reaches the highest possible count. At the end of the count, the counter goes back to 0000. The 1111 condition is decoded and the carry-out line goes high on the maximum count. This carry output level can be fed back (through an inverter) to the load input, enabling the preset values to be loaded again.

The counter counts from preset count to the maximum count, resets to the preset value, and begins counting a new cycle. The modulus is: $N - n_p$, where N is the natural modulus (maximum count) of the counter and n_p is the preset count. For example:

1. Preset counter to count 5 modulus: $n_p = 5$
   Natural modulus: $N = 16$
   Programmed modulus: $N - n_p = 16 - 5 = 11$
2. Preset counter to count 3 modulus: $16 - 3 = 13$

Figure 7-32 shows the 74161 programmed for modulus 11. The preset inputs are set to the difference between the natural modulus (16) N and the desired modulus (11). The presets must be set to +5 or gnd. $P_8 = 0$, $P_4 = 1, P_2 = 0, P_1 = 1$; where $+5V = 1$.

The fact that 0000 is not available except at the natural modulus can sometimes be a problem. These counters can be cascaded for larger counts as shown in Figure 7-33.

Problems

7. Program a 74161 counter for modulus 9, 10, and 12. List the missing binary combinations for each case.
8. The 74160 is a mod 10 version of the 74161. Look it up in the data book and show how to program it for a modulus of 6.
9. Make a drawing showing how to cascade two 74160 counters programmed for a modulus of 88.

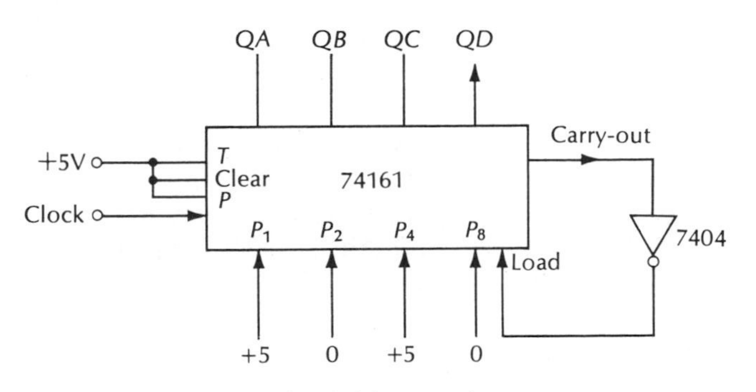

Figure 7-32 74161 Configured for a Mod 11

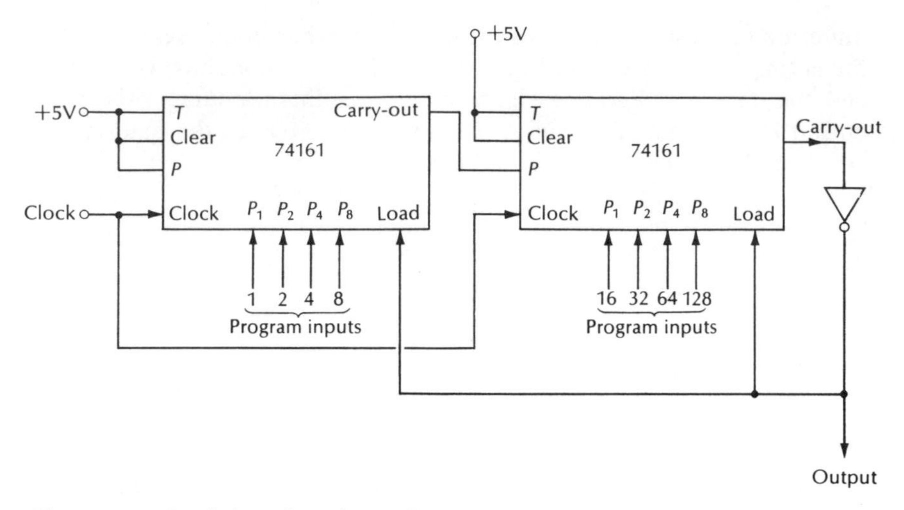

Figure 7-33 Divide-by-1-through-256 Counter

7-13 C-MOS Counters

There are three basic integrated circuit C-MOS counter types. C-MOS ripple counters are based on master-slave *J-K* flip-flops with circuits very similar to those used in TTL ripple counters. One version of synchronous C-MOS counters use type *D* flip-flops with circuits similar to those used in TTL synchronous counters. The third C-MOS counter is a form of synchronous shift register counter that we will examine in the next chapter.

C-MOS RIPPLE COUNTERS

Most C-MOS ripple counters require a positive clock pulse and count on the high-to-low transition (down-clock). The reset input is normally at ground for counting and is taken high for at least one-half a micro-second to insure that all flip-flops are reset and settled.

Typical current demands per C-MOS packages range from about 0.4 mA to 0.8 mA, except for the 4060 which features built-in clock circuitry. If the internal clock circuit is used, it requires an extra 2 mA.

Because C-MOS circuitry has been steadily improving during the last several years, it is necessary to consult the data manual for more exact data on the device in question. Table 7-4 lists typical parameters for common ripple counters.

C-MOS SYNCHRONOUS COUNTERS

Like C-MOS ripple counters, C-MOS synchronous counters tend to provide more counting hardware per package than similar TTL de-

Table 7-4 C-MOS Binary Ripple Counters

Type Number	Number of Stages	Divide by	Maximum Counting Frequency	Missing Outputs	Comments
4020	14	16384	@10V.,7MHz @5V.,2.5MHz	Stages 2 & 3 (2^2 & 2^3)	—
4024	7	128	@10V.,7MHz @5V.,2.5MHz	None	—
4040	14	4096	@10V.,6MHz @5V.,2MHz	None	—
4060	14	16384	@10V.,4MHz @5V.,1.75MHz	Stages 1, 2, & 3	Package contains internal clock generator for use as required.

vices. C-MOS counters are conservative with power consumption but have lower counting frequencies than their TTL counterparts. In cases where the lower counting rates of C-MOS counters can be tolerated, C-MOS may be preferred to TTL. Good compatibility between TTL and C-MOS makes it possible to mix the two logic forms in the same logic system.

Table 7-5 provides a summary of the characteristics of the most common synchronous C-MOS counters.

7-14 Decoding Counters

Most of the counters we have discussed so far produce output signals in standard binary or binary coded decimal (BCD). These counters can be decoded into a one output per binary group or seven-segment display format using standard MSI decoder packages. Some C-MOS counters have missing outputs because of limitations on the number of package pins and cannot be fully decoded. A standard decoder, such as the 74154, can be used to decode counters with non-standard count sequences by changing the output labels on the decoder. The 7492 mod 12 counter is one of those that require a change in the decoder output labels. In general, seven-segment decoders can be used only with counters that produce a normal binary or BCD count sequence.

Table 7-5 Synchronous C-MOS Counters

Type Number	Description	Number of Stages	Maximum Counting Frequency	Comments
4510B 4516B	BCD (Mod 10) Up/Down Presettable Binary Up/Down Presettable	4	@10V.,4MHz @5V.,2MHz	Clocks on positive-going edge of positive clock pulse. Up/down input is high for up-count and low for a down-count. Count preset is accomplished by momentarily taking preset inputs high. The counter can be reset by making the reset input high or by entering 0000 at the preset inputs.
4518B 4520B	Dual BCD (Mod 10) Up/counter Dual Binary (Mod 16) Up/counter	4	@10V.,6MHz @5V.,3MHz	Can be made to count either on the positive-going or negative-going edge of the positive clock pulse. The clock input can be grounded and the enable input can be used as a clock input when negative-edge clocking is required.
40102B 40103B	Two Decade BCD (Mod 10) Presettable Down/counter Binary Down/ counter. Presettable	8	@10V.,3.5MHz @5V.,1.5MHz	The only output from these counters occur at the end of the programmed down count.
40192B 40193B	BCD (Mod 10) Up/Down with Dual clock Reset input Presettable Binary (Mod 16) Up/Down with dual clock reset input presettable	4	@10V.,7MHz @5V.,3MHz	Separate control and clock inputs for up-counting and down-counting. A low on the preset enable input permits parallel presetting of the counter. During normal counting both carry and borrow outputs remain high. The carry output goes low when the highest count is reached. The borrow output goes low when the minimum count is reached. These counters can be cascaded.

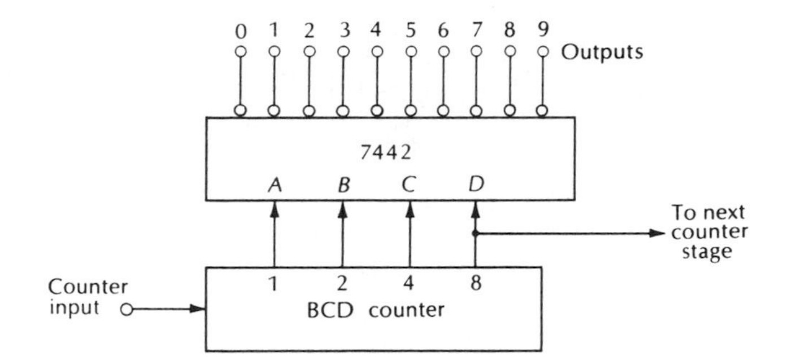

Figure 7-34 Decimal Counter and One-of-10 Decoder

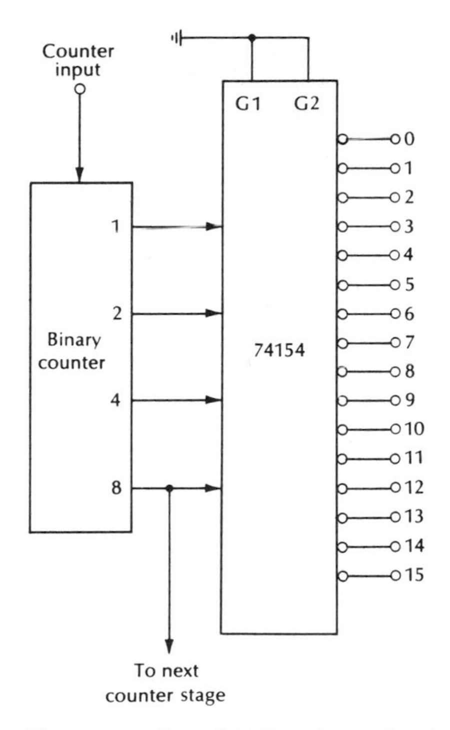

Figure 7-35 One-of-16 Decoder Used with a Binary Counter

Figure 7-34 shows the 7442 BCD-to-one of 10 decoder connected to a 7490 decade (mod 10) counter. Figure 7-35 shows a 74154 BCD-to-one of 16 decoder/demultiplexer used with a binary counter.

SEVEN-SEGMENT DECODER DRIVERS

Seven-segment decoder drivers generally have special circuitry to allow leading or trailing zeros to be blanked out when several decades

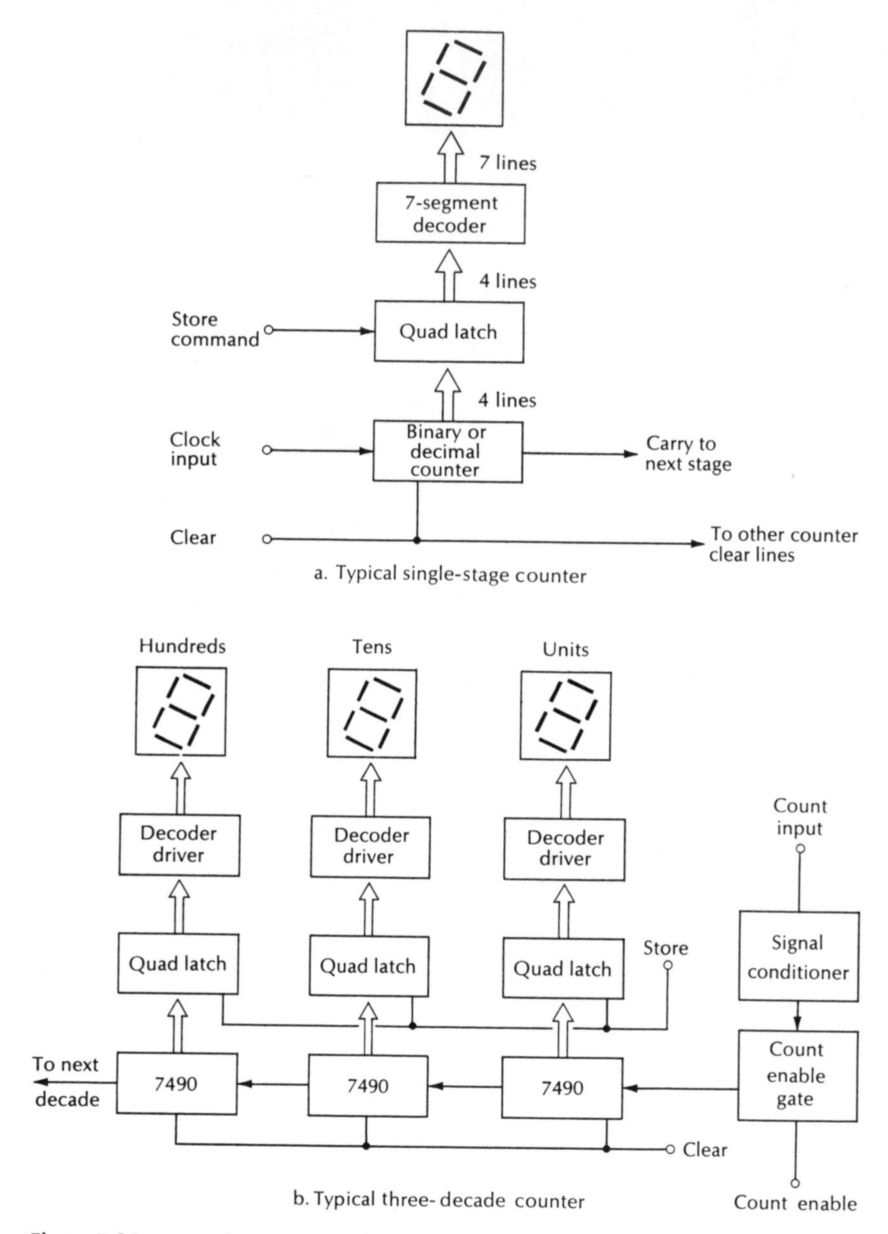

a. Typical single-stage counter

b. Typical three-decade counter

Figure 7-36 Typical Single-Stage Counter Arrangement and Three-Decade Counter

are involved. An override blanking input is usually provided to permit an entire digit to be blanked. This input is most often used to control the display brightness by pulse modulating the override blanking input. The override input is also used to time-multiplex displays. Time-multiplexing will be covered in a subsequent chapter. Most seven-segment decoder drivers also have a lamp-test input that lights all seven segments when activated.

DISPLAY MEMORY

Very often in counter display systems a memory in the form of a latch is employed to hold a display value while the counter is working on a new count. Without the latch the display would be a distracting blur during the counting period. The 7475 level clocking and 74100 quad latch are two popular ICs for this task. Note that this latch is level clocking, not edge clocking as some data sheets seem to imply.

For edge-clocking devices, the 74175 quad D and 74174 hex D latches are available. To use the level-triggered latches, a high is placed on the enable lines and the latches follow the inputs. The enable lines are taken to ground to hold the count on the display. To use the 74175 and 74174 edge-triggered latches, the clear is normally held high. The positive-going edge of a pulse enters data into the latch and holds it. The clear line is momentarily taken to ground to clear the latches. Latches are placed between the counter and the decoder.

Figure 7-36a shows a typical counter/latch/decoder circuit for a single decade. Figure 7-36b illustrates a typical three-decade counter circuit.

SUMMARY

IC counters are available in a variety of forms and with a variety of capabilities.

Counters are either ripple or synchronous or hybrid. Synchronous counters are faster but more complex. Bidirectional (up/down) counters are available in both synchronous and asynchronous (ripple) forms.

A counter can be made to count with any modulus by shortening the natural count. Programmable counters are variable modulus devices. The modulus is determined by setting conditions on the programming inputs.

In many applications, counters must be decoded to perform the desired function. A variety of IC decoders are available for this purpose.

Problems

10. Draw the logic block diagram for a three-stage binary ripple up-counter. What is its natural modulus?
11. Draw the logic block diagram for a binary ripple up/down counter.
12. Given a six-stage binary ripple counter using flip-flops with a delay of 30 nanoseconds per flip-flop, compute the maximum operating frequency for the counter.

13. Describe the basic techniques for obtaining a counter modulus that is not an integral multiple of 2.
14. How many equivalent propagation delays are there in a five-stage synchronous counter?
15. What kind of flip-flop should generally be used for stage one in the counter system in Figure 7-36b?
16. Draw the logic block diagram for a four-stage binary ripple up-counter using type *D* flip-flops.
17. Draw the logic block diagram for a four-stage binary ripple down-counter using type *D* flip-flops.
18. What is the purpose of a display memory? What is its most common form?
19. Define *modulus counter.*
20. Why must TTL reset (clear) pulses be 10 microseconds (or more) wide?
21. What is a programmable counter?
22. What is a presettable counter?
23. When a decoder is used with a counter, what is the purpose of controlling the decoder with a strobe pulse?
24. With reference to counters and decoders, what is a *glitch?*
25. Why is a ripple counter slower (lower counting rate) than a synchronous counter?
26. What is a *synchronous counter with ripple carry?* In what way is it a compromise between ordinary ripple and full synchronous counters?
27. What is the maximum count possible with 8 flip-flops?
28. Suggest a possible application where the reset-to-9 function of the 7490 decade counter could be useful.
29. Make a drawing showing a 74161 counter programmed for a modulus of 12.
30. Make a table of all presettable counters you can find in the data manuals. Look for both TTL and C-MOS types. (See Table 7-4 for an example.)
31. Look up the 7447 BCD-to-seven-segment decoder in a data manual. Explain the use of the RBO and RBI terminals on the device. Use a sketch combined with a brief written description to explain how and why these functions are used.
32. Given the circuit in Figure 7-35, change the labels on the output lines of the 74154 to accommodate the non-standard count sequence of the 7492 mod 12 counter. See the mode 1 truth table in Figure 7-29 for the count sequence.
33. Special Challenges
    a. Look up the 74191 up/down counter in the data manual. Add the necessary gate circuitry to convert it into a divide-by-10 counter.

b. Draw a logic diagram to convert the 74191 counter into a modulus 12 counter.
c. Draw a logic diagram showing the detailed gate structure to make a 74191 counter count up to ten and then automatically reverse itself and count back to zero.

CHAPTER 8

SHIFT REGISTERS AND SHIFT-REGISTER COUNTERS

Learning Objectives. *Upon completing this chapter you should know:*

1. *How to describe the shift register.*
2. *How flip-flop counters and shift registers differ.*
3. *How to define SISO, SIPO, PIPO, PISO.*
4. *How to explain the difference between true parallel loading and preset-only loading.*
5. *How to describe the characteristics of a left/right shift register.*
6. *How to draw the several functional external connections for the left/right shift registers.*
7. *How to describe the operating characteristics of the universal shift register.*
8. *How to draw and explain the operation of a shift-register ring counter.*
9. *How to identify the modulus of the ring counter by the number of stages.*
10. *How to draw the functional block diagram for a Johnson counter of length N.*
11. *How to draw the decoding circuitry for any Johnson counter.*
12. *How to define these terms: disallowed state, disallowed subroutine, self-start circuitry, self-correcting circuitry, modulus.*
13. *The Johnson counter's use in C-MOS.*
14. *The function of pseudo-random sequence generator.*

The shift register is a memory system consisting of flip-flops or MOS dynamic memory cells. The special feature of the shift register is that data can be transferred on command from cell to adjacent cell as many times as desired.

In the simplest type of shift register, data are entered serially one bit at a time until the desired number of bits have been entered and stored. Before each new bit enters the input, all of the previously entered bits are moved one stage (or cell) down the line to make room for the new bit. The data are shifted one stage at a time until all bits have arrived in turn at the output of the register. More complex registers provide outputs for each stage so that the entire contents of the register can be sampled in a single clock time. Parallel loading is also available on some shift registers where the entire register can be loaded in a single clock time.

The *universal* shift register is capable of right or left shift, both serial and parallel data entry, and output. Because of the number of output pins required for universal operation, it is generally restricted to shift registers with only a few stages.

Shift register packages are available with a wide variety of combinations of features and lengths. TTL shift registers are available with up to about 10 cells, and MOS devices with 2000 to 4000 stages are common.

In this chapter we will be primarily concerned with basic shift register operation and the application of small registers to special counting circuits. In Chapter 10 we will study some larger shift registers and their applications.

8-1 The Shift Register

Shift registers are synchronous systems using *J-K* flip-flops or type *D* flip-flops. Large MOS devices use either MOS flip-flops or dynamic MOS cells.

Figure 8-1 shows a basic shift register using *J-K* flip-flops. Unlike the ripple counter circuit, both Q and $\bar{Q}$ outputs are used and the circuit is *not* a frequency divider.

If there is a 1 in the first stage, Q_1 will be high. On the next clock pulse the high on Q_1 will transfer to Q_2. The first stage will either remain as it was before the clock or change states depending on the status of its J and K lines. If the first stage is in condition $\bar{Q}_1 = 1$, the next clock pulse will transfer that to stage 2 and make $\bar{Q}_2 = 1$.

In summary, the input conditions of each flip-flop at clock time T will

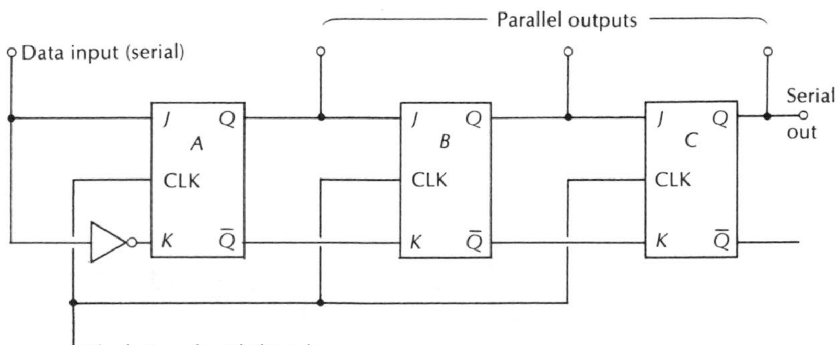

Figure 8-1 Using *J-K* Flip-Flops in a Shift Register

be transferred to the next stage at clock time $T + 1$. Notice that the first stage in Figure 8-1 uses a type D input circuit, a common configuration in shift registers. It is possible to construct an entire shift register using type D flip-flops as shown in Figure 8-2. Most of the latest shift registers use edge-triggered D flip-flops.

The J-K version in Figure 8-1 has a single data input in a D arrangement, called a *single-rail* input. If both J and K were available as separate data inputs, we would have a *double-rail* input. When both Q and $\bar{Q}$ are available as output, it becomes a double-rail output. If only Q is used as an output, it is a single-rail output.

Single-rail inputs and outputs are more of a packaging consideration than a logical one. Assume that an eight F-F shift register is to have all Q's and $\bar{Q}$'s available—this would require sixteen leads on the package with no leads available for inputs, clock, reset, GND, or V_{cc}. The register could only fit in a sixteen-lead package if fewer F-F's were used or the $\bar{Q}$ lines were simply not brought out of the package. In many applications the $\bar{Q}$ outputs can easily be left out or can be *derived* by using an inverter and the Q output.

Shift registers are also classified as serial or parallel *in* and serial or parallel *out*. Any shift register can be used in the serial-in/serial-out mode, but not all have parallel-in or parallel-out capability. Parallel in means that the data can be loaded into all flip-flops at the same time, while for a serial in, each bit (0 or 1) must be loaded into the input f-f and shifted over to make room for the next bit. The process continues until the register is loaded. When parallel out capability exists, all Q's are brought out of the package and the state of all flip-flops can be read at the same time. In a serial-out mode, the bits must be shifted out of the register one at a time to be used. The circuit in Figure 8-1 is a serial-in shift register with either parallel or serial out capability.

Shift-register classifications:

a. SISO: Serial-in/serial-out
b. SIPO: Serial-in/parallel-out
c. PISO: Parallel-in/serial-out
d. PIPO: Parallel-in/parallel-out

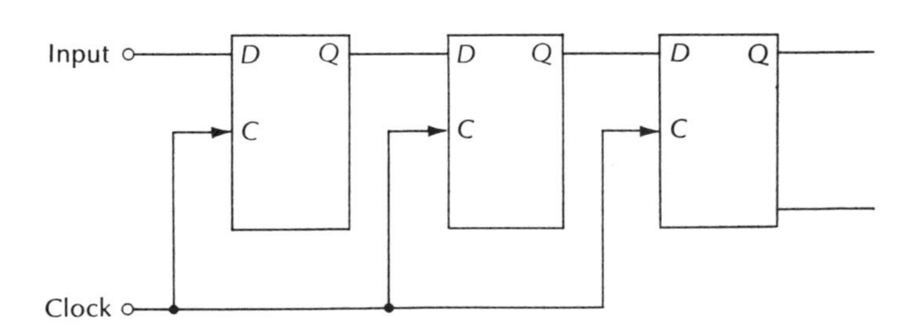

Figure 8-2 Type D Shift Register

All parallel-out registers are also capable of serial-out, and all parallel-in registers are capable of serial-in operation. The reverse of these two cases is not true.

PARALLEL LOADING CONSIDERATIONS

Some parallel load shift registers have preset-only capability. This means that an individual 1 in a given stage cannot be changed to a 0. The entire register must first be cleared to all zeros and reloaded with altered data.

Table 8-1 Common TTL and C-MOS Shift Registers

Length (Stages)	Type Number	TTL or C-MOS	Parallel Outputs	Direction	Parallel Load	Clear
8	7491	TTL	no	right	no	no
4	7494	TTL	no	right	preset only	yes
4	7495	TTL	yes	right/left	synchronous	no
5	7596	TTL	yes	right	preset only	yes
8	74164	TTL	yes	right	no	yes
8	74165	TTL	no	right	asynchronous	yes
8	74166	TTL	no	right	synchronous	yes
4	74194	TTL	yes	right/left	synchronous	yes
4	74195	TTL	yes	right	synchronous	yes

a. Common TTL shift registers

Type Number	Parallel Outputs	Number of Stages	Shift Direction	Parallel Load	Clear	Comments
4094	yes	8	right	no	no	tristate outputs
40100	no	32	right/left	no	no	recirculate capability
40104	yes	4	right/left universal	yes	no*	tristate outputs
40194	yes	4	right/left universal	yes	yes	

*The register can be cleared by taking both mode controls low and clocking the register.

b. Common C-MOS shift registers

True parallel load registers allow updating at any time without clearing first. Table 8-1 lists some common TTL and C-MOS shift registers and, among other information, states whether the register is a true parallel load or preset-only device.

Loading may be either asynchronous or clocked. In synchronous (clocked) load devices, loading always occurs with or at the fall of the clock pulse. Most TTL shift registers clock on the positive-going transition of the clock pulse.

8-2 The Right/Left Shift Register

One very useful capability in shift registers is that of bidirectional shifting—right shift or left shift. This mode generally requires that both parallel inputs and parallel outputs be available along with some added gates to control the shift direction. A shift register with all of these capabilities, and a few minor additions to increase the flexibility yields a configuration known as the *universal shift register*. See the functional block diagram for the 7495 left/right shift register in Figure 8-3.

The PE (parallel enable) controls the data entry mode (serial or parallel) and is frequently labeled *mode control*. A 0 input on PE enables all of the AND gates labeled 1. Both parallel and serial outputs are available, but when the parallel inputs are disabled by the AND gates labeled 2, it is in a serial-shift-right circuit in this mode. A 1 on the PE (mode control) input allows parallel data to enter and load the register.

After loading, the mode control can be switched back to serial-right shift and the data shifted *right*, through the register. In this way parallel data can be converted into serial form one bit at a time. Conversely, data can be converted from serial into parallel form by shifting the data into the register serially and taking the data from parallel outputs. The mode input should not be allowed to change states during clocking. Two clock inputs are provided so that parallel and serial entries can be made at different clock phases or rates. Proper timing of the two clocks can insure that the mode does not change too near the clocking edge of a clock pulse. A 10 ns delay is required for standard TTL after the clock pulse before mode switching can take place (although no delay is required between a mode change and next clock pulse). A 100 ns delay is required for the low-power TTL version of the left/right shift register.

The left/right shift register can be configured for a left/right serial input mode by wiring it as shown in Figure 8-4d.

UNIVERSAL SHIFT REGISTER

The universal shift register is similar to the right/left shift register just discussed except that extra gating circuits are provided so that all

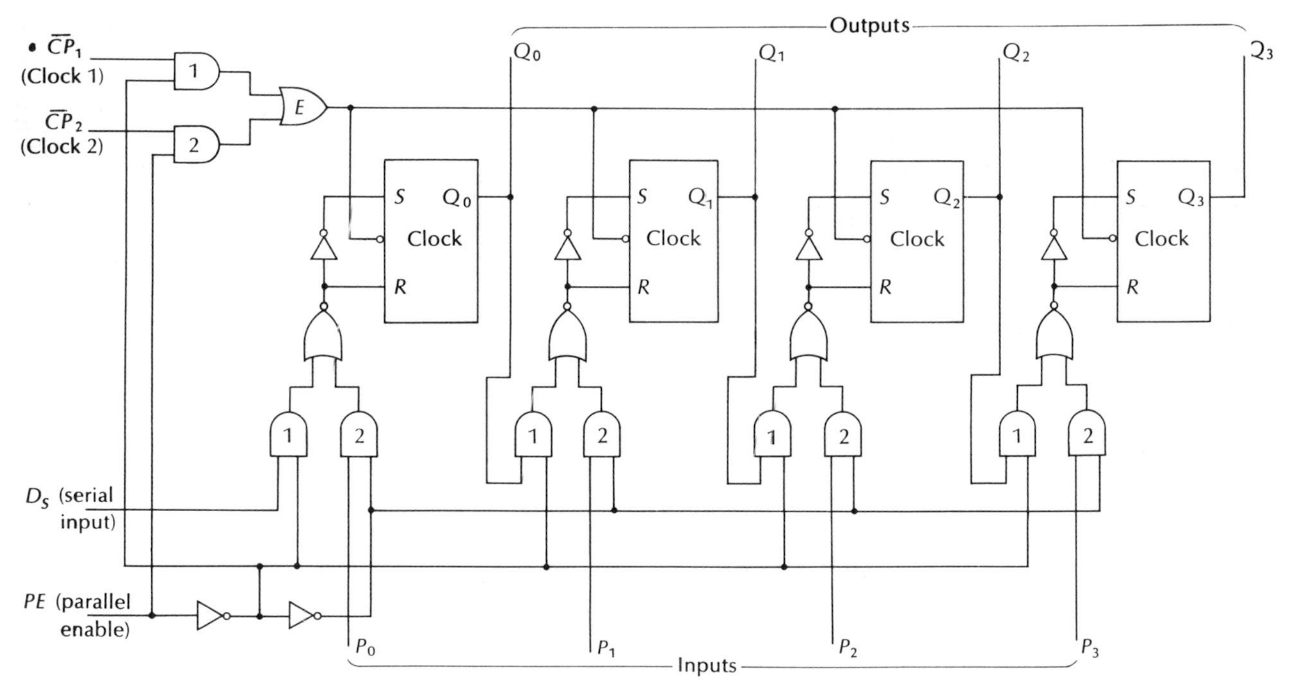

Figure 8-3 The 54/7495 Left/Right Shift Register

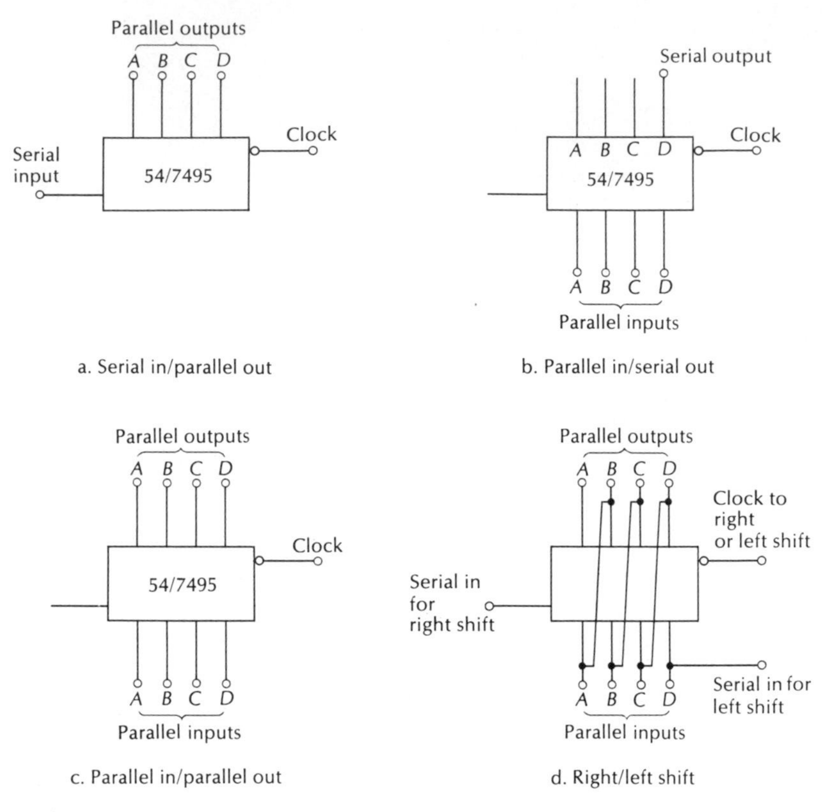

Figure 8-4 Operating Modes for the 54/7495 Shift Register

modes can be obtained upon command. No external wiring is required to get the right/left shift mode. A *load* input allows parallel loading when the input is at logical 1 and returns the register to a right/left shift mode with a logical 0 on the *load* input. The logic diagram of the 74194 universal shift register is shown in Figure 8-5.

Problems

1. Define *single-rail* and *double-rail inputs* and *outputs*.
2. What can be done to obtain a $\bar{Q}$ when $\bar{Q}$'s are not brought out of the IC package?
3. Define *serial in, serial out, parallel in,* and *parallel out*.
4. Answer the following with regard to shift registers with parallel-*out* capabilities:
   a. Are $\bar{Q}$'s always available?
   b. Can a parallel-*out* register always be used for serial-*out*?
5. Describe the capabilities of the universal shift register.

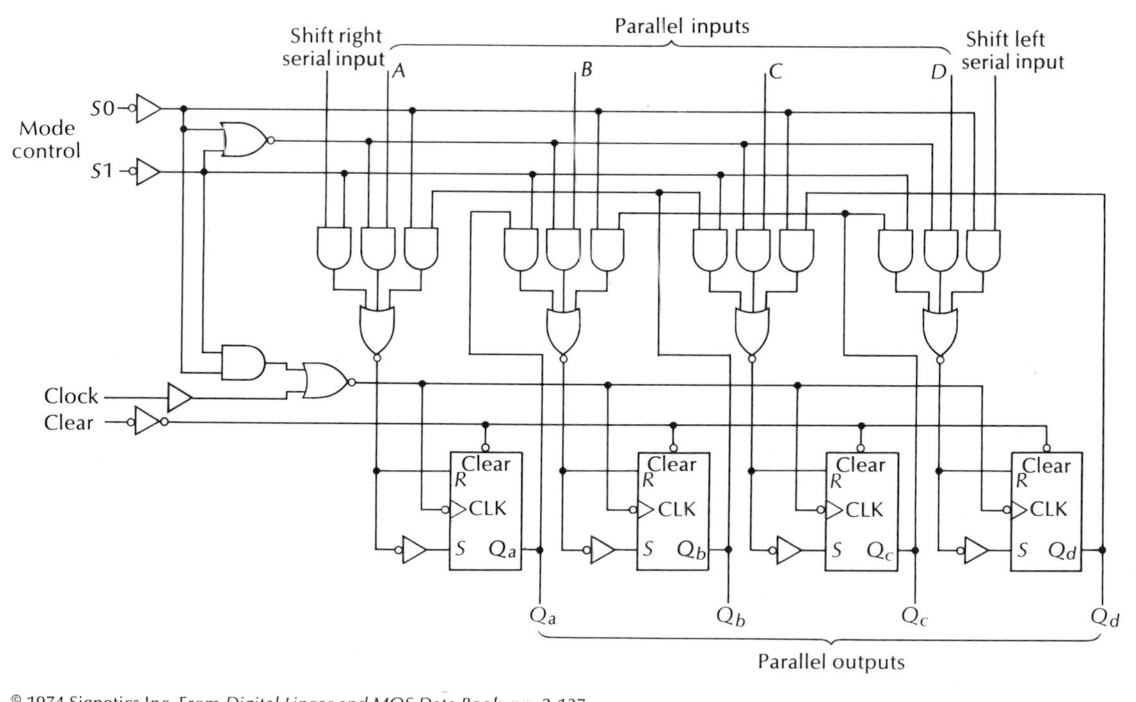

Figure 8-5 54/74194 Universal Shift Register

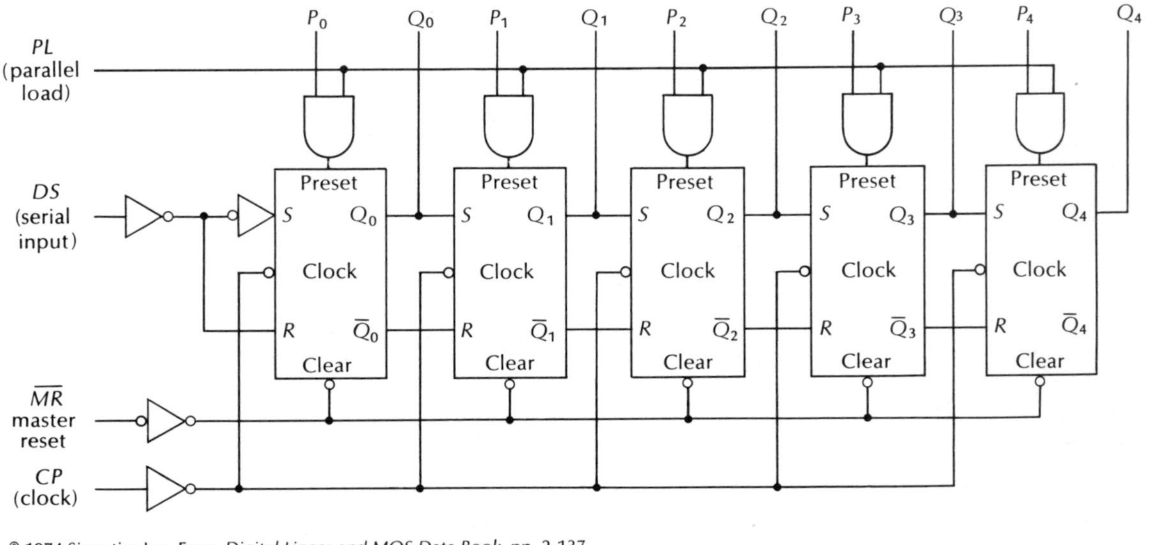

Figure 8-6 54/7496 Five-Bit Shift Register with Serial Parallel/In and Serial Parallel/Out (Right Shift Only)

8-3 Other Common Shift Register Configurations

Shift registers come in a variety of forms: serial-*in*/serial-*out,* parallel-*in*/parallel-*out,* serial-*in*/parallel-*out,* parallel-*in*/serial-*out* and configurations capable of any of these combinations. In addition, there are right-shift registers, left-shift registers, and bidirectional registers. Most of these varieties are available with several bit lengths, with four-, five-, and eight-bit lengths being the most common. Table 8-1 can be used as a shift register selection guide for 54/7400 series and C-MOS shift registers.

Figures 8-6 through 8-9 show the logic diagrams for some typical shift-register types.

Some shift registers, such as the 74165, are designed primarily for parallel-to-serial conversion. When the register is used for this purpose, data are loaded in parallel (all F-F's are set to 1 or 0 simultaneously), and the data are clocked out of their serial output one bit at a time. The data are marched in abreast and clocked out in single file.

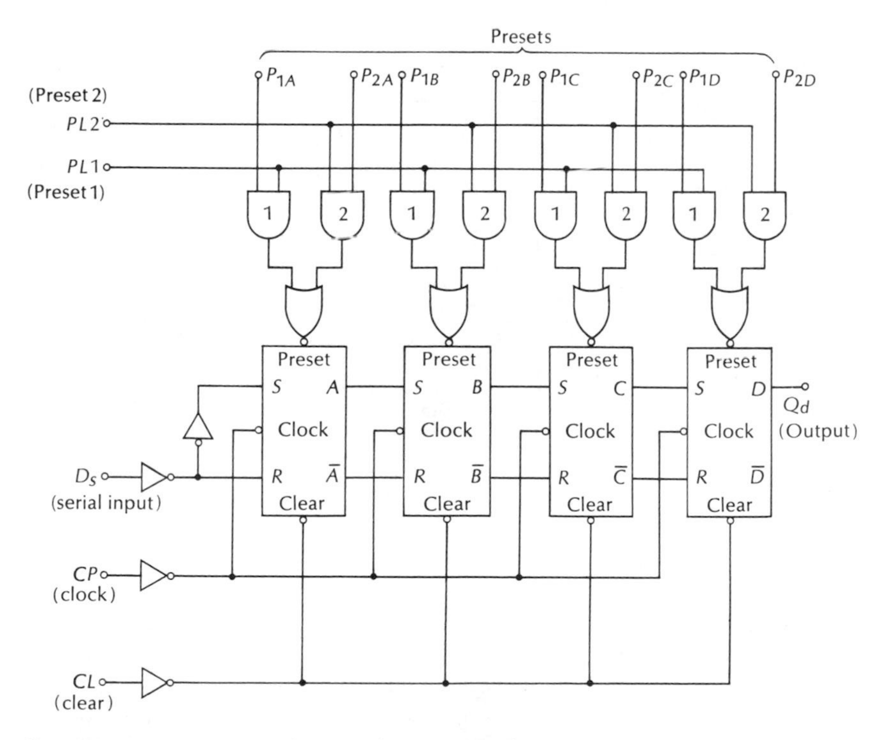

Figure 8-7 The 7494 Four-Bit Shift Register

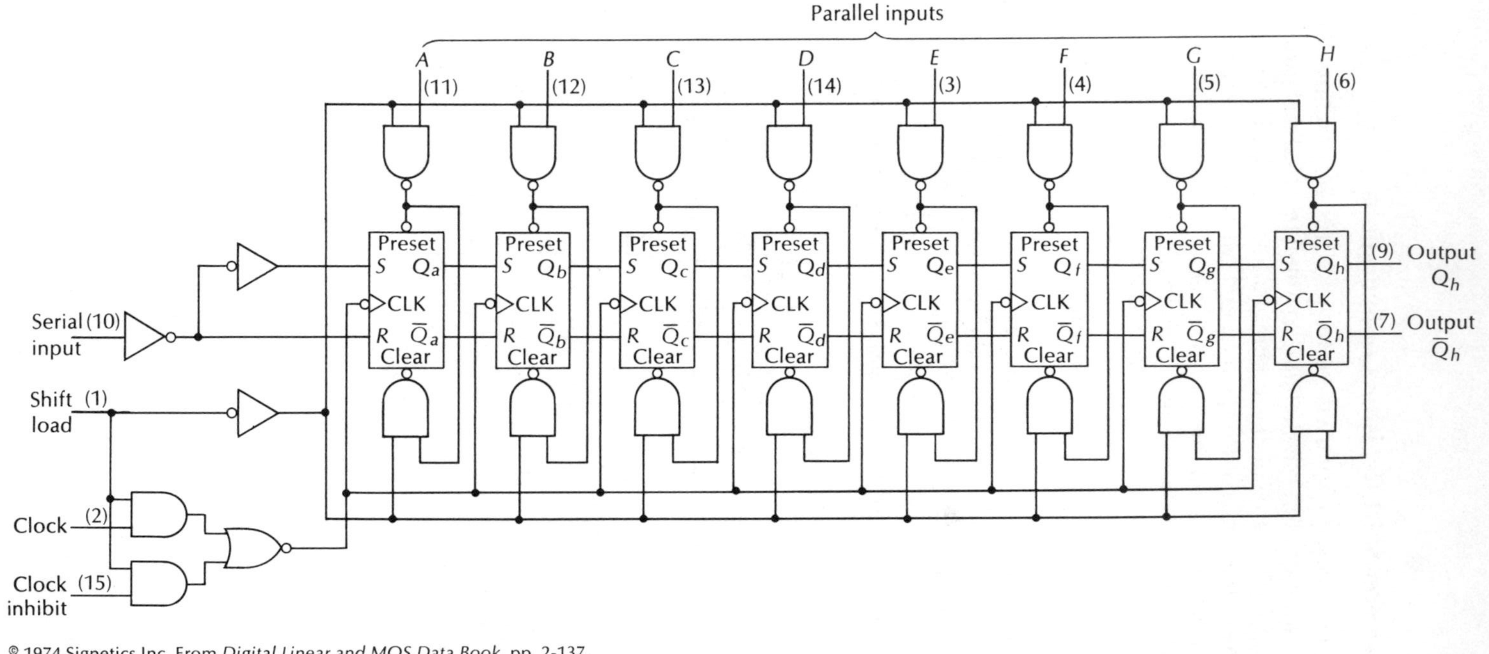

Figure 8-8 54/74165 Eight-Bit Shift Register

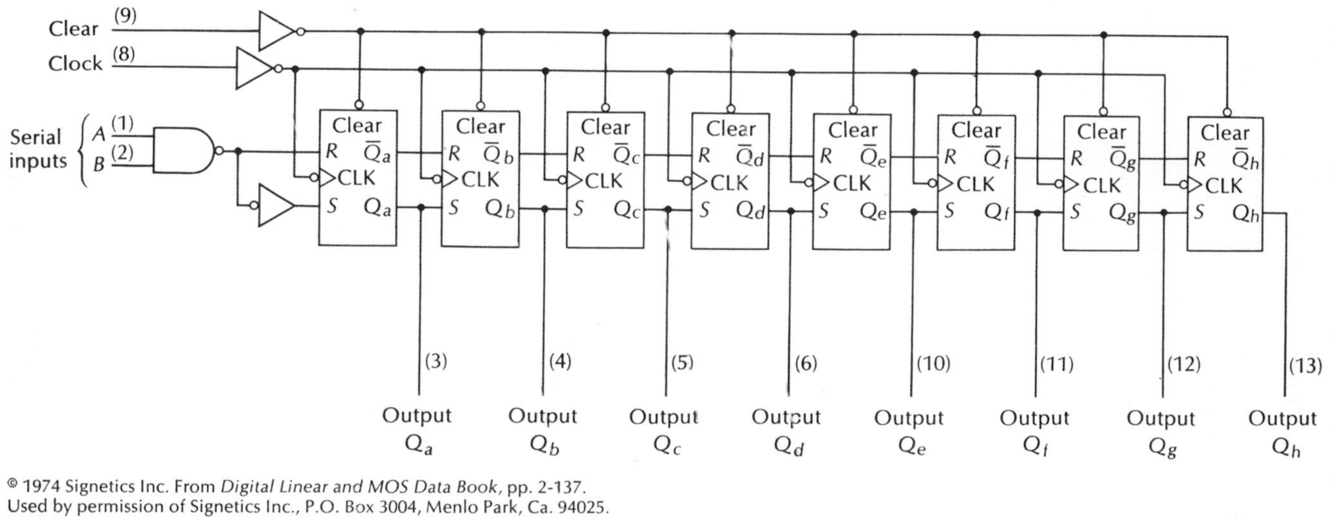

Figure 8-9 54/74164 Eight-Bit, Serial In-Parallel Out Shift Register

8-4 Four-Bit Data Selector/Storage Register

The 54/74L98 is not actually a shift register in the strictest sense, but it is very similar to one. As you will see by the logic diagram shown in Figure 8-10, it can be used as a shift register if properly connected. The *selected* word data are transferred to the F-F outputs on the negative-going edge of the clock pulse. Word one is *selected* by applying a logical 0 to the word-select input. A 1 on the word-select input selects word two. Two clock inputs are provided for different phases or clock rates for the two sets of data.

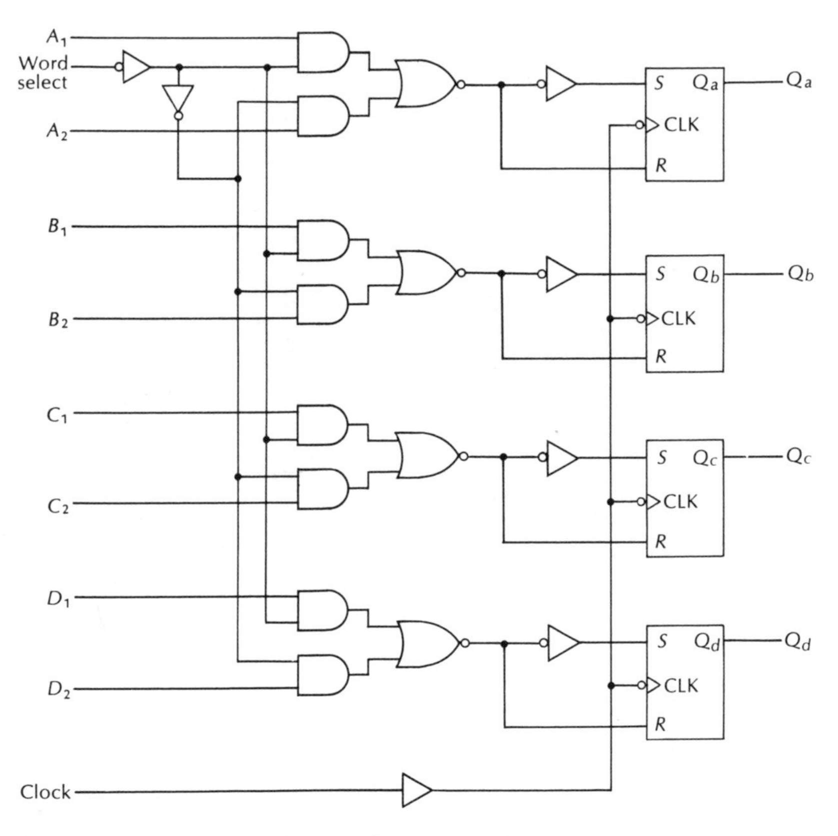

Figure 8-10 54/74L98 Data Selector/Storage Register

8-5 The Shift-Register Ring Counter

A ring counter can be formed by loading a 1 into the leftmost F-F of a shift register, with the serial output connected to the serial input, and clocking the 1 down the chain. After as many counts as there are F-F's, the 1 will have propagated out the end of the rightmost F-F and will have been returned to the one where it started. The ring counter is very inefficient in its use of flip-flops. For example, four flip-flops used in an ordinary binary counter can provide sixteen discrete outputs when decoded, while a ring counter can produce only four discrete states with four flip-flops. The key word here is *decoded*. The ring counter requires no decoding hardware, as opposed to the fairly complex decoding structures required for binary counters.

The binary counter gets 2^N counts, where N is the number of flip-flops, whereas the ring counter has a maximum count of N. Counters with a modulus that is not a power of 2 utilize flip-flops *less* efficiently than those with a power of 2, and *more* efficiently than the ring counter. However, the ring counter is the only common one that requires no decoding. Modulus counters may skip certain possible binary combinations, often by returning to zero before the natural maximum count has been reached. The combinations skipped are called *invalid* or *disallowed* states because, if the counter is working properly, they will never occur in the counting sequence. For example, in a decimal counter the binary equivalents of 10, 11, 12, 13, 14, 15 (base 10) will never occur because the counter uses only 10 states (0–9).

In the ring counter there are always $2^N - N$ disallowed states (N = the number of flip-flops). Figure 8-11 shows the functional block diagram and truth table of a four F-F ring counter. The *initiate* line resets all flip-flops but A to 0 and presets A to 1.

The ring counter must be cleared and loaded with a *single* 1. If noise, power supply shutdown, or some other factor causes more than a single 1 to be loaded (or no 1's at all), the counter will be loaded with a disallowed state and will start counting (when clocking begins) in a disallowed subroutine. For a four F-F ring counter there are five disallowed subroutines possible using the $2^N - N$ disallowed states. If $2^N = 16$ and $N = 4$, then $2^N - N = 16 - 4 = 12$ disallowed (or invalid) states. Once the counter is started with a disallowed state, it falls into a disallowed subroutine and will never count properly until the counter is forced by external influence to get to a state where *one and only one* F-F is at $Q = 1$. Any count sequence that includes a disallowed state is called a disallowed subroutine. In many cases it is necessary that the counter not only have just one $Q = 1$ in the chain, but also that the count begin with a specified F-F each time.

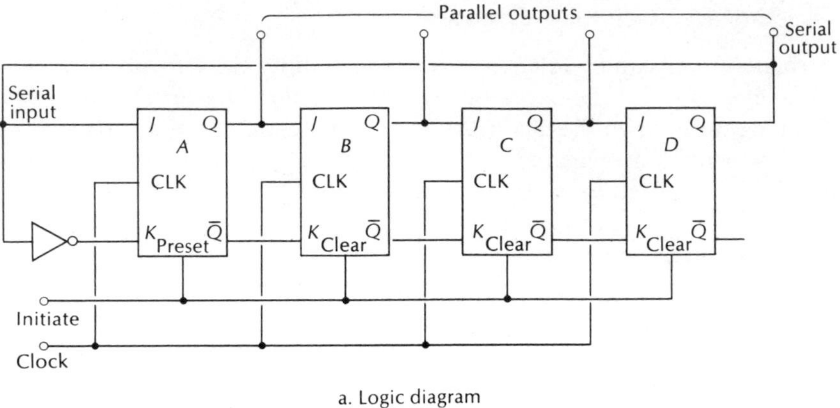

a. Logic diagram

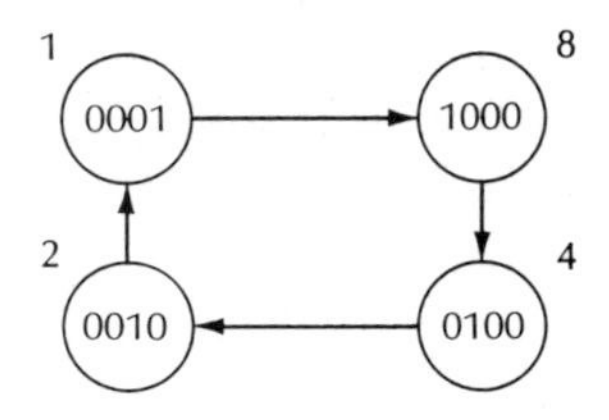

Desired counting routine

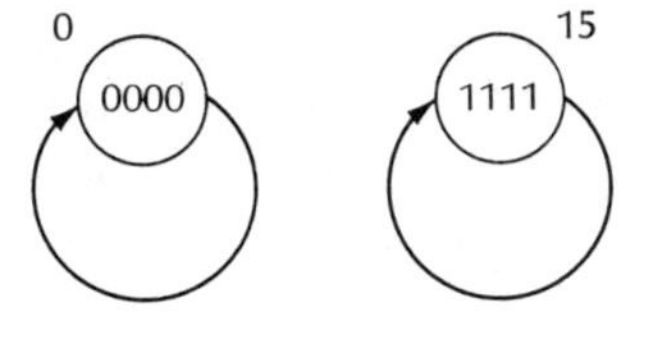

Disallowed subroutine

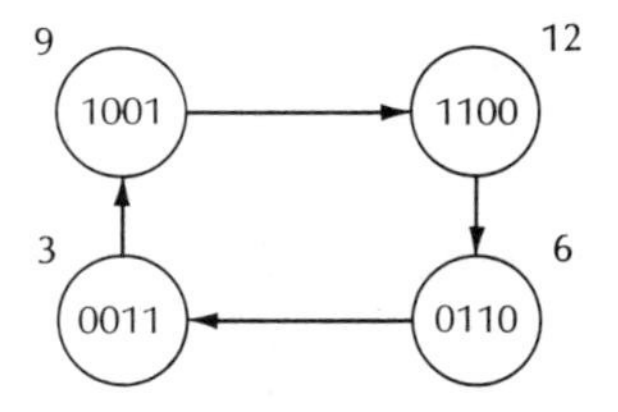

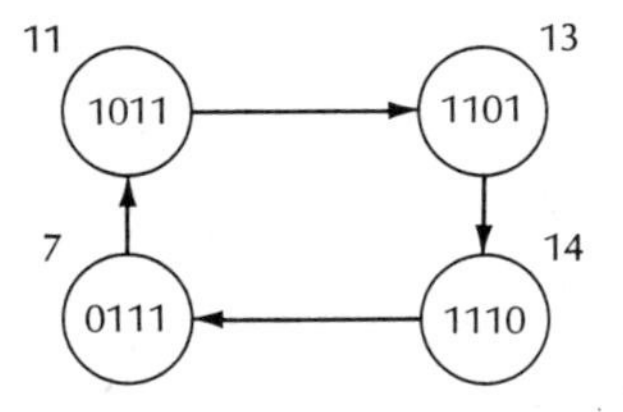

Disallowed subroutines

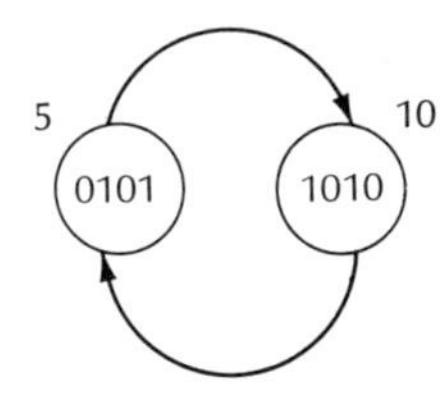

b. Disallowed states and disallowed subroutines for the ring counter

State	1	2	3	4	
1	1	0	0	0	Preloaded
2	0	1	0	0	
3	0	0	1	0	
4	0	0	0	1	
5	1	0	0	0	

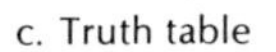
c. Truth table

Figure 8-11 Four-Flip-Flop Ring Counter

The circuit in Figure 8-11 provides an *initiate* input that sets all F-F's to 0 except *A*. *A* is set to 1 to prevent disallowed conditions. Figure 8-11b shows the disallowed states and possible disallowed subroutines for a four F-F ring counter.

SELF-CORRECTING FEEDBACK

The gate structure shown in Figure 8-12 can be used to automatically correct the counter when it gets locked into a disallowed subroutine. To use this circuit, you must be able to allow enough time (up to four counts) for the counter to make the correction. This configuration is self-starting even from 0000.

THE ZERO CIRCULATING RING COUNTER

Figure 8-13 shows a self-correcting, self-starting ring counter based on the 54/7496 shift register. The circuit is a five-bit ring counter that circulates a 0 rather than a 1. The NAND gate could be eliminated if the start of each count were preceded by loading 1's into all presets except *A* after clearing. All allowable combinations contain one 0 and four 1's. The NAND gate prevents disallowed combinations after a maximum of four counts.

8-6 The Johnson Counter

The Johnson counter is similar to the ring counter except that the *complement* of the output of the last flip-flop is fed back to the serial input of a shift register, resulting in a total count of $2N$, where N is the number of F-F's. The Johnson counter has $2^N - 2N$ disallowed states and can get locked into disallowed subroutines. The functional block

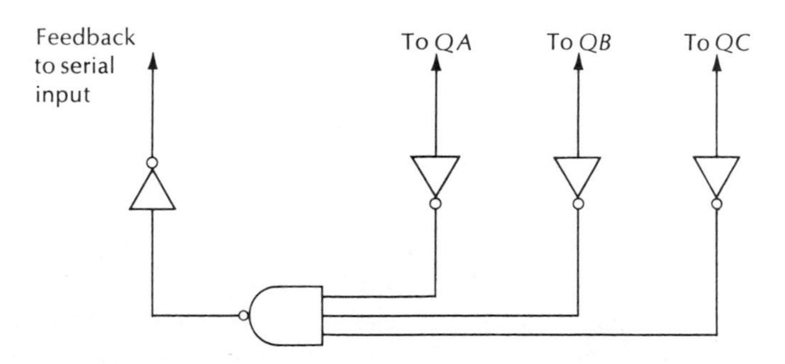

Figure 8-12 Feedback Circuit for a Four-Flip-Flop Ring Counter (Self-Starting and Self-Correcting)

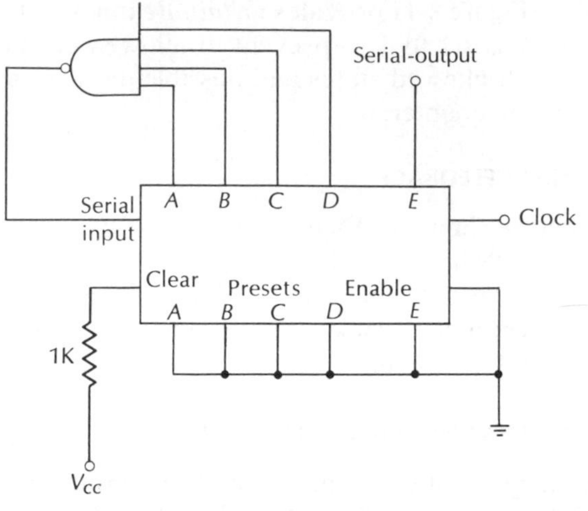

a. Logic diagram

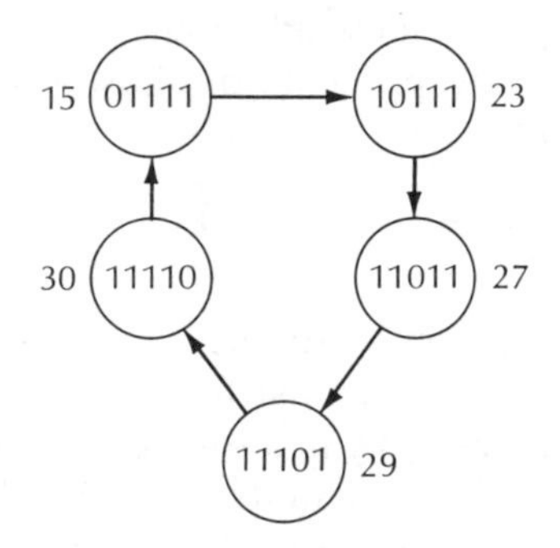

b. Allowed count routine

Figure 8-13 Zero-Circulating Ring Counter

diagram of the basic Johnson counter is shown in Figure 8-14, and the truth table for this device is shown in Table 8-2.

8-7 Decoding the Johnson Counter

Decoding the Johnson counter involves a two-input NAND gate for each decoded output. Each flip-flop drives two gate inputs. An evaluation of the truth table (Table 8-2) reveals that only one output changes states with any given clock pulse. Decoding is synchronous because only one change occurs at a time, and it happens with the clock. The output changes can be more easily visualized with the help of the timing

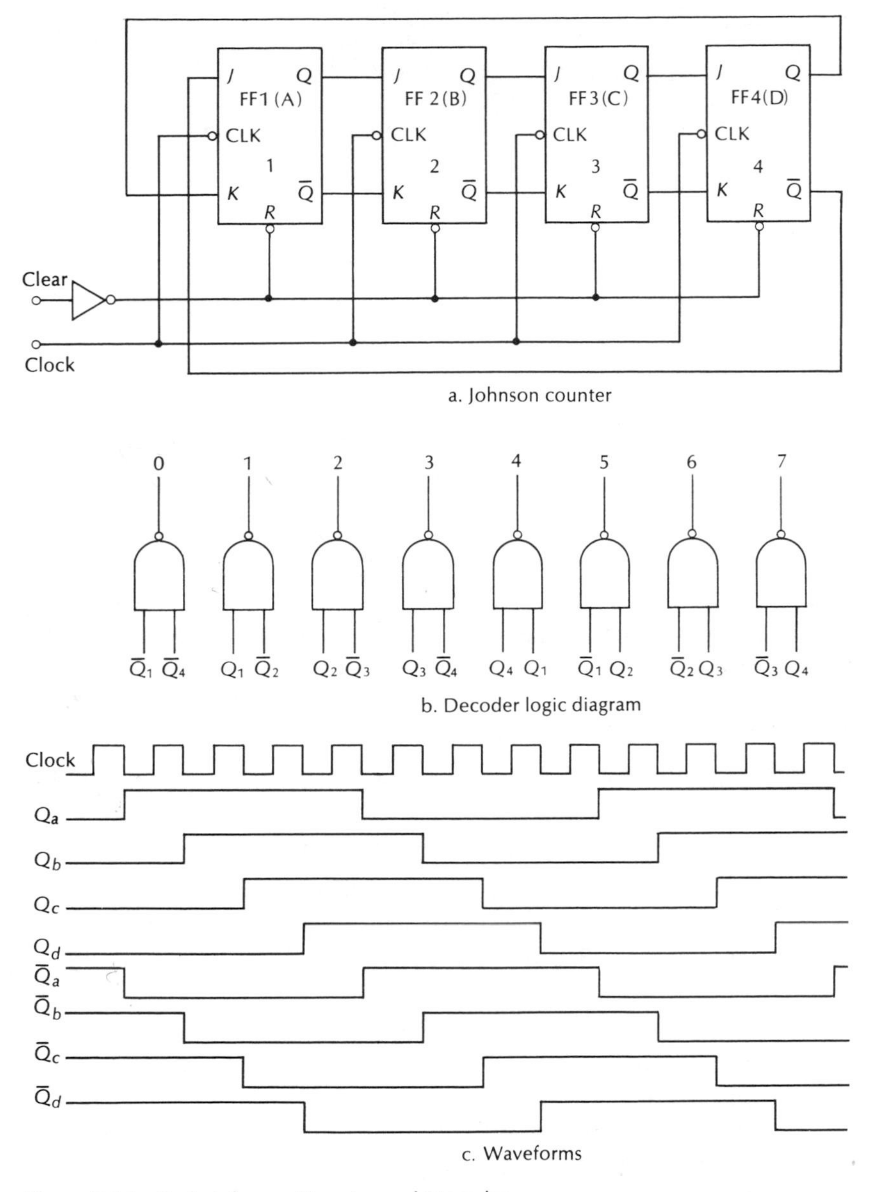

Figure 8-14 Basic Johnson Counter and Decoder

diagram in Figure 8-14. Notice that the Q outputs are shifted in phase by 360° where M is the counter modulus, and that the phase shift continues through the second 180° if we use the $\overline{Q}$ outputs. It is the symmetry shown in the timing diagram and truth table that makes

Table 8-2 Truth Table for the Four-Flip-Flop Johnson Counter

State	Flip-flop A	B	C	D
0	0	0	0	0
1	1	0	0	0
2	1	1	0	0
3	1	1	1	0
4	1	1	1	1
5	0	1	1	1
6	0	0	1	1
7	0	0	0	1
0	0	0	0	0

decoding a straightforward procedure. Figure 8-15 shows the decoding circuit for counters of modulus 3, 4, 5, 6, 7, 8, and 10. (Mod 9 is left as a student problem.)

If you study the even modulus decoder circuits in Figures 8-14 and 8-15 and the truth tables, you will observe the following pattern:

1. The Qs of flip-flops A and N are NANDed. N is the last flip-flop in the string.
2. The $\overline{Q}$s of flip-flops A and N are NANDed.
3. In each remaining row in the truth table there is a 1-0 pair adjacent to each other. These two entries define the inputs to the NAND gate for that row. No other entries in the row are used. Note these examples (see Table 8-2):

   Row 1: a 1 under $A = A$ and a 0 under $B = \overline{B}$ results in $f = A \cdot \overline{B}$.
   Row 2: a 1 under B is followed by a 0 under C. The NAND gate inputs are B and $\overline{C}$.
   Row 6: a 0 under B is adjacent to a 1 under C. The NAND gate inputs are $\overline{B}$ and C.

In the case of odd modulus number counters, one gate will be missing in the decoder as compared to that of the next higher even modulus counter. Otherwise all of the observations previously made are valid.

Problems

6. Write the truth table for a modulus 9 Johnson counter.
7. Draw the decoding circuit for a modulus 9 Johnson counter.

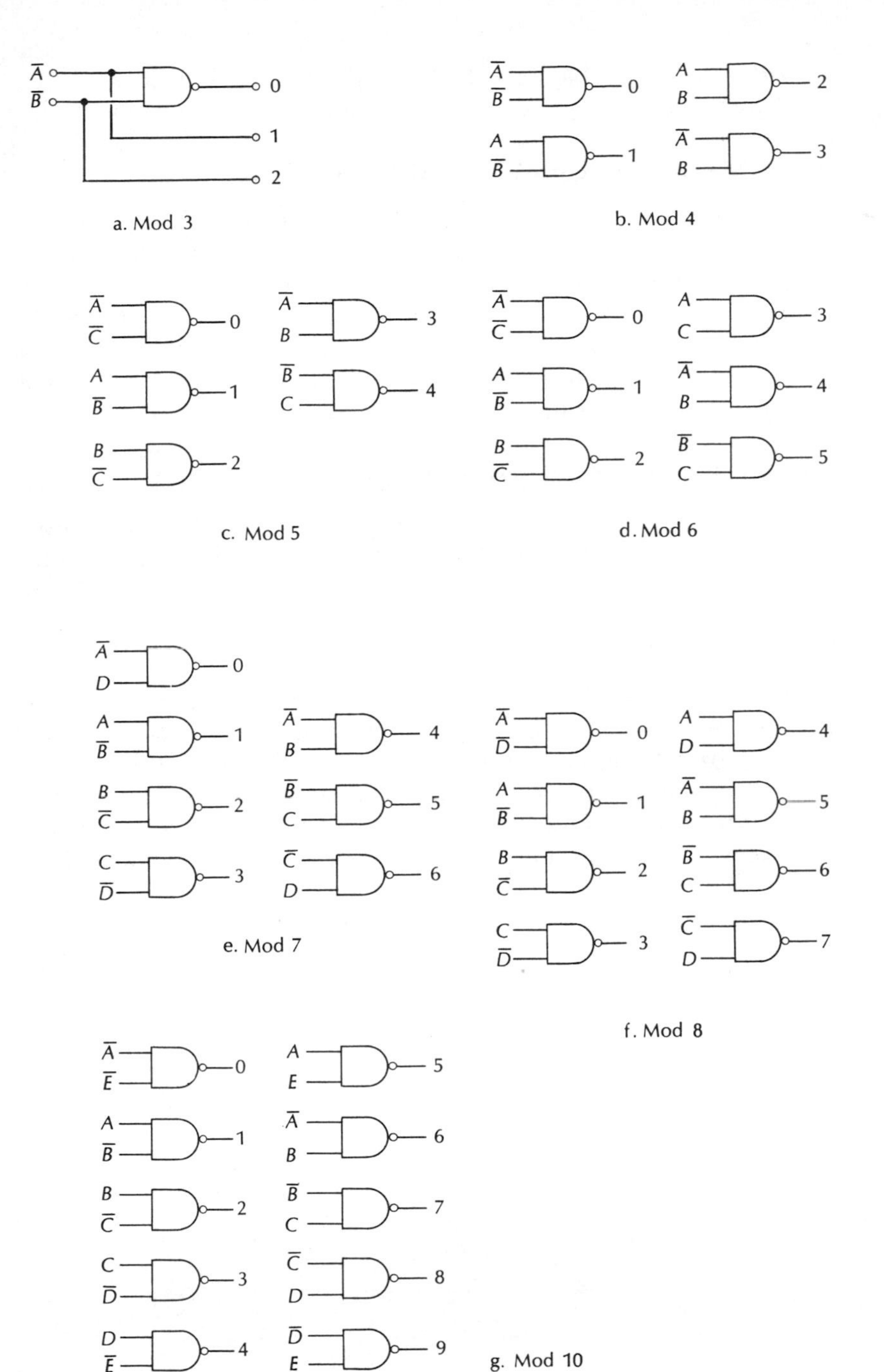

Figure 8-15 Johnson Counter Decoding

8. Write the truth table for a modulus 12 Johnson counter.
9. Draw the decoding circuit for a modulus 12 counter.

JOHNSON COUNTER USING 7495 SHIFT REGISTER

The Johnson counter can be implemented using the 7495 shift register. An enable pulse on the mode control and clock 2 lines loads the counter with zeros, acting as a reset-to-0 function. The functional block diagram is shown in Figure 8-16. The inverter is required because the Johnson counter demands that the complement of the output be fed back to the input. In the 7495, $\bar{Q}$ outputs are not available so an inverted Q_D is used. Refer to Figure 8-3 for the 7495 logic diagram.

SELF-STARTING, SELF-CORRECTING JOHNSON COUNTER

Figure 8-17 shows the functional block diagram of a four-F-F Johnson counter with additional gating to provide self-correction and self-starting in the desired counting routine.

ODD-LENGTH JOHNSON COUNTERS

So far we have been concerned with even-length (2, 4, 8, and so on) counters. An odd-length counter can be obtained by deriving the J input feedback from the $\bar{Q}$ of the next-to-last flip-flop in the chain. In many cases the $\bar{Q}$ outputs are not available, and an inverter must be used in conjunction with the Q output of the next-to-last stage. Figure 8-18 shows the configuration for Johnson counters of modulus 3, 5, 7, and 10.

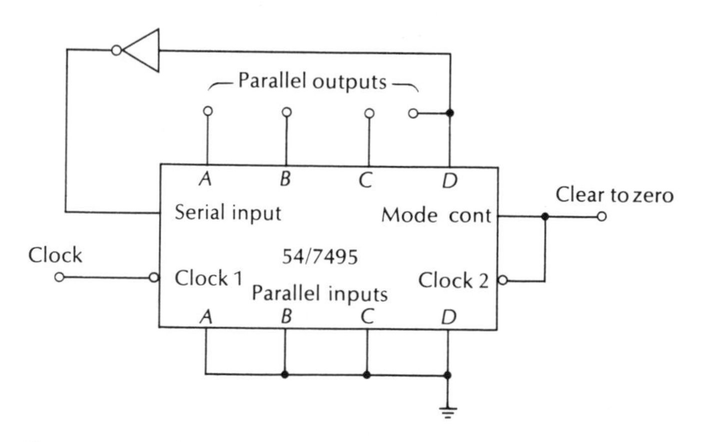

Figure 8-16 Johnson Counter Implemented with the 7495 Shift Register

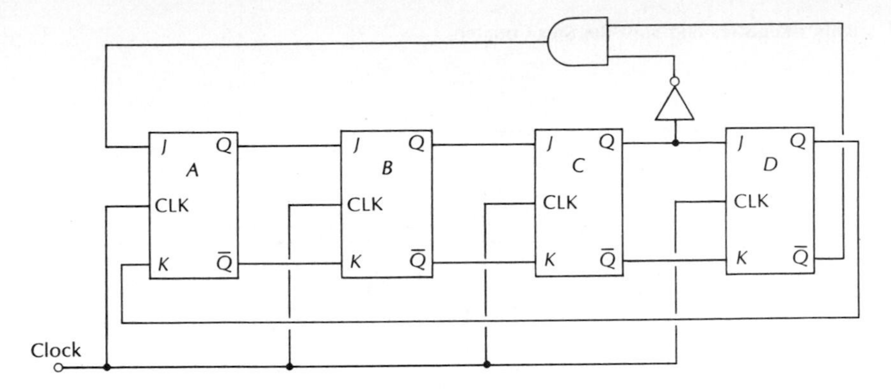

Figure 8-17 Four-Flip-Flop Johnson Counter with Self-Starting and Self-Correcting Gate

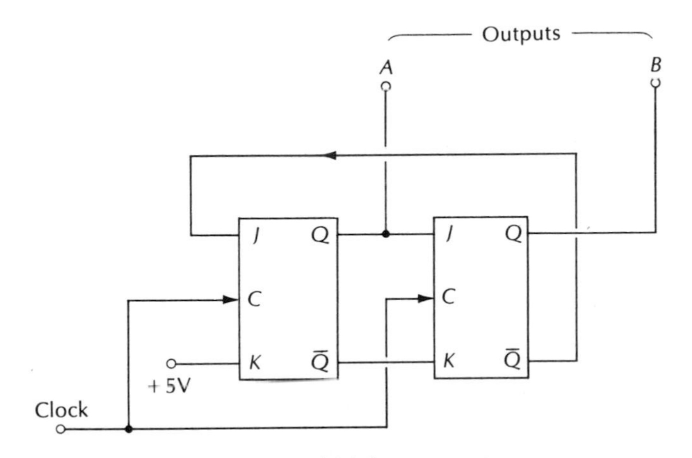

a. Mod 3 Johnson counter

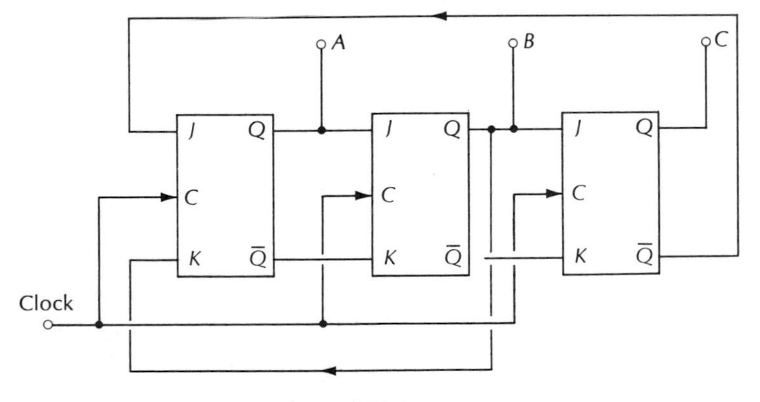

b. Mod 5 Johnson counter

Figure 8-18 Johnson Counters

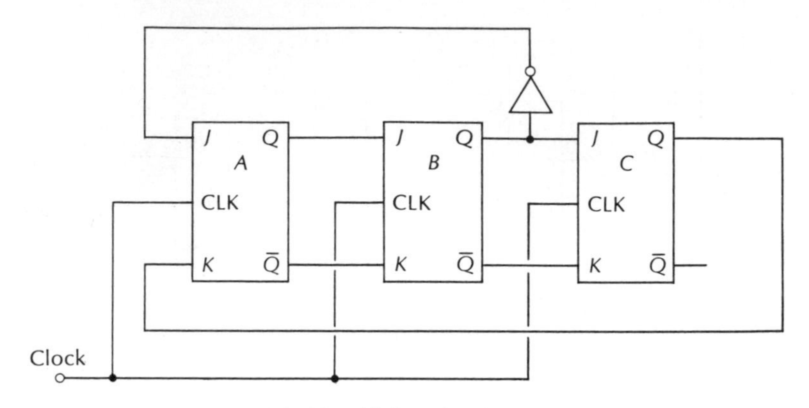

c. Mod 5 odd-length Johnson counter

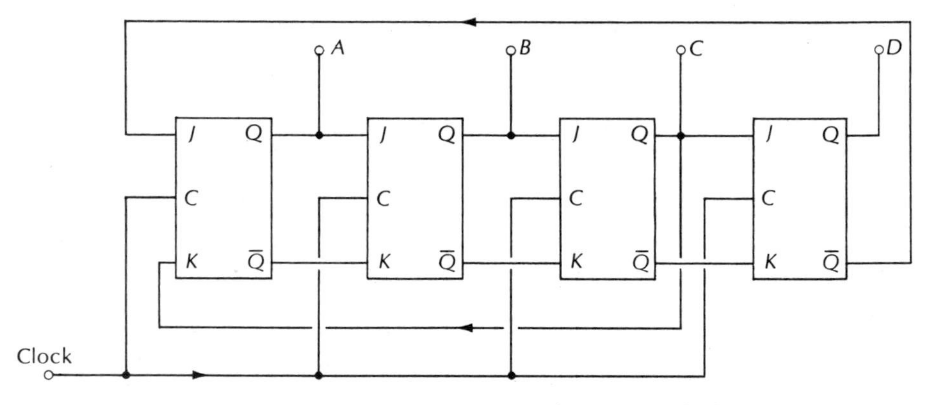

d. Mod 7 counter using 7473 for negative clocking

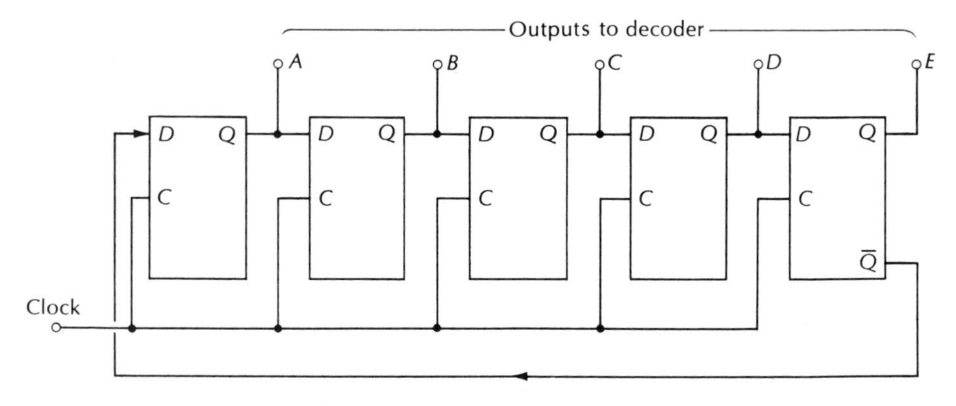

e. Mod 10 (decade) counter using type *D* flip-flops

Figure 8-18 continued

8-8 C-MOS Counters

The Johnson counter has become a popular circuit for C-MOS mod 10 (decade) BCD counters. Most of these C-MOS Johnson counters contain on-board decoder circuitry, either one-of-10 or seven-segment decoding. The Johnson counter is inherently synchronous, and although it is inefficient in its use of flip-flops, it does not require extra gate circuits to make it synchronous.

THE 4017 AND 4022 C-MOS COUNTERS

The 4017 and 4022 are both internally decoded Johnson type counters. The 4017 is a mod 10 (decade) counter, and the 4022 is a mod 8 (octal) counter. The 4017 produces one-of-10 outputs and the 4022 provides one-of-8 outputs.

Both counters count on the positive-going edge of the clock pulse (up-clock). Outputs are active-high. Reset and clock-enable inputs are held low for counting. The maximum counting frequency is 2.5 MHz with a supply voltage of 5 V, and 5 MHz with a V_{cc} of 10 V. The total unloaded package current at 1 MHz is 0.4 mA, at 5 V V_{cc}, and 0.8 mA at 10 V.

A special output terminal is provided. For outputs 0-4, the output terminal is high and goes low for counts between 5 and 9.

THE 4026 AND 4033 DECADE COUNTERS

The 4026 and 4033 decade counters are Johnson counters with internal decoding designed to drive seven-segment displays. Segment outputs are active-high and can provide up to 2.5 mA per segment with a V_{cc} of 5 V and 5 mA at 10 V. The output drive current is adequate for low-current LED displays and for fluorescent displays.

The maximum counting frequency for the 4026 and 4033 counters is 2.5 MHz with a V_{cc} of 5 V and 5 MHz with a supply voltage of 10.

The principal difference between the 4026 and the 4033 is that the 4033 has provisions for blanking leading and trailing zeros in multi-decade counters. The 4026 does not have the zero-blanking feature.

THE 4018B COUNTER

The 4018B is basically a divide-by-2-through-10 counter *only*. The counter can be programmed for any divide-by-count between 2 and 10, but individual count values cannot be conveniently decoded.

The count length is programmed by connecting an external feedback loop. Even-numbered divisions 2, 4, 6, 8, and 10 require only a single feedback line. All odd-numbered counts require a two-input AND gate in the feedback loop. Table 8-3 lists the proper feedback connections

Table 8-3 Feedback Table for the 4018B Counter

Divide By	Feedback to Input Terminal From:
2	$\overline{Q}1$
3	$\overline{Q}1$ and $\overline{Q}2$
4	$\overline{Q}2$
5	$\overline{Q}2$ and $\overline{Q}3$
6	$\overline{Q}3$
7	$\overline{Q}3$ and $\overline{Q}4$
8	$\overline{Q}4$
9	$\overline{Q}4$ and $\overline{Q}5$
10	$\overline{Q}5$

for count lengths 2 through 10. The output of the counter is taken from the feedback line or at the output of the feedback AND gate if one is used.

The reset input is held at ground for counting and taken high to reset the counter. The maximum counting frequency is 2.5 MHz with a V_{cc} of 5 V and 10 MHz at 10 V.

THE MC14553 THREE-DIGIT COUNTER

The MC14553 is a 3-digit decade counter with on-the-chip display memory (latches) and multiplexed seven-segment decoding for each of the three decades. This single 16-pin DIP can replace eight TTL packages.

ADVANCED TOPIC

8-9 The Pseudo-Random Number Generator

An interesting application of shift-register counting is the pseudo-random number generator. The numbers generated are random, but the sequence of random numbers repeats. For most configurations, the length of the count sequence is $2^N - 1$, where N is the number of shift-register stages. The circuit is similar to other shift-register counters except that exclusive-OR gates are used in the feedback network.

The output can be taken in parallel from the flip-flop, or as a train of pulses (serial) from the output of the final shift-register stage.

The pseudo-random number generator finds applications in cryptography, digital and analog test systems, games, and other statistical systems.

Figure 8-19 shows the logic diagram for a pseudo-random number generator with a count length of 1023. Table 8-4 provides feedback equations for other count lengths.

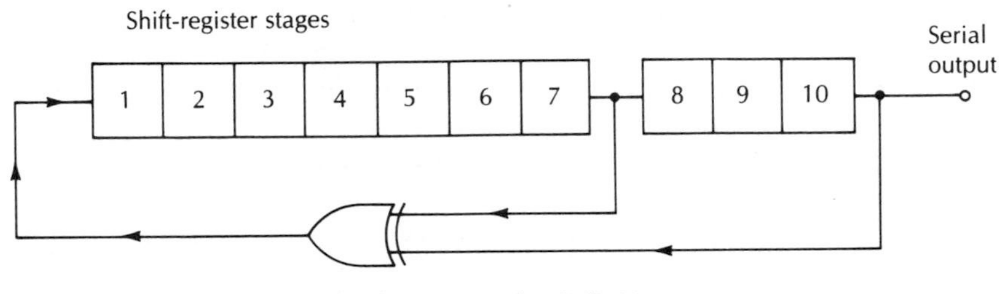

Figure 8-19 Pseudo-Random Sequence Generator Generates 1023 Random Numbers before Repeating

Table 8-4 Feedback Equations for Pseudo-Random Number Generators

Number of ffs	Sequence Length	Feedback Equation
8	255	$f = 5 \oplus 6 \oplus 7 \oplus 8$
9	511	$f = 5 \oplus 9$
10	1023	$f = 7 \oplus 10$
11	2047	$f = 9 \oplus 11$
12	4095	$f = 6 \oplus 8 \oplus 11 \oplus 12$
15	32,767	$f = 14 \oplus 15$
18	262,143	$f = 11 \oplus 8$
21	2,097,151	$f = 19 \oplus 21$
24	16,766,977	$f = 19 \oplus 24$
27	133,693,177	$f = 19 \oplus 27$
30	1,073,215,489	$f = 23 \oplus 30$

SUMMARY

Shift registers come in various lengths, capabilities, and limitations. A particular register is selected to fit an individual job. Shift registers are classified by direction of shift capability: right-shift, left-shift, or right/left shift. They are also categorized according to parallel or serial data entry and data exit.

Most shift registers have type *D* inputs to the first stage. All shift registers are capable of serial input and output, but only some are capable of parallel loading (input) or parallel output or both of them. Parallel loading is either preset-only, where changes in input data can be made only after clearing the entire register, or true parallel loading, where data can be changed at any time. The right/left shift register can be wired to perform any of the functions SISO, SIPO, PIPO, PISO, and right/left shift. They must, however, be connected according to the desired functions.

The universal shift register can be made to shift from one to another of all of the possibilities above by setting levels on control inputs.

Shift registers can be connected as synchronous counters by using appropriate feedback. Johnson counter decoding is simple and the counter produces clean symmetrical output signals. A number of C-MOS counters use the Johnson counter as the basic counter element with on-the-chip decoding.

Problems

10. What is meant by parallel-to-serial conversion (shift registers)?
11. What is the natural modulus of a five-bit shift register ring counter?
12. What are the primary advantage and the primary disadvantage of the shift-register ring counter?
13. What is meant by an invalid or disallowed state?
14. What is a disallowed subroutine?
15. What must be done to start a ring counter?
16. In a 5-bit ring counter, how many disallowed states exist?
17. Describe self-correcting feedback.
18. What is the difference between the 1 circulating and the 0 circulating ring counter? Does either form require decoding?
19. What is the purpose of decoding a counter? In light of your answer, why is it not necessary to decode a ring counter?
20. What is the basic circuit difference between a ring counter and the Johnson counter?
21. Compare the ring and Johnson counters with respect to the number of possible counts for *N* flip-flops?
22. How many disallowed states exist for a Johnson counter with *N* flip-flops?

23. Is the decoding circuitry for a Johnson counter more or less complex than for a binary ripple counter?
24. Draw the functional block diagram for a mod 8 Johnson counter using the 54/7495 shift register. (See Figure 8-3.)
25. Draw the functional block diagram for a mod 8 Johnson counter with self-start/self-correction gating using the 54/7495 shift register.
26. Draw the functional block diagram of a mod 10 (decade) Johnson counter using the 54/7496 shift register.
27. Look up the 4018B C-MOS counter in a manufacturer's data manual. Draw a diagram showing the 4018B programmed for each of the following counts: (a) 4, (b) 5, (c) 9.
28. State the range of maximum counting frequencies for the C-MOS counters covered in this chapter.
29. Assume that you are to design a tachometer to count revolutions per minute for a new turbine engine that has a maximum speed of 30,000 RPM. Would any of the C-MOS counters in this chapter be fast enough? Which ones?
30. (Extra problem) Draw the logic diagram of a pseudo-random number generator with a length of 4095.

CHAPTER 9

COMPUTER ARITHMETIC

Learning Objectives. *Upon completing this chapter you should know:*

1. *How to write numbers in signed magnitude and complement notations.*
2. *How to perform addition and subtraction with signed magnitude and radix-minus-1 and radix complement notation.*
3. *How to perform multiplication and division using computer methods.*
4. *How to trace data through the several kinds of adder circuits.*
5. *How to relate the data flow in adder circuits to the appropriate algorithm.*
6. *How to relate multiplication and division algorithms to the circuits that perform them.*

All mathematical operations from the simplest to the most complex can be reduced to processes of simple addition and subtraction, and subtraction can also be reduced to addition. Addition could even be reduced to the level of simple counting and the manipulation of pebbles.

The adder in most computers performs the full range of mathematical operations. Even a relatively simple problem may involve a sizable number of steps. The computer works so fast that a great many simple steps can be taken in a very short time. The instructions that provide the computer with step-by-step directions for solving a problem are called an *algorithm*. The computer program can be viewed as a very long compound algorithm.

Numbers are represented either in a form called *signed magnitude* or in a form called *complement notation*. These representations have been chosen because they require simpler circuitry to implement than other methods of number representation.

In this chapter we will study the methods of number representation, and how arithmetic is performed in each system. Then we will examine the circuitry that performs addition and subtraction in each system of notation. Finally, we will see how the adder can be used to perform multiplication and division.

DEFINITIONS OF ARITHMETIC TERMS

Minuend: The number from which another is to be subtracted.
Subtrahend: The number to be subtracted from the minuend.

Addend: A number to which another is to be added.
Augend: The number to be added to the addend. Because addition is commutative, the terms *addend* and *augend* are occasionally used interchangeably.
Commutative: Because $A + B = B + A$ addition is commutative and since $A \times B = B \times A$, multiplication is commutative. Because $A - B \neq B - A$, subtraction is not commutative. Division is not commutative because $A/B \neq B/A$.
Multiplicand, Multiplier, and *Product:*

$$\begin{array}{rl} 5 & \text{multiplicand} \\ \times 4 & \text{multiplier} \\ \hline 20 & \text{product} \end{array}$$

Quotient, Dividend, and *Remainder:*

$$\begin{array}{rrl} & 21 & \text{quotient} \\ \text{divisor} & 7\overline{)150} & \text{dividend} \\ & \underline{147} & \\ & 3 & \text{remainder} \end{array}$$

9-1 Algorithms

The arithmetic section of a computer is generally little more than a memory and logic system capable of performing binary addition. Subtraction is performed by a procedure known as *complement arithmetic,* in which the subtrahend is complemented and the difference found by addition. Multiplication can be accomplished by successive additions, and division can be performed by successive subtractions.

A computer must have detailed instructions telling it when to add, when to complement, where to store partial and finished results, and where to find numbers and new instructions for the next operation. Such instructions are called a *program*. There are three kinds of programs: hardwired, firmware, and software. Software is the most versatile of the three because it can be anything that can be conceived, within the limitations of the machine. It consists of a written detailed list of all of the operations a computer is to perform and the order in which they must be performed, along with the memory locations of data to be stored and retrieved during the problem.

These instructions are then put on punch cards, tape, or some other medium whereby they can be loaded into the computer's memory. The data to be worked with is also loaded into the memory when the *go* button is pushed; the machine then follows the program stored in its memory, *fetching* new data and further instructions at its own rate without human intervention until the problem is finished.

An algorithm is a special kind of program that instructs the computer in how (step-by-step) to use the adder to perform subtraction, multiplication, and division, to extract roots, and so on. Algorithms were not developed for computers, however. The example that we will study here, known as Newton's method, was developed by Isaac Newton. It is a commonly used algorithm in modern electronic computing.

The difference between an algorithm, or computer program, and most other lists of instructions—a recipe, for example—is that the number of steps in an algorithm is not known in advance. Another important aspect of an algorithm is the concept of varying degrees of correctness. An algorithm may be iterated (repeated a number of times) until the solution reaches the answer that is correct to some predetermined value. The program is complete only when the proper degree of correctness is reached.

Suppose, for example, that we wish to extract an accurate square root using a pocket calculator that is capable of only add, subtract, multiply, and divide functions. Let us assume the problem is to find the square root of 125. Here is the algorithm for extracting the square root (Newton's method):

a. Call the square root r.
b. Let X = the number for which the square root is to be found.
c. Let S = 0.00001 (some predetermined small value).

Algorithm Step	*Instruction*
1.	Set: $r = 1$
2.	Compute: $r = \frac{1}{2}(x/r + r)$
3.	Evaluate: Is $(r^2 - X) \lesseqgtr S$? a. If yes, stop. The problem is finished. b. If no, return to step 2 and repeat.

The first step in the algorithm establishes a starting point. Setting $r = 1$ as a first approximation is done because the computer cannot easily make a reasonable rough approximation of the square root of a number. If a fair approximation can be made, it takes fewer steps to get to an answer; where we start is otherwise unimportant. Extra steps are of little consequence at computer speeds, but for our example let us start by setting $r = 10$. $10^2 = 100$, which is not far from 125, but is enough to demonstrate the method.

Considering the limitations of the calculator we are using for the example, suppose we break steps 2 and 3 down into the appropriate substeps. We will retain the notation n^2, but we will have to implement it by multiplying n by itself (assuming the calculator has no n^2 key). Here is the algorithm for a calculator:

Algorithm Step	*Instruction*
Step 1.	Set r = (first approximation of x); in this case it is 10. Let S equal 0.0001.
	a. Set r equal to 10.
	b. (1) $x/r =$
	(2) $+r =$
	(3) $\div 2 = r$ (the revised approximation)
	c. Evaluate
	(1) $r^2 =$
	(2) $-x = S$ stop or reiterate
First iteration	
Step 2.	a. $x/r = 125/10 = 12.5$
	b. $+r = 22.5$
	c. $\div 2 = 11.25$
Step 3.	a. $r^2 = 11.25 \times 11.25 = 126.5625$
	b. $-x = 1.5625$
	$1.5625 > S$
Second iteration	
Step 2.	a. $x/r = 125/11.25 = 11.11111$
	b. $+r = 22.36111$
	c. $\div 2 = 11.18056$
Step 3.	a. $r^2 = 11.18056 \times 11.18056 = 125.00492$
	b. $-x = 0.00492$
	$0.00492 > S$
Third iteration	
Step 2.	a. $x/r = 125/11.18056 = 11.180119$
	b. $+r = 22.360679$
	c. $\div 2 = 11.180339$
Step 3.	a. Evaluate
	(1) $r^2 = 124.99998$
	(2) $-x = 0.00002$
	$0.00002 < 0.0001$
End.	

Newton's method is an excellent example of an algorithm suitable to mechanization because it consists of the repetition of a few very simple steps. From a human standpoint (doing it by hand at any rate), this method is tedious, but it is ideally suited for digital systems.

Problems

Using Newton's method, find:

1. $\sqrt{14}$

2. $\sqrt{490}$
Assume $S = 0.0001$.

BINARY REPRESENTATIONS FOR ARITHMETIC OPERATIONS

There are three common methods used to represent numbers in binary computers. In most cases, both positive and negative numbers must be handled. The choice of either signed magnitude, radix complement, or radix-minus-one complement notation affects the logical design of the system.

9-2 Signed Magnitude Representation

In signed magnitude representation part of the computer word is used for the absolute value in binary. Another part of the word is reserved for the sign using 0 to represent + and 1 to represent −. For example, assume an eight-bit word with the leftmost bit reserved for the sign.

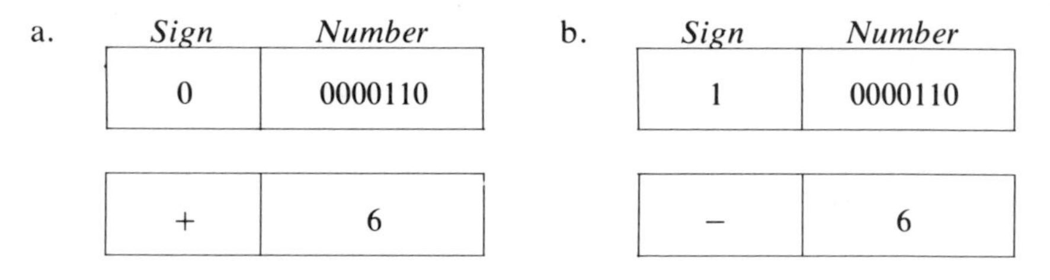

Problems

Express the following decimal numbers in signed magnitude binary representation:

3. −64 4. 125 5. −125 6. 14

The algorithms used for arithmetic operations must include rules for dealing with signs:

Addition algorithm (signed magnitude)
1. To add numbers with like signs, add the magnitudes and affix the common sign.
2. To add numbers with unlike signs, find the *difference* of the magnitudes and use the sign of the number with the largest magnitude.

Subtraction algorithm (signed magnitude)
1. Change the sign of the subtrahend.
2. *Add* the subtrahend, with the changed sign, to the minuend.

9-3 The Radix Complement

Because complement arithmetic has many advantages in terms of ease of mechanization, it is the most popular form for representing binary numbers. The radix complement of a computer number is defined as:

$$\text{Radix complement} = R^n - N$$

where R is the radix
n is the number of bits (binary digits) in the computer word
N is the number to be complemented

Example

If $R = 10$ and $n = 5$, find the radix complement of 976:

$$10^5 - 976$$
$$100000 - 976 = 99024$$

The radix-minus-1 complement is one less than the radix complement and can be found as follows:

Find the radix-minus-1 complement of 976:

99999	
−976	
99023	radix-minus-1 complement
+1	adding 1
99024	we get the radix complement

9-4 Addition and Subtraction

BINARY ADDITION

The following defines the addition of binary digits A and B:

$A + B$	*Sum*	*Carry*
0 + 0	= 0	0
0 + 1	= 1	0
1 + 0	= 1	0
1 + 1	= 0	1
1 + 1 + 1	= 1	1

The last case includes a carry from the previous column.

Examples

$$\begin{array}{r} 1 \\ +1 \\ \hline 10 \end{array} \qquad \begin{array}{r} 0 \\ +1 \\ \hline 1 \end{array} \qquad \begin{array}{rl} 111 & \text{carry} \\ 1101 & \\ +0111 & \\ \hline 10100 & \end{array}$$

(1., 3., 5.)

$$\begin{array}{r} 1 \\ +0 \\ \hline 1 \end{array} \qquad \begin{array}{r} 0 \\ +0 \\ \hline 0 \end{array} \qquad \begin{array}{r} 1010 \\ +1101 \\ \hline 10111 \end{array}$$

(2., 4., 6.)

BINARY SUBTRACTION

Subtraction follows the same pattern as that used in decimal subtraction except that a *borrow* involves borrowing a 2 from the next higher order column instead of a 10 as in decimal subtraction.

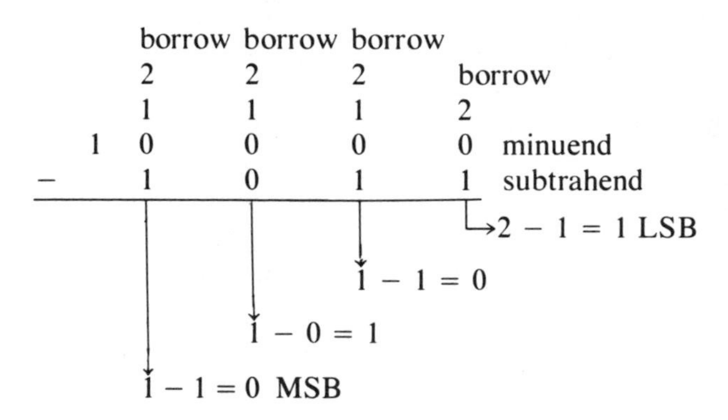

Problems

Add the following binary numbers:

$$\begin{array}{r} 10110 \\ +11101 \\ \hline \end{array} \qquad \begin{array}{r} 00111 \\ +10101 \\ \hline \end{array}$$

(7., 9.)

$$\begin{array}{r} 100110 \\ +010100 \\ \hline \end{array} \qquad \begin{array}{r} 010111 \\ +111001 \\ \hline \end{array}$$

(8., 10.)

9-5 Binary Complement Arithmetic

In the binary system complementing is easily mechanized using an inverter or an exclusive-OR true complement gate. The radix-minus-1 complement requires only that all ones be changed to zeros and all zeros to ones.

The radix complement can then be obtained by simply adding 1. The add 1 operation can be performed by the existing adder hardware, and no subtraction operation is required to get either the one's or two's complement.

Find the one's and the two's complement of each of the following binary numbers:

1. 010110 the number
   101001 the one's complement
   +1 adding 1
   101010 the two's complement

2. 1011000 the number
   0100111 the one's complement
   +1 adding 1
   0101000 the two's complement

Problems

Find the one's complement and the two's complement of the following binary numbers:

11. 10101101
12. 11100011
13. 110101011
14. 000111
15. 0100011
16. 001000

TWO'S COMPLEMENT NOTATION

In two's complement notation the leftmost bit is reserved as a sign bit. A positive number is written as a normal binary number with a 0 as a sign bit. A negative number is expressed as the two's complement with a 1 for the sign bit. Consider the following examples:

a. One's complement representation of +5 and −5 using an eight-bit register:

b. The two's complement representation of +5 and −5 using an eight-bit word.

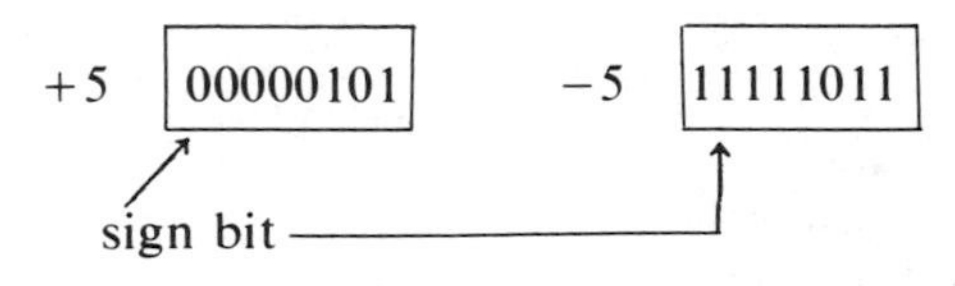

Complement representation is easy to implement in digital hardware, and addition and subtraction can be performed without regard for the sign of the number. The following examples illustrate typical addition operations using two's complement notation. The boxes represent registers.

1. *decimal* / *binary* (sign bit = leftmost bit)

decimal		binary	
13		00001101	addend (13)
+6	+	00000110	augend (6)
19		00010011	sum (19)

2. *decimal* / *binary* (sign bit = leftmost bit)

decimal		binary	
−13		11110011	addend (−13)
+ 6	+	00000110	augend (6)
− 7		11111001	sum (−7)

3. *decimal* / *binary* (sign bit = leftmost bit)

decimal		binary	
−13		11110011	addend (−13)
+(−6)	+	11111010	augend (−6)
−19	1	11101101	sum (−19)

register overflow discarded (the 1)

4. *decimal* / *binary* (sign bit = leftmost bit)

decimal		binary	
13		00001101	addend (13)
+(−6)	+	11111010	augend (−6)
7	1	00000111	sum (7)

register overflow discarded (the 1)

SUBTRACTION

Subtraction is performed by taking the two's complement of the subtrahend and *adding* it to the minuend. The process is illustrated by the following examples.

1. *decimal* / *binary*

decimal		binary	
6		00000110	minuend (6)
−(−13)	−	11110011	subtrahend (−13)
19		00000110	minuend (6)
	+	00001101	complement of subtrahend
		00010011	difference (19)

2. *decimal* / *binary*

decimal		binary	
6		00000110	minuend (6)
−13	−	00001101	subtrahend (13)
− 7		00000110	minuend (6)
	+	11110011	complement of subtrahend
		11111001	difference (−7)

3. *decimal* / *binary*

decimal		binary	
13		00001101	minuend (13)
− 6	−	00000110	subtrahend (6)
7		00001101	minuend (13)
	+	11111010	complement of subtrahend
	1	00000111	difference (7)

register overflow discarded (↑ 1)

4. *decimal* / *binary*

decimal		binary	
−13		11110011	minuend (−13)
− 6	−	00000110	subtrahend (−6)
−19		11110011	minuend (−13)
	+	11111010	complement of subtrahend
	1	11101101	difference (−19)

register overflow discarded (the 1)

Problems

Perform the following additions using two's complement notation:

17. 22 + 11
18. 16 + (−8)
19. −6 + 12
20. −14 + 29
21. (−66) + (−100)
22. 49 + 68

Perform the following subtractions using two's complement notation:

23. (−22) − (−14)
24. 22 − (−14)
25. (−22) − 14
26. (−16) − 64
27. 64 − 16
28. 44 − 72

9-6 Floating Point Notation

Floating point notation is the computer representation of numbers written in scientific notation. In fixed point notation the proper positioning of the decimal point requires significant effort on the part of the programmer. Scientific notation (floating point) is easier for the programmer and has the specific advantage that nonsignificant zeros need not be stored. The part that is normally the units part (mantissa) is carried as a fraction. The value $.642136 \times 10^{10}$ could be stored as:

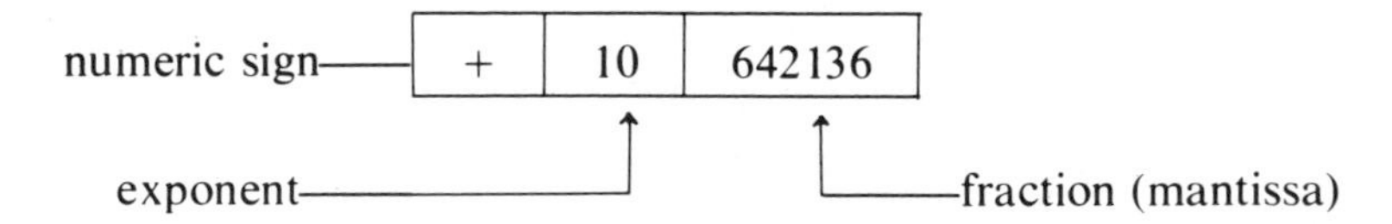

Because the exponent may require a range of from +99 to −99, some way of indicating the sign of the exponent is necessary. The most common approach is to *bias* the exponent by adding to it some large value such as 64. If the exponent part is less than 64, it is negative.

If the bias is 64, the number 0.642136×10^{-3} would be represented and stored as:

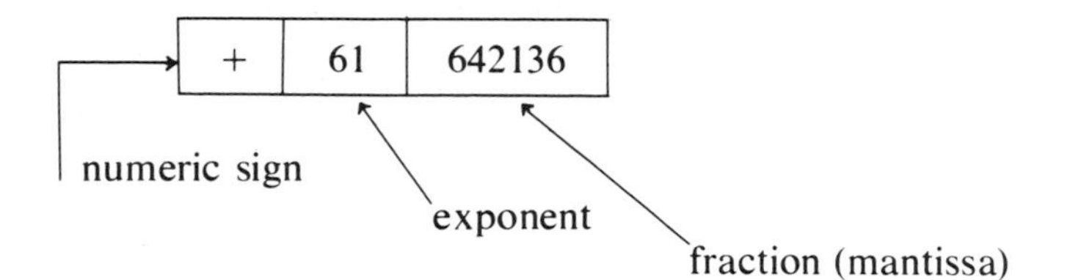

The value −419632 would be represented as:

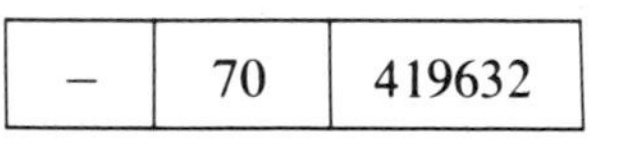

When the representation is in binary, as is normally the case, a 16-bit floating point word would be represented and stored as:

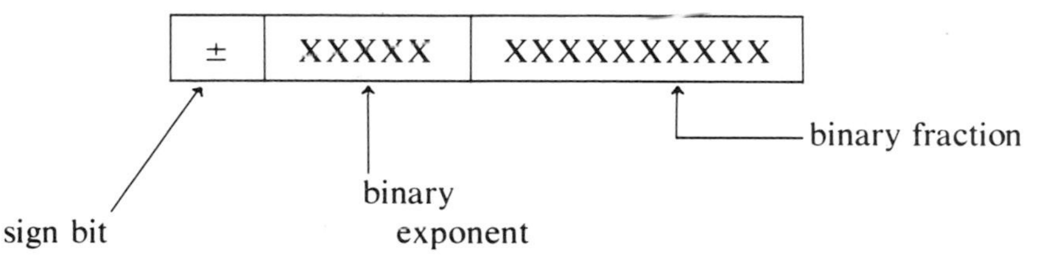

The leftmost digit indicates the bias figure, in this case 10000 (decimal 16). The binary exponent represents a power of 2.

for: $10101.10_2 = .1010110 \times 10^{101}$,
where $10_2^{101} = 2^5_{10}$

The *radix number* in *any* radix is written as 10. The number would be represented and stored in a register as:

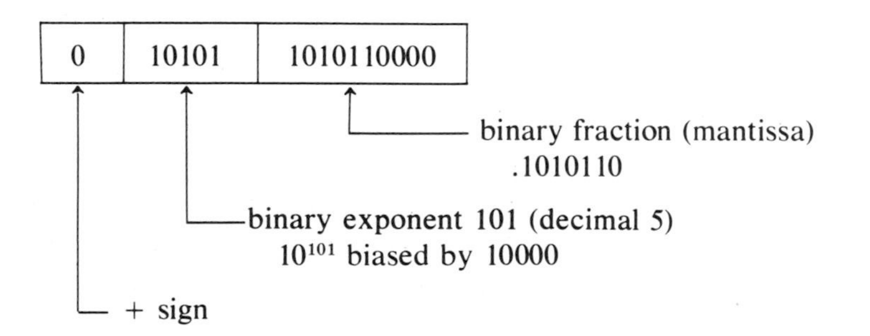

Problems

Write each of the following numbers in (a) sign magnitude notation, (b) two's complement notation, and (c) one's complement notation. Write all results in binary.

29. 26_{10}
30. 19_{10}
31. 11_{10}
32. -26_{10}
33. -19_{10}
34. -11_{10}
35. -44_{10}

9-7 Binary Addition and the Half Adder

The several binary combinations and their sums are:

0	0	1	1	augend (A)
+0	+1	+0	+1	addend (B)
0	1	1	10	sum (S)
			↑ (leftmost 1)	carry (C)

Table 9-1 shows the binary addition table and the same table written in truth table form. An examination of minterms yields the equation $S = \overline{A} \cdot B + A \cdot \overline{B}$. You may recognize this as the exclusive-OR function $S = A \oplus B$.

The truth table for the carry function is identical to the truth table for the AND gate. Figure 9-1 shows two representations for the implementation of a circuit to perform binary addition.

The circuit shown in Figure 9-1 is called a *half adder* because it does not have a provision for adding a carry from a preceding addition. The functional block diagram for the half adder is shown in Figure 9-2.

Table 9-1 Binary Addition

$A+B$ = Sum	Carry
0+0 = 0	0
0+1 = 1	0
1+0 = 1	0
1+1 = 0	1

a. Addition table

A	B	Sum	Minterms
0	0	0	
0	1	1	($\overline{A}$ B)
1	0	1	(A $\overline{B}$)
1	1	0	

Sum equation: sum = $\overline{A}B + A\overline{B}$
This is the exclusive-OR function and
and can be written as sum = $A \oplus B$.

b. Sum truth table

A	B	Carry	Minterms
0	0	0	
0	1	0	
1	0	0	
1	1	1	($A \cdot B$)

c. Carry truth table

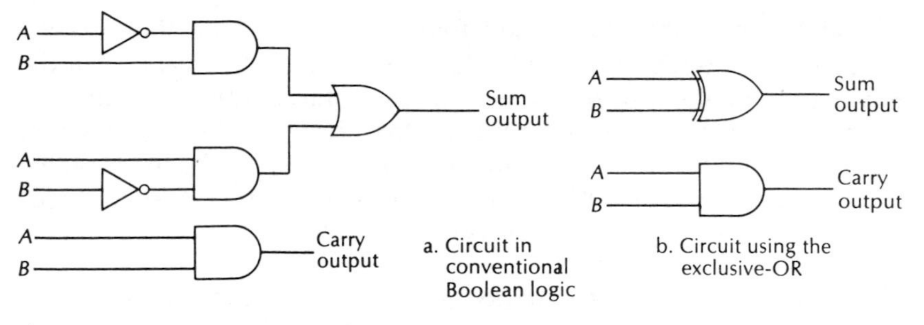

Figure 9-1 Logic Circuit to Perform Binary Addition

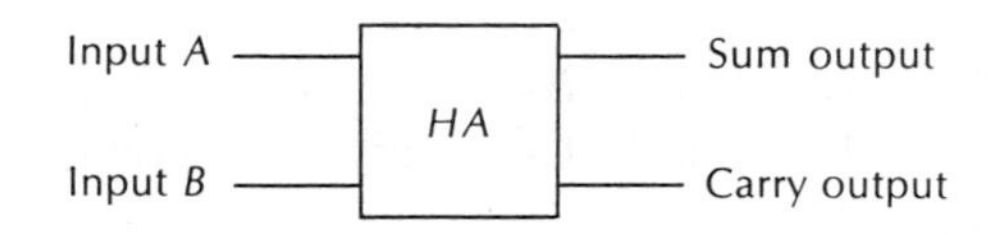

Figure 9-2 Half Adder Block Diagram Symbol

Addition Examples

1.
```
   111  —— carry
  1011  —— addend
 +0101  —— augend
 -----
 10000
 carry
```

2.
```
 11    —— carry
 0110  —— addend
 0010  —— augend
 ----
 1000
```

9-8 The Full Adder

Consider the following addition problem:

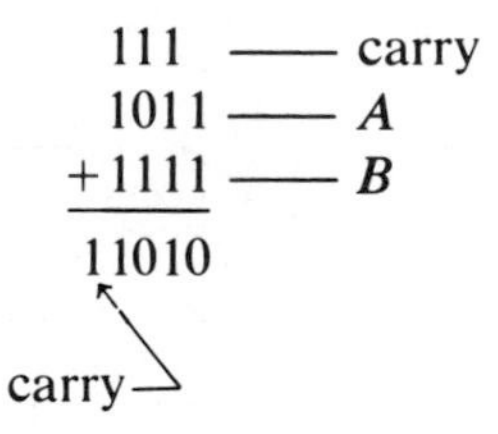

Notice that in the second and fourth columns (from the left) we ended up with three ones to be added as a result of a carry from the preceding addition. A full adder involves three variables: the addend, the augend, and the carry from the previous addition. Because the bits may be added in any order, we can call them *A*, *B*, and *C* on the truth table in any order. We must now expand our set of addition rules to include the possibility of a carry.

ADDITION COMBINATIONS WHEN A CARRY IS TO BE ADDED

A	*B*	*C*	*Sum*	*Carry*
			2^0	2^1
0 +	0 +	0	= 0	0
0 +	0 +	1	= 1	0
0 +	1 +	0	= 1	0
0 +	1 +	1	= 0	1
1 +	0 +	0	= 1	0
1 +	0 +	1	= 0	1
1 +	1 +	0	= 0	1
1 +	1 +	1	= 1	1

A—Addend
B—Augend
C—Carry from previous addition

The bottom row involves the addition of three units, but there is no three-weight digit in binary so it must be written as 11 ($2^1 + 2^0$). Table 9-2 shows the truth table version of the addition table including the carry from a previous addition. If we combine the minterms for the sum equation we get:

$$S = ABC_{n-1} + A\bar{B}\bar{C}_{n-1} + \bar{A}\bar{B}C_{n-1} + \bar{A}B\bar{C}_{n-1}$$

Table 9-2 Full Adder Truth Table

m	A	B	C	Sum	Carry
0	0	0	0	0	0
1	0	0	1	1	0
2	0	1	0	1	0
3	0	1	1	0	1
4	1	0	0	1	0
5	1	0	1	0	1
6	1	1	0	0	1
7	1	1	1	1	1

and for the carry equation we get:

$$C_n = \overline{A}BC_{n-1} + A\overline{B}C_{n-1} + AB\overline{C}_{n-1} + ABC_{n-1}$$

We can plot both equations on the Karnaugh map in Table 9-3 to see if they can be simplified. It is apparent that the sum equation cannot be simplified. The carry equation reduces to:

$$C_n = BC_{n-1} + AC_{n-1} + AB$$

The logic circuits for the full adder sum and carry functions are shown in Figure 9-3.

9-9 A Full Adder Composed of Two Half Adders

In this section we will introduce a useful shortcut, practice some important algebraic manipulations, and see how two half adders can be combined to form a full adder (Figure 9-4 shows the circuit for doing so).

We will use the letters A, B, and C as variables. Table 9-2 shows what is expected from any full adder, including one composed of two half adders. An examination of the diagram in Figure 9-4 shows that the

Table 9-3 Karnaugh Maps for the Full Adder

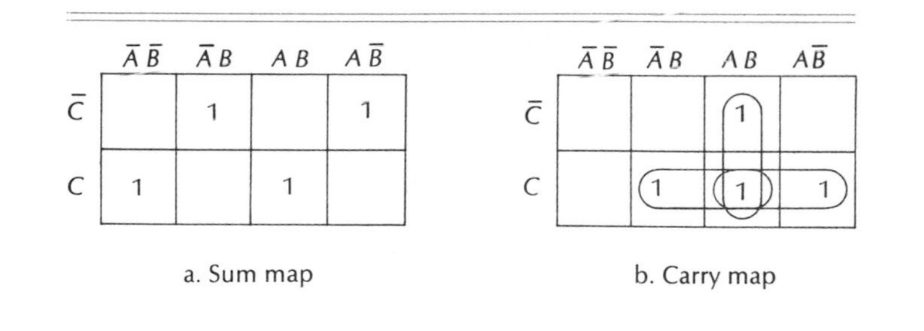

a. Sum map

	$\overline{A}\,\overline{B}$	$\overline{A}B$	AB	$A\overline{B}$
$\overline{C}$		1		1
C	1		1	

b. Carry map

	$\overline{A}\,\overline{B}$	$\overline{A}B$	AB	$A\overline{B}$
$\overline{C}$			1	
C		1	1	1

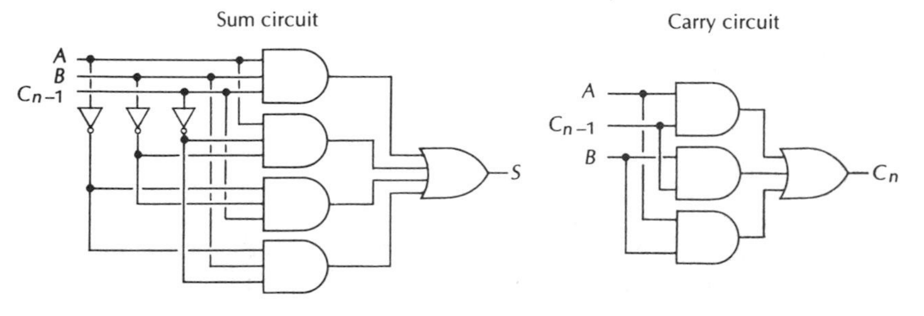

Figure 9-3 Full Adder Logic Diagram

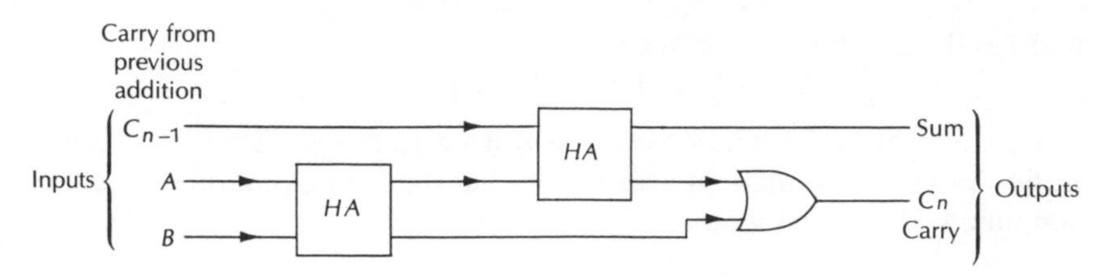

Figure 9-4 Full Adder Composed of Two Half Adders

output of half adder 1 forms one input of half adder 2. The minterm equation is:

$$S = (\bar{A} \cdot \bar{B} \cdot C) + (\bar{A} \cdot B \cdot \bar{C}) + A \cdot \bar{B} \cdot \bar{C}) + (A \cdot B \cdot C)$$

Call one input to the second half adder R (the useful shortcut) and write the sum equation for the second half adder, using C and R to set up the initial equation. The sum equation for the second half adder can be written first in terms of C and R. The more complex sum output, which feeds into it, can be substituted after the basic equation has been set up. The equation is as follows:

$$S = (\bar{C} \cdot R) + (C \cdot \bar{R}), \text{ (or } S = C \oplus R)$$

Substituting the output of the first half adder for R, where

$$R = (A \cdot \bar{B}) + (\bar{A} \cdot B)$$

we get:

$$S = \bar{C} \cdot [(A \cdot \bar{B}) + (\bar{A} \cdot B)] + C \cdot [\overline{(A \cdot \bar{B}) + (\bar{A} \cdot B)}]$$

Multiplying the first complex term through by $\bar{C}$, we get:

$$(\bar{C} \cdot A \cdot \bar{B}) + (\bar{C} \cdot \bar{A} \cdot B)$$

The second term also requires multiplying out by C, but first clear the long complement bar.

$$C \cdot [(\bar{A} + B) \cdot (A + \bar{B})]$$

Multiplying through by C, we get the following products:

$$C \cdot \bar{A} \cdot A$$
$$C \cdot \bar{A} \cdot \bar{B}$$
$$C \cdot B \cdot A$$
$$C \cdot B \cdot \bar{B}$$

The first (and last) term drops out, because (A and $\bar{A}$ and B and $\bar{B}$) cannot coexist, or, stated another way, $A \cdot \bar{A} = 0$ and $C \cdot 0 = 0$. This leaves us with:

$$(C \cdot \bar{A} \cdot \bar{B}) + (C \cdot B \cdot A)$$

Combining the results of the two simplified sum terms, we get a final equation (1), and by rearranging we get (2).

$$(1)\ S = (\overline{C} \cdot A \cdot \overline{B}) + (\overline{C} \cdot \overline{A} \cdot B) + (C \cdot \overline{A} \cdot \overline{B}) + (C \cdot A \cdot B)$$
$$(2)\ S = (\overline{A} \cdot \overline{B} \cdot C) + (\overline{A} \cdot B \cdot \overline{C}) + (A \cdot \overline{B} \cdot \overline{C}) + (A \cdot B \cdot C)$$

The second equation (2) is the same equation we derived from Table 9-2.

Now analyze the carry circuit for the full adder composed of two half adders. Table 9-2 is the truth table for the carry circuit of any 3-bit binary full adder. The minterm equation is:

$$C_n = (\overline{A} \cdot B \cdot C) + (A \cdot \overline{B} \cdot C) + (A \cdot B \cdot \overline{C}) + (A \cdot B \cdot C)$$

The simplified equation is:

$$C_n = (A \cdot B) + (A \cdot C) + (B \cdot C)$$

Examine the circuit in Figure 9-4. The OR gate has $A \cdot B$ as one input. The second input is taken from the number 2 half adder's carry output. The second half adder's inputs are C and R, where

$$R = (\overline{A} \cdot B) + (A \cdot \overline{B})$$

The carry equation is:

$$C_n = A \cdot B + C \cdot R$$

Substituting $\overline{A} \cdot B + A \cdot \overline{B}$ for R, we get:

$$C_n = A \cdot B + C \cdot (\overline{A} \cdot B) + (A \cdot \overline{B})$$

Multiplying by C in the second term, we get:

$$C_n = A \cdot B + (\overline{A} \cdot B \cdot C) + (A \cdot \overline{B} \cdot C)$$

Expanding the first term,

$$C_n = A \cdot B \cdot C + A \cdot B \cdot \overline{C} + \overline{A} \cdot B \cdot C + A \cdot \overline{B} \cdot C$$

which, in a slightly different order, is the same equation as derived from Table 9-2.

SUBTRACTION USING THE RADIX AND RADIX-MINUS-1 COMPLEMENT

Subtraction in parallel machines is nearly always performed by adding either the radix or the radix-minus-1 complement of the subtrahend to the minuend. The method of subtracting through addition with complements can be illustrated by some examples.

Radix Complement Example

Subtract 124_{10} from 125_{10} by means of the radix complement.

1. Generating the complement of 124_{10}, we have:

$$\begin{array}{r} 1000_{10} \\ -124_{10} \\ \hline 876\ \text{(radix 10)} \end{array}$$

2. Adding 125_{10} to the complement of 124_{10}, we have:

$$\begin{array}{r} 125_{10} \\ +876_{10} \\ \hline 1\ 001 \text{ (radix 10)} \end{array}$$

3. The 1 generated at the far left is considered an overflow and is not part of the result. It does, however, provide one important bit of information. The fact that the overflow 1 was generated indicates that the result of the subtraction is a positive number.

Example Subtract 125_{10} from 124_{10} by means of the radix complement.

1. Generating the radix complement of 125_{10}, we have:

$$\begin{array}{r} 1000 \\ -125 \\ \hline 875 \text{ (radix 10)} \end{array}$$

2. Adding 124_{10} to the radix complement of 125_{10} gives us:

$$\begin{array}{r} 875 \\ +124 \\ \hline 999 \text{ (radix 10)} \end{array}$$

3. In this case there is no overflow, which indicates that the answer is a negative number and that the final result will be obtained after this result has been complemented. Complementing the answer for step 2 yields:

$$\begin{array}{r} 1000 \\ -999 \\ \hline 0001 \text{ (radix 10)} \end{array}$$

The absence of an overflow in step 2 indicates a negative sign or a 0. Thus the final result is -1_{10}.

In all three previously discussed systems of binary signed number representation, negative numbers are written in complement form and the result would be correct at the end of step 2, and the recomplementing operation would *not* be performed.

THE RADIX-MINUS-1 COMPLEMENT

If the radix-minus-1 complement is to be used, the computer must "know" when to add 1 to the result and when to withhold it. When the result of a subtraction is a positive number, an overflow is generated.

This tells the computer to add 1, and the operation is known as an end-around carry. Since an overflow is not generated when the result of a subtraction is negative, there is no end-around carry. The absence of the overflow is, in a sense, an instruction to complement again and to affix a negative sign to the result.

Let us repeat the two previous examples using the $R - 1$ complement. Subtract 124_{10} from 125_{10} using the $R - 1$ complement.

1. Finding the $R - 1$ complement of 124:

$$\begin{array}{r} 999 \\ -124 \\ \hline 875 \end{array} \text{ (radix 10)}$$

2. Adding 125 to the complement of 124:

```
  125
 +875
 ----
 1000
 └──→1     end-around carry
 ----
 +001      result (radix 10)
```

The fact that an end-around carry was generated indicates that the result is a positive number.

Now subtract 125_{10} from 124_{10} using the $R - 1$ complement.

1. Finding the $R - 1$ complement of 125:

$$\begin{array}{r} 999 \\ -125 \\ \hline 874 \end{array} \text{ (radix 10)}$$

2. Adding 124 to the complement of 125:

$$\begin{array}{r} 124 \\ +874 \\ \hline 998 \end{array} \quad \text{result (radix 10)}$$

No end-around carry is generated; the sign is negative and the result must be recomplemented to get the absolute value.

3. Taking the $R - 1$ complement of 998:

$$\begin{array}{r} 999 \\ -998 \\ \hline -001 \end{array} \text{ (radix 10)}$$

The result is -1.

Problems

Using the radix complement, carry out the following subtractions (all radix 10 numbers):

36. 1257 − 1134
37. 59 − 67
38. 4293 − 3971
39. 3971 − 4293

Perform the following subtractions using the diminished radix complement (all radix 10 numbers):

40. 1257 − 1134
41. 59 − 67
42. 4293 − 3971
43. 3971 − 4293

9-10 Addition and Subtraction Using Sign and Magnitude Binary Representation

The following examples illustrate addition and subtraction procedures for binary numbers represented in sign and magnitude notation.

a. Adding two positive numbers

sign

$+11_{10}$ = 0 01011 —— magnitude

$+13_{10}$ = 0 01101

+24 = 0 11000

To add two positive numbers, add the magnitudes and use the common sign (0 = +, 1 = −).

b. Adding two negative numbers

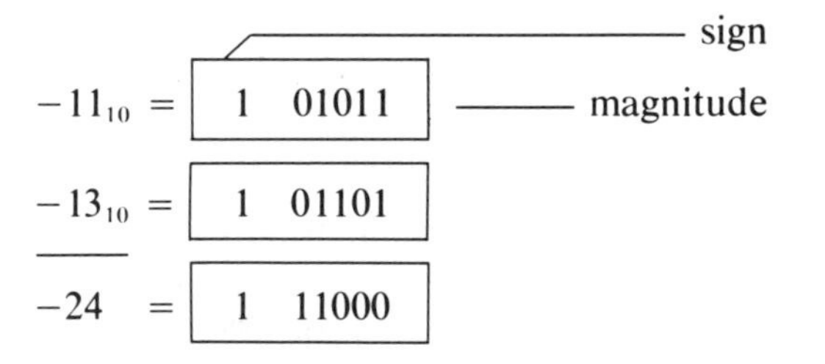

To add two negative numbers, add the magnitudes and use the common sign.

c. Adding two numbers with unlike signs
   (1) Augend magnitude larger

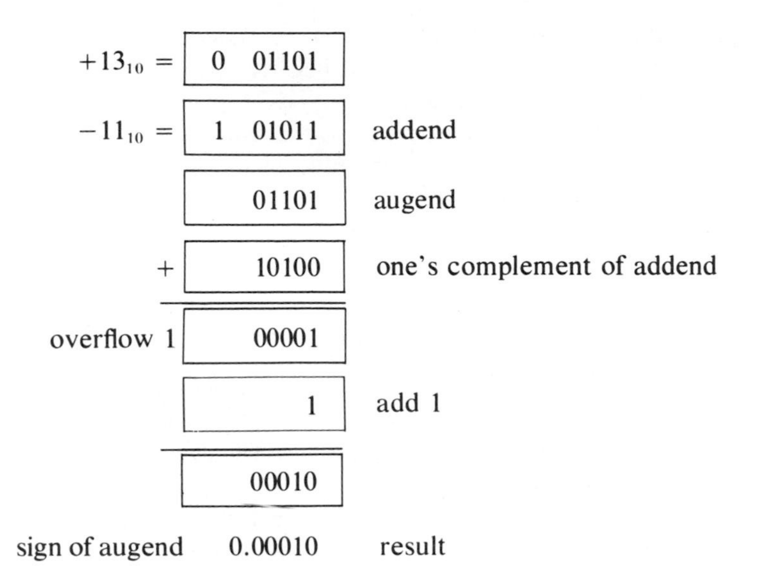

To add two numbers with unlike signs, add the augend to the complement of the addend. Then add the end-around carry to the sum. Affix the sign of the augend to the final sum (the sign of the number with the largest magnitude).

(2) Addend magnitude larger or equal

$+11_{10}$ = 0 01011

-13_{10} = 1 01101

magnitude

	01011	augend
+	10010	one's complement of addend
	11101	sum—no end-around carry generated
	00010	complement sum
	1. 00010	affix sign of addend to sum

To add two numbers with unlike signs, add the augend to the complement of the addend. There is no end-around carry, so the sum must be complemented. Affix the sign of addend (the sign of the number with the largest magnitude).

SUMMARY OF SIGNED MAGNITUDE ALGORITHMS

a. Adding two positive or two negative numbers
   (1) Add magnitudes.
   (2) Affix the common sign to the sum.

b. Adding two numbers with unlike signs and with the *augend* having the *larger* magnitude
   (1) Add the augend magnitude to the complement of the addend magnitude.
   (2) Add the end-around carry to the sum (magnitude part).
   (3) Affix the sign of the number with the larger magnitude, in this case the sign of the augend.

c. Adding two numbers with unlike signs and with the *addend* having the *larger* magnitude
   (1) Add the augend magnitude to the complement of the addend magnitude.
   (2) No end-around carry is generated; complement the sum.
   (3) Affix the sign of the number with the larger magnitude, in this case the sign of the addend.

9-11 The Logical Implementation of Signed Magnitude Addition/Subtraction

The 7483 is a typical TTL IC adder package. It uses ripple-through (serial) carry internally wired through the four full adders in the package.

The circuit uses DTL logic for inputs and high fan-out TTL logic with a high-speed Darlington pair output circuit. The use of DTL and Darlington output circuitry reduces the carry delay and minimizes the need for more complex look-ahead carry circuitry. The logic diagram for the 7483 is shown in Figure 9-5. It should be noted that the actual logic circuit for the 7483 may differ from one manufacturer to another.

LOOK-AHEAD CARRY CIRCUITS

Faster adder circuits have a feature called *look-ahead carry,* which eliminates the speed limitations caused by the cumulative carry propagation time as it ripples through the adders. The look-ahead carry adder has additional gates to sample the input and decode the condition for each stage that will produce a carry so that the carry does not have to ripple through a string of adders. The look-ahead gates deliver the carries to all stages at the same time. In this system the carries are handled in parallel, while they are serial in ripple carry adders. Figure 9-6 shows the logical implementation of a signed magnitude adder/subtracter. The exclusive-OR gates on input and output lines function as controlled inverters.

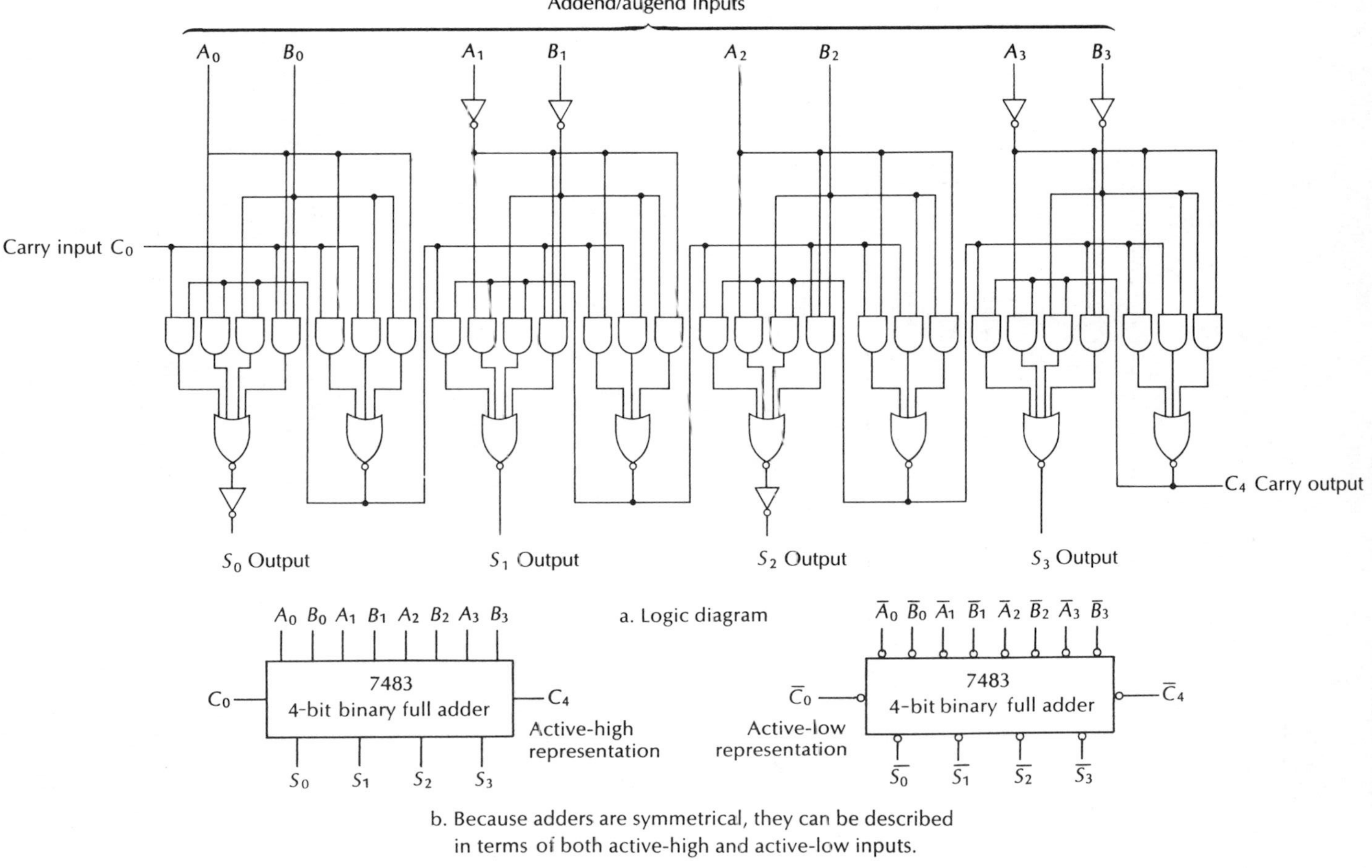

a. Logic diagram

b. Because adders are symmetrical, they can be described in terms of both active-high and active-low inputs.

Figure 9-5 Logic Diagram for the 7483 Four-Bit Ripple Carry Adder

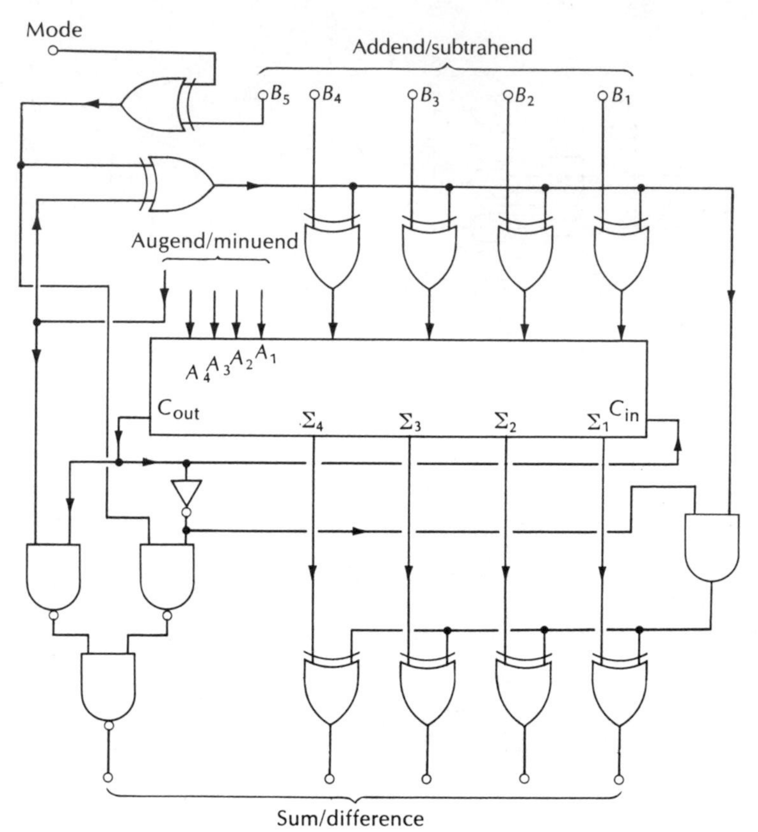

Figure 9-6 Logical Implementation of Sign and Magnitude Addition/Subtraction Addend/Subtrahend

With a constant 0 on input A, a 1 on B produces a 1 output, and a 0 on B produces a 0 output. This is the non-inverted condition. With a constant 1 on input A, a 0 on input B produces a 1 output, and a 1 on B produces a 0 output. This is the inverting mode. See the truth table in Figure 9-7b. Figure 9-7c shows a 4-bit true/invert (true/complement) circuit. If only four bits are required (including the sign bit), $A4$ is connected to V_{cc} and $B4$ is connected to GND to enable carry propagation.

Problems

Write the following decimal numbers in sign-magnitude binary form and then add them.

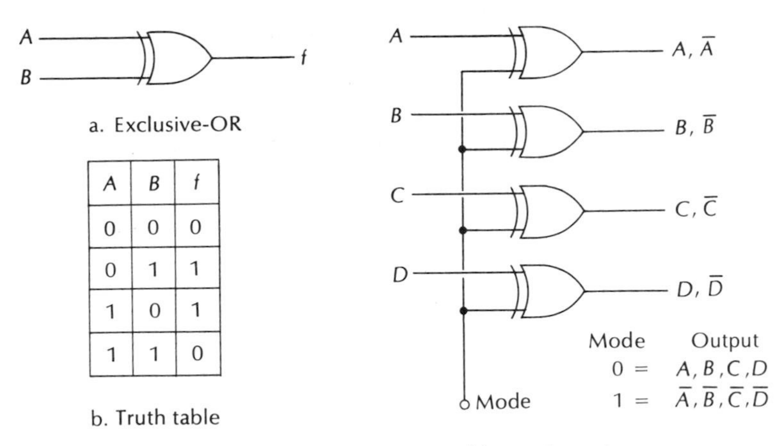

A	B	f
0	0	0
0	1	1
1	0	1
1	1	0

Figure 9-7 True/Complement Generator

44. $\begin{array}{r} +49 \\ \underline{+37} \end{array}$ 45. $\begin{array}{r} -64 \\ \underline{+28} \end{array}$ 46. $\begin{array}{r} -39 \\ \underline{-42} \end{array}$ 47. $\begin{array}{r} +64 \\ \underline{-28} \end{array}$

9-12 One's Complement Addition/Subtraction

The following examples illustrate the addition of signed numbers in one's complement notation:

a. Adding two numbers that have the same sign
 (1) two positive numbers

$$\begin{array}{rl} +13_{10} & = 0.01101 \\ \underline{+11_{10}} & = \underline{0.01011} \\ 24 & = 0.11000 \end{array}$$

Add numbers, including the sign bit.

 (2) two negative numbers

$$\begin{array}{rrl} & 1 \text{ carry} & \\ -13 = & 1.10010 & \text{augend} \\ \underline{-11} = & \underline{1.10100} & \text{addend} \\ -24 & \longrightarrow 1 & \text{add end-around carry} \\ & 1.00111 & \text{sum} \end{array}$$

Add the two numbers including sign bit and then add the end-around carry. Note that both numbers are written in one's complement form.

b. Adding two numbers with opposite signs

(1) positive answer

```
        1 11
 -11 =  1.10100  augend
 +13 =  0.01101  addend
 ---    -------
 + 2 = |0.00001
       └──────→ 1
        -------
        0.00010
```

(2) negative answer

```
          1
 -13 =  1.10010  augend
 +11 =  0.01011  addend
 ---    -------
 - 2 =  1.11101  no end-around carry
```

Add the two numbers including sign bit. If the numbers are written in proper one's complement form, the result is in proper one's complement form.

c. Subtraction

For subtraction in one's complement notation, complement the subtrahend, including the sign bit, and follow addition procedures.

Problems

48. Summarize the algorithms for addition in one's complement notation.

Write the following decimal numbers in one's complement notation and then add them.

```
49. +14      50. -14      51. +23      52. -69
    + 6          - 6          -19          +49
    ---          ---          ---          ---

53. -72      54. -23      55. +54
    -53          +19          -72
    ---          ---          ---
```

Write the following decimal numbers in one's complement notation and then subtract them.

```
56. +46      58. -14      60. +23
    +22          -47          +19
    ---          ---          ---

57. -63      59. -22
    +39          -37
    ---          ---
```

LOGIC FOR IMPLEMENTING ONE'S COMPLEMENT ADDITION AND SUBTRACTION

The logic diagram of a system for performing one's complement addition and subtraction is shown in Figure 9-8. Figure 9-9 shows the 74H87 true/complement (1/0) element and its truth table.

9-13 Two's Complement Addition/Subtraction

The following examples illustrate the algorithms for adding and subtracting in two's complement notation.

a. Addition
   (1) Adding two positive numbers

	1111	carry
+13 =	0.01101	augend
+11 =	0.01011	addend
+24 =	0.11000	sum

Add the two numbers including sign bit.

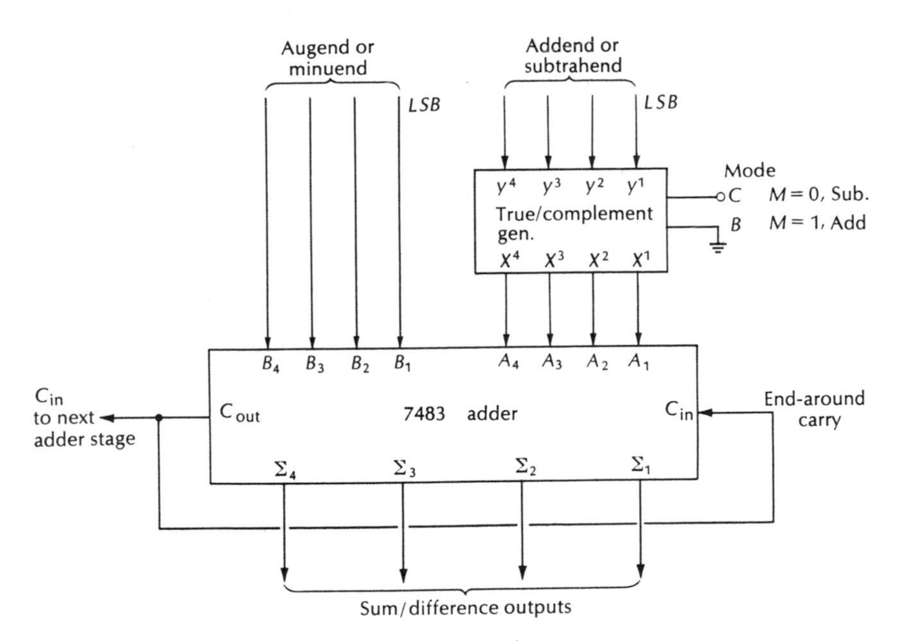

Figure 9-8 Typical One's Complement Adder/Subtracter

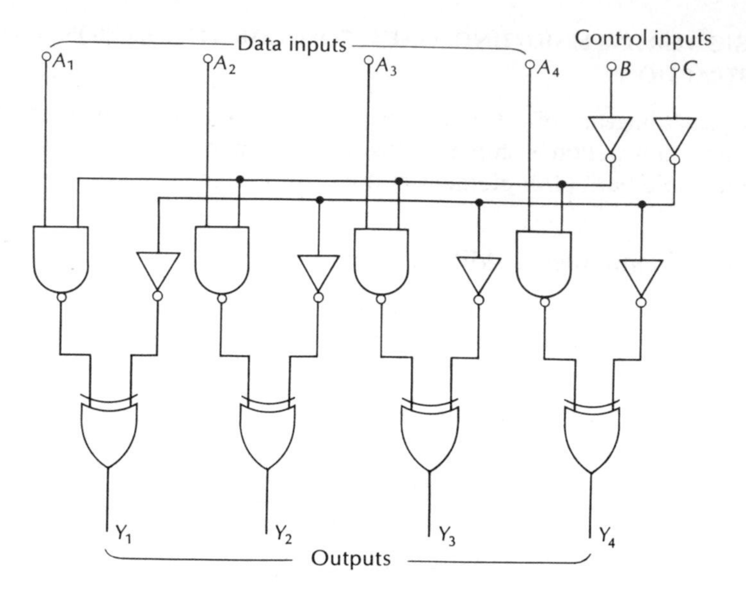

a. Logic diagram

Control inputs		Outputs			
B	C	Y_1	Y_2	Y_3	Y_4
L	L	$\bar{A}_1$	$\bar{A}_2$	$\bar{A}_3$	$\bar{A}_4$
L	H	A_1	A_2	A_3	A_4
H	L	H	H	H	H
H	H	L	L	L	L

b. Truth table

Figure 9-9 Logic Diagram and Truth Table for the True/Complement (1/0) Element

(2) Adding two negative numbers

$$\begin{array}{rrl} & 1\ \ 111 & \text{carry} \\ -13 = & 1.10011 & \text{augend} \\ \underline{-11} = & \underline{1.10101} & \text{addend} \\ -24 = & 1.01000 & \text{sum} \end{array}$$

carry discarded → 1

Add the two numbers including the sign bit, ignoring the carry from the addition of the sign bit column. Numbers are written in proper two's complement form.

(3) Adding numbers with opposite signs

```
                      1 11 1    carry
            -11  =  1.10101    augend
            +13  =  0.01101    addend
            ---     -------
            + 2  =  0.00010    sum

carry       → 1
discarded

                      1 11 1    carry
          +13₁₀  =  0.01101    augend
          -11₁₀  =  1.10101    addend
          -----     -------
carry       + 2     0.00010    sum
discarded ───────→ 1
```

Add the two numbers and ignore all carry outputs from the sign bit column.

b. Subtraction

```
(1)      +13  =  0.01101   minuend
      -(+11)  =  0.01011   subtrahend
      ------
        + 2

                          1 11 1    carry
                         0.01101    minuend
                         1.10100    complement of subtrahend
                               1    add 1
                         -------
carry discarded → 1      0.00010    difference

(2)      +13  =  0.01101   minuend
      -(-11)  =  1.10101   subtrahend
      ------
        +24

             1111    carry
          0.01101    minuend
          0.01010    complement of subtrahend
                1    add 1
          -------
          0.11000    difference
```

Add the one's complement of the subtrahend to the minuend. Add 1 to the result to restore the sum of the two's complement form. Ignore any carry from the sign bit column.

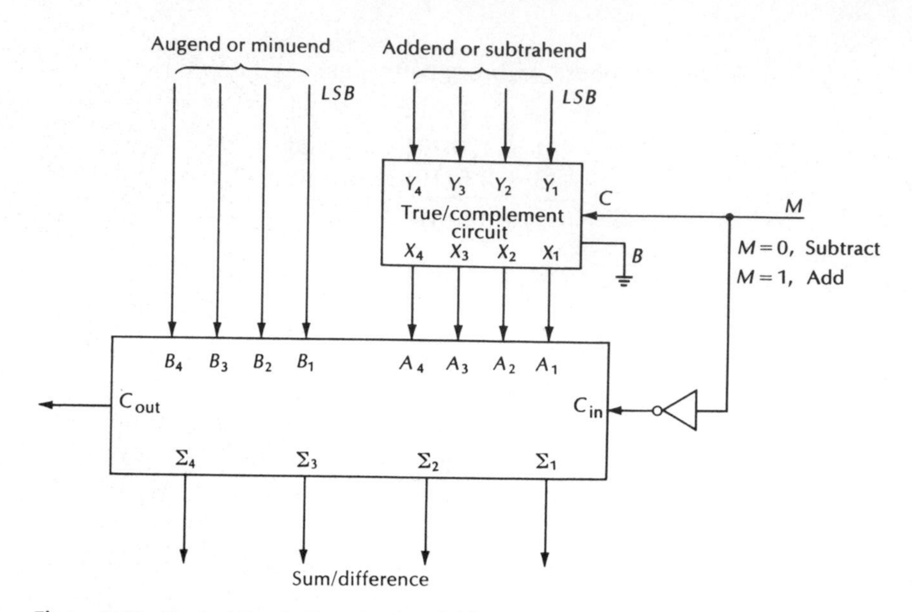

Figure 9-10 Typical Two's Complement Adder/Subtracter

9-14 Implementing Two's Complement Addition and Subtraction

The functional block diagram for a two's complement addition and subtraction circuit is shown in Figure 9-10. The circuit has no end-around carry provision because the carry from the sign bit column is always discarded. In the subtract mode the subtract command is inverted and fed into the carry input ($C_n - 1$), effectively adding a 1 to the result, to restore the *difference* to the proper two's complement form. The two's complement can always be formed by adding 1 to the one's complement.

9-15 Binary-Coded Decimal (BCD) Addition and Subtraction

Many calculators and other small-scale digital systems display the results directly on numerical readouts. For this kind of system a decimal code, generally the natural BCD (8421) or the excess-3 code, is often the most practical implementation of the necessary computing hardware.

BINARY-CODED DECIMAL SYSTEM

In a BCD number system the original positional decimal structure is retained, but each decimal digit is represented by a four-bit binary number:

9	5	4	Decimal number
1001	0101	0100	BCD equivalent

Four binary digits is the minimum number by which all decimal digits 0 to 9 can be represented. Table 9-4 shows the structure of the BCD system. Look at the following example:

Write 234_{10} in BCD.

2	3	4	Decimal
0010	0011	0100	BCD

Further examples are shown in Table 9-5.

When BCD numbers are used, conversion to display becomes simple, since it can be done one digit at a time where no digit is ever larger than 9. Arithmetic is actually done within the machine in binary, and therein lies a problem. Since radix 2 and radix 10 are not related, the carry generated by adding two digits will not occur under the same conditions for decimal digits and BCD digits. The following example will illustrate this problem:

Decimal	*BCD*		
	1	0	position
26	0010	0110	
+38	+0011	1000	
64	0101	1110	

Table 9-4 Structure of the BCD System

3	2	1	0	Position
Thousands	Hundreds	Tens	Units	Position value
$2^3\,2^2\,2^1\,2^0$	$2^3\,2^2\,2^1\,2^0$	$2^3\,2^2\,2^1\,2^0$	$2^3\,2^2\,2^1\,2^0$	BCD value in exponent form

Table 9-5 Decimal Numbers and Their BCD Equivalents

Decimal numbers	BCD equivalent
001	0000 0000 0001
123	0001 0010 0011
546	0101 0100 0110
879	1000 0111 1001

Note that in adding $26_{10} + 38_{10}$ a carry is produced in position 0 ($1110_2 = 14_{10}$). In a sense the result in BCD is correct (0101 + 1110 = 50 + 14), but the maximum allowable weight number for position 0 has been exceeded. If the decimal-position notation scheme is to be preserved, the result must be expressed as 60 + 4 (0110 – 0100). The following will help to illustrate the point:

Decimal		*BCD*	
1		0001	
+2		+0010	
3	no carry required	0011	result correct
3		0011	
+4		+0100	
7	no carry required	0111	result correct
5		0101	
+5		+0101	
10	carry required	1010	result not correct, no carry generated
6		0110	
+5		+0101	
11	carry required	1011	result not correct, no carry generated
9		1001	
+7		+0111	
16	carry required	1 0000	carry generated, but result not correct

In this last case a carry is generated in the BCD addition, but position 0 is incorrect. In the decimal system the first carry occurs at 10. In the BCD system the first carry occurs at 16; that is, 15 is the largest decimal number that can be expressed in four binary digits. The difference between 10 and 16 is 6, and notice that 6 is the number required to make the BCD sum correct.

A brief study of the system will indicate that 6 must be added to any sum 10_{10} or greater to maintain the decimal structure. This implies that the machine must be able to recognize when a number 10_{10} or larger has been generated.

For sums between 10 and 15 the NBCD error detector examined in Chapter 6 can be used. For sums larger than 16 a *natural* carry is generated. The outputs from the error detector and from the adder-carry output can be used to activate a separate add-6 or 0 (only) circuit.

Example BCD Addition

Add 259_{10} and 378_{10}

Decimal	*BCD*			
	1	1	carries	
259	0010	0101	1001	
+378	+0011	0111	1000	
637	0110	1101	0001	sum
	+	0110	0110	add
	0110	0011	0111	BCD result
	6	3	7	decimal equivalent of BCD results

1. Add each BCD group as an ordinary pair of binary numbers.
2. If the decimal equivalent of the sum produced is less than 10, the result is correct.
3. If the decimal equivalent of the sum is 10 or greater, add binary 6 and add any overflow beyond four binary digits to the LSB column of the next BCD group to the left.

Problems

Write each of the following decimal numbers in BCD and then add them.

61. 398 + 297
62. 525 + 425
63. 2081 − 1097
64. 5700 + 479

BCD ADDER CIRCUIT

The BCD adder circuit is shown in Figure 9-11. The two adders are four-bit binary full adders of the same kind used in the previous adder circuits. The two AND and one OR gates form the BCD error detector (decoder) that instructs the correction adder to add 0000 for sums less than 10 and to add 0110 for sums between 10 and 15. The carry output from the adder is also connected to the input of the OR gate to implement the add-6 correction adder for sums larger than 15. The output of the add-6 decoder gate also provides the proper *decimal* carry to the next BCD adder decade. The correction adder can be the same device

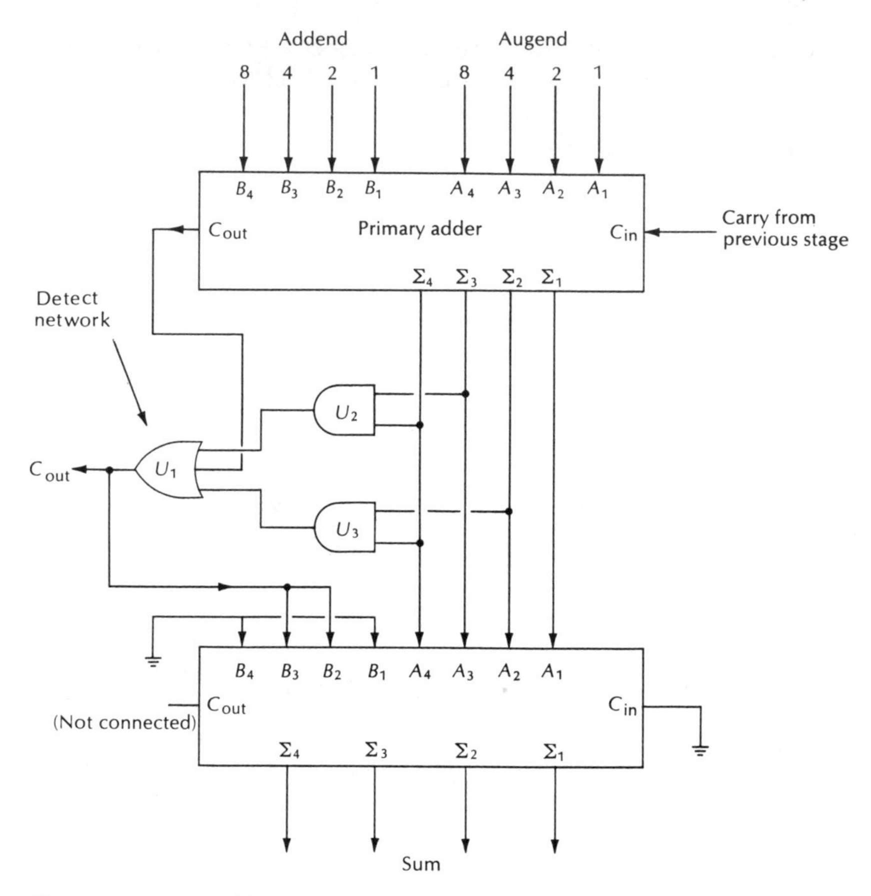

Figure 9-11 BCD Adder Circuit

type as the primary adder. To add 0110, $B1$ and $B4$ can be kept at a constant 0 logic level.

BINARY-CODED DECIMAL SUBTRACTION

Figure 9-12 shows a BCD subtracter circuit that uses the one's complement method. The minuend and the one's complement of the subtrahend are added by the primary adder. The result is transferred through a true/complement circuit (four exclusive-OR gates or a 74H87 or similar true/complement circuit) to the correction adder. The correction adder then adds either 0000 or 1010 depending upon the sign of that particular decade and the sign of the total result.

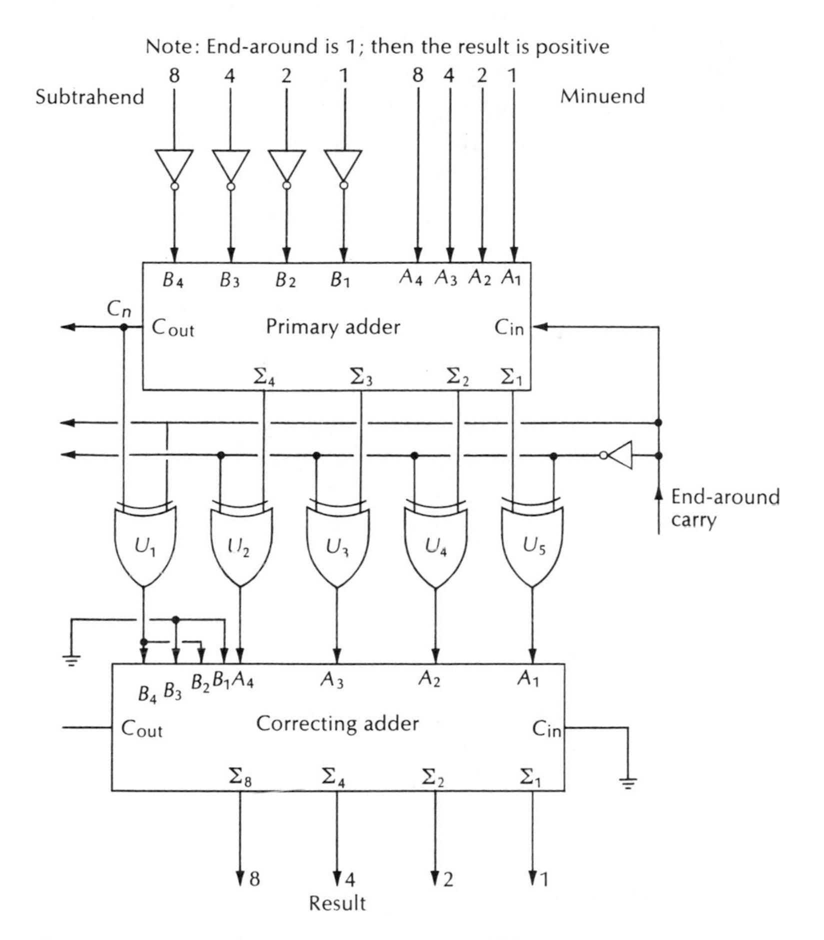

Figure 9-12 BCD Subtracter Stage One's Complement Type.

ALGORITHMS

1. True/complement circuit
   a. If the end-around carry is 1, the sign is +. *Transfer true results* to the correcting adder.
   b. If the end-around carry is 0, the sign is −. *Transfer the complement* of the output of the primary adder to the correcting adder.
2. The correcting adder
   a. End-around carry = 1
      $C_n = 1$ correcting adder adds 0000
      $C_n = 0$ correcting adder adds 1010

b. End-around carry = 0
   $C_n = 1$ correcting adder adds 1010
   $C_n = 0$ correcting adder adds 0000
c. All carries from the correcting adder are ignored.
   These rules can be summarized in Boolean notation as:

$$\text{Add } 1010 = C_{eac} \cdot \overline{C_n} + \overline{C_{eac}} \cdot C_n$$

where C_{eac} is the end-around carry.

This proves upon inspection to be the exclusive-OR function and can be written as:

$$\text{Add } 1010 = C_{eac} \oplus C_n$$

An examination of Figure 9-12 indicates that an exclusive-OR gate is used to control the correcting adder.

BCD SUBTRACTION USING THE NINE'S COMPLEMENT

Because the BCD system is a coded decimal system, subtraction can be carried out using the nine's complement. Figure 9-13 shows a special nine's complement circuit and Figure 9-14 shows a complete BCD nine's complement subtraction functional block diagram. An adder can be used to extract the nine's complement. The adder nine's complement generator adds 1010 (the complement of 0101) to the subtrahend.

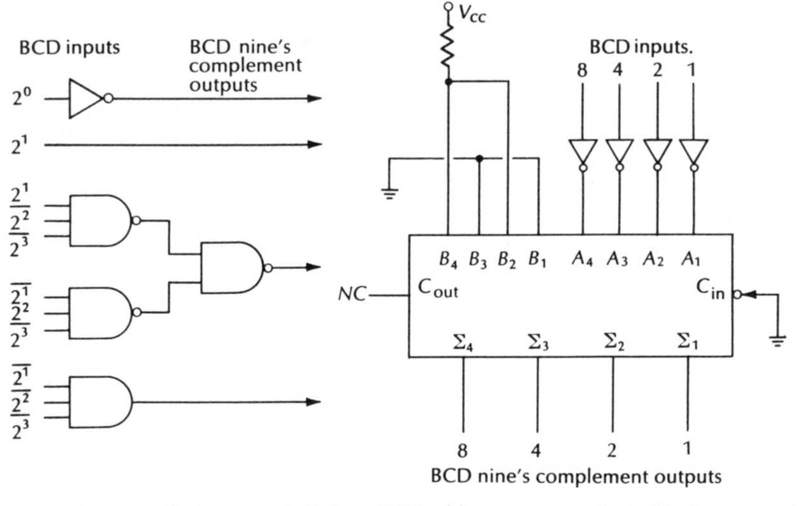

Figure 9-13 Two Nine's Complement Generators

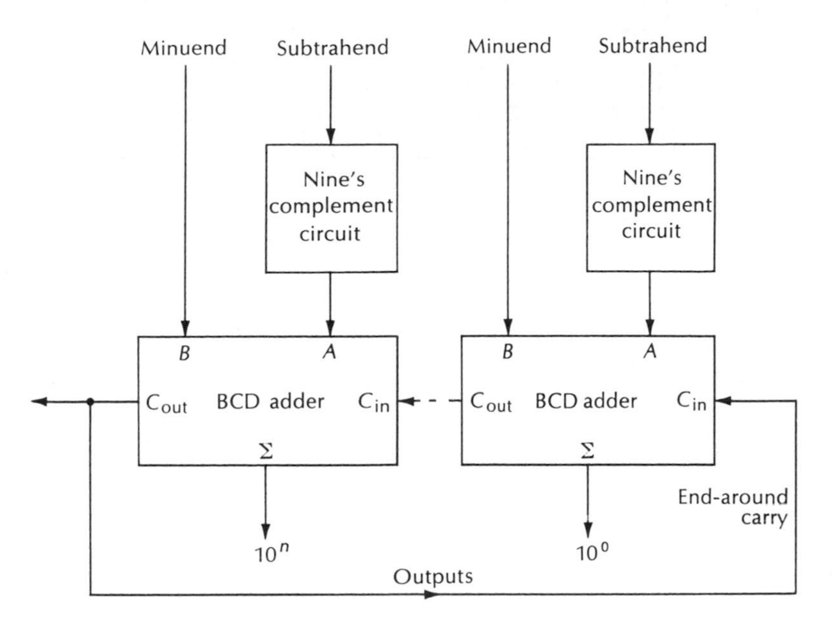

Figure 9-14 Typical Nine's Complement BCD Subtracter

9-16 Excess-3 Addition and Subtraction

The excess-3 code is the most popular of several self-complementing codes. In this context self-complementing means that the one's complement is identical to generating the nine's complement of the decimal number.

The simple inversion of each bit in the XS-3 code yields the nine's complement. For subtraction the subtrahend can be complemented by a true/complement circuit instead of a special nine's complement generator. The functional block diagram of an XS-3 adder is shown in Figure 9-15. Table 9-6 provides XS-3 equivalents for decimal digits and nine's complement values.

The addition algorithm for XS-3 addition is shown below. Add the two numbers.

a. If an end-around carry is produced, add 0011 (3_{10})
b. If *no* end-around carry is produced, subtract 0011 or add its complement, 1100.

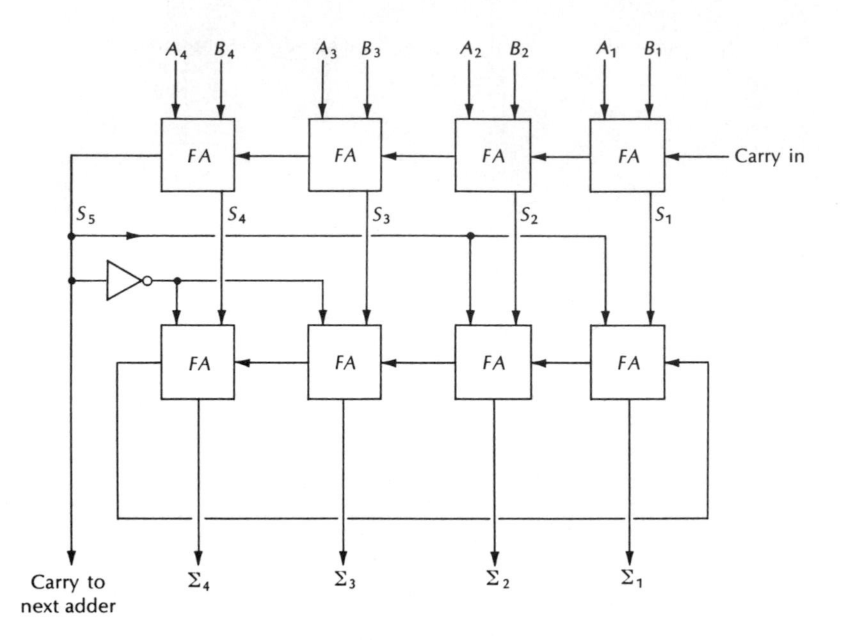

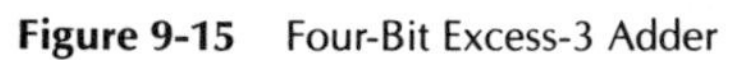
Figure 9-15 Four-Bit Excess-3 Adder

Table 9-6 Decimal Equivalents in the XS-3 Code

Decimal number	Excess-3 Equivalent	Nine's Complement
0	0011	1100
1	0100	1011
2	0101	1010
3	0110	1001
4	0111	1000
5	1000	0111
6	1001	0110
7	1010	0101
8	1011	0100
9	1100	0011

Examples

a. Add 879_{10} and 132_{10}.

Decimal	*Excess-3 BCD*			
	1	1	carry	
879	1011	1010	1100	
+132	+0100	0110	0101	
1011	1 0000	0001	0001	uncorrected sum

This result is correct for BCD, but 3 must be added to each column to restore the number to the excess-3 code.

0001	0000	0001	0001	uncorrected sum
0011	0011	0011	0011	add 3
0100	0011	0100	0100	corrected sum

b. Add 231_{10} and 122_{10} in excess-3.

Decimal	*Excess-3 BCD*			
231	0101	0110	0100	
+122	+0100	0101	0101	
353	1001	1011	1001	uncorrected sum

There was no overflow so 3 must be subtracted. This is accomplished by adding the complement of 0011, 1100.

11	1111	carry	
1001	1011	1001	uncorrected sum
1100	1100	1100	add 1100
0110	1000	0101	
		1	end-around carry
0110	1000	0110	corrected sum

17 Half and Full Subtracters

Although subtraction through the use of complement notation is the most widely used method, some machines use a subtracter circuit that is very similar to the common adder circuit. Table 9-7 is the truth table for binary subtraction. The *difference* equation written from Table 9-7 is $D = \bar{A} \cdot B + A \cdot \bar{B}$. This is the exclusive-OR function and can be written as $D = A \oplus B$. The *borrow* equation as written from the truth table is $B = \bar{A} \cdot B$. These two equations differ but little from the half adder equations:

$$\text{Sum} = A \oplus B$$
$$\text{Carry} = A \cdot B$$

Table 9-7 Subtraction Truth Tables

A	B	Difference	Minterms
0	0	0	
0	1	1	$(\bar{A}B)$
1	0	1	$(A\bar{B})$
1	1	0	

a. Subtraction difference truth table

b. Difference equation: $D = \bar{A}B + A\bar{B}$

This is also the exclusive-OR function: $D = A \oplus B$

The sum and difference equations are identical

A	B	Borrow	Minterms
0	0	0	
0	1	1	$\bar{A}B$
1	0	0	
1	1	0	

c. Subtraction borrow truth table

d. Borrow equation: $B = \bar{A} \cdot B$

Summary of equations

Addition:

Sum $= A \oplus B$

Carry $= A \cdot B$

Subtraction rules

$A-B$	=	Difference	Borrow
$0-0$	=	0	0
$0-1$	=	1	1
$1-0$	=	1	0
$1-1$	=	0	0

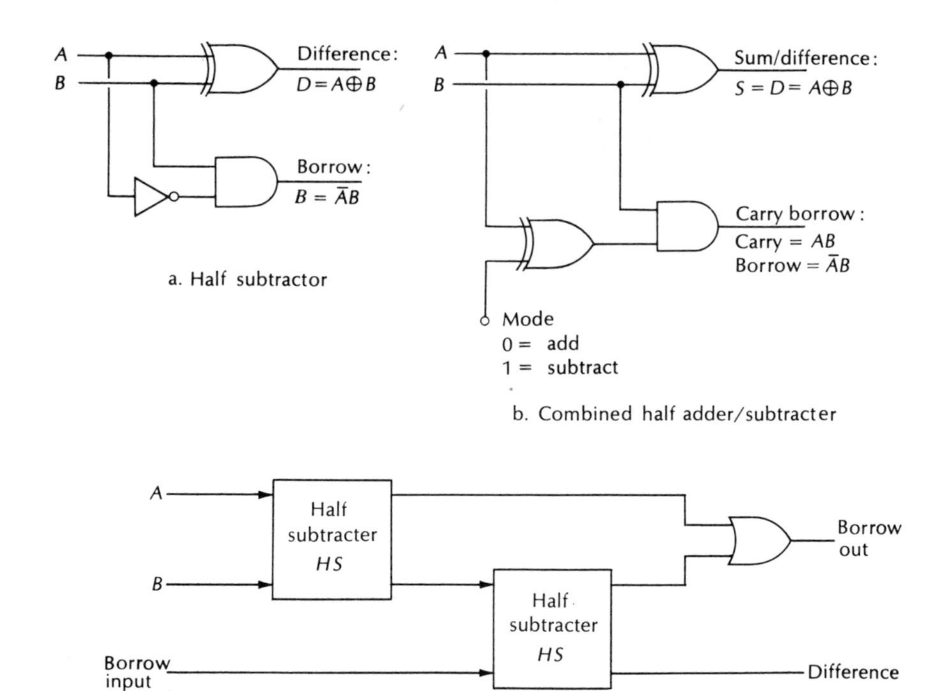

Figure 9-16 Half Subtracter, Half Adder/Subtracter, and Full Subtracter

The difference is simply an inverted A for the subtracter and *not* inverted for addition. Figure 9-16 shows a half subtracter circuit, a combined adder/subtracter circuit using an exclusive-OR gate as a *controlled* inverter, and the method of combining two half subtracters to form a full subtracter. Figure 9-17 shows a 4-bit subtracter functional block diagram.

9-18 Serial Addition

Figure 9-18 shows a serial adder scheme where a full adder and flip-flop perform serial binary addition. For active-*high* operands, the carry

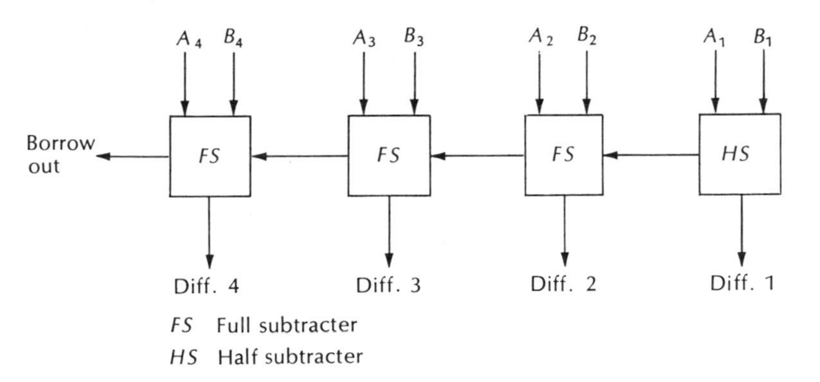

Figure 9-17 A Four-Bit Binary Subtracter

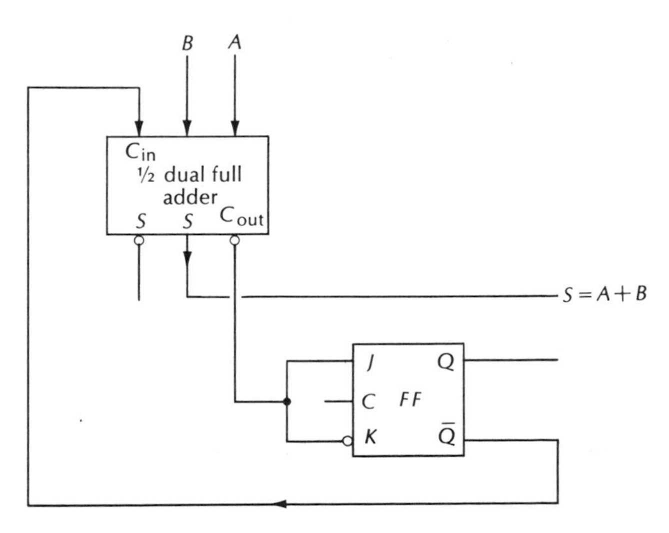

Figure 9-18 Serial Adder

flip-flop must be set when the least significant bit is applied. For active-*low* operands, the flip-flop must be reset when the least significant bit is applied.

Serial addition/subtraction can easily be accomplished using two full adders and a flip-flop. The upper adder in Figure 9-19a serves as a conditional inverter. Exclusive-OR gates could be used. This scheme requires either a second pass for end-around carry or that the carry flip-flop starts out set for add, reset for subtract (with active-*high* operands, but the opposite with active-*low* operands). This second pass can be avoided by using two exclusive-OR gates in the data path, thereby effectively using the adder with active-*high* operands in one mode and active-*low* operands in the other.

For both addition and subtraction, the carry flip-flop must start out set for active-*high* operands, reset for active-*low* operands (see Figure 9-19b).

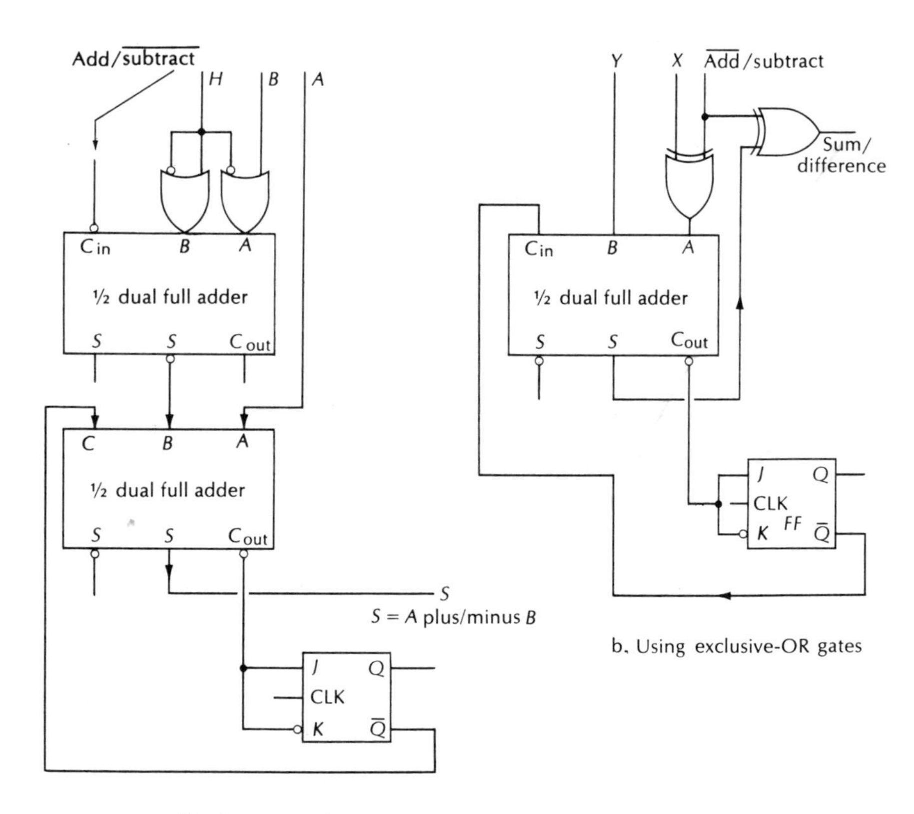

a. The first approach

Figure 9-19 Serial Addition and Subtraction

9-19 Binary Multiplication

Multiplication is performed as successive additions. The notation 5×7 implies a shortcut method of adding seven fives:

$$5 \times 7 = 5 + 5 + 5 + 5 + 5 + 5 + 5 = 35$$

The algorithm used in human calculation involves multiplying the multiplicand by each digit in the multiplier and adding the partial products. We also use the device of shifting the partial product one position to the left, which in the decimal system is equivalent to multiplying by 10 (conversely a shift of one position to the right would be the equivalent to dividing by 10).

In binary notation shifting one position to the left is equivalent to multiplying by 2. Binary multiplication can be mechanized using an approach similar to the longhand method because binary multiplication (one pair of bits at a time) is such a simple process. In binary there are only two possible results for the multiplication of any pair of bits. Note the following examples:

$$(1)\ \begin{array}{r} 0 \\ \times 0 \\ \hline 0 \end{array} \qquad (2)\ \begin{array}{r} 1 \\ \times 0 \\ \hline 0 \end{array} \qquad (3)\ \begin{array}{r} 1 \\ \times 1 \\ \hline 1 \end{array}$$

and for multibit numbers:

$$\begin{array}{r} 1101 \\ \times 0 \\ \hline 0000 \end{array} \qquad \begin{array}{r} 1101 \\ \times 1 \\ \hline 1101 \end{array} \qquad \begin{array}{l} \text{multiplicand} \\ \text{multiplier} \\ \ \end{array}$$

The rules are as follows:

a. When a binary number is multiplied by 0, the result is 0.
b. When a binary number (multiplicand) is multiplied by 1, the result is the original number (the multiplicand).

The following example illustrates the procedure that is likely to seem most obvious because it is very much like the way we normally do multiplication. In this example, all of the partial products are found and then added simultaneously. Successive addition of each partial product is as follows:

$$\begin{array}{r} 111 \\ 101 \\ \hline 111 \end{array}$$

If the least significant bit of the multiplier is a 1, the multiplicand is transferred to the output as a partial product.

0111	If the next least significant bit is a 0,
0111	the product is shifted right one position.
111	If MSB is a 1, the product is shifted right
100011	one position and added to the multiplicand.

We will use this example to illustrate the mechanics of the process in Figure 9-20, parts a through f.

a. This part shows the initial problem loaded into the appropriate registers. The multiplicand register will not change during the operation. The multiplier is loaded into the MQ register. The output of the MQ register is connected to the adder, but it does *not* supply bits to be added. Its output state is actually an instruction that tells the adder either to add or not to add (1 = add, 0 = don't add). This particular input on the adder is simply an add enable input.

b. The 1 on the add enable input enables the adder to add the contents of the accumulator to the multiplicand.

000	accumulator
+111	multiplicand
111	new accumulator contents

Part b shows the status of the registers after the addition.

c. Shift the accumulator and MQ registers right one position. This puts a zero in the rightmost cell of the MQ register, which instructs the adder *not* to add, thereby inhibiting addition. Part c shows the contents of all registers at this stage of the process.

d. Again, the accumulator and MQ registers are shifted right. The 1 in the rightmost cell of the MQ register enables the adder, and the contents of the accumulator are added to the multiplicand.

11	carry
001	accumulator
+111	multiplicand
1 000	new accumulator contents

Notice that a carry-out of the adder is produced. This is stored in the delay element for use after the next shift right.

e. Shift right. This part shows the contents of all registers after the shift. The 1 in the rightmost cell of the MQ register enables the adder. However, in this case there is a 1 stored in the carry delay cell. Gating circuitry, not shown in Figure 9-20, inhibits the trans-

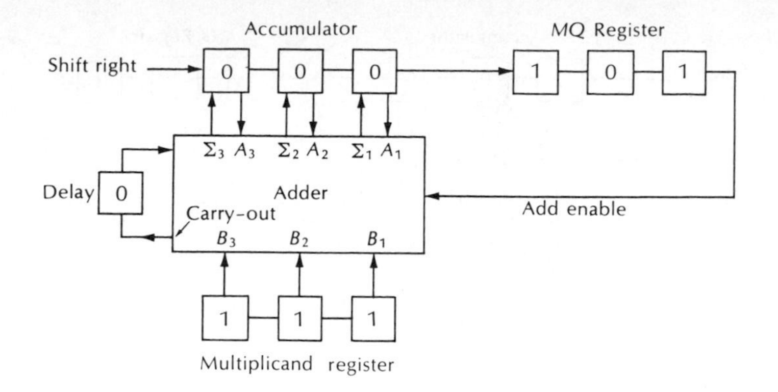

a. First step

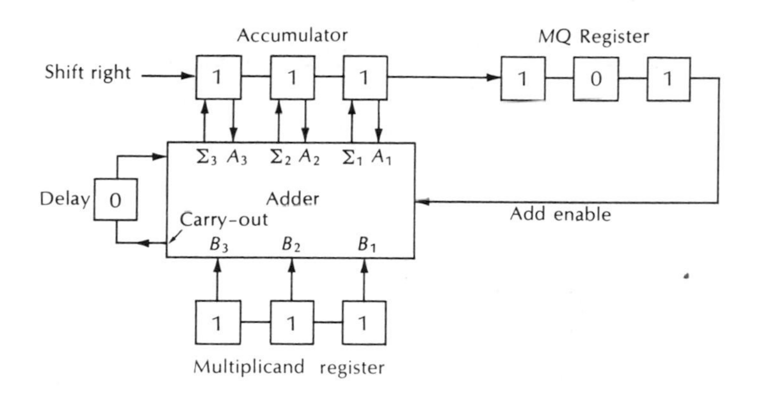

b. Second step

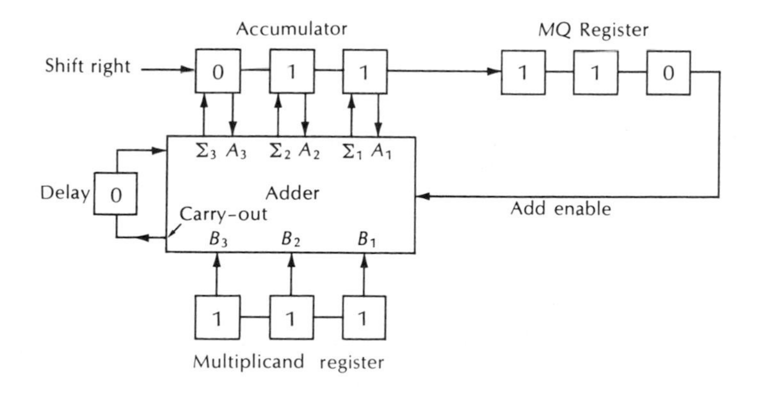

c. Third step

Figure 9-20 Binary Multiplication

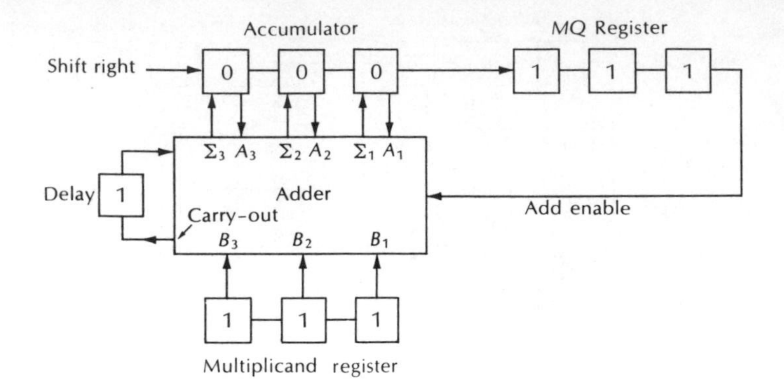

d. Fourth step

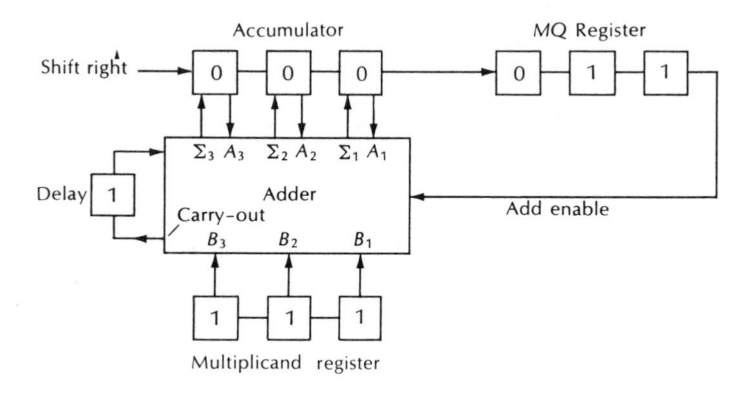

e. Fifth step

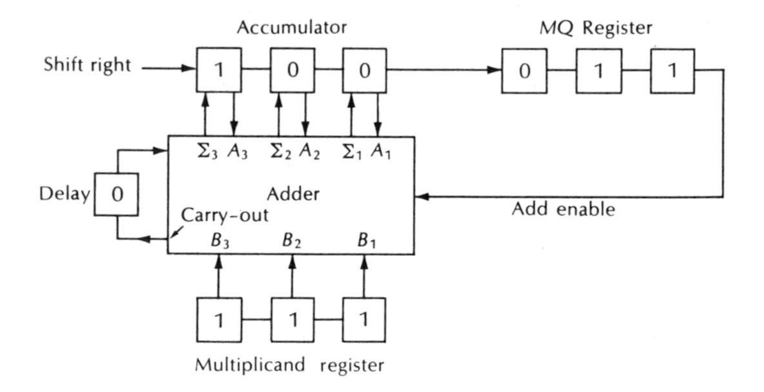

f. Sixth step

Figure 9-20 continued

fer of data from the multiplicand register so that only the carry is added to the contents of the MQ register.

f. This part shows the final contents of all registers.

BCD MULTIPLICATION

BCD multiplication is handled very much like binary multiplication except that BCD groups are carefully arranged in the registers to maintain their essentially decimal character. This method of listing partial products is slightly different from the usual longhand procedure in order to keep down the number of registers.

Example

a. *Conventional*

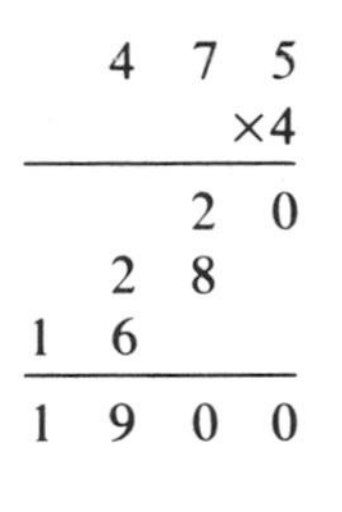

b. *Diagonal*

```
4 7 5
  ×4
-------
1 2 2
 6 8 0
-------
1 9 0 0
```

The diagonal arrangement requires only two registers.

LOOK-UP TABLES

As read-only memories (ROM) go down in price, the use of look-up tables becomes increasingly attractive. Individual digit products can be placed permanently in memory and retrieved when needed. This eliminates the need for multiplication and reduces the multiplication process to one of simply looking up the partial products and adding them as shown in example b.

9-20 Binary Division

Division is the opposite of multiplication and can therefore be accomplished by a series of repeated subtractions, as illustrated below:

Example

a. *Conventional*

	5	quotient
divisor 10	50	dividend

b. *By subtraction*

50	
−10	
40	first subtraction
−10	
30	second subtraction
−10	
20	third subtraction
−10	
10	fourth subtraction
−10	
00	fifth subtraction − quotient = 5

The process can be extended to a larger figure but the number of steps becomes so large that the division operation becomes quite slow. To divide 1800 by 6 would require 300 subtractions. The shifting process can be used to drastically reduce the number of subtractions. The following example approximates one common computer method. It differs only in the fact that we can determine the proper number of zeros behind the 6 by inspection, where the computer must use a trial-and-error procedure to make the determination. The quantity added to the quotient at each step is the equivalent of a computer register left shift (10 = one-place shift, 100 = two-place shift, and so on), as shown below:

	Add to quotient	*Quotient total*
300		
6 | 1800		
−600		
1200	100	100
−600		
600	100	200
000	100	300 quotient

BINARY DIVISION ALGORITHM (RESTORING METHOD)

a. Subtract divisor from dividend. If result is a positive number, put 1 in rightmost bit of quotient register; if result is a negative number, add divisor back to dividend.
b. Shift quotient left by one bit and shift divisor right by one bit (or dividend left).
c. Repeat steps a and b.
d. Continue steps a through c until a subtraction yields a difference of all 0 bits, until required accuracy is obtained, or until all available bit positions in quotient register are filled.

Example in Applying the Division Algorithm*

1. Subtract
2. Shift and subtract
3. Add
4. Shift and subtract
5. Shift and subtract

	Contents of quotient register
1011	
1010 ⟌ 1101110	
1010	0000
001111	0001
1010	0010
11011	
1010	
01111	0010
1010	0100
01010	0101
1010	1010
0000	1011 final quotient

Problems

Write the following in BCD and add:

65.	66.	67.	68.
49	546	421	997
+63	+194	+315	+894

* For a comprehensive discussion of division processes see Flores, *The Logic of Computer Arithmetic,* Prentice-Hall, 1963.

Write the following in binary signed magnitude, radix complement, and radix-minus-1 complement notations, and perform the indicated operations:

69. (−42) − 36
70. (−36) − 42
71. (−64) + 29
72. (−29) + 64
73. (−64) + (−29)
74. (−64) − (−29)
75. 64 − 64
76. (−64) + (−64)
77. 43 − 96
78. 43 − (−96)

BINARY MULTIPLIERS

The National Semiconductor DM7875A/B forms a two-package binary multiplier system. The multiplier can multiply two 4-bit numbers in 36 nanoseconds, considerably faster than the successive addition approach required when an arithmetic logic unit (ALU) is used to multiply binary numbers. The 7875A/B features tristate outputs. Binary multipliers are less common than ALU devices, and most machines do not use them. There are, however, systems where the use of the separate multiplier circuit and its associated registers and control circuitry is justified.

9-21 The Arithmetic Logic Unit (ALU)

The arithmetic logic unit (ALU) is the generally preferred device for most arithmetic circuits. The ALU contains all the necessary circuitry for performing simple addition and subtraction, along with 14 other useful add/subtract functions and 16 common logic functions. Each function is selected by programming a set of four *select* inputs with the appropriate binary combination. A *mode* control determines whether the ALU operates in its arithmetic or its logic mode.

The device can be used as either an active-high or an active-low system. Table 9-8 lists the logic and arithmetic functions available in the 74S181 Schottky ALU. Other available ALUs, including the C-MOS 40181B, feature logic and arithmetic functions almost identical to those of the 74S181 ALU.

The 74S181 has about 75 gates on the chip, making it a fairly complex device. Each package can process two 4-bit binary words, and stages can be cascaded for longer binary words. Where speed is not critical, ripple-carry propagation from stage to stage can be used. When higher speed operation is required, the companion look-ahead carry generator, the 74S182, can be used.

The 74S181 can perform the addition of two 8-bit numbers in 19 ns when used with the 74S182 look-ahead carry generator. This is about half the add-time required by the older standard TTL 74181 ALU.

The 40181B C-MOS ALU has the same logic and arithmetic capabil-

Table 9-8 ALU Functions

Function Select				Active-High Data		
				Mode-High Logic Functions	Mode-Low: Arithmetic Operations	
S3	S2	S1	S0		C_n = High (No Carry)	C_n = Low (With Carry)
L	L	L	L	f = $\overline{A}$	f = A	f = A plus 1
L	L	L	H	f = $\overline{A + B}$	f = $A + B$	f = $(A + B)$ plus 1
L	L	H	L	f = $\overline{A}B$	f = $A + \overline{B}$	f = $(A + \overline{B})$ plus 1
L	L	H	H	f = 0	f = minus 1 (2's compl)	f = zero
L	H	L	L	f = $\overline{AB}$	f = A plus $A\overline{B}$	f = A plus $A\overline{B}$ plus 1
L	H	L	H	f = $\overline{B}$	f = $(A + B)$ plus $A\overline{B}$	f = $(A - B)$ plus $A\overline{B}$ plus 1
L	H	H	L	f = $A \oplus B$	f = A minus B minus 1	f = A minus B
L	H	H	H	f = $A\overline{B}$	f = $A\overline{B}$ minus 1	f = $A\overline{B}$
H	L	L	L	f = $\overline{A} + B$	f = A plus AB	f = A plus AB plus 1
H	L	L	H	f = $\overline{A \oplus B}$	f = A plus B	f = A plus B plus 1
H	L	H	L	f = B	f = $(A + \overline{B})$ plus AB	f = $(A + \overline{B})$ plus AB plus 1
H	L	H	H	f = AB	f = AB minus 1	f = AB
H	H	L	L	f = 1	f = A plus A	f = A plus A plus 1
H	H	L	H	f = $A + \overline{B}$	f = $(A + B)$ plus A	f = $(A + B)$ plus A plus 1
H	H	H	L	f = $A + B$	f = $(A + \overline{B})$ plus A	f = $(A + \overline{D})$ plus A plus 1
H	H	H	H	f = A	f = A minus 1	f = A

ities as the 74S181 except that it is much slower, with an add-time of 200 ns. The 40182B is a C-MOS look-ahead carry generator for the 40181 ALU. The industry type MC14581 is another common C-MOS ALU with characteristics similar to the 40181B.

Figure 9-21 shows the logic diagram of the 74S 182 look-ahead carry generator. Figure 9-22 shows how the 74S 181 ALU can be cascaded with either ripple-carry or look-ahead carry propagation.

SUMMARY OF ALGORITHMS

1. Addition by sign and magnitude

   Two numbers of same sign

   To magnitude of augend, add magnitude of addend. Use common sign.

   Two numbers of opposite sign—augend magnitude larger

   To magnitude of augend, add one's complement of addend magnitude. End-around carry. Affix sign of augend.

   Two numbers of opposite sign—addend magnitude larger (or equal)

   To magnitude of augend add one's complement of addend magnitude. There is no end-around carry. Complement result and affix sign of addend.

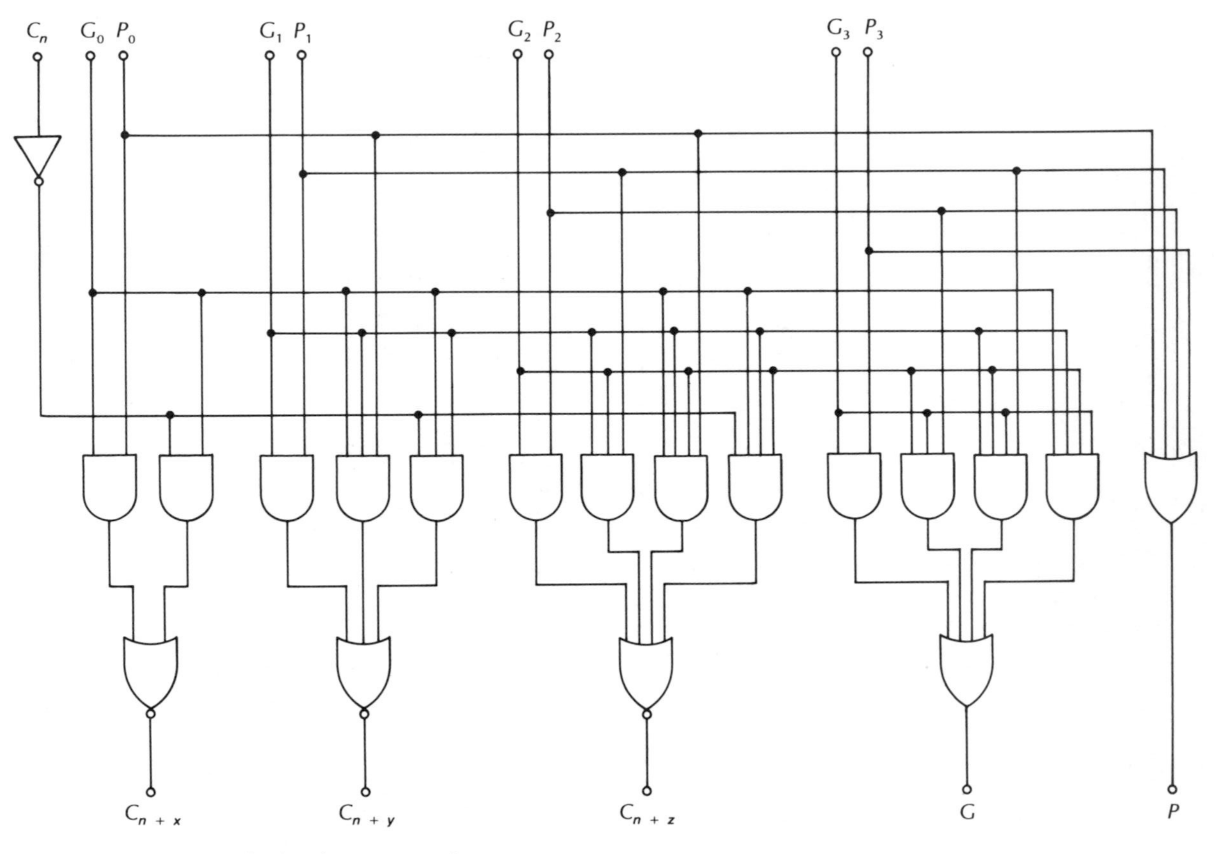

Figure 9-21 74182 Look-Ahead Carry Logic Diagram

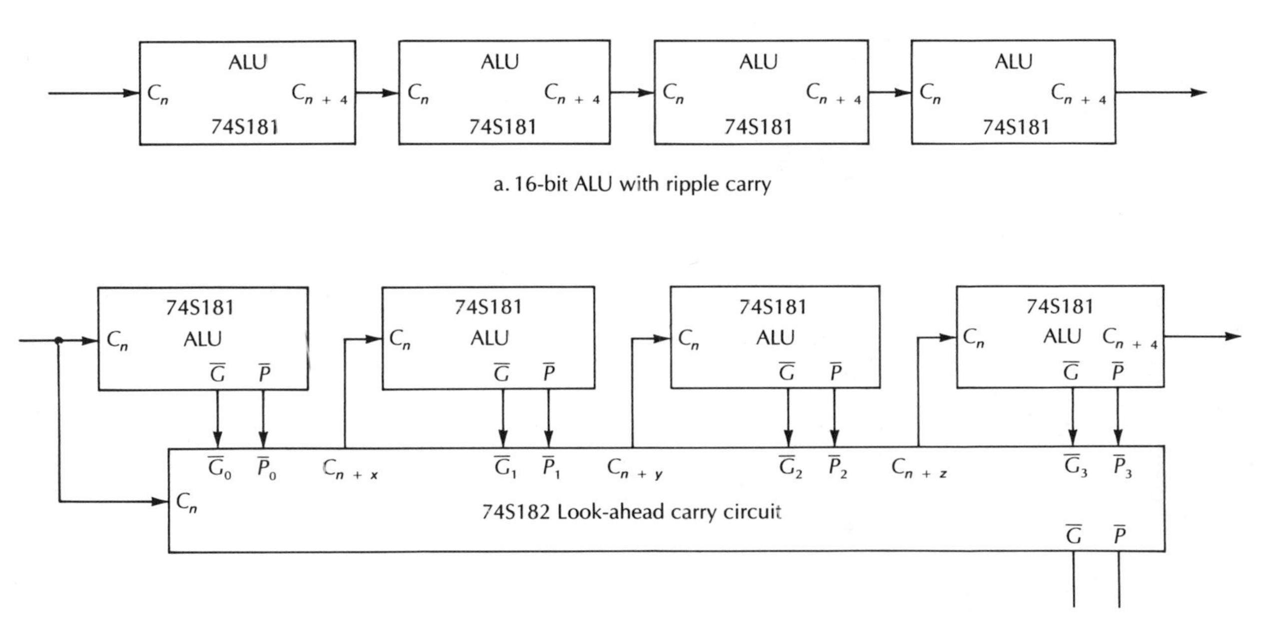

b. 16-bit ALU with look-ahead carry circuit

Figure 9-22 Cascading ALUs

2. Addition using one's complement notation

   Two numbers of same sign

   To augend magnitude, add addend magnitude. Use common sign. To one's complement of augend magnitude, add one's complement of addend magnitude. Add end-around carry. Affix common sign (answer in one's complement form).

   Two numbers of opposite sign

   To magnitude of augend (in one's complement form, if negative), add magnitude of addend (in one's complement form, if negative). Add end-around carry.

3. Addition using the two's complement

   Two numbers of same sign

   To augend magnitude, add addend magnitude. Use common sign. To two's complement of augend magnitude, add two's complement of addend magnitude and ignore carry. Affix common sign.

   Two numbers of opposite sign—augend magnitude larger

   To magnitude of augend, add two's complement of addend including sign bits. Ignore carry from sign bits. To two's complement of augend magnitude, add magnitude of addend; add sign bits also. Ignore carry from sign bits.

   Two numbers of opposite sign—addend magnitude larger (or equal)

   To two's complement of augend magnitude, add magnitude of addend including sign bits. Ignore carry from sign bits.

PROBLEMS

Given Table 9-8, state the values (H or L) required at the *select* and *mode* terminals for the 74S181 ALU to perform the following functions:

79. Add A to B.
80. Subtract 1 from A.
81. Add $A + B + 1$.
82. The Boolean equation; $f = A + B$.
83. Compare two 4-bit numbers to determine whether the two are equal. *Hint:* If you can't figure it out from the table, look up the 74S181 in the data manual and see whether you can find the clue there.
84. Table 9-8 lists the function AB (multiply $A \times B$). Can this function be used to multiply two 4-bit numbers in a single operation? If not, why not?

CHAPTER **10**

MEMORY SYSTEMS

Learning Objectives. *Upon completing this chapter you should know:*

1. How to classify (according to use and access time) the following memory types:
   - *a. Core*
   - *b. MOS LSI memories*
   - *c. MOS LSI shift-register memories*
   - *d. Bipolar scratch pad and buffer memories*
   - *e. Cassette drives*
   - *f. Core memories*
   - *g. Punched paper tape and cards*
   - *h. Computer tape*
   - *i. ROM*
   - *j. RAM*

2. How to describe core memory construction and principle of operation.

3. How to draw the schematic diagrams for: a static MOS, a dynamic MOS, and a bipolar typical memory cell.

4. How to explain the operation of each of the memory cells (in 3).

5. How to define DRO and know where it applies.

6. How to describe both bit and word organization in RAMs.

7. The output drive capabilities of typical MOS memories.

8. The kinds of ROMs commercially available and the special characteristics of each.

9. What kinds of (recirculating MOS) shift registers are typically available.

10. What kinds of output configurations are available in MOS memory chips.

11. MOS timing problems and their relationship to troubleshooting.

12. ROM programming and erasing techniques.

Automatic processing of data in a computer depends upon being able to store a list of operational commands called the *program,* along with the numerical or other data to be operated on. The computer must have access to instructions and data at computer speeds.

Any computer contains several kinds of memories, each of which is generally allocated the kind of memory task best suited to it. The use of several memory forms is dictated by a trade-off in cost per bit of memory and memory speed. As a general rule, the faster the access to a memory type, the higher the cost per bit.

MAIN MEMORY

The computer main memory is random-access, which means that the time required to retrieve data from the memory is the same regardless of the memory location. This is in contrast to magnetic tape, for example, where many unwanted memory locations may have to be passed through before arriving at the desired location. In the case of magnetic tape, the number of unwanted locations that have to pass under the head before the desired data could be retrievable can vary considerably depending on its relative location on the tape. Consequently, the time required to retrieve a given set of data is variable. For main memory purposes, constant access time is important to the overall computer timing cycle. Access time must be predictable and constant. Random-access memories are generally magnetic core or semiconductor arrays. Main memories are moderately fast but generally slower than some of the smaller semiconductor support memories.

REGISTERS AND BUFFER MEMORIES

Register memories are support memories often dedicated to specific tasks such as storing immediate data for input to an arithmetic logic unit (ALU) or for retaining the output results of an ALU. Registers are generally flip-flop arrays, often in shift register form. Small but fast buffer memories, often called *scratch pad,* are coming into increasing use. These are often random-access memories where blocks of data can be transferred from the main memory just prior to the time that the machine requires the data. This allows the computer to access data at the faster rate of the scratch pad instead of the slower rate of the main memory.

MASS MEMORIES

Mass memories are capable of storing a very large quantity of data inexpensively, but with very slow access times. Mass memories include magnetic tape, punched paper tape, punched cards, and magnetic disc and drum memories.

Magnetic-bubble memories (MBMs) and charge-coupled devices (CCDs) are also used for mass storage with the advantages of greater flexibility, higher access speeds, smaller size, and no moving parts.

Mass memories are not normally accessed directly. Data are transferred from mass storage into main memory and wait there until the computer requires them. Data are then generally transferred into special registers for immediate access. These memory transfers allow the computer to operate at its own speed rather than having to slow down to accommodate relatively slow memory systems.

RANDOM-ACCESS MEMORIES

Until fairly recently the magnetic core memory was the choice for main memory applications. However, semiconductor memories promise to largely displace core memories. The cost per bit of core memory has stabilized and, barring some breakthrough, is not likely to become less expensive. Semiconductor memories have already become cheaper for medium-sized memories and are faster and cheaper than core memories.

10-1 Memory Mediums

PUNCHED PAPER CARDS

Punched paper cards were used by Joseph Marie Jacquard as early as 1801 for controlling the weaving of textile patterns on an automated loom. (See Figure 10-1.) By computer standards, the punched card is a very slow access memory, but it is capable of a theoretically infinite storage capacity at a very low cost per bit of storage.

The most common punched card format has 80 columns. Each column can represent either the decimal digits 0–9, letters of the alphabet, or special symbols depending upon the arrangement of punches in three special zones. The punched card format is shown in Figure 10-2a and Table 10-1 shows the punched card code. Figure 10-2b shows a card punched to contain the following: DIGITAL TECHNOLOGY-647 SMITH, SHAKESPEARE. #46975 & L310@$23.00/UNIT.

PUNCHED PAPER TAPE

Punched paper tape is commonly used in conjunction with teletype (TTY) systems. Many modern machines use the American Standard Code for Information Interchange (ASCII or US ASCII), but there are several other codes in use, including the older five-level Baudot code. The complete ASCII code can be found in Table 6-3. Figure 10-3 shows a common ASCII tape format, and Figure 10-4 shows the format of the five-level code. The arrow in Figure 10-3 indicates the direction of tape travel. The tape is read against the arrow.

Figure 10-1 Jacquard Loom (Smithsonian Institution)

MOVING MAGNETIC MEMORY TRACKS

Magnetic track recording methods using reel-to-reel tape, cassettes, magnetic discs (both rigid and flexible or "floppy"), and drums have become common methods for storing large volumes of data. Tape is capable of storing indefinitely large amounts of data, but its sequential

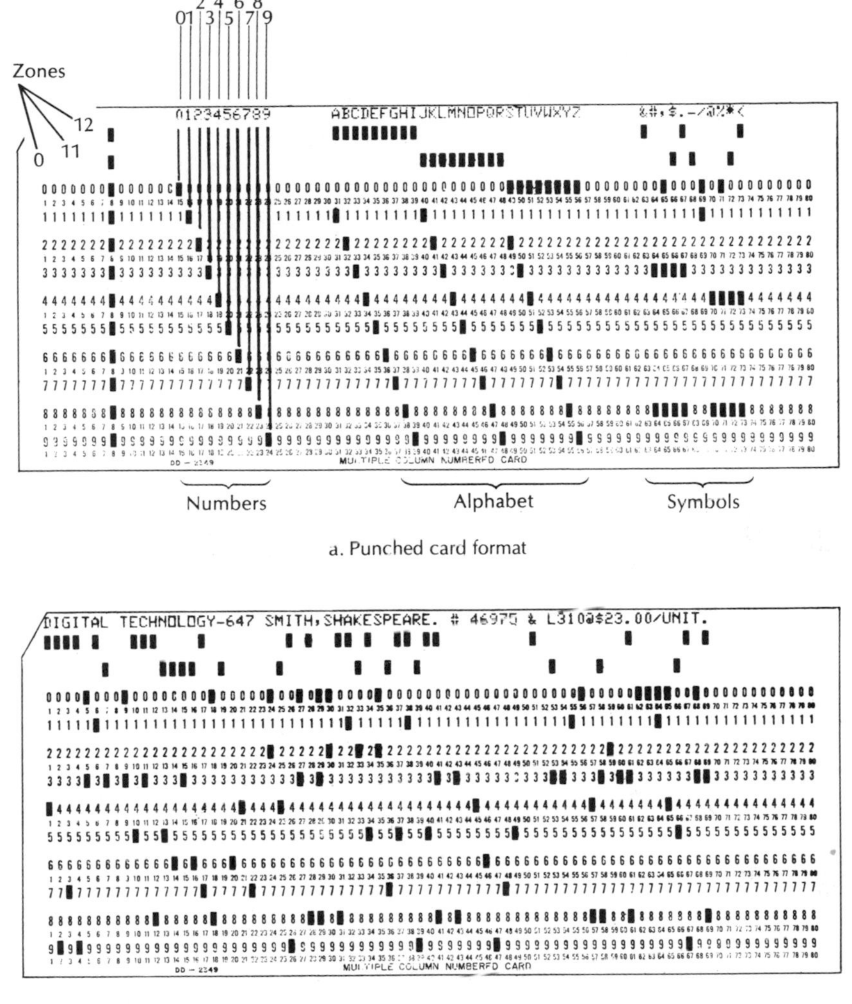

a. Punched card format

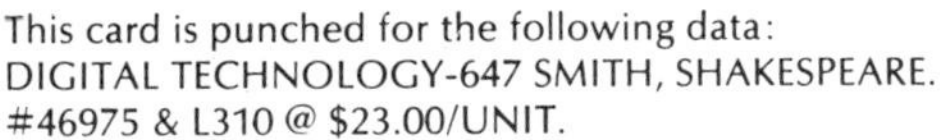
This card is punched for the following data:
DIGITAL TECHNOLOGY-647 SMITH, SHAKESPEARE.
#46975 & L310 @ $23.00/UNIT.

b. Sample punched card

Figure 10-2 Punched Paper Cards

(serial) nature and mechanical speed limitations make it a slow access storage medium. Sophisticated electronically controlled tape transport machines are used in larger computer systems. Entertainment cassette machines have recently become popular for use with programmable calculators, microcomputers, and remote terminals.

Table 10-1 Punched Card Code

Numerical Row Only	Zone 12 Plus Numerical Row Below	Zone 11 Plus Numerical Row Below	Zone 12 Plus Numerical Row Below
0=0			
1=1	1=*A*	1=*J*	
2=2	2=*B*	2=*K*	2=*S*
3=3	3=*C*	3=*L*	3=*T*
4=4	4=*D*	4=*M*	4=*U*
5=5	5=*E*	5=*N*	5=*V*
6=6	6=*F*	6=*O*	6=*W*
7=7	7=*G*	7=*P*	7=*X*
8=8	8=*H*	8=*Q*	8=*Y*
9=9	9=*I*	9=*R*	9=*Z*

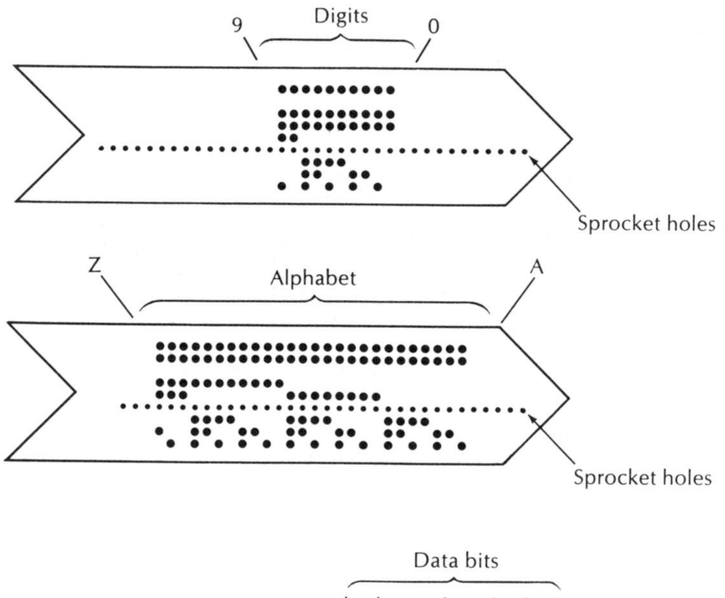

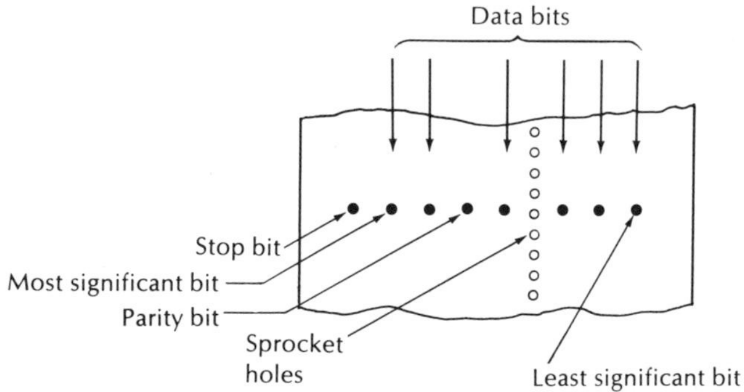

Figure 10-3 ASCII Coded Punched Paper Tape

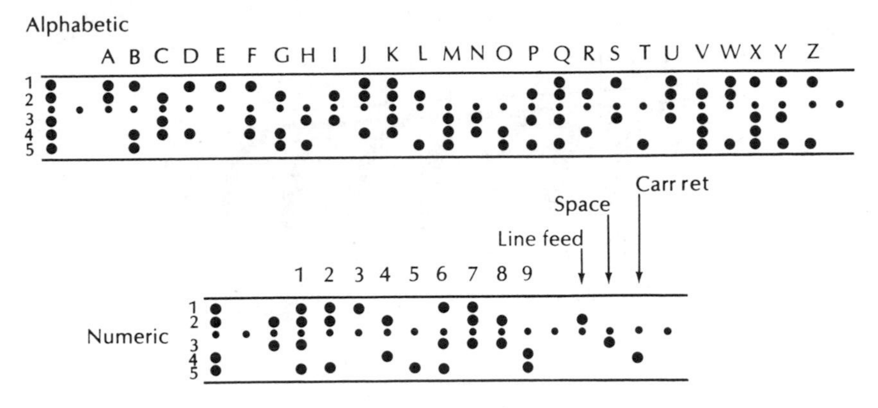

Figure 10-4 Five-Level Punched Paper Tape Code

Magnetic discs and drums use the same magnetic recording principle as that of magnetic tape except that the recording track is a specific *band* on the surface of the disc or drum. Each drum or disc contains a number of bands and often has several read and write heads. Multiple heads permit parallel reading or writing of several bands and allow a string of information to be repeated several times around the track to shorten the access time. Many of these drums and discs turn more than 10,000 RPM and require that write and read heads be floated on a cushion of air to minimize friction and wear.

Although a given band on a drum or disc has a short finite length compared with tape, it still can store a considerable amount of data. Because the velocity of the head moving along the track can be so much greater than practical tape velocities, drum and disc storage has a much shorter access time, although this method of storage is still far slower than the internal electronic working speed of even the slowest digital machine. As a result, its principal use is as a speed buffer between keyboard, punched cards, tape, and so on and the high-speed internal main memory.

Figure 10-5 shows the typical organization of a magnetic drum. Figure 10-6 shows the magnetic tape format.

DISC MEMORIES

Disc memories using stacks of solid metal discs with an oxide coating have been in use for some time. These machines are fast but bulky and expensive. A slower but very inexpensive disc memory system uses inexpensive, easily stored, flexible plastic discs. The floppy-disc memory is cheap enough to be used with small systems such as microcom-

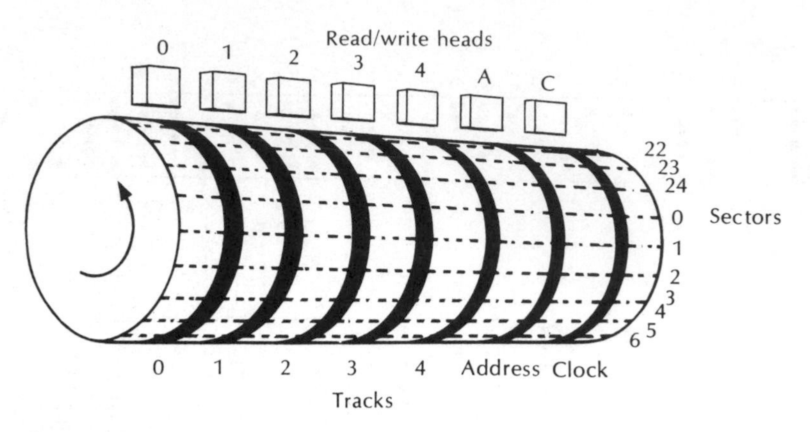

Figure 10-5 Magnetic Drum Memory

Figure 10-6 Typical Magnetic Tape Format

puters. Discs can store several million bytes on a plastic disc. Floppy discs have a rotational speed of 360 RPM and have an average access time of less than 300 ms. There are generally 70 or more tracks which are selected by moving the read/write head across the disc. Figure 10-7 shows the basic features of the floppy-disc drive mechanism. Figure 10-8 shows how data are organized on a typical floppy disc.

NONMECHANICAL MAGNETIC MEMORIES

Several memory schemes use magnetized domains located at fixed spots in a matrix. The movement from one magnetic spot to another is accomplished through electronic switching instead of mechanical motion. The most common form of this class is the *core* memory. It consists of an array of tiny (≈.050″ OD), toroidal (doughnut-shaped) cores threaded on hairlike wires, forming a rectangular grid. Each core is a

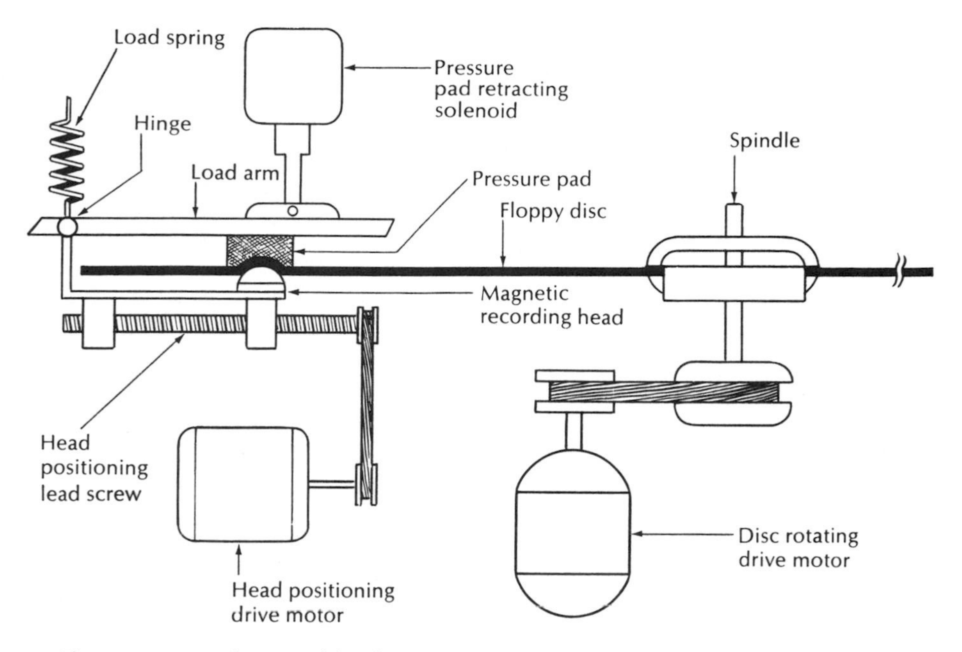

Figure 10-7 Mechanism of the Floppy-Disc System

memory cell that can store either 0 or 1. Individual cores are either wound as a few turns of ultra-thin magnetic ribbon or made by pressing a magnetic oxide and a binder into a dense toroid (doughnut) shape. The pressed (ferrite) core is the most common. Core memory has been the favored main memory element for several years. It requires power only when writing or reading out data, but read/write currents are quite high—100 mA to 1 A. Core memory cost per bit is fairly low but it is not likely to become appreciably cheaper.

Plated wire and magnetic film memories also use fixed position magnetic domains but not discrete cores. These memories have promised denser storage at lower cost than core memories, but have not yet lived up to the projected potential and have proved in practice to be slower than cores. At this moment it is questionable that further improvements in plated wire and magnetic film memories will make them truly popular forms.

Cryogenic memories are still in the research and development stage. These memories are based on the fact that certain materials at critical temperatures a few degrees above absolute zero exhibit zero resistance. A current induced in a tin (or lead) loop at cryogenic temperatures will continue to circulate indefinitely unless the temperature is changed, the current loop is opened, or an additional magnetic field is used to alter

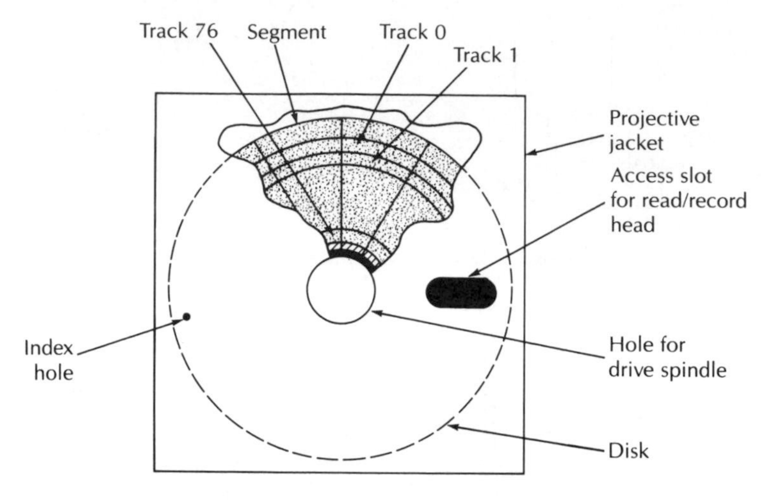

a. Disc organization

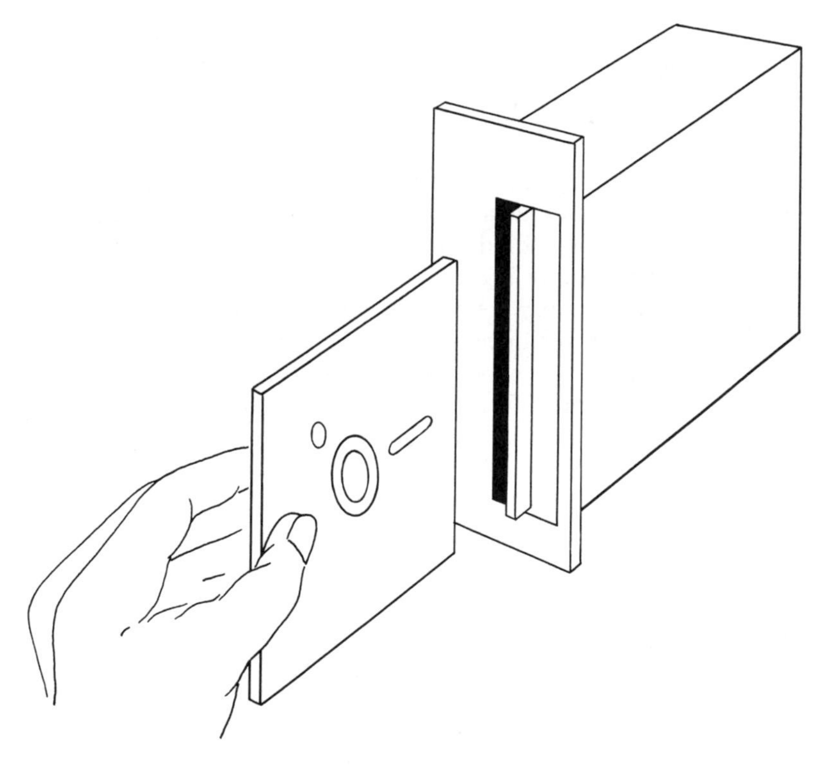

b. Inserting floppy-disc cartridge into machine

Figure 10-8 The Floppy Disc

the loop's zero resistance threshold. In an experiment at Massachusetts Institute of Technology started in 1956, a current of several hundred amperes was induced in a lead ring immersed in liquid helium. It was still flowing in 1965 when the experiment was terminated. There are still, however, a number of practical problems to be solved in this type of memory.

The IBM Corporation is currently developing a cryogenic memory based on the Josephson junction. The experimental device consumes only 10 microwatts of power, about 1 percent of the power required by semiconductor MOS memories and is 100 times faster than modern MOS memories. A family of Josephson junction OR, AND, NAND, and NOR gates is also under development. The drawback to cryogenic memories is the requirement that the devices be maintained at a temperature only 4.2° (C) above absolute zero.*

MAGNETIC-BUBBLE MEMORIES (MBMs)

Magnetic-bubble memories promise to replace some of the mechanical disc, tape, and drum memories in many systems. MBM is a fairly new technology, but one-megabit modules are available with access times of 25–50 milliseconds. Higher speeds and larger memory capacities are on the horizon. MBMs are compact and contain no moving parts.

CHARGE-COUPLED DEVICES (CCDs)

Until recently, charge-coupled devices, consisting of a single MOS cell per stage shift-register memory, were in direct competition with bubble memories. Thus far, CCDs have proved to be more difficult to manufacture, more costly than anticipated, and prone to soft fail. *Soft fail* is caused by the background radiation of alpha particles, primarily emitted by the materials used to encase the integrated circuit. Alpha particles do no damage to the memory cells but can alter the data stored in them, resulting in errors.

The problems in CCDs can probably be solved, but there is now little incentive to do so because conventional MOS memory devices have become so cheap and cell densities have become so high that most of the projected advantages of CCDs have been wiped out. Conventional MOS memories are inherently more flexible and faster than CCD devices.

Soft fail is also, to a lesser extent, a problem with conventional MOS memories and is a problem that *will* be solved. It may be too early to write off CCDs. Magnetic bubble memories lay dead in the water for

---

* Absolute zero is −273°C.

two years before several small, but critical breakthroughs made the technology viable.

Long-Chain Shift Registers

MOS shift registers with 1000 or more stages serve as serial memories. Many of these long-chain shift registers have provisions for recirculating the data from the output back to the input. These long-chain shift registers normally have no parallel outputs, so data that must be read out more than once must be pushed back into the register for later use.

SEMICONDUCTOR MEMORIES

Semiconductor memories are the workhorses of the digital world. Static memories use flip-flops as the basic memory cell and can use MOS, TTL, ECL, I^2L, or C-MOS technology. Bipolar devices (TTL, ECL, and I^2L) produce memory cells with shorter access times than memory cells based on MOS or C-MOS. *Access time* is the time required for the memory cell flip-flops to settle into a stable condition. Data on the output lines of a memory are *valid data* only after the flip-flops have had time to settle into a stable condition.

MOS and C-MOS devices generally have longer access times than bipolar devices but use less power. Power consumption per memory cell is significant since an 8-bit microprocessor, for example, may use over a half-million semiconductor memory cells in its memory system. Larger computers require even larger memories.

The selection of a particular semiconductor memory type generally involves a trade-off between short access time (high speed) and power consumption.

DYNAMIC MEMORIES

Dynamic memories are MOS devices. There is no bipolar counterpart. Dynamic memory cells store data in the form of a charge on a small capacitor. Dynamic memory is the cheapest type, and its simple cell structure allows more cells to be put on a single chip than static (flip-flop) designs.

Dynamic memory is far more difficult to use than static memory because of its critical timing requirements and the need to refresh the memory every few milliseconds. Refreshing is required because the charge on the memory capacitors tends to leak off. Dynamic memories are also more susceptible to soft-fail problems than static memories.

READ-ONLY MEMORIES

Read-only memory (ROM) is intended primarily as a permanent data store. We have already examined ROMs as used to generate complex logic functions, a more or less secondary application. ROMs are normally programmed outside the computer by a special programming unit. Once installed in the computer, the ROM data can be read by the computer, but the computer cannot alter the data stored in the ROM.

ROM can be used to store mathematical tables, frequently used subroutines, and so on. ROMs can be programmed to tell the computer what kind of machine it is to be. ROMs that determine the computer's orientation are often called *personality modules*. The personality module may orient the machine in a way that makes it most efficient as a data processor, as a controller for a certain kind of industrial process, or as a chess partner.

An important application of ROMs is that of making it easier to program computer hardware. Internally, the computer speaks only binary, while human programmers would much rather use something close to their own language.

For example, a simplified 4-bit educational computer used by the author requires that the binary instruction 0010-0111-0001-0001-0100-0011 perform the operation 7+1=?. A built-in ROM allows the student to type in 7+1=?. The ROM translates that human-oriented instruction into the machine-required instruction 0010-0111-0001-0001-0100-0011. A ROM is normally used to translate symbols typed on a standard typewriter keyboard into ASCII code groups (which are in binary form); a binary-form ASCII translation ROM at the computer output converts the code groups into something that prints out in ordinary English or standard mathematical symbols. There are a multitude of other ROM applications.

Some ROMs can be erased and reused and some cannot. Because ROM is also a random access memory in terms of address organization, there is a current trend to refer to RAM as Read/write memory.

10-2 Core Memory

Magnetic core memories consist of many toroidal cores strung in a rectangular grid as illustrated in Figure 10-9.

The cores have a square loop magnetization curve as shown in Figure 10-10. This kind of curve describes the behavior of a magnetic material that has a critical point at which the flux density produced by a secondary field causes the core to become permanently magnetized. At flux densities below this critical value, the magnetic material relaxes to its original state as soon as the external magnetic field is removed. If the external field is produced by current flowing through a wire, there is some critical current at which the core will remain permanently magnetized after the current ceases to flow through the wire. In Figure 10-9, slightly more than half the critical current is passed down line Y_2

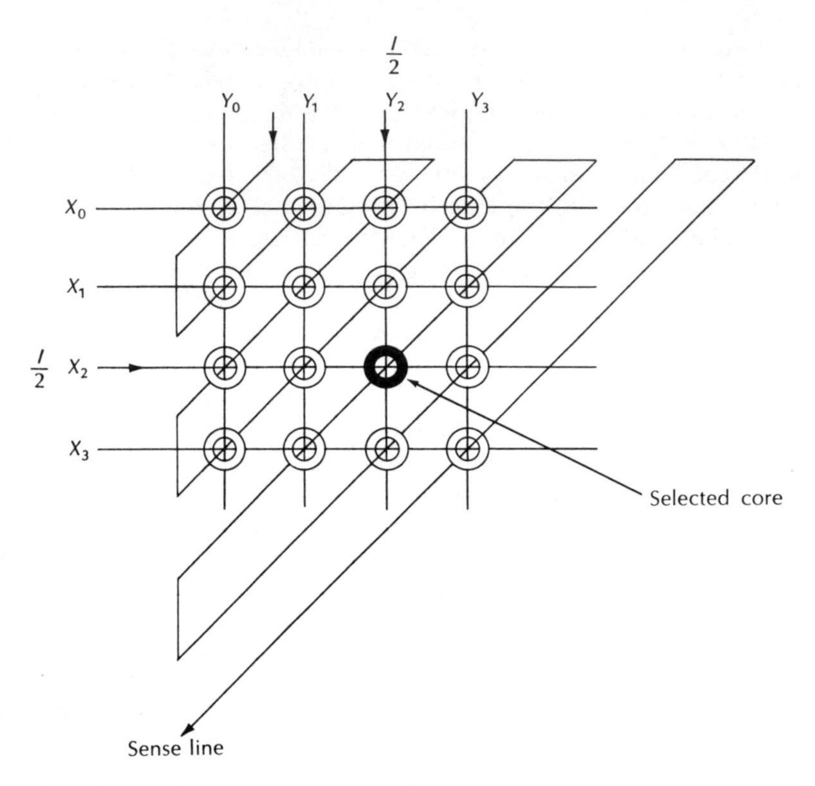

Figure 10-9 Magnetic Core Memory Plane

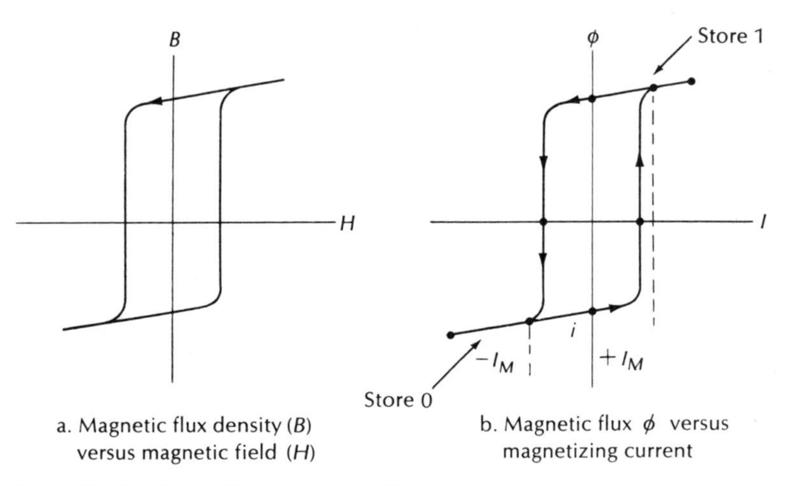

a. Magnetic flux density (B) versus magnetic field (H)

b. Magnetic flux ϕ versus magnetizing current

Figure 10-10 Ferrite-Core Hysteresis Curves

and a similar amount through line X_2. The two fields combine to magnetize the core at the intersection of Y_2, X_2. All other cores receive only about half enough current-produced magnetic field to switch them into a permanently magnetized condition. Thus, one and only one (selected) core is *switched* as a result of the two half-critical currents.

Figure 10-9 shows a sense line threaded through all cores. The line is threaded back and forth to effectively cancel induced currents in the sense line as a result of magnetic domain movement in unselected cores. To read out of the memory, the status of desired cores is determined by selecting each core by passing half-critical currents through the appropriate X and Y lines in a direction opposite that used for writing into the core. As the selected core is switched from one magnetized direction to the opposite direction, its collapsing field induces a voltage in the sense winding. Figure 10-11 illustrates current flow and magnetic field direction for storing either zero or one. This method of core selection is called *coincident current* selection. Notice that only 8 lines are required to select 16 cores (Figure 10-9). 1024 cores can be selected by 64 lines. Two lines, a combination of one X and one Y, out of the 64 select one of the 1024 cores.

The output voltage from the sense line is quite small and, because of the way the sense line is routed through the cores, may produce either a positive or a negative pulse when a selected core is sensed. A special sense amplifier is used to establish proper levels for the rest of the system.

As a matter of construction convenience, when core assemblies are organized in stacks, the X and Y select lines have their directions alternated as shown in Figure 10-12 to allow for the addition of an inhibit line.

Computer data are generally organized into groups of bits called a *word*. It is common practice to have a stack of core planes in which there are as many planes as there are bits in the particular computer's word. A stylized sketch of a core plane memory stack is shown in

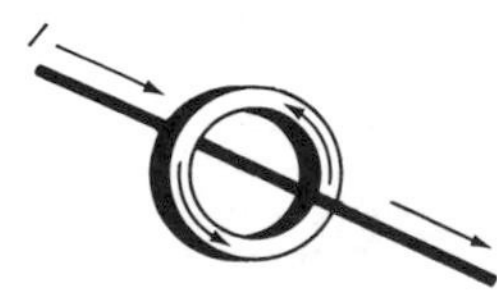

a. Current is applied; core stores a 1.

b. Core is magnetized; core remembers a 1.

c. Current is reversed; core reverses its magnetic state and stores a 0.

Figure 10-11 Current Flow and Magnetic Field Directions in Magnetic Cores

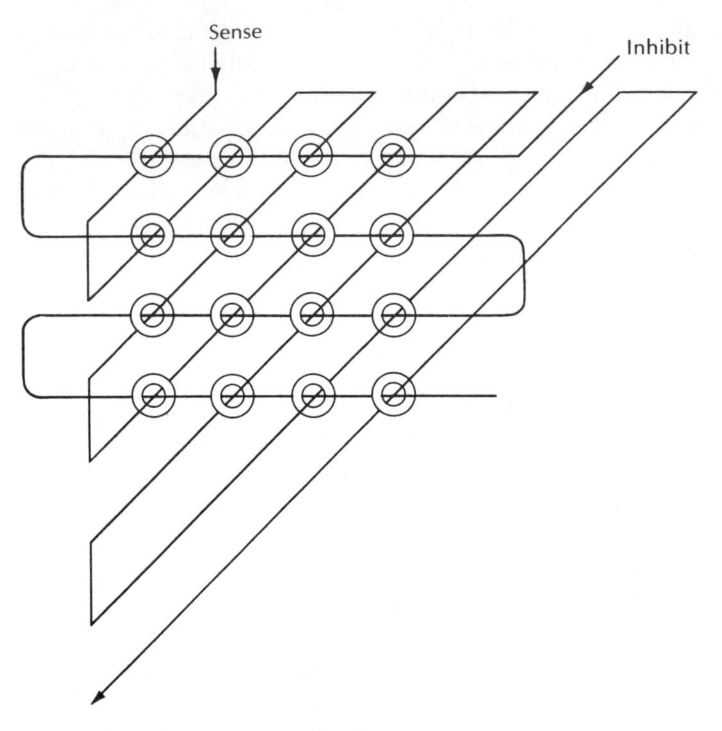

Note: Select lines are omitted for clarity.

Figure 10-12 Core Plane with Sense and Inhibit Lines

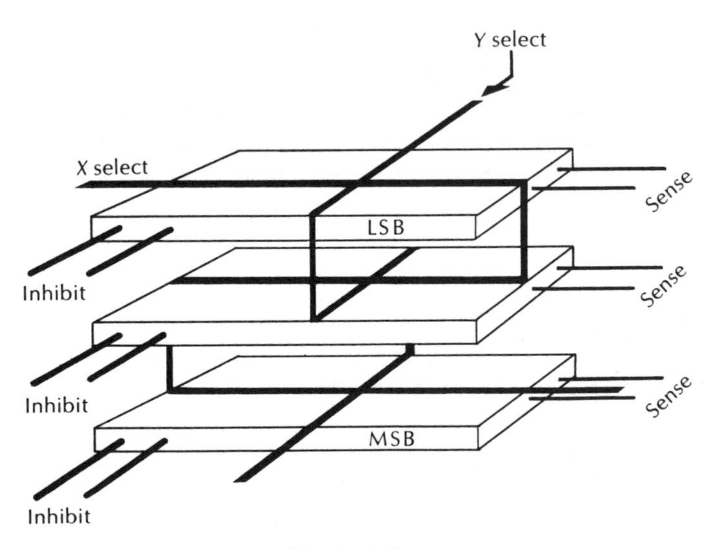

Figure 10-13 Core Array of Stacked Planes

Figure 10-13. This organization permits the same X and Y select lines to address all of the planes in the stack by inhibiting an entire plane when a 1 is not to be written at a particular x-y intersection of a given plane. In planes where a 1 is to be written at that location, the plane is not inhibited. This arrangement greatly reduces the number of select lines required and the associated drive hardware. A memory of 1024 20-bit words (20,480 bits) requires 32 X lines, 32 Y lines, and 20 inhibit lines; thus, 84 lines can address 20,480 bits.

In order to read out of a core, the core's magnetic state must be changed. The readout is said to be destructive because the original data are lost in the process of reading it. (Destructive readout is abbreviated DRO.) The destruction of data in the process of reading it is not generally tolerable. In order to overcome this deficiency in the core memory system, data are stored briefly in flip-flops and written back into the memory. Figure 10-14 shows the scheme for accomplishing nondestructive readout. The read cycle becomes a combined read/write cycle and proceeds as follows:

1. Clear all flip-flops.
2. Select cores to read—all selected cores set to the zero state. Data transferred to flip-flops for temporary storage.
3. Write pulse gate flip-flop states into core driver and inhibit amplifiers. The $\bar{Q}$ lines from the flip-flop control the inhibit lines.

0-3 Semiconductor Random Access Memories

Static semiconductor random access memories (RAMs) are composed of flip-flop cells using MOS or bipolar transistors. RAM organizations and functions are the same whatever the transistor type. The different transistor and circuit types yield different access times and have varying power needs. Table 10-2 lists some common memory chips and

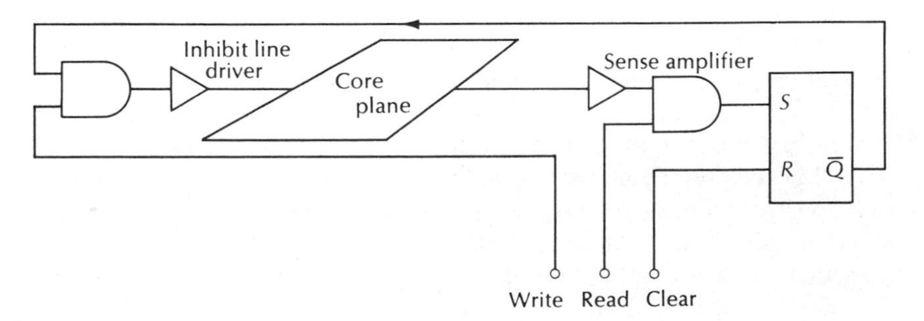

Figure 10-14 Core Memory Stack Showing Nondestructive Readout System

Table 10-2 Comparison of Typical Static RAM Memory Chips

Type Number	Organization	Transistor and Circuit Type	Access Time (nanoseconds)	Power Required (milliwatts)
2101*	256 × 4 bits	N-MOS	500	150
2102*	1K × 1 bit	N-MOS	500	150
2115	1K × 1 bit	N-MOS	45–70	656
2115AL	1K × 1 bit	N-MOS	45–70	395
2115AH	1K × 1 bit	N-MOS	15–20	600
2147	4K × 1 bit	N-MOS	55–70	900
2147H	4K × 1 bit	N-MOS	35–45	945
4801	1K × 8 bits	N-MOS	55–90	656
6147P	4K × 1 bit	C-MOS	55–70	75
93471A	4K × 1 bit	Bipolar TTL	45–60	893
93L471	4K × 1 bit	Bipolar TTL	45–60	472
93H471	4K × 1 bit	Bipolar TTL	30	893
10415	1K × 1 bit	Bipolar ECL	20	780

*These older, relatively slow chips are still common.

compares their organization, speed (access time), and power requirements.

Memory cells are often arranged in an x/y coordinate grid. A particular cell is selected by a row (x) signal and a column (y) signal. The cell to be written into, or read out of, is the cell at the intersection of the x and y signals. A decoder selects one line at a time on the x (row) axis for a specific binary combination on the input of the decoder. A second decoder selects the column (y) line. The binary numbers applied to inputs of the decoders are called *memory addresses*. The decoders are included on the memory chip.

There are two basic memory organizations, bit-organized and word-organized. In a bit-organized memory a given address selects a single cell capable of storing a single bit. The bit-organized memory will have a single output and a single input, or a single combined input/output line. The input is called the *write line* and the output is called the *read* or *sense line*.

In a bit-organized memory a given address selects a single memory cell. In a word-organized memory a given memory address selects several cells at the same time, the number of cells depending on the word length. The word-organized memory has as many inputs/outputs as there are bits in the word.

Figure 10-15 shows a 64-bit, bit-organized memory and an 8×8 bit word-organized memory. A *word* may typically be anything from 4 bits to 36 bits. An 8-bit word is a byte and 8-bit memories are sometimes called byte-organized memories.

ADDRESS INPUTS

A memory needs a number of address inputs to select the proper cell or word in the memory array. The total number of address inputs depends on the number of addresses available in the memory array. In general a memory requires n address lines, where 2^n is the number of addresses in the array. An address would select a single bit in a bit-organized memory, and an entire word in a word-organized memory.

Example A memory is organized as $1K \times 1$ bit. How many memory locations are there, and how many address lines are required?

The description $1K \times 1$ bit is read as 1K words of 1 bit each. A $1K \times$ 4-bit memory would be composed of 1K words of 4 bits each. Actually, the term 1K is a shorthand way of saying 1024 bits or 2^{10}. The actual number of addresses is always an integral power of two. For memories of fewer than 1024 addresses, the number is specified: 64, 128, 256, and so on. The abbreviation 1K is used for 1024 (2^{10}), 2K for 2048 (2^{11}), 4K for 4096 (2^{12}), and so on.

The first part of the memory description, 1K in this example, states the number of addresses. If there are 1024 addresses, we need 10 address lines because $2^{10} = 1024$.

A typical 8-bit microprocessor uses an 8-bit (1-byte) word length and has 16 address lines available for 8-bit word memory-address selection. The machine is, therefore, capable of selecting any one of 65,536 memory locations, each of which can store one 8-bit word ($2^{16} = 65,536$). All

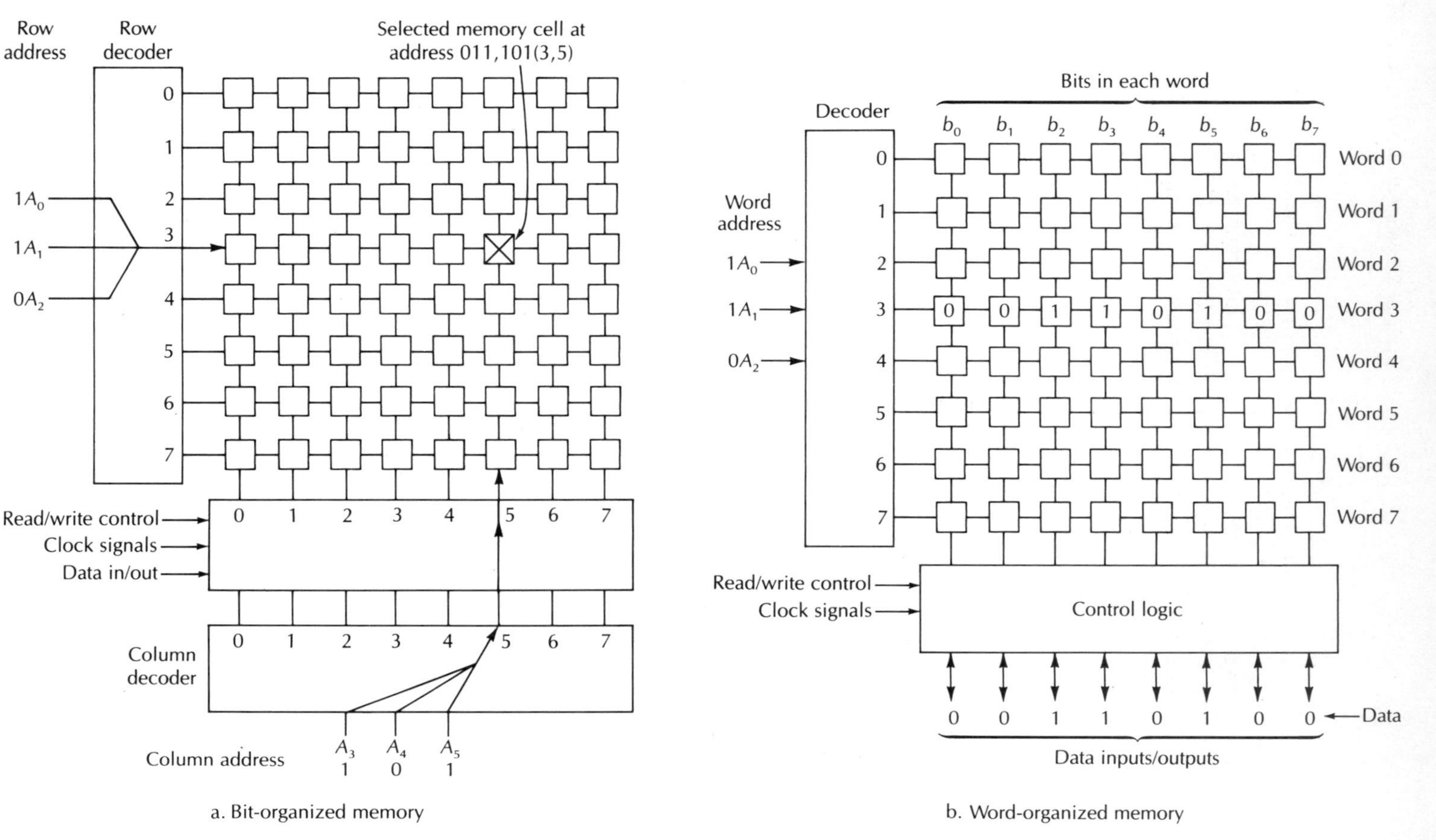

Figure 10-15 Bit- and Word-Organized Memories

65,536 memory locations are generally not devoted exclusively to RAM addresses. Some addresses are reserved for input/output equipment and some are used for addresses in read-only memory. The proportion of memory devoted to RAM and ROM is dependent on the job for which the microcomputer is designed. Sometimes there is very little RAM required but a lot of ROM. In other cases most of the memory is RAM. A given microcomputer may use only a part of the memory capacity it is capable of addressing.

OTHER INPUTS AND OUTPUTS

In addition to the address inputs that select the cell or group of cells to be written into or read out of, the memory requires other inputs and outputs. Bit-organized memories require a data input and a data output. The *write* input is the data input and the *read* output is the data output. In bit-organized memories the Read and Write lines may be either separate or combined into a read/write line. An additional mode control called read/write (R/W) input is required, if they are combined, to determine the function of the R/W line at any given time.

Word-organized memories more often combine read and write lines. The word-organized memory requires a read output and a write input for each bit in the word. Combining the read and write lines reduces the number of pins on memory package.

CHIP SELECT

The chip-select (CS) input is a *chip-enable* input and is sometimes designated CE. The designation is more often CS to reflect the fact that a given memory chip is selected from within a memory consisting of many chips through the use of this input. Most modern IC memories use 3-state outputs and are arranged so that the entire chip is effectively disconnected from the system until activated by the CS input. Some memory chips have several CS inputs to facilitate the use of the chip in large memory systems. We will examine multichip memories a little later in this chapter.

10-4 The Register Concept

A common way of looking at word-organized memories is to view a section of memory as a separate register. An address selects a particular register word to write into or read out of. The entire register is considered a stack of words one word wide by N-words deep. Thus far, we have used the word *register* in conjunction with shift registers. Shifting capability is not an inherent capability of RAM-type registers. However, the computer central processing unit can generally move data around in RAM memory in a way that simulates shift register behavior when such behavior is required.

The register concept may seem a trivial idea when viewing a single memory chip but becomes a very useful notion when an entire memory system is under consideration.

For example, in a complete memory system certain blocks of RAM are often reserved for specific functions. These blocks can be designated as x, y, or z registers (or whatever describes the function), even though the register is actually only a group of memory addresses in a sea of RAM memory addresses. A group of memory addresses that has been given a register *title* can be shown on a block diagram as a separate entity, thus greatly facilitating the understanding of how the system works. Somewhere in the manual for the system there will be a list of addresses for each register when that detail is needed.

10-5 Random Access Memory Specifications

Unfortunately memory chips have not been as well standardized in terms of pinout and specifications as users would like. Timing and speed specifications are particularly difficult to interpret.

All too often a certain chip will appear to be a direct replacement for another, and the manufacturer may even indicate that that is the case. It may turn out that the replacement chip doesn't even have the same pinout as the original, or that subtle timing differences preclude its use in some systems.

Timing problems can best be evaluated by studying the manufacturer's timing diagram. The published *access time* must be considered only a rough guide to memory speed and timing. In normal memory operation, the chip select (CS) input is held in the inactive state until the flip-flop cells have been selected and the *access* time has elapsed, allowing the flip-flops to settle and the data to be considered *valid*. After the chip select input has been activated, it takes a finite time for the data to become available. This time period may be as long as the access time. The system clock must produce pulses in the proper time relationship to insure correct memory operation.

Figure 10-16 shows a typical memory timing diagram, but it is only an example. Each memory product will have a slightly different timing diagram. Slight differences can be important.

POWER DOWN IN MOS MEMORIES

Some of the more recent MOS memories automatically reduce their power demand when not being written into or read out of. There is always a finite time required to bring the power level up to normal for

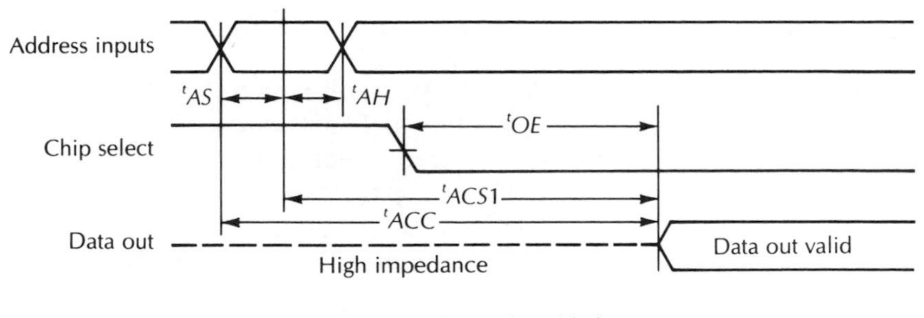

a. Read cycle (read enabled)

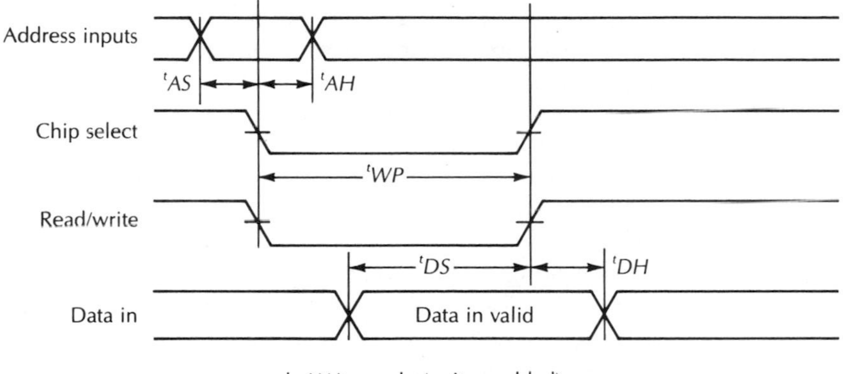

b. Write cycle (write enabled)

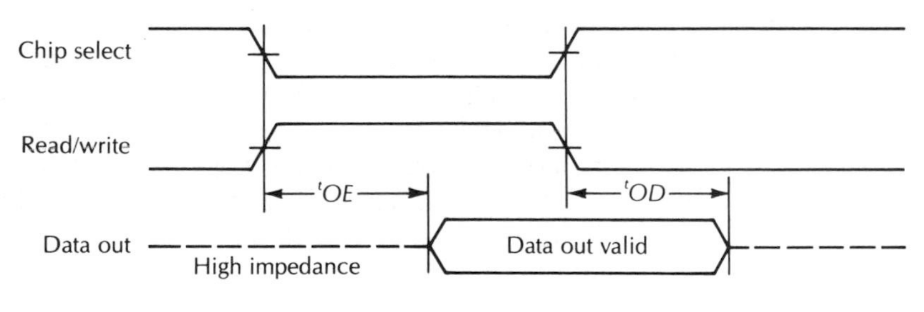

c. Output enable/disable

Figure 10-16 Typical Memory Timing Diagram

read or write operations. This time delay makes the access time (^{t}acc) less meaningful and alters the timing diagram for these devices. At this writing bipolar memories do not have this feature, and C-MOS devices do not need it. C-MOS devices draw power only when they are being written into or read out of.

GENERAL COMMENTS

If you will refer back to Table 10-2, you will notice that most of the memory devices listed have access times of 55 ns or less. This is a little misleading because most of the devices selected for the table are the latest available at this writing. There are two well-defined market segments for memory devices: those with access times of 100 ns and greater fill the needs of one market segment, and those with 55 ns or less fill the needs of the other.

The older 2102 has long been an industry standard for the lower speed segment of the market and is available from a number of manufacturers.

The 2147 has become a recent industry standard for medium to higher speed requirements, and has become a small family of devices, available from several manufacturers.

Bipolar ECL memories have dominated the highest speed end of the market, but Schottky TTL devices are fast approaching them in speed.

An examination of Table 10-2 seems to reveal no great advantage of one type of transistor technology over another. Indeed manufacturers committed to a given technology are doing their best to compete.

However, at this writing N-MOS seems destined to become the leader of the pack, with C-MOS the choice for systems where low power consumption is primary. This situation could change next week with an announcement of some new breakthrough.

The principal reason MOS technology is most promising at this time is economic. Currently, MOS technology can put more memory cells on a chip than other technologies. As a general rule, the memory chips are the cheapest part of the memory system. Circuit boards, interconnecting devices (connectors, cables, and so on), and power supplies represent the bulk of the cost. The total cost of a memory system tends to go up exponentially with the number of memory chips while the cost of larger memory chips increases at an almost linear rate. The ideal would be to have the entire RAM system on a single chip. A typical 8-bit microcomputer would require a memory chip with 65K bytes. This density seems to be a goal for the future, perhaps the distant future. For the present, MOS technology can yield a memory system with fewer chips and therefore has the cost advantage. This estimate of the future is, of course, only a projection of current trends. Memory technology is the most rapidly changing area in the digital field, which makes any predictions risky at best.

10-6 Static RAM Cell Structures

The following RAM cell structures are intended to provide *typical* cell circuits. There are a number of variations among manufacturers and new variations are constantly being devised. More often than not, these modifications have little to do with circuit performance but can make manufacturing processes more efficient or reduce manufacturing cost. In any event, a proper evaluation of cell circuitry cannot be made without considering both circuit behavior and the manufacturing process.

Figure 10-17 shows typical RAM cell circuits.

10-7 Dynamic MOS Memory

The operation of dynamic cells depends upon the nearly infinite input impedance and the capacitive nature of the MOS FET transistor. Data are stored as a charge on the input (or an auxiliary capacitor) capacitance. Because any real capacitor has some leakage, the dynamic memory will gradually "forget," and must be periodically refreshed. A precharge method simplifies the refreshing operation but does not eliminate the problem.

There are several basic random-access memory cell structures used in MOS technology. Figure 10-18 shows three typical RAM cell circuits. In part a, transistor Q_2 is the storage cell and C_1 is the input

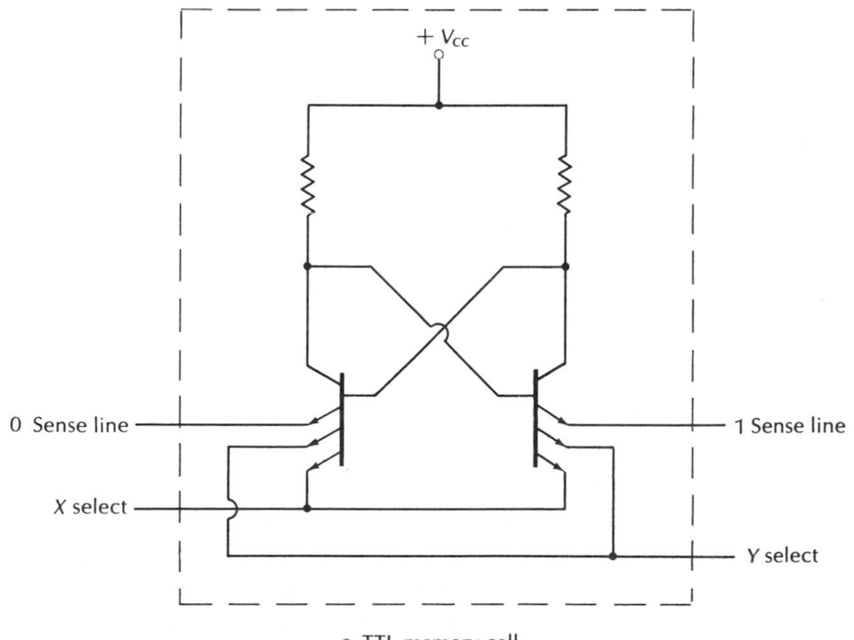

a. TTL memory cell

Figure 10-17 Memory Cell Structures

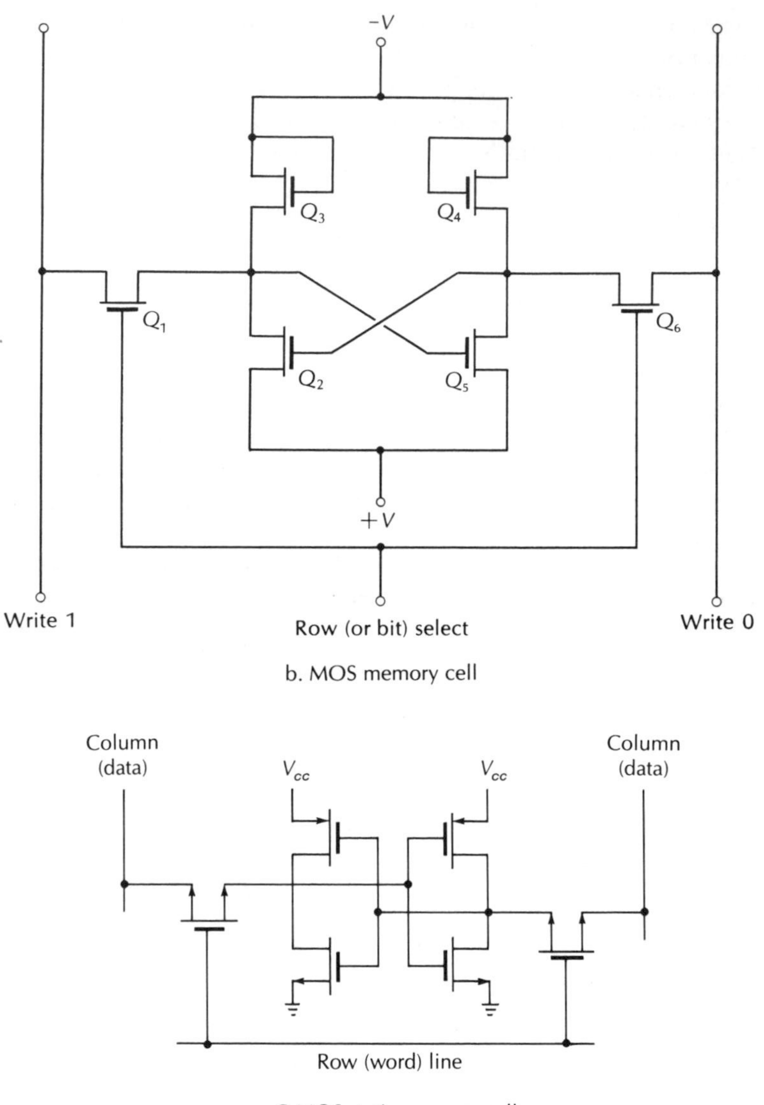

b. MOS memory cell

c. C-MOS static memory cell

Figure 10-17 continued

capacitance of Q_2. To write in data, Q_1 is switched to a low impedance state discharging C_1. Transistor Q_3 is a combination active-load resistor and transfer gate that transfers the output of the memory cell (Q_2) to the read data line upon a command from the read select line.

The circuit in part b is basically the same as that in part a except that

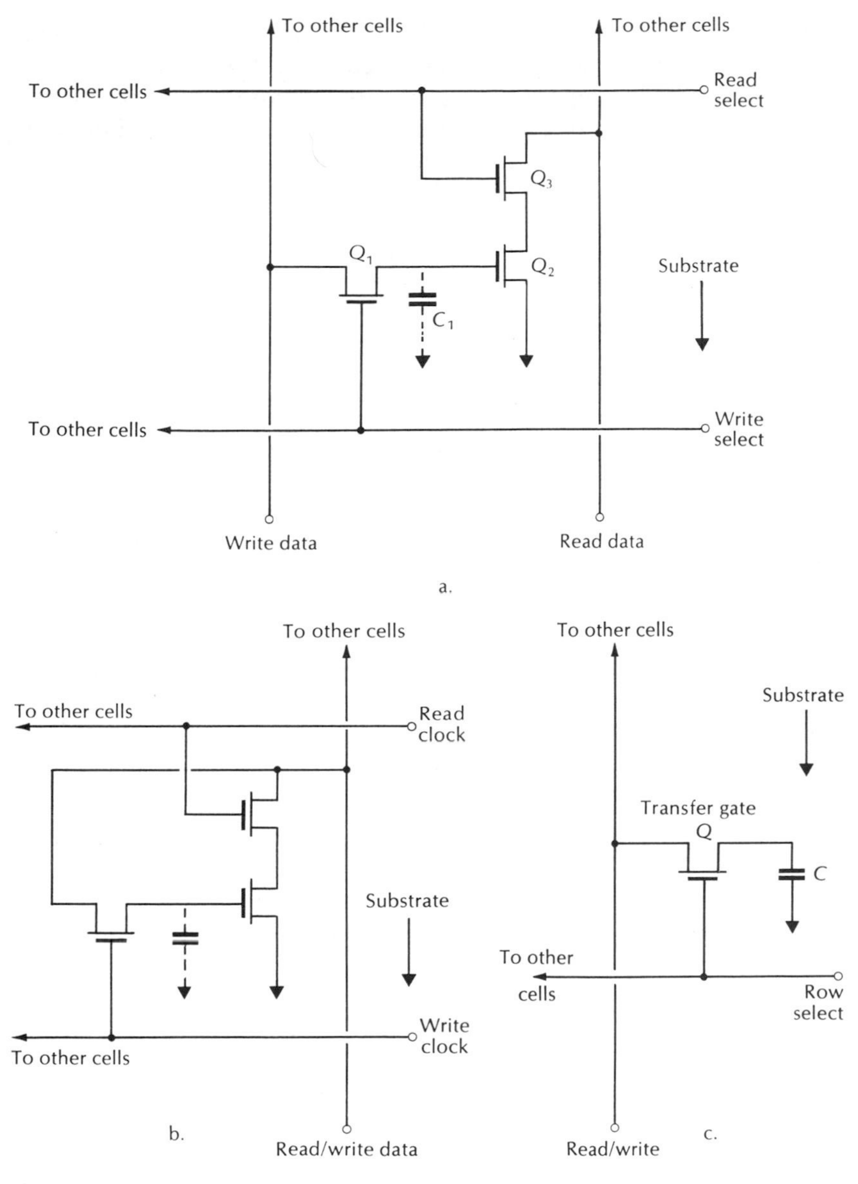

Figure 10-18 MOS Dynamic RAM Cells

read and write lines are combined to reduce the amount of interconnecting wiring on the chip. A two-phase clock controls the read/write cycles in the device. In part c the transistor Q is used as a transfer gate, and a passive on-the-chip capacitor is used as the storage element.

The memory capacitors are precharged to represent a *zero*. A *one* written into a cell discharges it. Periodically a refresh charge current must be delivered to all memory capacitors to make up for capacitor leakage in the fully charged capacitor. The discharged capacitors also receive a small charge but not enough to change the logic level, and that charge leaks off before the next refresh cycle.

Dynamic memories have proven attractive because their simple cell structure allows more cells to be placed on a chip. This feature makes dynamic memories cheaper than static systems. However, dynamic memories have some disadvantages. The need for refreshing the memory every few milliseconds requires very critical timing if refresh demands are not to interfere with other operations in the computer system.

Dynamic memory cells are also more prone to soft fail and memory crashes due to electrical noise transients.

Newer devices such as the Motorola MCM 6664, 64K × 1-bit dynamic RAM, have put most of the refresh and timing control circuits on the memory chip. With the addition of a special control chip such as the 3242, eight MCM 6664 chips can form a 64K × 8-bit memory system that "appears" to the computer to be a static memory system. Efforts have also been made to reduce the memory's susceptibility to soft fail and noise-induced memory crashes. These new 16 to 64K dynamic memories may or may not prove highly competitive with static devices.

10-8 Read-Only Memories (ROMs)

Read-only memories are a permanent data storage device. In the course of normal operation, they are not volatile and cannot be written into. Permanently held data can be read out during processing. These memories can be used to store lookup tables and microprograms such as algorithms to replace complex combinatorial logic circuits, to convert from one code to another, and to generate characters for video display units.

ROMs are organized in the same fashion as RAMs, either bit or word organized. They are designed to be compatible with RAMs so that they can be integrated into the same memory system. From a system standpoint ROMs differ from RAMs only in that they cannot be written into in normal operation. Internally ROM cells are different from RAM cells. Their simpler structure permits more cells to be put on a single chip.

ROMs use much simpler cells than RAM memories. Theoretically, a cell need be no more than a transistor biased to a permanently off state or, in the case of MOS, left off the chip during manufacture. There are three common varieties of ROMs: mask programmed, field programmed, and erasable.

Mask-programmed ROMs are programmed to customer order or to certain standard programs during manufacture by controlling transistor base (or gate) potential or by leaving out specific cells. Figure 10-19 illustrates one way this could be accomplished.

In the manufacturing process the resistor of the base of each transistor (cell) is made to be either a low resistance, causing the transistor to conduct heavily and thus produce a low (logical 0) output, or it can be made high and produce a logical 1 output. The cells are not always quite as simple as that shown in Figure 10-19 because loading factors must be taken into account, cell select provisions must be provided, and very often a chip select input is also provided. However, the basic principle involved is correctly illustrated in Figure 10-19. Once programmed, masked programmed ROMs cannot be altered.

Field programmable units allow us to program our own ROMs for short-run or prototype use. Sophisticated programming units are available if much in-house programming is to be done. The principle of operation and the nature of the programming in bipolar devices are similar to those of mask programmed devices except that the base resistances are provided in the form of fusable nichrome (nickel

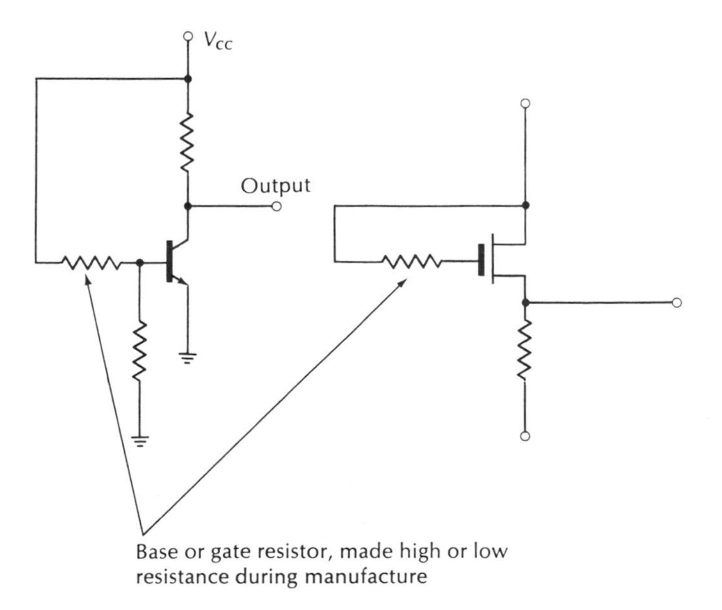

Figure 10-19 Examples of Masked Programmed ROM Memory Cells

chromium) links. As delivered to the customer, all links are low resistance. The customer programs the unit by selecting a particular cell in the same way it would normally be selected, and applying an abnormally high voltage (and current) that burns the fusible link open, causing the output of the transistor (cell) to go high. Once the link is opened, that particular cell is forever at a logical 1 and cannot be altered. Each cell is selected in turn, and the link is either fused open or left intact depending upon whether a 0 or a 1 is to be stored in that particular location.

The MOS erasable ROM (EROM) is similar to the dynamic cell used in RAMs; however, the gate is left open and glass encapsulated. The memory is programmed by applying a relatively high voltage (compared to operating voltages). The charge is trapped at the gate-glass interface and is stable unless irradiated by a short wavelength ultraviolet light or some similar radiant energy. The memory is erased by removing the protective cover and exposing the chip to ultraviolet light. The chip must then be completely reprogrammed. This program-erase cycle can be repeated several times before the chip must be replaced.

Some of the newer PROMs can be programmed or reprogrammed electrically, but not in the process of normal computer operation. Like other ROMs, once they are installed in the system, the computer can read the data but cannot write data into the ROM.

Sophisticated ROM programming units are available that allow the ROM program to be loaded into computer RAM, tested for accuracy, and then automatically loaded permanently into a ROM. The programmed ROM can then be installed in the computer memory system.

Figure 10-20 shows a typical mask programmed MOS ROM cell circuit.

10-9 Memory Systems

Digital systems, more often than not, require more memory than can currently be put on a single chip. Multiple chip memories can be configured for any required word length and with virtually any number of words. The computer in which the memory is used generally dictates the word length and the *maximum* number of words in the memory. Most modern memories are bus-oriented, and both random-access and read-only memories can be connected to the bus to provide a memory system that is a mixture of RAM and ROM.

THE BUS

Bus architecture is a feature of most small computers and some larger ones. There are two major and one or two minor busses associated with

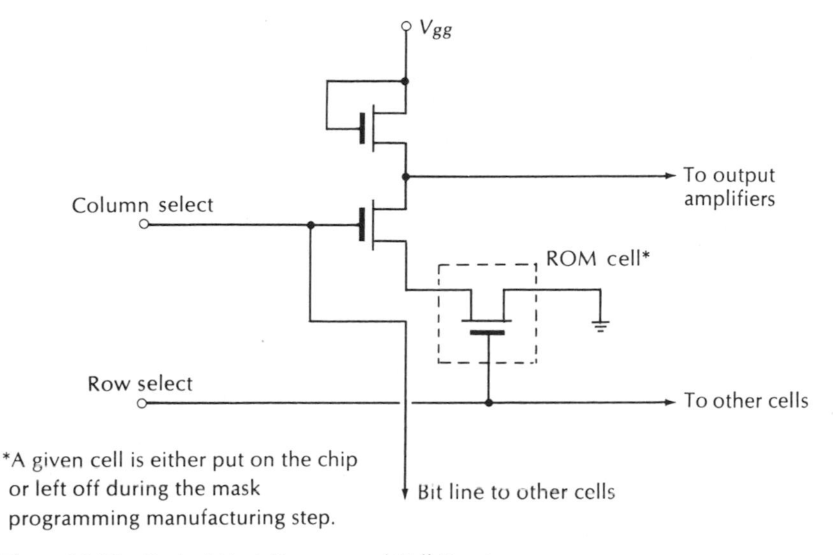

Figure 10-20 Typical Mask Programmed Cell Structure

small computer memory systems. The typical 8-bit microprocessor will have an 8-bit (8-line) data bus, a 16-bit (16-line) address bus, and one or two control busses. The bus is actually a set of lines connected to all the devices on the bus. Data can flow in either direction on the data bus, and flow is under the direction of the central processing unit (CPU). The central processor determines whether data are to be stored in memory or read out of memory. It also puts a memory address on the address bus to write the data into a particular memory location or to read out the data stored there. Certain memory addresses are assigned to peripheral devices such as displays, keyboards, and printers. A decoder connected to the bus will recognize a special address code and connect the printer, for example, to the data bus to receive the data being sent from the CPU. The bus can generally have only two devices connected to it at a given time, one device to transmit the data and one to receive the data. All other devices on the bus are disabled; if tristate output circuits are used, the disabled devices are in the high impedance mode and are effectively disconnected from the bus.

Figure 10-21 illustrates a simplified bus-oriented microcomputer system. The central processor contains a number of specialized working registers, control circuits, an arithmetic logic unit (ALU), and other circuits depending on the features of the microprocessor unit (often called MPU instead of CPU).

Normally data are transferred to or from the microprocessor unit and to or from memory or peripheral devices, such as printers and

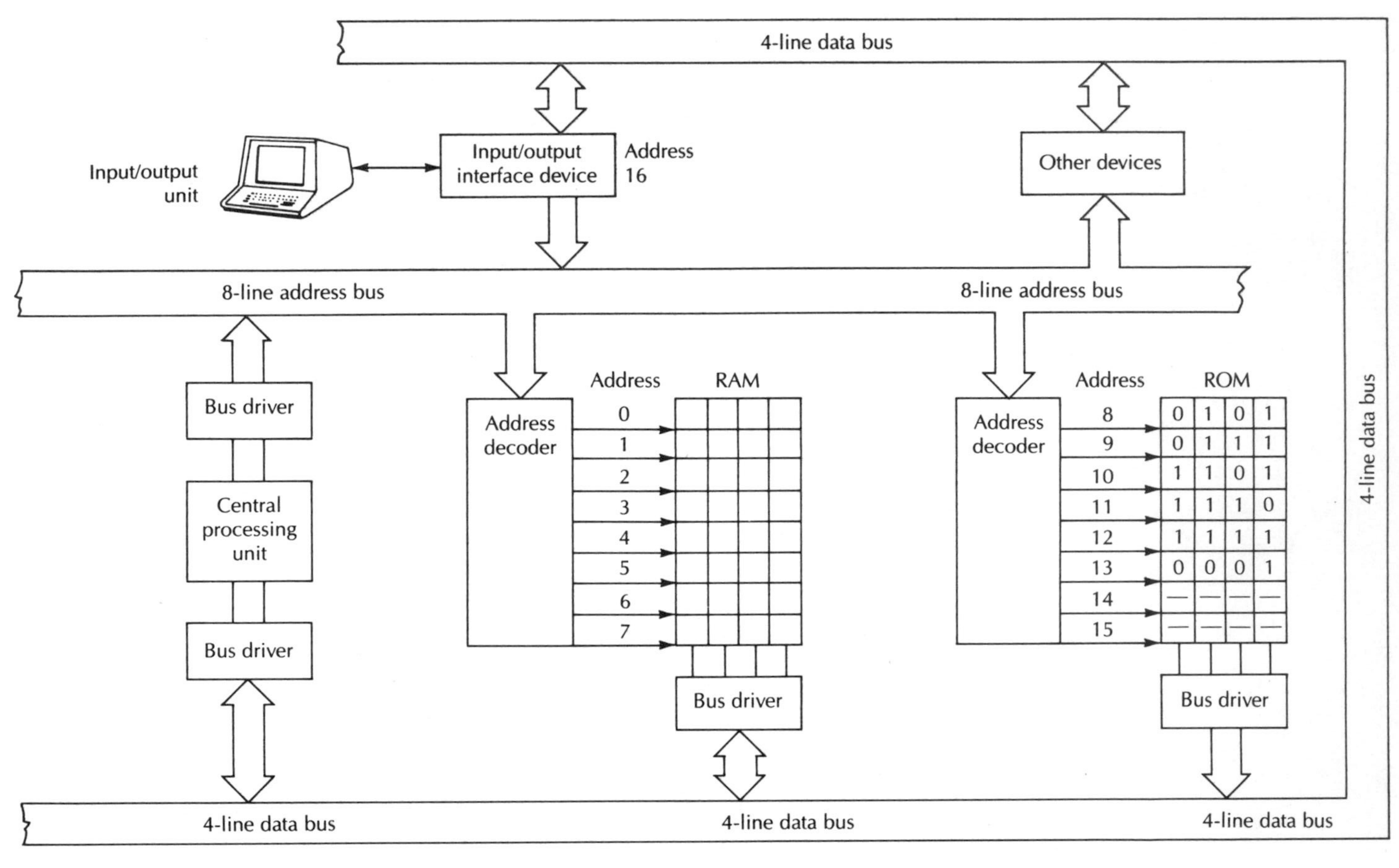

Figure 10-21 Bus Organized System

keyboards. Data to be loaded into external RAM (on the bus) are generally stored temporarily in a dedicated microprocessor memory (register) and then transferred along the bus to a particular address in the RAM. The data stored in the on-the-bus RAM will wait until the processor again puts the same address on the address bus and reads the data out of RAM via the data bus. The data will then flow from the RAM to a specific microprocessor register where, within a few microseconds, the processor will perform some operation on the data or translate the binary information into a command for its next operation.

Figure 10-21 is a 4-bit machine; that is, its standard word is 4 bits long. The data bus would need four lines (wires or copper traces on a circuit board). The address bus has eight lines, and because $2^8 = 256$, the processor can address 256 distinct memory locations or input/output devices. If there were only one address assigned for input/output, the sample machine in Figure 10-21 could store 255 4-bit words at any given time.

The arrows on lines connecting the various devices to the bus indicate the direction of information flow. Because the central processor initiates the memory addresses, address information flows out of the processor and into the various address decoders. As we shall see shortly, the chip select (CS) input to the memory chips is often a part of the address and may be connected to the address bus.

In most cases access to the data bus is bidirectional: data can flow from processor to memory or from memory to processor. Exceptions are ROMs, which cannot be written into, and peripheral devices limited to either inputting or outputting data.

Bus drivers may be on the memory chip or separate units. They can be bidirectional or unidirectional and are most often tristate devices. Large-scale MOS memory or CPU chips are often capable of driving only one TTL load or equivalent, which may not be adequate to drive the bus system.

The input/output interface device in the example (Fig. 10-21) is located at memory address 16 and would contain a decoder to detect the fact that the input/output unit is being addressed by the microprocessor. The processor can then send data to or receive data from the input/output unit just as though it were a memory device. The I/O interface also changes the data rate to make the slower I/O device compatible with the higher speed of the processor. The interface device may also be required to make appropriate code conversions and to change voltage and power levels to suit the requirements of the particular input/output devices.

The block to the right of the I/O interface simply represents the addresses available for the addition of more memory or other I/O devices.

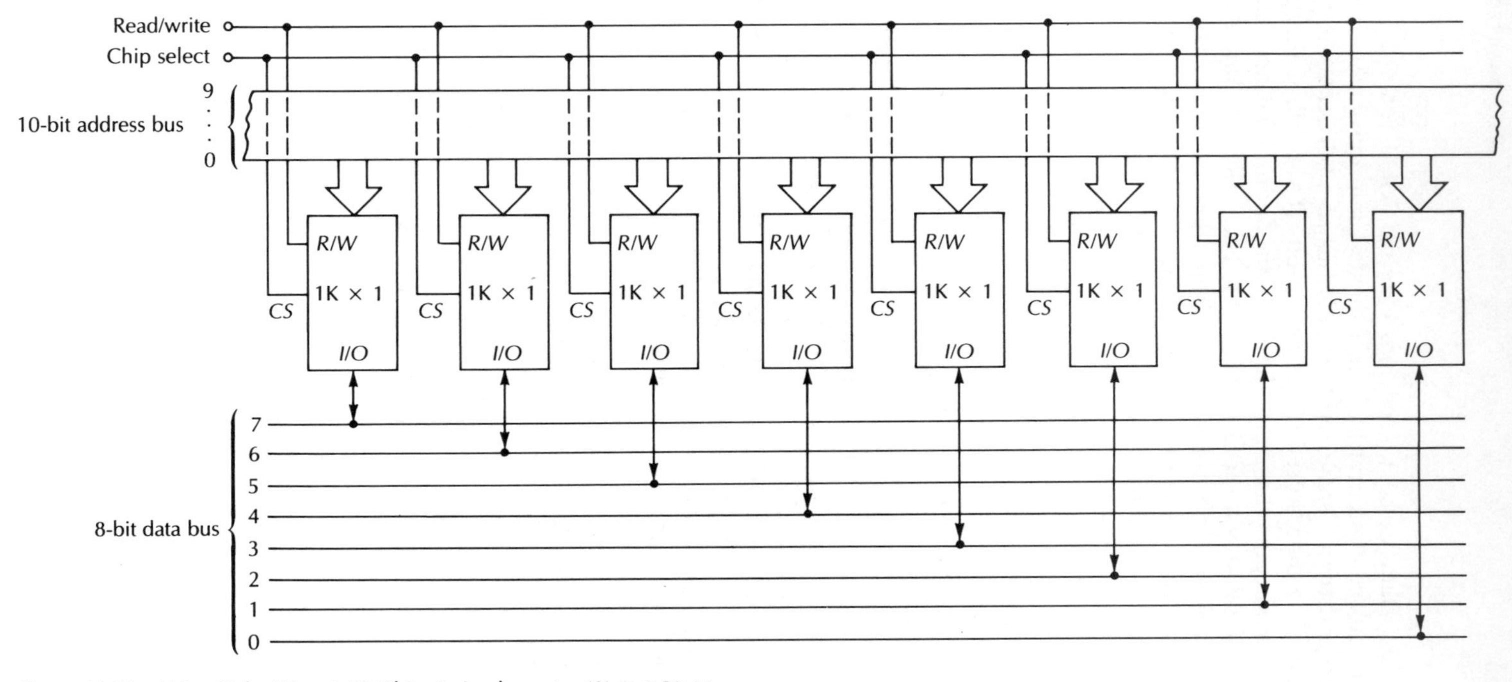

Figure 10-22 Using Eight 1K × 1-Bit Chips to Implement a 1K × 8-Bit Memory

The ROM is shown with binary data permanently stored in its memory cells. The words at addresses 14 and 15 contain dashes indicating that the ROM had a little more storage space than required by the required data. In reality, addresses 14 and 15 would contain all zeros or all ones and therefore generally could not be assigned to an I/O or other device. They are simply unused addresses. The programmer will not use those addresses, and the processor will not put those addresses (14 and 15) on the bus. Sometimes, rather than waste the memory space, the ROM designer will fill an unused ROM address with some mathematical or other data that might find only occasional use.

COMBINING MEMORY CHIPS TO MAKE LARGER MEMORY SYSTEMS

Figure 10-22 shows how to combine eight 1K × 1 bit memory chips into a 1K × 8-bit memory array. Each chip in Figure 10-22 is capable of storing 1024 1-bit words. A 10-bit address bus addresses bit 1, for example, in all eight of the memory chips. The data inputs/outputs are connected in such a way that one bit of the 8-bit word on the data bus is written into (or read out of) each of the chips.

Figure 10-23 shows how two 16 × 4-bit chips can be connected to form a 32 × 4-bit memory array. In this case the chip select ($\overline{CS}$) inputs are part of the address circuitry. Bit four in the memory address selects

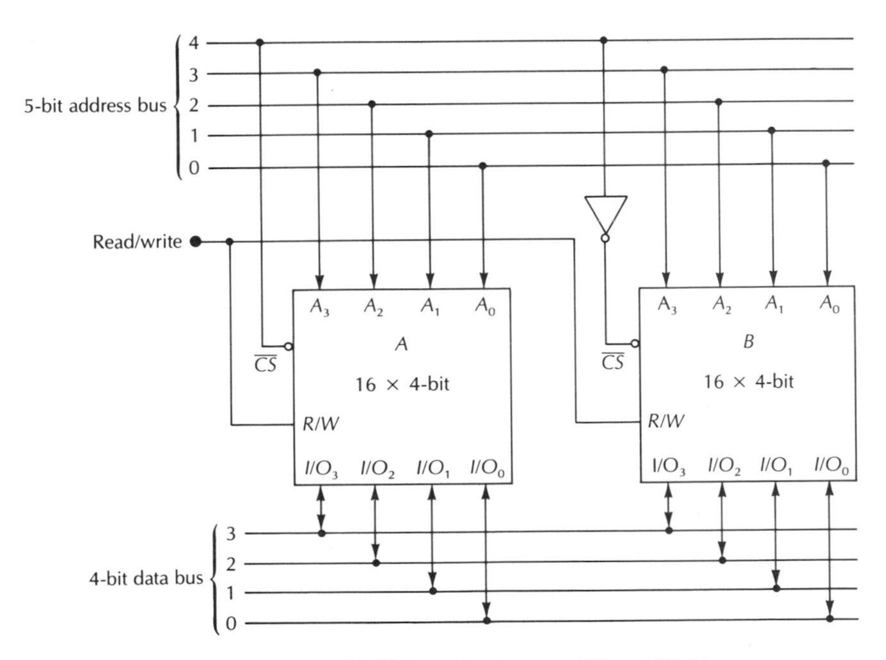

Figure 10-23 Using Two 16 × 4-Bit Chips to Implement a 32 × 4-Bit Memory

the appropriate chip. Chips A and B are both addressed by address lines 0, 1, 2, and 3, but only one of the chips will be put on the bus depending on the status of bit four in the address. A low on bit four will activate chip A and address the first 16 words. A high on bit four will keep chip A off the bus, but the inverter connected to the $\overline{CS}$ input of chip B will put chip B on the data bus to select the second group of 16 words.

Figure 10-24 is a more elaborate version of the technique used in Figure 10-23. Two inputs of a 3-line/8-line decoder are used to select one of the four 256 × 8-bit memory chips. The address lines zero through seven are connected to all four memory chips. Memory address bits 8 and 9 are used to select the one appropriate chip out of the four.

The addition of one address line and the use of the entire decoder would allow the use of eight 256 × 8-bit chips for a memory capacity of 2048 8-bit words. By combining the techniques used in Figure 10-22 with those illustrated in Figures 10-23 and 10-24, memory systems of almost any word length and practical size can be implemented.

10-10 Shift-Register Memory

MOS shift registers that can store from 100 to several thousand bits are available for a multitude of tasks. These are generally serial-in/serial-out (SISO) and are available in both static and dynamic forms. TTL registers with lengths of more than 8–16 bits are not common.

STATIC DEVICES

Static MOS shift-register cells are modified MOS flip-flops. Figure 10-25 shows a typical MOS static shift-register cell. Basically the circuit consists of two MOS inverters, using transfer gates in the feedback circuits. The circuit is basically a flip-flop. The two cells are isolated from each other by the transfer gate transistors Q_3 and Q_5. Data transfer into the cell and are coordinated from one half of the cell to the other by a three-phase clock. The three-phase clock signals are normally generated by circuitry on the memory chip. The three-phase signals are derived on the chip from a standard TTL single-phase clock signal input. TTL interface input/output and similar circuitry is also included on the memory chip, making it fully TTL compatible.

The circuit in Figure 10-25 operates as follows: when Ø1 is *low,* data are transferred through Q_1 to C_1. At this time Ø2 and Ø3 are *high* and C_2 is isolated from the left half of the cell. When the TTL system clock goes *low* (for active-high units), Ø1 goes *high* and Ø2 goes *low*. When Ø1 is *high,* C_1 is isolated from the input terminal. Ø2 *low* causes Q_3 to

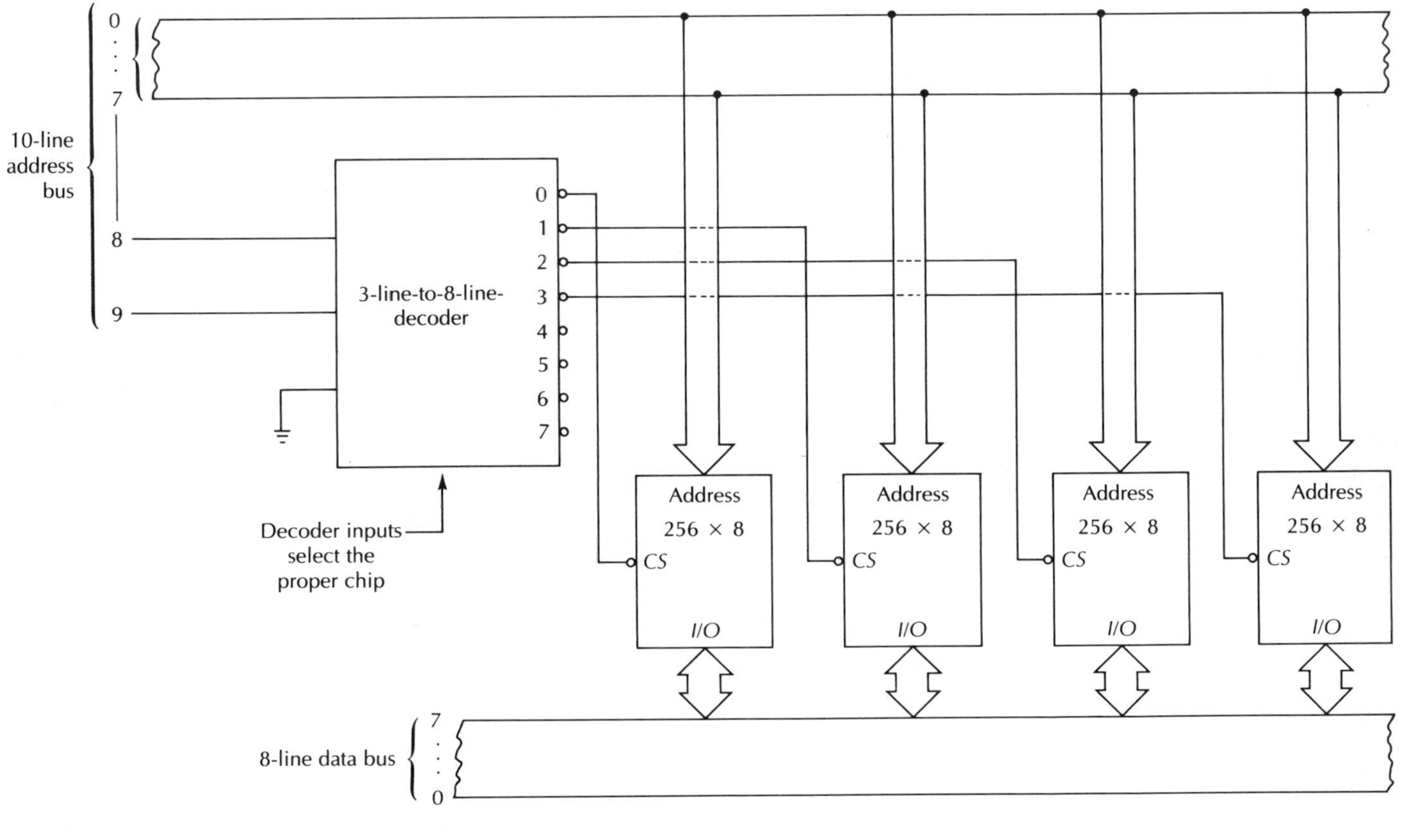

Figure 10-24 Using Four 256 × 8-Bit Chips to Implement a 1K × 8-Bit Memory

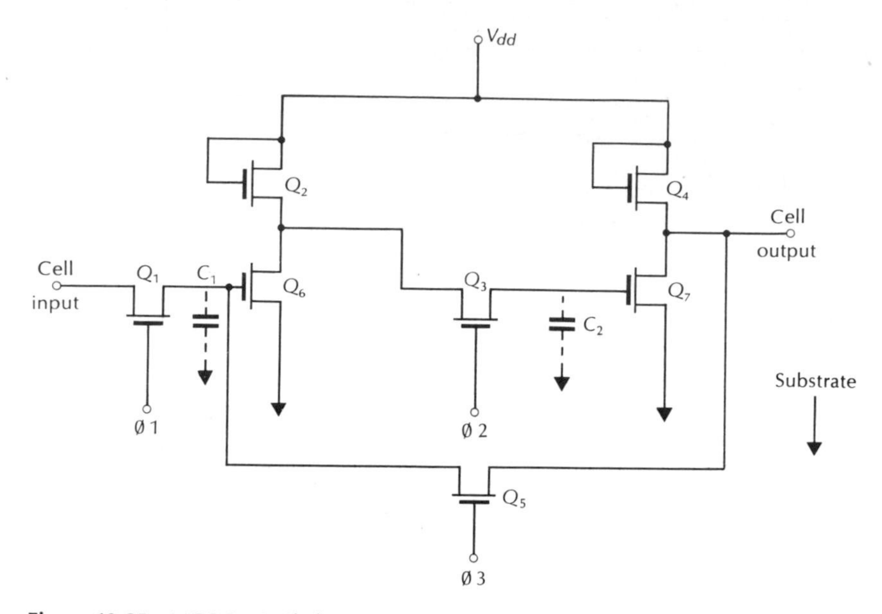

Figure 10-25 MOS Static Shift Register Cell

conduct, transferring data to C_2. Phase 3 requires about 5 microseconds to go *low,* to provide sufficient latch-up time. The feedback signal through Q_5 latches up the cell until the next TTL clock pulse arrives and, via the three-phase internal clock, causes new data to shift into the cell, while the cell outputs the previous data for the next cell in line. The system of Ø1 and Ø2 clocks provides behavior similar to that of a bipolar master-slave flip-flop and, like the master-slave, eliminates the race problem. Ø3 is necessary to control the latch-up function. The Ø3 clock imposes a limitation on the maximum system clock pulse width, which can be found in the data sheet for the particular device being used.

DYNAMIC SHIFT-REGISTER CELLS

The dynamic shift-register cells use the stored charge principle. A two-phase, non-overlapping clock is required, but no third phase is necessary because there is no latch-up feedback. Capacitor refreshing is required, thus imposing a limitation on the minimum clock speed, typically a few hundred hertz. Figure 10-26 shows the schematic diagram of the dynamic shift-register cell. The circuit works as follows:

When Ø1 (in) is *low,* the data stored as a charge on C_1 are transferred through Q_3 to C_2. When Ø1 (in) goes *high,* Ø2 (out) goes *low,* transferring data out of the cell. The *high* on Ø1 (in) isolates C_2 from the input. Data are transferred from one cell to the next with each pair of clock pulses.

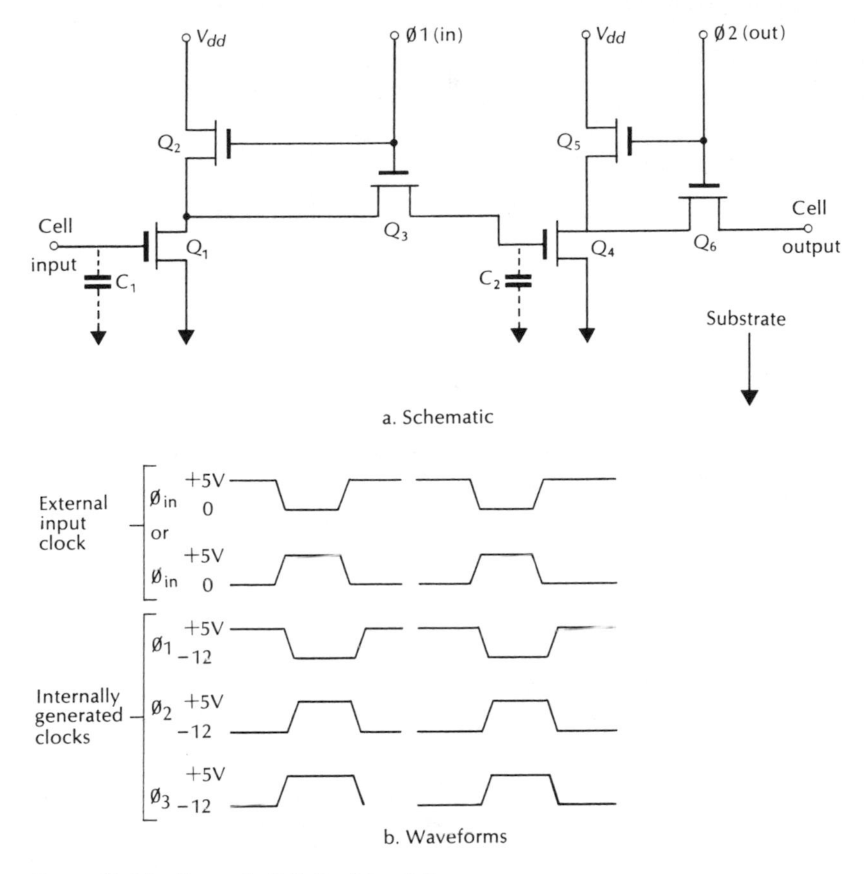

Figure 10-26 Dynamic Shift Register Cell

Q_2 and Q_5 are turned on by the clocks. As a result, power consumption varies with the clock duty cycle. Refreshing is accomplished in this case by simply shifting data frequently enough, which means that there is always a minimum useful clock rate for this kind of shift register.

RECIRCULATING SHIFT REGISTERS

Data stored in a shift register can be taken from the output, fed back to the input, and shifted around the loop with the addition of some logic. Recirculating the data can make the readout nondestructive if data are returned to the first stage at the same time they are transferred to the output. In the case of dynamic registers, recirculating is necessary if data must be held for more than a millisecond because of the dynamic logic's refresh requirement.

The recirculating logic is often included on the chip in MOS devices, but outboard logic can also be used. Figure 10-27a shows a typical

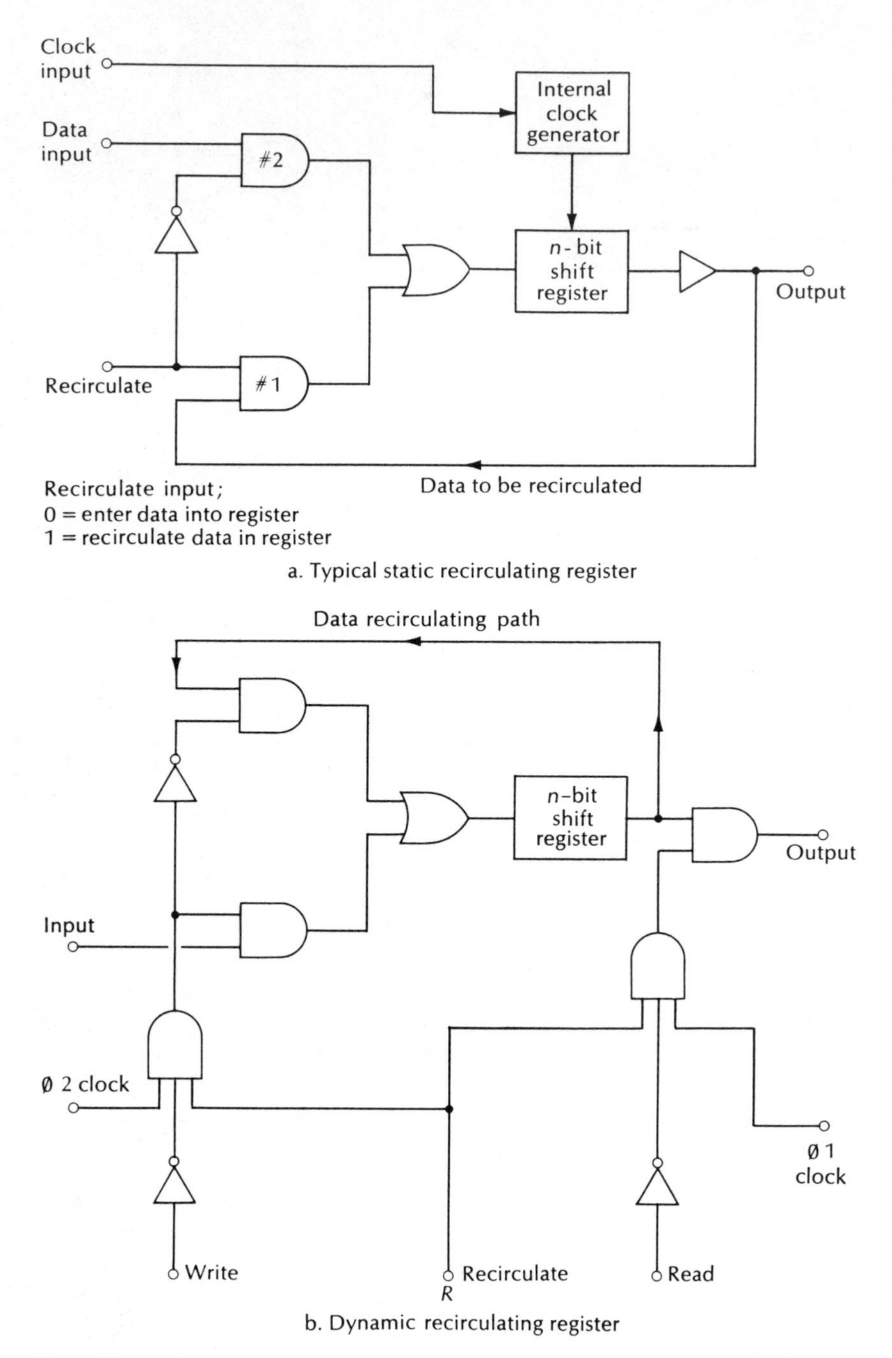

a. Typical static recirculating register

b. Dynamic recirculating register

Figure 10-27 Recirculating Registers

Truth table

Write	Read	Function
0	0	recirculate, output is 0
0	1	recirculate, output is data
1	0	write mode, output is 0
1	1	read/write mode, output is data

Recirculate input: $R = 0$, data recirculates;
$R = 1$, see truth table

c. Dynamic recirculating register truth table

Figure 10-27 continued

MOS static register with recirculating logic. In the circuit in Figure 10-27a output data is always available. A recirculate input at gate 1 determines whether new or recirculated data will enter the *n*-bit shift register. With a 0 on the recirculate input, gate 1 is disabled and recirculated data is locked out. The inverter applies a *high* to the lower input of gate 2, allowing new data to enter the register. With a 1 on the recirculate input, recirculated data is allowed to enter the register and new data is locked out. Most MOS static shift registers have on-chip circuitry to develop the required clock signals from the system clock.

The recirculating gate circuit for dynamic registers in Figure 10-27b is similar to the one used for static registers, except that the output is gated and the two-phase clock is applied to the recirculating control and output gates. The truth table in Figure 10-27c tabulates the gate control functions for the dynamic register circuit.

SUMMARY

SEMICONDUCTOR MEMORIES

The phenomenal progress made in large capacity semiconductor memory techniques in recent years has begun a revolution in digital technology. In recent years 16K-bit (or more) memories on a chip have become common and inexpensive items.

Cost

Cost is a very important factor because of the relatively large number of storage cells required for even a small memory system. The availability of inexpensive memory systems, mostly of the semiconductor variety, is responsible for the current boom in pocket calculators, microprocessors, special remote terminals with auxiliary computing capability (called *intelligent terminals*), digital instrumentation, and industrial process control systems. The availability of inexpensive semiconductor memories has even made it possible for the automotive industry to project the use of a small computer to control fuel mixtures, brakes, transmission, and a variety of other automotive functions—all at a projected cost of less than $50 per car. Memory cost is defined in terms of cents per bit.

STATIC RAM MEMORY

Static MOS memories use flip-flops as the basic memory element (cell) and are available in both bipolar and MOS versions. Static RAMs are slower than dynamic RAMs but can be operated without complex clock signals or, in most cases, in an asynchronous (no clock) mode. Because flip-flops are used to store data instead of the charged capacitor approach used in dynamic MOS, there is no need for refresh circuitry.

MOS RAMs are available with both static and dynamic cells. The static cell is basically a flip-flop latch similar to that used in bipolar memories. The dynamic cell, on the other hand, has no counterpart in bipolar technology. In this type of cell the transistor gate capacitance is charged and the stored charge controls the source-to-drain current. Because of the small capacitance and normal leakage currents, the gate capacitors must be recharged (refreshed) every few milliseconds. The reading out of a memory location automatically refreshes the charge at that location, but this process cannot be relied upon in practice. It is unlikely that each cell will be read frequently enough in the course of normal processing operations. Although the dynamic cell may be faster than the static MOS cell, the time that must be allowed for refresh is a definite drawback. The increased complexity and timing accuracy of the clock also adds to the problem. In addition, some of the potential operating speed is lost as a result of the more complicated gating involved in handling the more complex clock sequence. These peripheral speed losses narrow the speed margin between dynamic and static structures.

SHIFT-REGISTER MEMORIES

There is a variety of both dynamic and static MOS shift-register memories that can be used when SISO data storage is available. These memories are relatively inexpensive and easy to implement. All shift-register memories can be used as recirculating registers, but not all of them have on-the-chip recirculation logic.

Shift registers can serve as low-cost sequential memories, large bit-number buffer memories, CRT display memories, delay line replacement, and a number of other functions.

Bipolar shift registers are, of course, available but the smaller lengths on each chip make them less suitable for the kinds of applications just mentioned.

READ-ONLY MEMORIES (ROMS)

The read-only memory is used for permanent storage such as frequently used control routines, tables of data, and many other computer housekeeping chores. ROMs are available in up to 1K–2K bits in bipolar forms. Bipolar units are faster but more expensive than MOS devices. These devices are available as factory programmed or field (fusible) programmable units (PROMs).

MOS devices are available as factory programmed (mask programmed) and erasable field programmable units.

Problems

1. Compare the following in terms of approximate access time:
   a. disc — d. core
   b. drum — e. semiconductor main memory
   c. tape — f. semiconductor scratch pad
2. For what purpose is each of the following memories used?
   a. core
   b. large-scale integrated circuit memories
   c. bipolar (LSI) RAM
   d. cassette
   e. drum/disc
   f. computer tape
   g. punched cards and tape
   h. ROM
3. What is a cryogenic memory?
4. Is core memory destructive or nondestructive?
5. Describe a stack of core planes. What are the advantages of this configuration?
6. What is a bipolar RAM?
7. Describe the two popular RAM organizations.
8. What is the basic memory cell in bipolar RAMs?
9. Are bipolar RAM cells classified as static or dynamic? Why?
10. Is the readout of bipolar RAM cells a destructive readout? Why?
11. What is the purpose of memory decoding? (Respond in detail.)
12. What kind of organization is involved in a 16×32-bit memory?
13. A 1K memory has an actual bit count of ______?
14. Explain the value of tristate outputs on memory chips.

15. Why is MOS preferred to TTL type circuitry for larger memories?
16. Compare static and dynamic MOS in terms of complexity and cost.
17. Draw the schematic diagram of an MOS RAM static cell and briefly describe how it works.
18. Draw and explain the operation of an MOS dynamic memory cell.
19. Explain the term *refresh.* Where is it used and why?
20. Draw the schematic diagram of a two-stage MOS dynamic shift register and explain how it works.
21. Why are MOS shift registers preferred over the TTL type for memory applications?
22. What is a read-only memory (ROM)?
23. What are ROMs used for?
24. List the ROM types commonly available.
25. Of what does a bipolar ROM cell consist? An MOS ROM cell?
26. What is a field-programmable ROM?
27. How is a field-programmable ROM programmed?
28. What is an erasable ROM? How is it programmed? How is it erased?
29. What is a recirculating shift register?
30. Write out the information found on the following punched card.

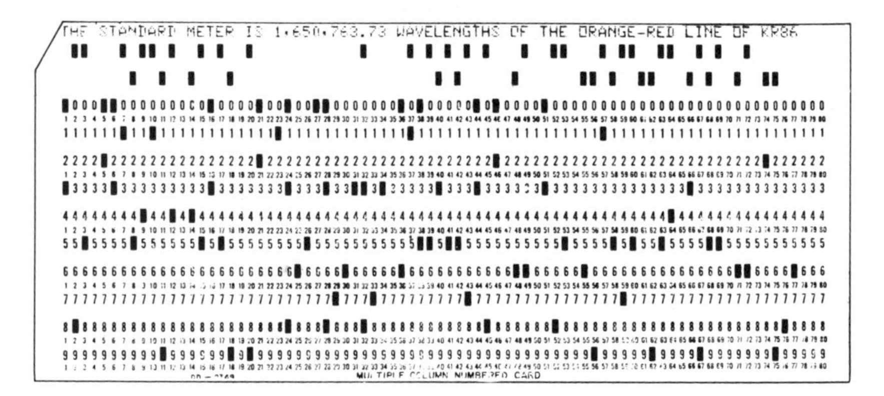

31. Write out what is encoded on the sample ASCU coded punched paper tape below.

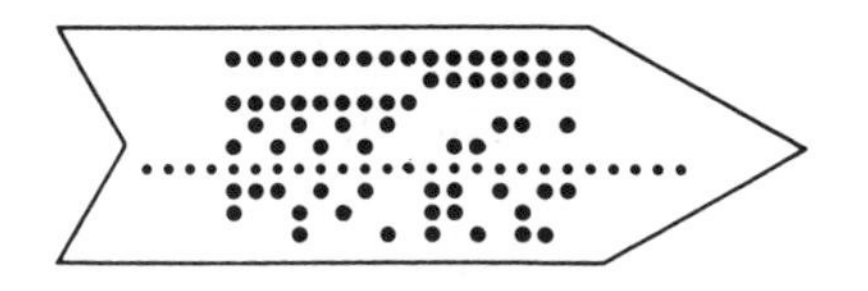

32. In a bus-organized memory system, is the data bus generally bidirectional? If so, are there cases where data travel in only one direction on the bus? Give an example to support your answer.
33. A microprocessor has the capability of addressing 2K of memory. How many address lines are required?
34. What is the *exact* number of bits a 4K × 1-bit memory can store?
35. There are 12 address lines in a memory system. What is the largest number of memory locations that can be addressed?
36. *Access time* is the primary memory parameter. Is the access time parameter adequate to compare two memory chips when you are trying to find a substitute? Explain.
37. C-MOS memory is generally slower than other MOS technology. What makes C-MOS memory attractive for some systems?
38. Explain briefly how a bit is stored in a dynamic memory cell and how *refreshing* operates.
39. List some advantages and disadvantages of dynamic memory as compared to static memory.
40. Define *soft fail.*
41. Can the computer, in normal operation, write into ROM?
42. Why is it generally possible to put more ROM than RAM on a given chip?
43. In a computer system, where do memory addresses originate?
44. Why is a bus driver often used in conjunction with a memory chip?
45. Draw a block diagram showing how a group of 2K × 4-bit memory chips can be used to construct a 4K × 8-bit memory system.

CHAPTER 11

INTERFACING

Learning Objectives. *Upon completing this chapter you should know:*

1. *How to describe the basic properties of the operational amplifier.*
2. *How to write the voltage gain equations for the inverting and non-inverting op-amp circuits.*
3. *How to state the approximate input impedances for the inverting, non-inverting, and voltage follower op-amp circuits.*
4. *How to explain the operation of the comparator circuit.*
5. *How to define hysteresis in comparator circuits.*
6. *How to describe the operation of the 555 timer in the monostable mode.*
7. *How to describe the operation of the 555 timer in the astable mode.*
8. *How to explain the function of a digital clock.*
9. *How to define master clock and subordinate clock.*
10. *How to define the term* analog.
11. *The function of a transducer.*
12. *How to define the acronymn DAC.*
13. *How to define the acronymn ADAC.*
14. *How to explain the operation of the summing amplifier DAC.*
15. *The purpose and advantages of the ladder-type summing network.*
16. *How to define accuracy and resolution in digital-to-analog converters.*
17. *The basic parallel (simultaneous) analog-to-digital converter.*
18. *The basic techniques used in sequential analog-to-digital converters.*
19. *The techniques and purpose of time multiplexing digital-to-analog and analog-to-digital converters.*

Interfacing circuits are necessary whenever digital circuits must transmit to or receive from devices that are not compatible with the digital circuits used. Ultimately digital circuits must communicate with display devices, relays, telephone lines, lamps, printers, and other similar devices.

These outside-world devices generally require voltage, current, and power levels that differ from standard logic levels. Outside-world devices also tend to be slower than the internal logic. Further, outside-world devices often operate in some binary code that differs from straight binary.

Within a logic system, both MOS and TTL may be used. Connecting some MOS devices to TTL requires a logic level translation.

The circuits used to alter voltage, current levels, speed, or codes to make two devices compatible are called *interface circuits*. There are IC interface devices available to solve many, but not all, of the interface problems normally encountered.

A special case of interfacing is converting analog information into digital data and digital data into analog information. This interface is particularly important in instrument and control systems. One of the most common examples of analog-to-digital conversion is the digital voltmeter. Voltages in the real world do not occur in convenient digital-like steps. They exist in a continuous range of values and must be converted into equivalent digital values before the digital circuitry can deal with them.

11-1 The Operational Amplifier

The operational amplifier (op-amp) is basically a linear (analog) device, but it is important in many analog/digital interface systems.

The operational amplifier is also used in several strictly digital applications, where it is given the special title of *voltage comparator*. In its comparator mode, the op-amp is used in analog/digital interface circuits, timing circuits for digital clocks, data transmission-line systems, and a number of other digital applications.

OPERATIONAL AMPLIFIER CHARACTERISTICS

The operational amplifier gets its name from its original use in analog computers. By definition the operational amplifier must have the following characteristics:

1. High input impedance
2. Low output impedance
3. Very high open-loop (without feedback) voltage gain (100,000 or more)
4. Provision for external feedback to set the voltage gain at any desired value
5. Very good stability
6. Capable of frequency response from DC to some high frequency
7. One 180-degree phase shift (−) input and one zero-degree phase shift (+) input

Most modern IC operational amplifiers are multi-stage differential amplifiers, generally including an emitter-follower output stage to provide a low impedance output. Figure 11-1 shows the operational amplifier symbol.

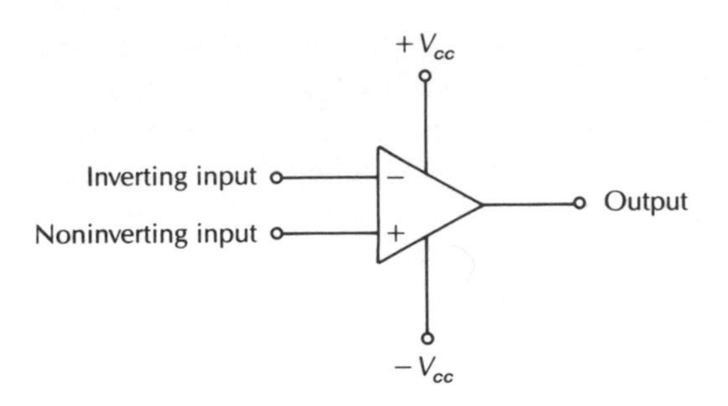

Figure 11-1 The Operational Amplifier Symbol

SOME COMMON OP-AMP CIRCUITS

The operational amplifier can be configured in hundreds of ways. We have only enough space in this book to examine briefly a few of those configurations that are often used in conjunction with digital circuits and systems.

The Inverting Amplifier

Figure 11-2 shows the circuit configuration, voltage gain, and input impedance equation for the inverting op-amp configuration. The large amount of (negative) feedback being returned to the inverting input lowers the input impedance between point A and ground to practically zero ohms—a virtual ground. For all practical purposes the input impedance is simply the value of R_{in}.

The voltage gain is governed by the ratio of R_f to R_{in} and is normally independent of all amplifier parameters, providing the amplifier qualifies as an op-amp.

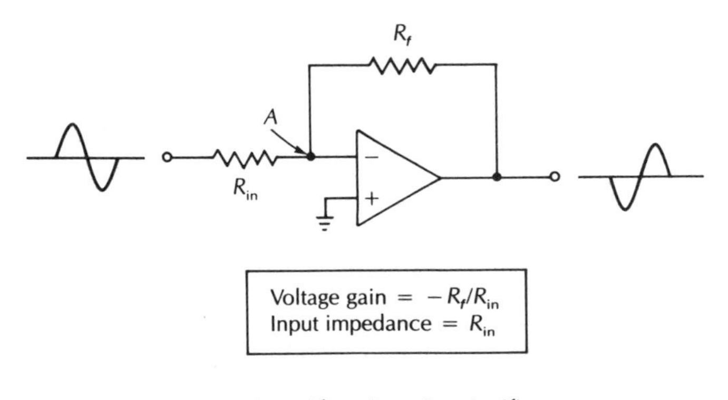

Figure 11-2 The Inverting Amplifier

The Non-Inverting Amplifier

The non-inverting op-amp circuit provides a high input impedance, again because of the negative feedback. The voltage gain is controlled by the ratio of R_f to R_{in} and is independent of the amplifier characteristics. Figure 11-3 shows the non-inverting op-amp configuration.

The Summing Amplifier

The summing amplifier in Figure 11-4 is a flexible and accurate way in which to *add* analog voltages. Because the inverted input (with feedback) is a virtual ground, it provides an ideal summing node.

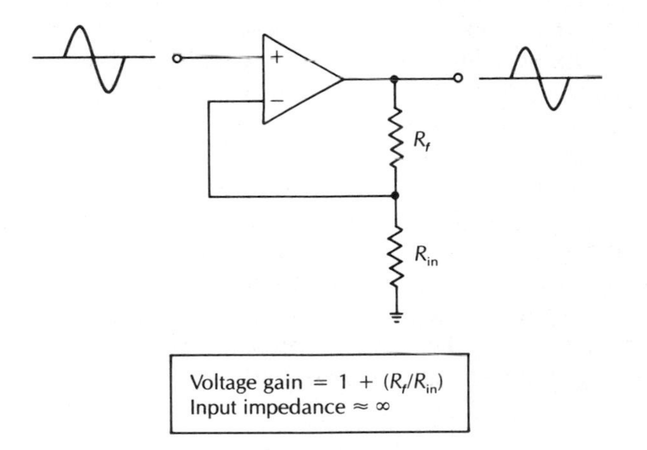

Figure 11-3 The Non-Inverting Amplifier

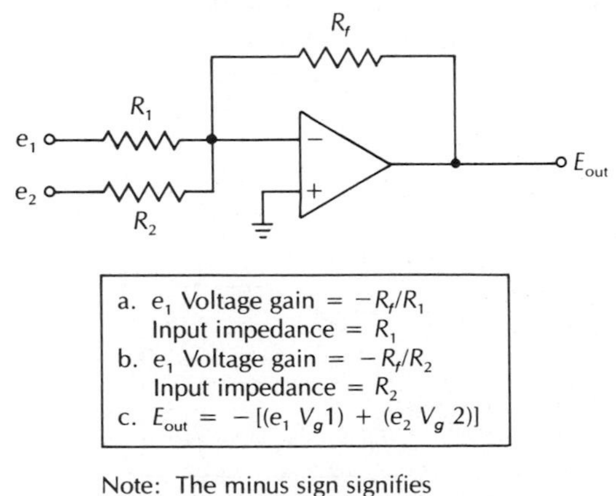

Figure 11-4 The Summing Amplifier

High Input, Low Output Impedance Amplifier

The circuit in Figure 11-5 is a voltage follower circuit with a voltage gain of unity (1), nearly infinite input impedance, and a very low output impedance. The circuit is similar to the transistor-common collector circuit (emitter-follower), but has more nearly ideal characteristics.

The Comparator

The voltage comparator is the only common op-amp circuit that is used without negative feedback. The comparator takes advantage of the high open-loop gain of the amplifier to make a sudden switch when the input voltages on the inverting and non-inverting inputs are equal.

Because of the very high voltage gain, less than a millivolt difference between inverting and non-inverting inputs will drive the amplifier to one power supply rail (voltage) or the other. When the two input voltages are equal within a small fraction of a millivolt, the comparator suddenly switches to zero volts output. The comparator may be operated with two power supplies to provide an output voltage range from some negative to some positive voltage. Some operational amplifiers may also be operated with a single power supply to provide an output voltage range from zero to plus 5 volts, for example.

Any op-amp can be used as a comparator, but there are IC op-amps available designed especially for comparator service. In addition to fast switching capabilities, these special op-amps often include special features such as enable/disable control inputs.

HYSTERESIS

In some cases it is desirable to have a comparator with some built-in hysteresis to provide the combined action of a comparator and Schmitt trigger. As in the Schmitt trigger, *positive* feedback is used to provide a lower trip-point, an upper trip-point, a dead-band between, and regenerative action to insure fast rise and fall times.

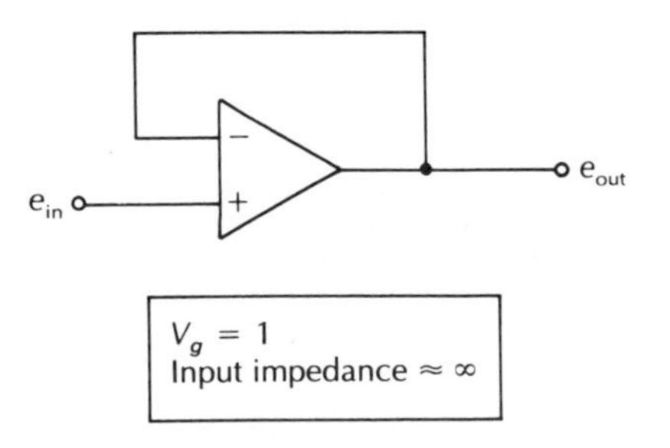

Figure 11-5 The Voltage Follower

The price paid for hysteresis and the regenerative action is a tendency for the comparator to produce some overshoot (or undershoot) and ringing. The ringing tendency can increase the settling time of the comparator. In many cases the need for hysteresis is dictated by noise on data transmission lines, which are relatively slow anyway. In such cases, the longer settling time presents no problem.

Figure 11-6 shows three common comparator symbols along with a symbolic summary of comparator behavior. Figure 11-7 shows the positive feedback connection to provide hysteresis in a comparator circuit.

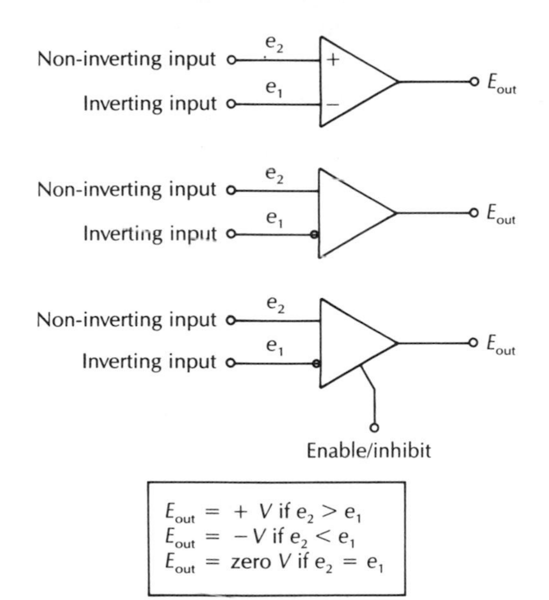

$E_{out} = +V$ if $e_2 > e_1$
$E_{out} = -V$ if $e_2 < e_1$
$E_{out} =$ zero V if $e_2 = e_1$

Figure 11-6 The Comparator

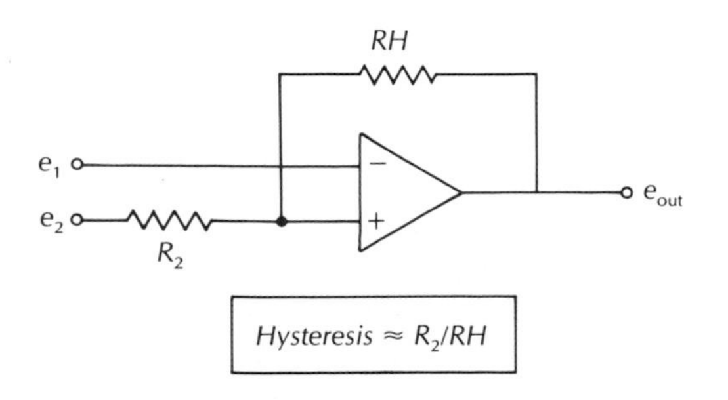

Hysteresis $\approx R_2/RH$

Figure 11-7 Comparator with Hysteresis

11-2 The 555 Timer

The 555 timer is a stable, highly flexible timing circuit. It can be used to provide clock pulses or time delay for digital circuits. It is also used in a number of digital interface circuits and is common in analog timing circuits as well.

For the following discussion please refer to Figure 11-8. The resistors R_1, R_2, and R_3 form a voltage divider that provides a different reference voltage to one input of each of the two comparators. Comparator B has a reference voltage of two-thirds of the power supply voltage, and comparator A has a reference voltage of one-third of the power supply voltage. Most timing devices that depend on an R-C time constant vary their timing period along with any variations in power supply voltage. R-C timing circuits are also subject to timing inaccuracies due to the non-linear capacitor charge curve. The 555 timer design circumvents both of these problems. The voltage divider makes the charge range of the capacitor independent of the absolute voltage and ensures that the

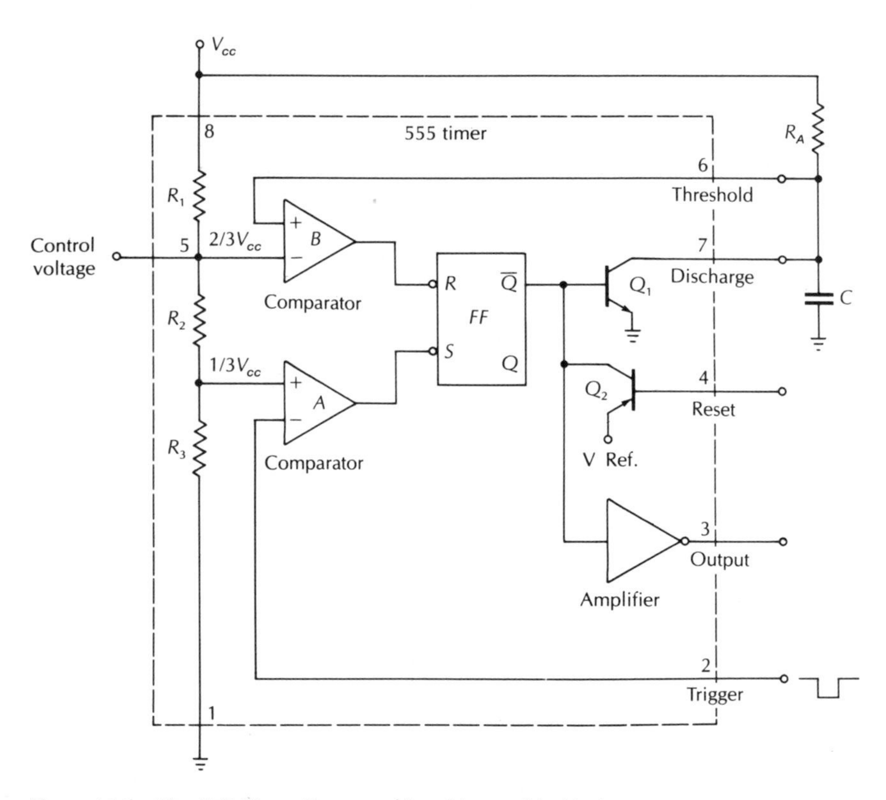

Figure 11-8 The 555 Timer Connected in a Monostable Mode

device operates only on the most linear part of the capacitor charge curve. The 555 timer has two basic operating modes; astable and monostable. In the astable mode, the timer is a free-running TTL compatible-pulse generator that can serve as a clock (not the timekeeping kind) for digital circuits. In the monostable mode, a trigger pulse turns the device on, and it stays on for a time period set by an external resistor and capacitor. The output pulse is TTL-compatible and the pulse width has no relationship to the pulse width of the trigger pulse. The monostable configuration provides a time delay or functions as a "pulse stretcher."

The best way to understand the 555 circuit is to examine its operation in its two principal modes.

MONOSTABLE OPERATION

In the monostable circuit in Figure 11-8, the monostable cycle begins with transistor Q_1 turned on, holding timing capacitor C in the discharged condition. The output voltage is *low*. The arrival of a negative trigger pulse at pin 2 (trigger) sets the flip-flop. When the flip-flop is set, the transistor Q_1 is turned off and the output (pin 3) goes *high*. The capacitor C starts charging toward two-thirds V_{cc}, at which point the comparator resets the flip-flop. Transistor Q_1 is again turned full-on and rapidly discharges C. The output (pin 3) goes low.

The comparator is a high-gain differential amplifier. One differential input on comparator A is connected to a voltage of one-third V_{cc} and the other to the trigger input. The comparator provides an abrupt transition in its output voltage at the point where the voltages to its two inputs are equal.

The device triggers on a negative-going input signal when the level reaches one-third V_{cc}. Once triggered, the circuit will remain in this state until the set time is elapsed, even if it is triggered again during this interval. The time that the output remains in the high state can be determined by the nomogram in Figure 11-9. The charge rate and the threshold level of the comparator are both directly proportional to supply voltage, and the timing interval is independent of slow variations in power supply voltage. However, the circuit is sensitive to supply voltage variations that are faster than the timing period. V_{cc} decoupling is frequently required. The frequency modulation input (pin 5) is connected to ground through a .01 μF capacitor. When this input is not needed, it is good practice to use this capacitor to decouple the V_{cc} line.

Applying a negative-going pulse simultaneously to the reset terminal (pin 4) and the trigger terminal (pin 2) during the timing cycle discharges the external capacitor and causes the cycle to start over. The

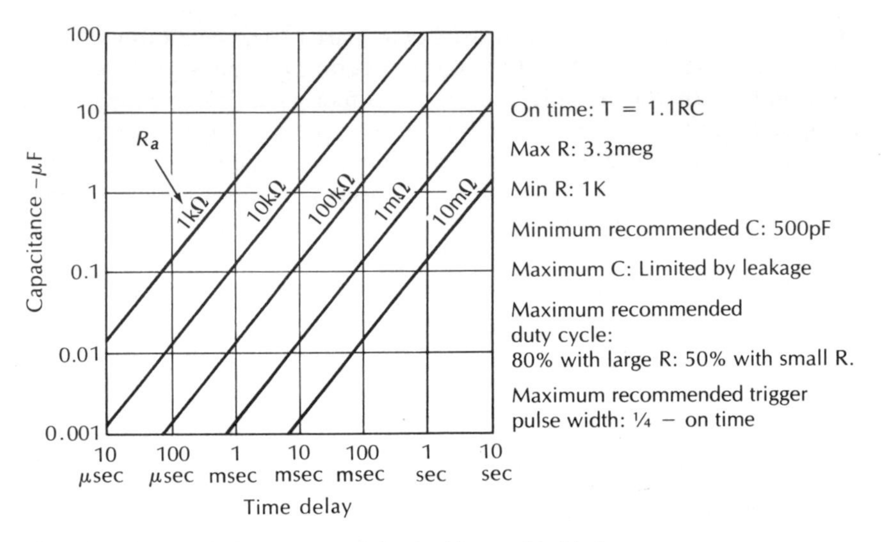

Figure 11-9 Time-Delay Nomogram for the Monostable Mode

timing cycle will now begin on the positive edge of the reset pulse. During the time the reset pulse is applied, the output is driven low.

When the reset function is not in use, it should be connected to V_{cc} to avoid the possibility of false triggering.

ASTABLE OPERATION

Figure 11-10a shows the timer connected to operate in an astable mode. In this mode the timer will trigger itself and operate as a free-running multivibrator.

The external capacitor C charges through R_A and R_B and discharges through R_B alone. As a result, the duty cycle (ratio of on-to-off time) is controlled by the ratio of resistors R_A and R_B. When the timer is connected as an astable oscillator, this capacitor C charges and discharges between one-third and two-thirds V_{cc} as it is in the triggered mode. Because the charge and discharge points are at specific fractions of V_{cc}, the timing is independent of slow variations in V_{cc}.

The basic astable circuit in Figure 11-10a cannot produce a symmetrical (square) wave pulse. When it is necessary to generate a pulse train in which on and off times are equal, the circuit can be modified as shown in Figure 11-11. With the addition of diodes D_1 and D_2, the total period is $0.7(R_A + R_B)C$. If $R_A = R_B$, the on-time or off-time pulse width is $0.7R_BC$. The frequency of the square wave pulse is the reciprocal of the total period.

The following equations define the timing and frequency relation-

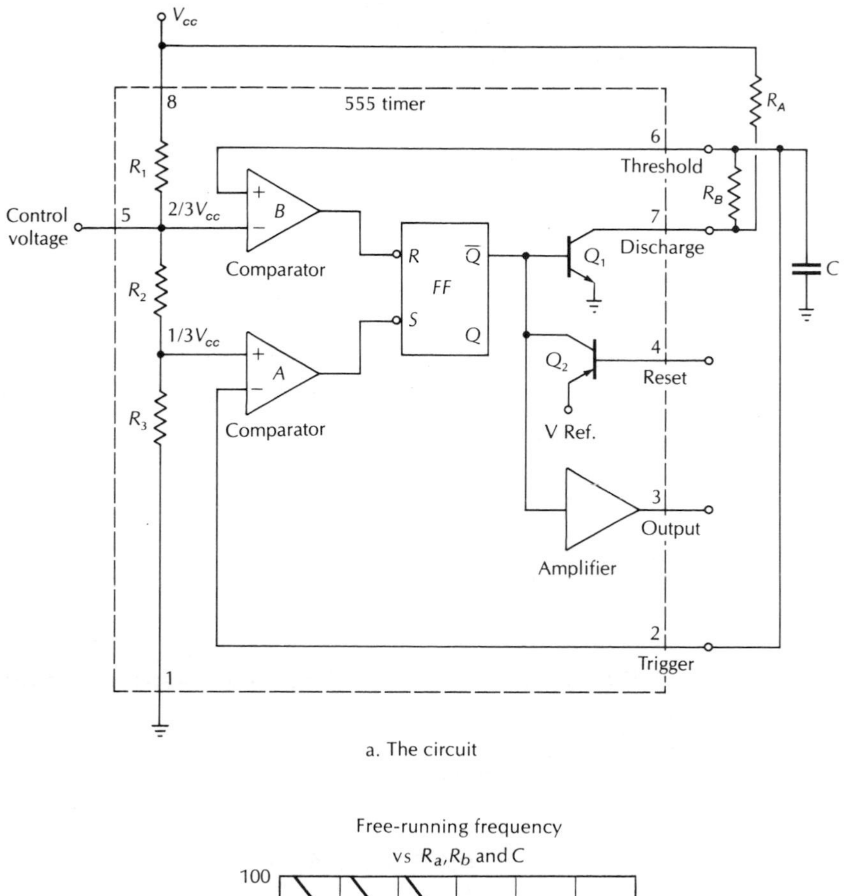

a. The circuit

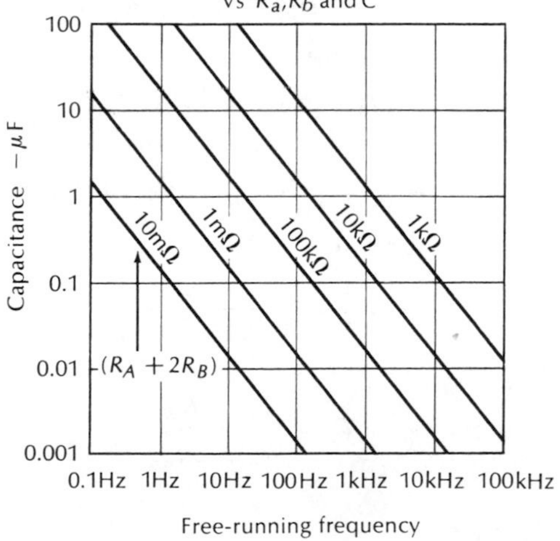

b. Free-running frequency nomogram for the 555 astable mode

Figure 11-10 The 555 in the Astable Mode

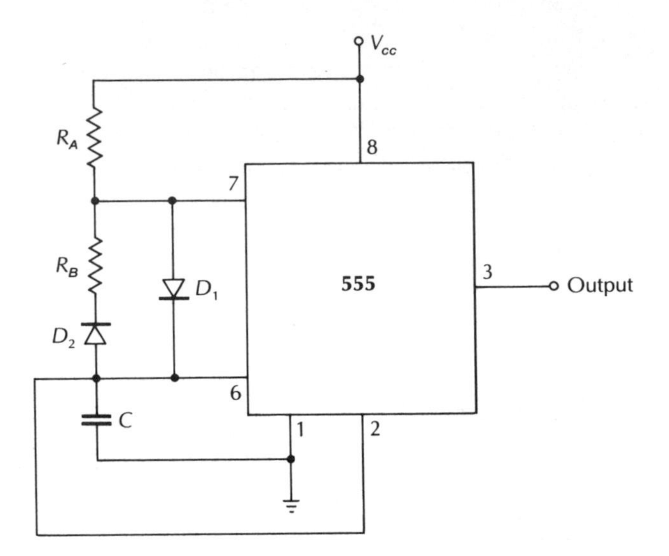

Figure 11-11 The 555 Astable Circuit Modified to Produce a Symmetrical Waveform

ships. Figure 11-10b shows a free-running frequency nomogram for the astable configuration.

TIMING AND FREQUENCY EQUATIONS

1. Charge time (output *high*)

$$t_1 = 0.693\,(R_A + R_B)\,C$$

2. Discharge time (output *low*)

$$t_2 = 0.693\,(R_B)\,C$$

3. Total period

$$T = t_1 + t_2 = 0.693\,(R_A + 2R_B)\,C$$

4. Frequency of oscillation

$$f = \frac{1}{T} = \frac{1.44}{(R_A + 2R_B)C}$$

5. Duty cycle

$$D = \frac{R_A + R_B}{R_A + 2R_B}$$

FREQUENCY DIVIDER

The timer can be used as a frequency divider by connecting the circuit as a monostable and adjusting the timing. The timer is triggered by the first incoming pulse and starts its timing cycle. Because it cannot be retriggered during the timing cycle, it is immune to further pulses until the cycle is complete. Thus it can be set to time-out every 2, 3, or 4 pulses, and so on. Synchronization with the input signal results from the fact that the timer begins each new cycle only when initiated by the input frequency signal.

MISSING PULSE OR BURST DETECTOR

The configuration shown in Figure 11-12 provides for the detection of a missing pulse in a continuous train or it can detect the start and end of a burst of pulses.

The timing cycle is continuously reset by the pulse train. A change in frequency, a missing pulse, or the termination of the train (end of a burst) allows the timer to time-out causing a change in output level. The time delay is set to be slightly longer than the normal time between pulses.

PULSE WIDTH MODULATOR

Figure 11-13 shows how the monostable configuration can be used for pulse width modulation. The timer is triggered by a continuous pulse train and the threshold voltage of the comparator is modulated by the

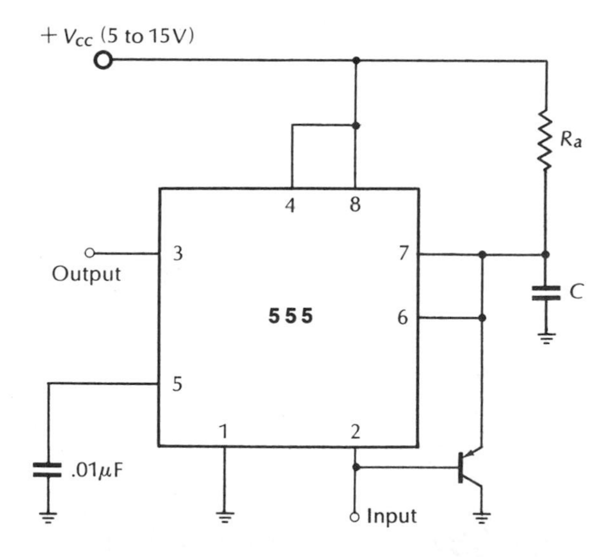

Figure 11-12 The 555 Timer Used as a Missing Pulse Detector

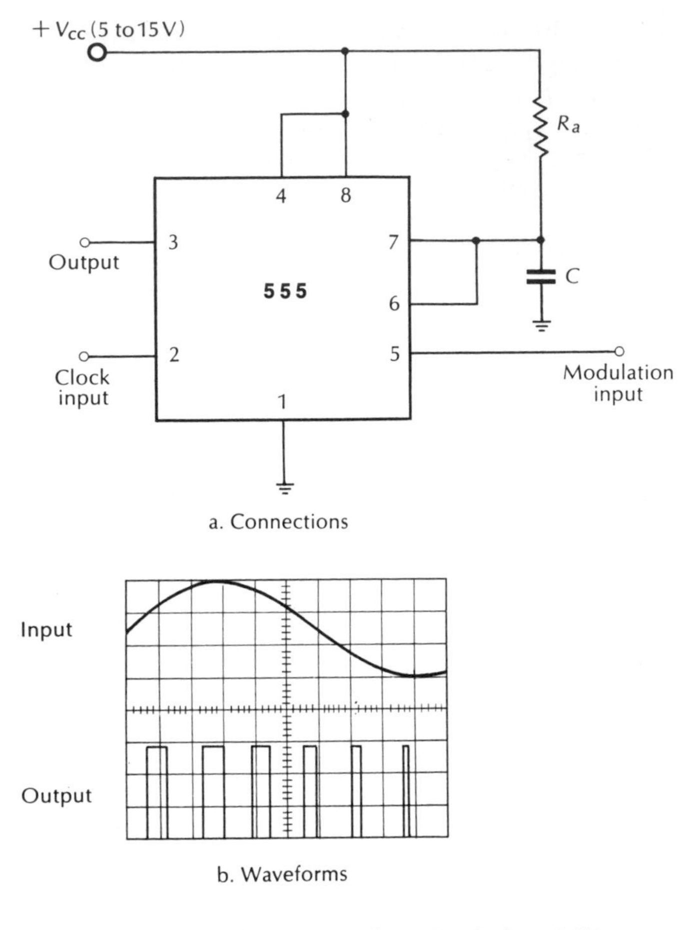

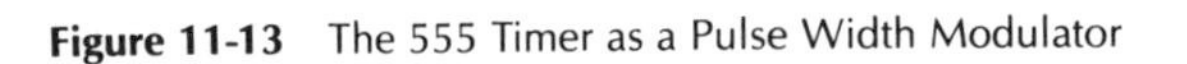

Figure 11-13 The 555 Timer as a Pulse Width Modulator

signal applied to the control-voltage terminal (pin 5). The pulse width is modulated by varying control voltage. Figure 11-13b shows typical waveforms.

PULSE POSITION MODULATION

In the circuit in Figure 11-14 the timer is used as a pulse position modulator. The modulation signal is applied to the control-voltage terminal as in the pulse width modulator. In this case, however, the timer is configured for an astable mode instead of a monostable mode.

The 555 is completely compatible with TTL when used with a 5 volt power supply. The practical upper frequency limit is about 500 kHz.

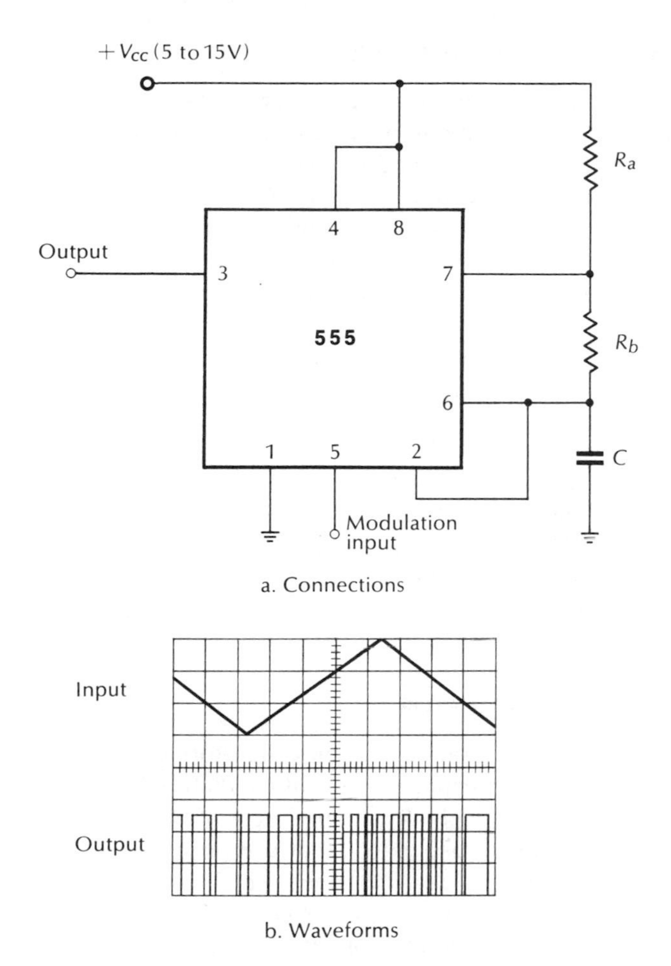

Figure 11-14 Pulse Position Modulation with the 555 Timer

11-3 Other Clock and Timing Circuits

Nearly all digital circuits require some kind of clock-pulse generator. In many cases a simple astable circuit is adequate and the 555 timer in its astable mode is a popular choice. There are other simple clock generators in common use as well. We will briefly examine some of those in this section. When more precise timing is required, a crystal-controlled oscillator is generally used.

Integrated circuits are available that contain all necessary clock circuitry except for the crystal and one or two other small components. A number of microprocessor chips also contain on-chip clock circuitry that requires only the external crystal or *R-C* timing components. Here we will be concerned with separate clock circuits.

MASTER AND SUBORDINATE CLOCKS

Whatever the source, the master clock provides a train of pulses that controls all of the operations in the system, determines the speed at which the system operates, and determines how long it takes to perform an operation—for example, addition.

Clock pulses are often required that are shifted in phase from the master clock or have pulse widths or repetition rates that differ from the basic pulse train. The circuits that provide these master clock-related pulses are called *subordinate clocks*. As the name implies, subordinate clocks derive their output pulses by performing some operation on the pulses generated by the master clock. They are *not* independent pulse generators.

In many systems where TTL and MOS are used together, the master clock is a simple TTL single-phase clock. The multiphase clock pulses required by memory circuits and so on are derived from and are subordinate to the TTL master clock. The current trend is to include subordinate clock circuitry on the memory chip when it is necessary. Special IC chips are available for subordinate clock functions when separate circuitry is required.

LOGIC GATE RING OSCILLATOR

There are a number of practical clock generators based on logic gates. Any odd number of inverting gates will oscillate when connected in a ring, as shown in Figure 11-15. The principle of operation differs, however, from conventional oscillators that use positive feedback at some frequency (f_0) that produces a 180° phase shift. In the case of logic gate oscillators, the transition of a gate through its linear region requires a very short time. As a result, conventional phase shift explanations become difficult. A better viewpoint is to consider the gates as ideal switches with built-in delays. The oscillator then can be viewed as logic 1 following itself around the ring. Each time the 1 arrives and leaves at the gate connected to the output, an output pulse appears.

Frequency is a function of the total delay time. Additional elements

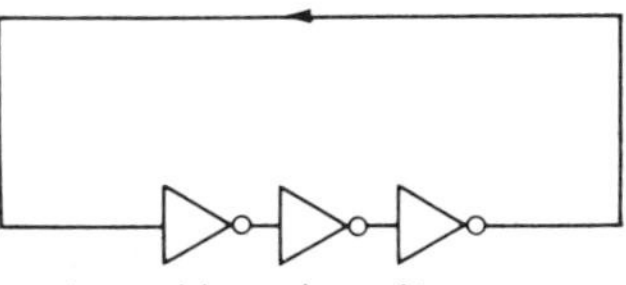

Any odd number of inverters

Figure 11-15 The Basic Logic Gate Ring Oscillator

such as RC time constant circuits can be used to increase the delay and lower the frequency. The maximum frequency is determined by the intrinsic gate delays. Figure 11-16 shows a practical version of a logic gate ring oscillator using TTL 7403 open collector NAND gates. The circuit is moderately stable and the frequency is controlled by the values of R and C. Figure 11-17 shows a C-MOS version using 4069 C-MOS inverters. The equation for the approximate frequency of the C-MOS circuit is:

$$f \approx \frac{1}{2C(0.405R_{eq} + 0.693R_1)}$$

where
$$R_{eq} = \frac{R_1R_2}{R_1 + R_2}$$

C-MOS CRYSTAL OSCILLATOR

A logic gate oscillator with crystal control can easily be constructed from a C-MOS gate. TTL gates are not as practical for use in crystal-

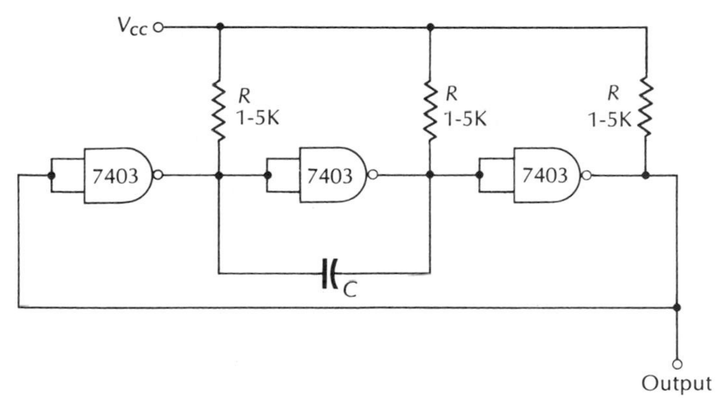

Figure 11-16 The Practical Ring Oscillator

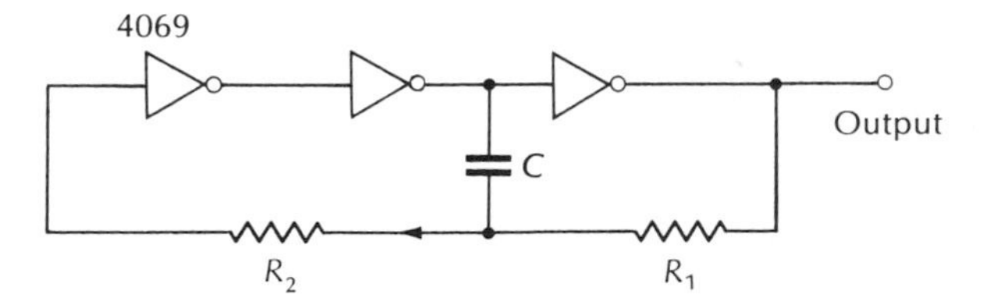

Figure 11-17 The C-MOS Ring Oscillator

controlled logic gate oscillators because their low input impedance loads the crystal, lowers the Q, and impairs stability. The high input impedance of C-MOS makes it ideal for crystal-controlled (or LC) oscillators. Figure 11-18 shows a simple but practical C-MOS crystal-controlled logic oscillator.

11-4 The Integrated Circuit One-Shot

The 74121 in Figure 11-19 is a complete monostable multivibrator except for an external timing resistor and capacitor. It is a highly stable, temperature-compensated circuit. There are three gated inputs: A_1, A_2, and B. The input gate equation for triggering is $T = (\bar{A}_1 + \bar{A}_2)B$, where B is a Schmitt trigger gate input.

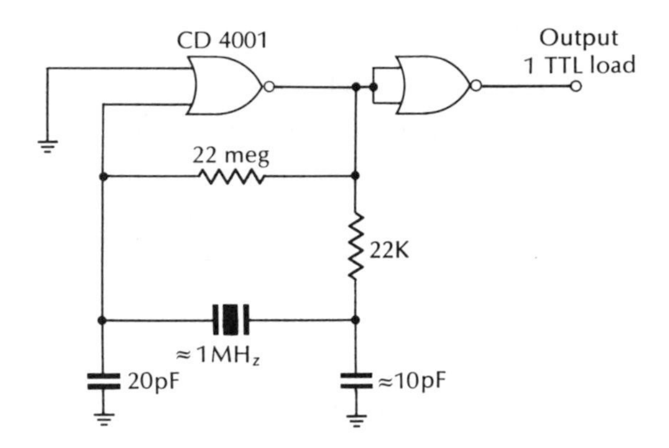

Figure 11-18 The C-MOS Crystal Oscillator

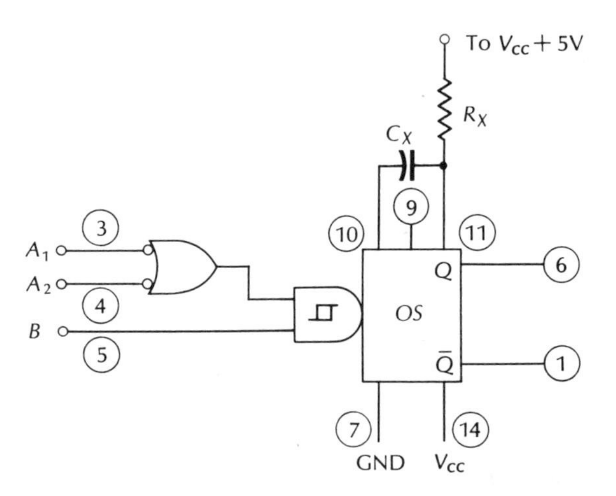

Figure 11-19 The 74121 One-Shot (OS) Monostable

This allows the B input to respond to a slowly going positive input voltage, for example a low-frequency sine-wave input. The B input will respond to voltage changes as slow as 1 volt per second. (The A inputs require rates of change in voltage greater than 1V/μs.) The B input fires at between 1.5 and 2 volts. The hysteresis voltage is about 0.8V. The pulse out of the one-shot is initiated by the Schmitt trigger, but the pulse width is determined by the one-shot time constant. The B input is enabled when either A_1 or A_2 is low. Possible stable output pulse widths range from 40 ns to 40 seconds. The minimum delay of $\approx$ 40 ns uses no external capacitance and only the internal timing resistor. To use the internal timing resistor, pin 11 must be connected to V_{cc}. Figure 11-20 shows graphs for selecting external timing components.

Retriggerable one-shots are also available. The 74122 is an example. Figure 11-21 shows how 74121 one-shots can be used to delay and shorten a pulse.

11-5 Digital-to-Analog and Analog-to-Digital Conversion

Most of the quantities in the real world that we wish to measure or control electronically are analog in nature. As the day warms up, for example, the temperature does not stay at 70 degrees for a period and then suddenly jump to 71 degrees; it rises gradually. Between 70 and 71 degrees there is an infinite number of temperatures.

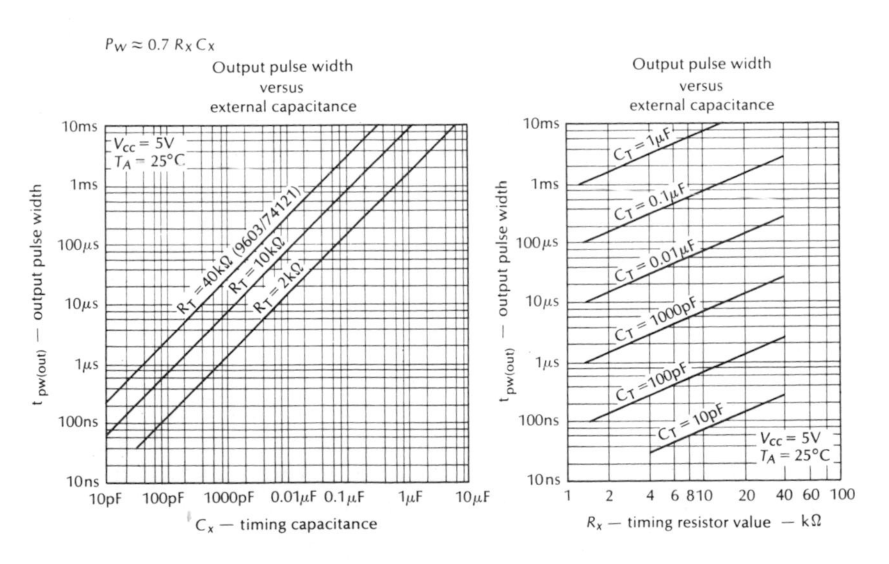

Figure 11-20 External Timing Component Graphs for the 74121

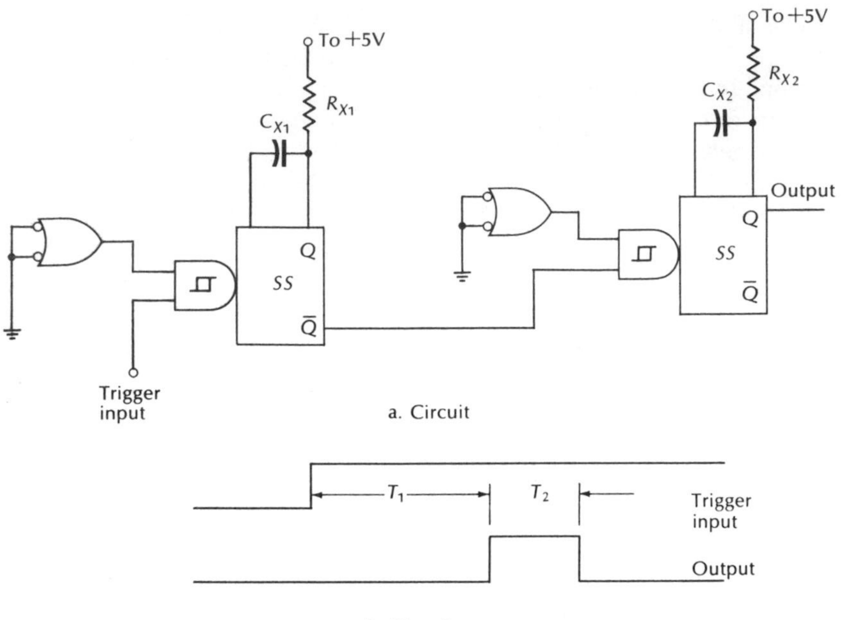

Figure 11-21 Delaying and Shortening a Pulse with 74121 One-Shots

When we measure a physical quantity, such as temperature, we use a transducer that converts temperature into a voltage analog. The output voltage of the temperature transducer is proportional to the temperature being measured, but the output voltage is still an analog quantity. If the transducer produces an output voltage of one volt at 70 degrees and two volts at 71 degrees, there is still an infinite number of voltages between one and two volts.

When we wish to convert an analog voltage into a digital representation, we convert the range of output voltages of the transducer into discrete steps. In the process of converting from analog to digital we lose everything between the steps. In our previous example the transducer produces an output voltage of one volt at 70 degrees and two volts at 71 degrees or one volt per degree. If our analog-to-digital converter converts in one-volt steps it can display temperatures of 70, 71, 72 degrees but cannot resolve the difference between, say, 70.2 degrees and 70.3 degrees. Both 70.2 degrees and 70.3 degrees would be displayed as 70 degrees. The step size is called the *resolution*. The resolution here is one volt and, in terms of temperature, one degree.

In this example, if we need to be able to distinguish between 70.2 and 70.3 degrees we must reduce the step size of the analog-to-digital converter to tenths-of-volts. In no case will the ADC (analog-to-digital converter) be capable of infinite resolution. In every case we must select a resolution appropriate to the task at hand, keeping in mind that the cost of the converter tends to increase with the resolution.

The other side of the coin is converting data in digital form to data in analog form. Digital control circuits are far cheaper and have a far broader range of capabilities than currently available analog hardware. Many devices, such as temperature-regulated ovens and speed-controlled motors, require varying analog voltages to alter temperature or speed. If these devices are to be controlled by digital circuitry, digital output data must be converted into analog voltages. The device used to make the conversion is called a *digital-to-analog converter* (*DAC*).

In many industrial process control systems both ADCs and DACs are required. For example, in a temperature-controlled oven the temperature must be measured, compared to some reference, and corrected when it deviates from the desired value. Figure 11-22 shows a block diagram of such a closed-loop system.

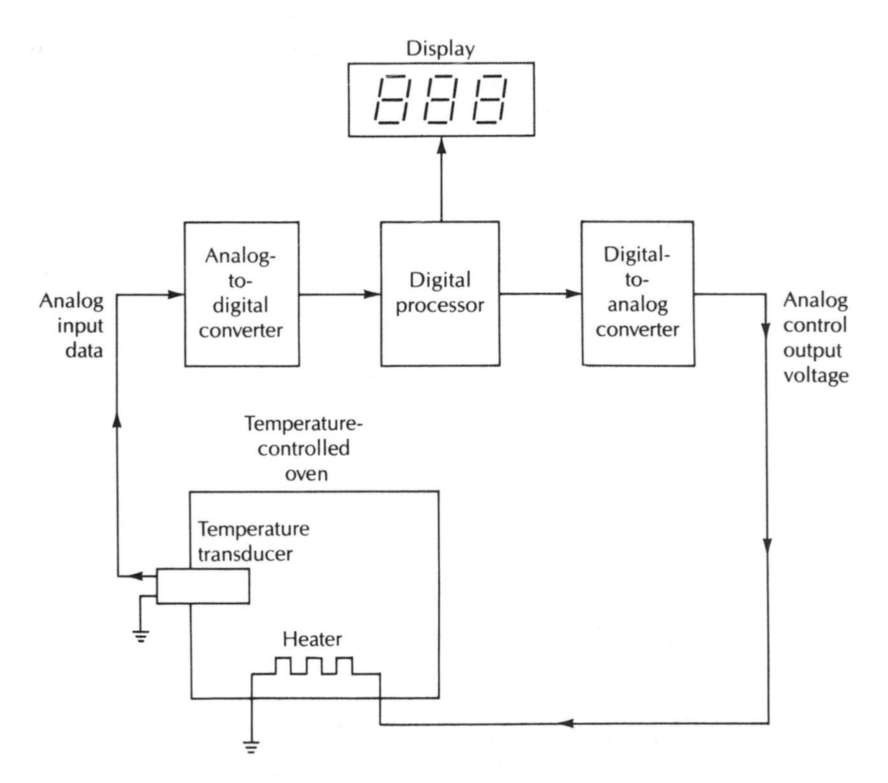

Figure 11-22 Closed-Loop Digital Control System

11-6 Digital-to-Analog Conversion Techniques

Digital-to-analog conversion hardware is generally simpler than analog-to-digital hardware. In addition, many analog-to-digital converters include a digital-to-analog converter as part of the converter system. For these reasons we will look at digital-to-analog conversion techniques first.

ACCURACY AND RESOLUTION

Two important parameters of digital-to-analog converters are accuracy and resolution. *Accuracy* is a measure of how close the actual analog output voltage is to the designed (or computed) output voltage. It is dependent largely on the precision of ladder resistors, precision power supply, and other components that have an inherent tolerance. *Resolution* is a measure of the smallest increment that can be resolved and is totally dependent on the number of bits in the input to the converter.

THE SUMMING AMPLIFIER DAC

Figure 11-23 shows a summing amplifier digital-to-analog converter. The gain for each input is weighted by using appropriate values of input resistors. The feedback resistor (R_f) value is selected to provide the desired output voltage steps. The following equations define the weighting of each bit when its input is at 5 volts.

1. The general equation: $E_0 = -E_{\text{in}}\ (R_f/R_{in})$
2. Input D: $E_0 = 5(2K/10K) = -1$ volt
3. Input C: $E_0 = 5(2K/5K) = -2$ volts
4. Input B: $E_0 = 5(2K/2.5K) = -4$ volts
5. Input A: $E_0 = 5(2K/1.25K) = -8$ volts

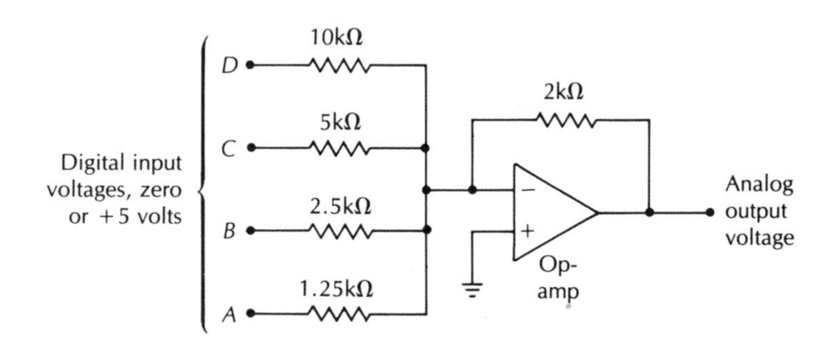

Figure 11-23 Digital-to-Analog Converter Using an Op-Amp Summing Amplifier

Table 11-1 Input and Output Conditions for the DAC

Input Code				Output Volts	Summed Output Voltage
MSB			LSB		
A	*B*	*C*	*D*	*A* *B* *C* *D*	
0	0	0	0	0 + 0 + 0 + 0	0
0	0	0	1	0 + 0 + 0 + 1	1
0	0	1	0	0 + 0 + 2 + 0	2
0	0	1	1	0 + 0 + 2 + 1	3
0	1	0	0	0 + 4 + 0 + 0	4
0	1	0	1	0 + 4 + 0 + 1	5
0	1	1	0	0 + 4 + 2 + 0	6
0	1	1	1	0 + 4 + 2 + 1	7
1	0	0	0	8 + 0 + 0 + 0	8
1	0	0	1	8 + 0 + 0 + 1	9
1	0	1	0	8 + 0 + 2 + 0	10
1	0	1	1	8 + 0 + 2 + 1	11
1	1	0	0	8 + 4 + 0 + 0	12
1	1	0	1	8 + 4 + 0 + 1	13
1	1	1	0	8 + 4 + 2 + 0	14
1	1	1	1	8 + 4 + 2 + 1	15

Note: Negative signs omitted for ease of understanding.

Table 11-1 summarizes the operation of the DAC for the 16 possible binary input conditions for four bits. The drawing in Figure 11-24 shows a counter connected to a summing amplifier and the digital wave-form output as the counter goes through its count sequence. The outputs of a counter or register are fed to the inputs of level amplifiers, and the outputs of the level amplifiers are fed into an amplifier using weighted input resistance values. The level amplifiers are required because typical TTL (or MOS) output levels are not precise or predictable enough. The approach in Figure 11-25 is essentially the same except that the summing network has been transformed into a ladder form. This resistive arrangement, often called the *R*-2*R* ladder, has a number of advantages over the summing network shown in Figure 11-24, particularly when a large number of binary outputs are involved. An examination of the resistance values reveals that each resistor value decreases by a power of 2 as the number of binary outputs increases. The result is a different loading value on each level amplifier, eventually resulting in an unacceptable current in summing resistors for higher order binary outputs.

The kind of ladder network shown in Figure 11-25 has long been popular in constant-input/constant-output audio and RF attenuators.

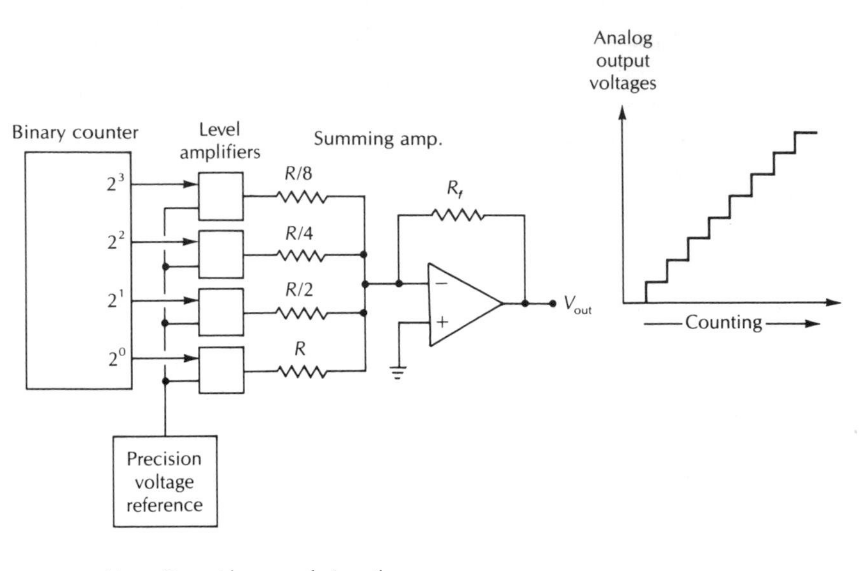

Note: To avoid any confusion, the fact that the Op-amp is an inverting amplifier has not been indicated.

Figure 11-24 Digital-to-Analog Converter with Level Amplifiers

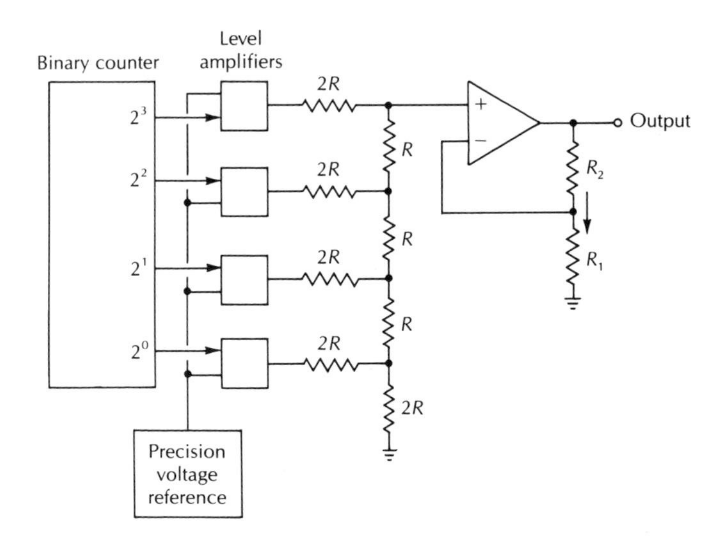

Figure 11-25 Digital-to-Analog Converter with a Ladder-Type Summing Network

One of its important characteristics is that the equivalent resistance as seen from any node is R (using R-$2R$ ratios) regardless of the state (0 or 5 V) of individual inputs to the ladder. The assumption is made that the output resistance of the level amplifiers closely approaches zero ohms at both high and low levels. This requirement dictates a low output impedance for the precision power supply that serves the level amplifiers.

The analog output voltage for each binary input can be determined as follows (assume voltage gain of unity for the amplifier):

$E_0 =$

MSB	V/2
2nd MSB	V/4
3rd MSB	V/8
4th MSB	V/16
.	.
.	.
.	.
nth MSB	$V/2^n$

where V is the output voltage of all level amplifiers. This voltage can be either zero volts or V, (some arbitrary specified value). The total output voltage from the ladder is the sum of the output voltages generated by all individual weighted-bit outputs.

Example

$$\text{MSB} = \frac{10}{2} = 5 \text{ V}$$

$$\text{2nd MSB} = \frac{0}{4} = 0 \text{ V}$$

$$\text{3rd MSB} = \frac{10}{8} = 1.25 \text{ V}$$

$$\text{LSB} = \frac{0}{16} = 0 \text{ V}$$

Sum 6.25 V analog output voltage

There are other techniques used in integrated circuit converters, but most of them are variations of the circuits we have discussed. The most common variation uses constant current sources in conjunction with a current-summing amplifier instead of a voltage-summing amplifier. Another rarely used technique uses a voltage-controlled oscillator whose output frequency is proportional to an applied analog input voltage.

11-7 Analog-to-Digital Converters

There are two broad classifications of analog-to-digital converters: simultaneous (or parallel) and sequential (or serial) converters. The simultaneous (parallel) converter is straightforward, simple in concept, and the faster of the two types. Unfortunately, it also uses more hardware and is more expensive than the sequential (serial) converter.

The parallel converter requires $2^n - 1$ comparators to produce an n-bit digital output. It also requires a precision voltage divider and encoding logic. Because each comparator is a fairly complex circuit in itself, it is an expensive approach.

Sequential (serial) analog-to-digital converters are all based on a common principle, but there are several circuit variations intended to improve the conversion speed and precision.

11-8 The Simultaneous (Parallel) Analog-to-Digital Converter

The basic circuit for the parallel ADC is shown in Figure 11-26. The precision voltage divider divides the precision reference voltage into ten equal parts. The reference voltage determines the step size (resolution). A reference voltage of one volt would yield a resolution of one-tenth of a volt; a 0.1 volt reference voltage would yield a resolution of 0.01 volt, and so on.

In Figure 11-26, the analog input voltage is applied to a voltage-follower buffer amplifier with unity voltage gain. The output of the buffer amplifier is fed to one input on each comparator.

If we assume a reference voltage of 10 volts, the lower comparator (1) will switch when the analog input voltage is 1 volt. Comparator (2) will switch when the analog input voltage is 2 volts, and so on. The outputs from the comparators is digital but essentially decimal, not binary. The encoding circuitry is necessary to convert the decimal into binary form. Encoders are covered in Section 4-1.

11-9 Sequential Analog-to-Digital Converters

Figure 11-27 is representative of most modern sequential analog-to-digital converters. Variations and improvements on the basic technique are used to increase the conversion speed or to improve the precision of the converter.

HOW IT WORKS

Please refer to Figure 11-27 for the following explanation.

1. A signal at the *start conversion* input resets the flip-flop to Q = high. When Q goes high, the AND gate (AG) allows pulses from the free-running clock to start advancing the counter.

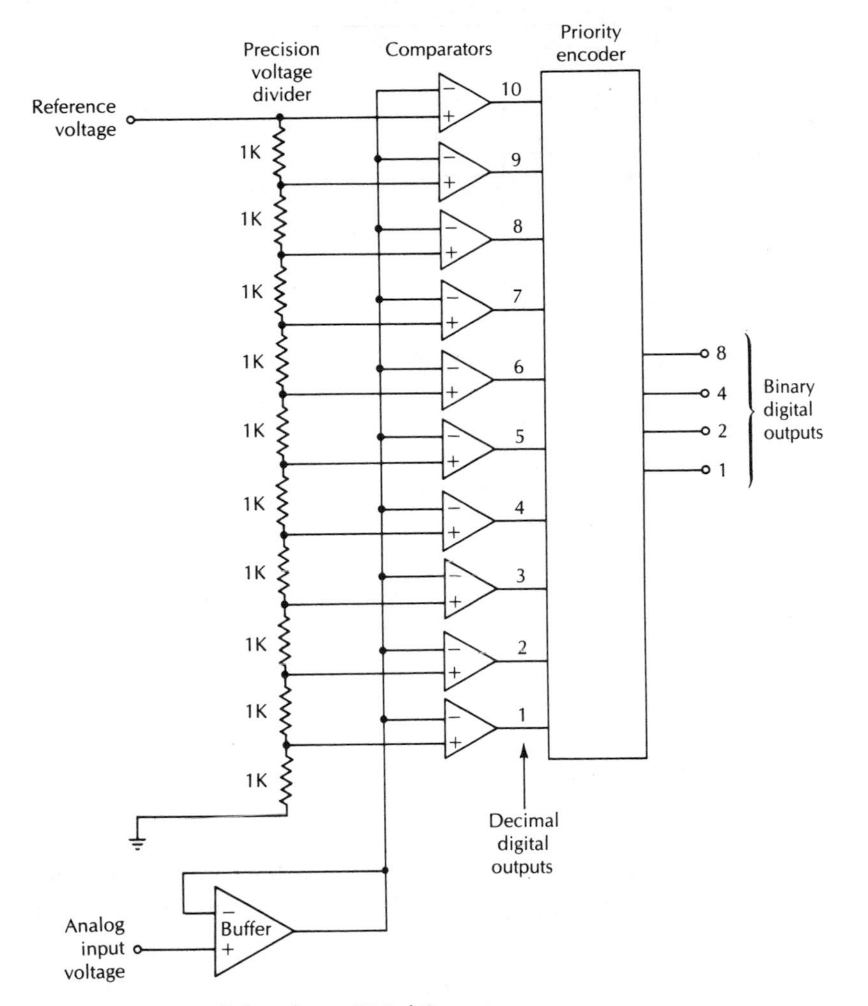

Figure 11-26 Parallel Analog-to-Digital Converter

2. The output of the counter is routed to a digital display or other appropriate digital circuitry and to a digital-to-analog converter.

3. The digital-to-analog converter produces a rising staircase voltage as shown in the waveform inset in the figure. This stairstep voltage is fed to one input of the comparator.

4. The analog input voltage is applied to one input of the comparator. The output voltage of the DAC (feedback signal) is applied to the other input of the comparator. The counter continues to count and the DAC continues to produce a rising stairstep voltage until the DAC output voltage equals the analog

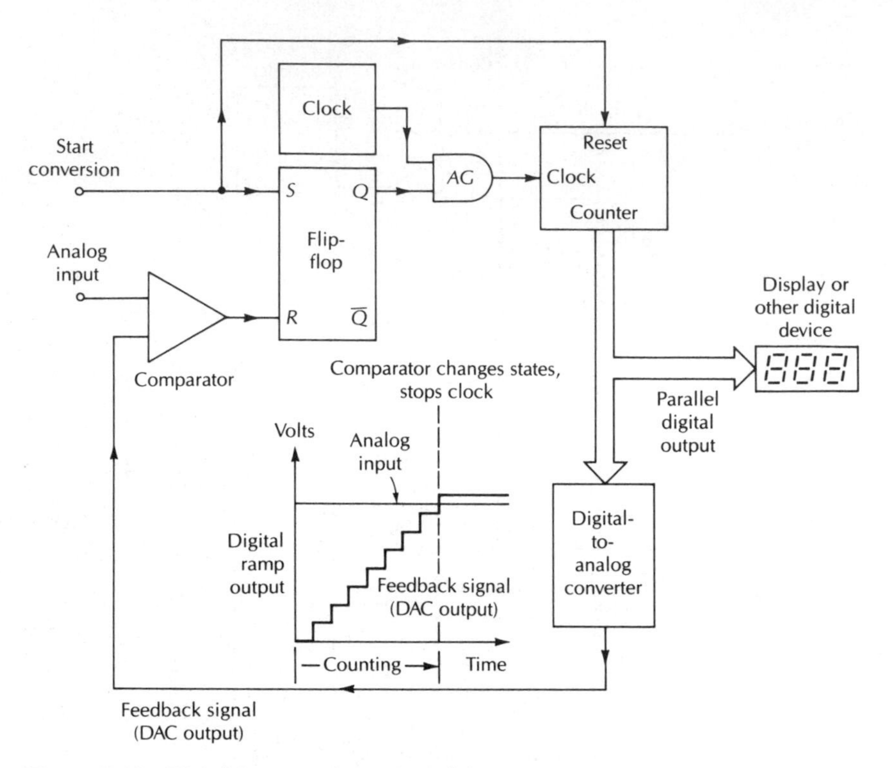

Figure 11-27 Digital Ramp Analog-to-Digital Converter

input voltage. When the DAC feedback voltage equals the analog input voltage, the comparator changes state and resets the flip-flop.

5. When the flip-flop resets, Q goes low and clock pulses are no longer delivered to the counter. The counting stops and the count held in the counter is the digital equivalent of the analog input voltage.

11-10 The Successive Approximation Analog-to-Digital Converter

The simple ramp-type converter suffers from two disadvantages, a long conversion time and a conversion time that increases as the analog input voltage increases.

The successive approximation converter improves the conversion speed by a factor of 100 or more and makes the conversion time independent of the absolute analog input voltage. The conversion is carried out by entering a 1 into a flip-flop representing the most significant bit

(MSB) and comparing the entire register contents (one FF for each bit) to the analog input voltage. If the entire register contents represent a value larger than the analog input voltage, the FF is reset. If the register contents represent a value less than the analog voltage, the flip-flop holds the 1. A 1 is then tried in the next most significant bit. If the register contents represent a value less than the analog voltage, the next most significant bit is held at 1. If the register equivalent is greater than the analog voltage, the flip-flop is reset to zero. This process continues down to the last significant bit.

When all bits have been tested in this fashion, the register contains the binary digital equivalent of the analog input voltage. The procedure is essentially the same as a person might use to convert a decimal value into its binary equivalent.

Figure 11-28 shows a simplified block diagram of a 4-bit successive approximation analog-to-digital converter. The circuit operates as follows:

1. A *start conversion* pulse sets F-F *F* to Q = 1, presets F-F *E* to Q = 1, resets all flip-flops in the 4-bit shift register and the 4-bit register to zero.

2. When Q on F-F *F* goes high, AND gate 9 delivers clock pulses to the ring counter consisting of the 4-bit shift register and flip-flop *E*.

3. During the first clock pulse the output of the comparator will be *high* if the analog voltage is not beyond the range of the converter. The *high* on QD and the *high* output of the comparator will set flip-flop *D* to Q = 1.

4. The digital-to-analog converter will produce an output corresponding to the total value stored in the register.

   If the output voltage of the DAC exceeds the analog input voltage, the comparator will switch to *low* and reset flip-flop *D* to Q = 0. If the DAC output voltage is less than the analog input voltage, the comparator output voltage will remain *high* and flip-flop *D* will remain in the Q = 1 condition.

5. The next clock pulse advances the ring counter to QC (f-f *C*) where the same sequence that occurred at flip-flop *D* is repeated.

6. The ring counter advances through the remaining bits in the same fashion.

7. On the fifth clock pulse, Q on flip-flop *E* goes low, resetting flip-flop *F* and stopping the clock pulses. The register now contains the binary digital equivalent of the analog input voltage.

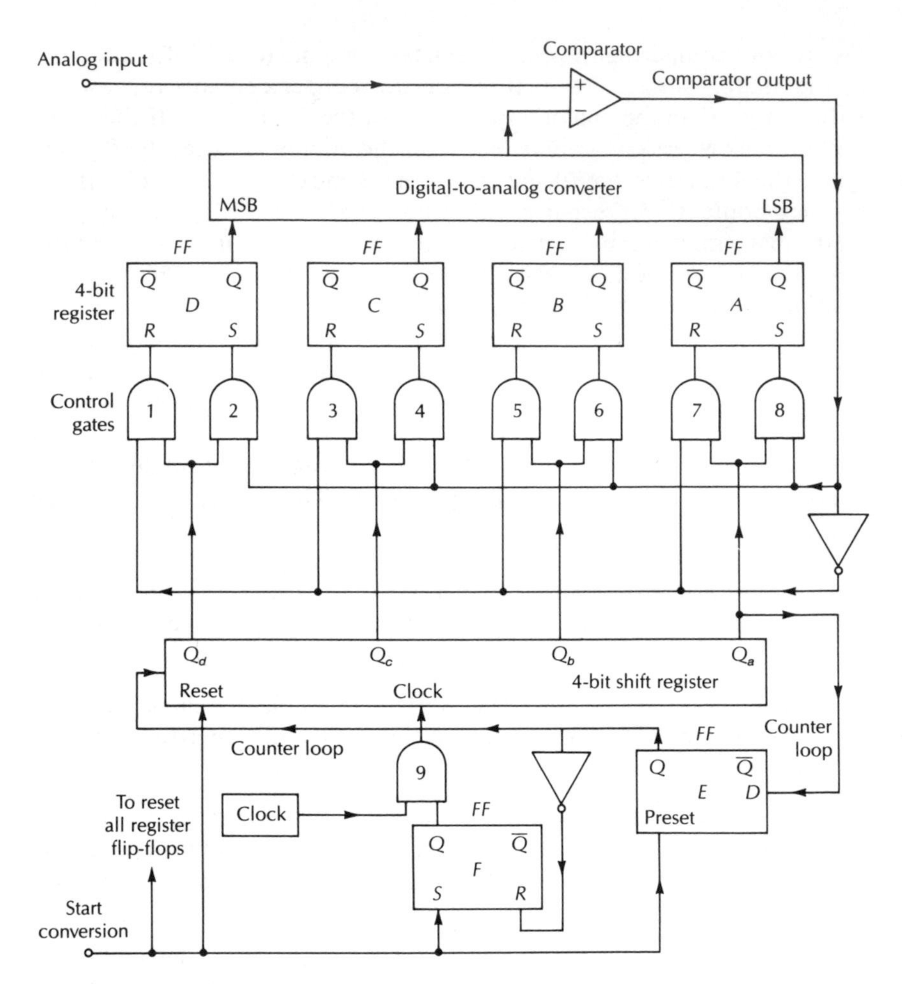

Figure 11-28 Successive Approximation Analog-to-Digital Converter

Table 11-2 illustrates the register operation for converting an 11-volt analog voltage into its binary digital equivalent.

11-11 Time-Sharing Multiplexing

Most analog systems are fairly slow compared to digital processing speeds. For example, it takes time for an oven to change temperature, for a motor to change speeds, or for a liquid or gas flow rate to change. Because of this difference in speed between real-world systems and digital circuits, it is common practice to use a single digital system to control several real-world systems. Multiple control of real-world systems is accomplished by time-sharing multiplexing.

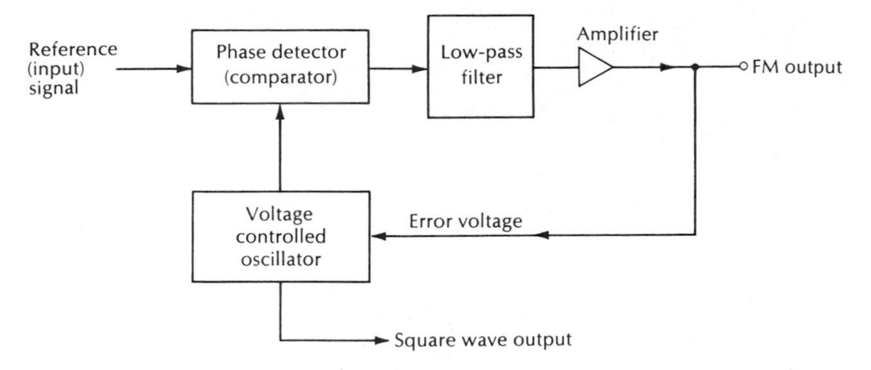

Figure 11-34 Phase-Locked Loop Block Diagram

get the sum and different frequencies 201 kHz and 1 kHz. The low pass filter removes the sum frequency. A DC voltage proportional to the frequency difference is fed back to the VCO, changing its frequency to more nearly that of the input signal. The correction continues until the two signals are phase locked. The error-signal output serves as a demodulated FM output. The circuit is designed to make the error voltage a linear function of the deviation of the input signal. The *capture range* is the range over which the loop can acquire lock. The lock range is generally wider than the capture range.

For use as a frequency multiplier, the VCO can be set to run at a harmonic of the input frequency and it will still lock on the input signal. For larger multiplication factors there is a more satisfactory approach. The VCO output square wave is TTL compatible. Figure 11-35 shows the block diagram of a commercial PLL.

PLL FREQUENCY MULTIPLICATION

One important application for the PLL is digital frequency multiplication. To accomplish frequency multiplication, the loop is opened and a digital counter (divider) is inserted in the loop as shown in Figure 11-36. The VCO is set to run at the frequency that it is desired to multiply to. For example, to multiply a 10 kHz signal to 100 kHz, a divide-by-10 counter is placed in the loop and the VCO is set (by the external timing R and C) to run at 100 kHz. The digital counter divides the 100 kHz VCO frequency down to 10 kHz so that VCO frequency (phase) can be compared with the input signal frequency for each cycle. This results in a precision not obtainable with harmonic-locking techniques; it requires no tuned circuits and can track the input frequency over the range of about an octave.

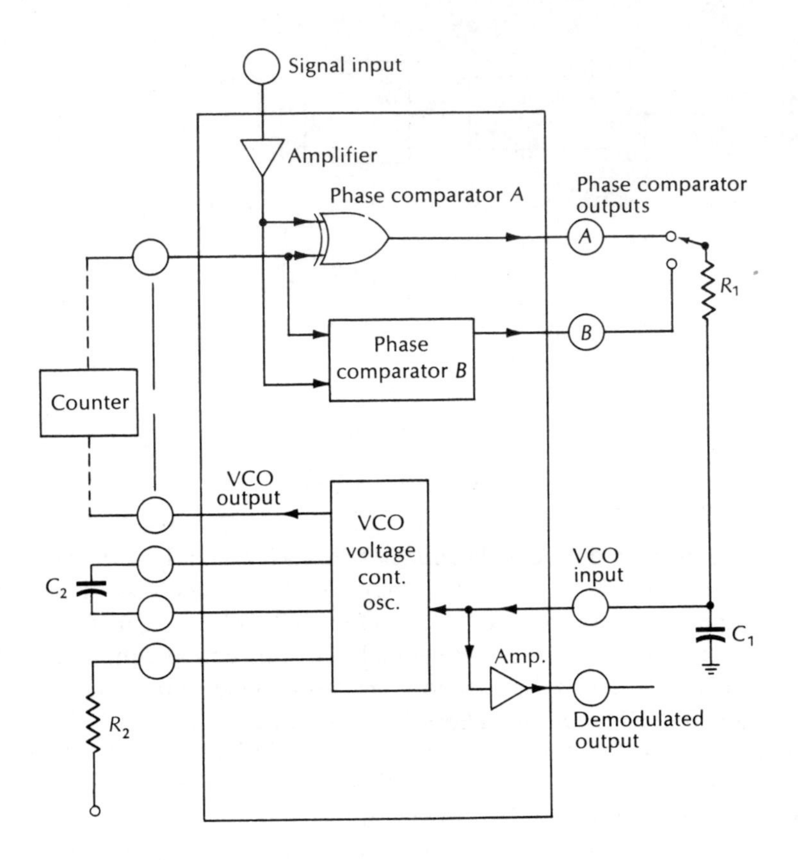

Note: C_2 and R_2 are VCO frequency determining components; R_1 and C_1 form a low-pass filter network.

Figure 11-35 Typical Commercial Phase-Locked Loop

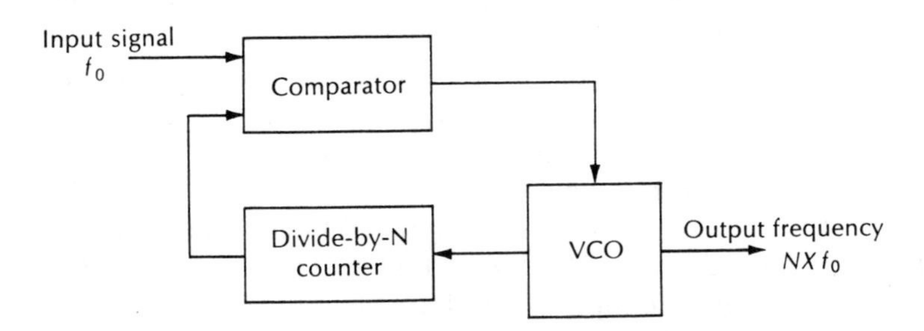

Figure 11-36 Frequency Multiplication with Phase-Locked Loop

FSK (FREQUENCY SHIFT KEYING)

FSK is a form of digital frequency modulation in which a carrier is shifted to one frequency for a logical zero and to another for a logical one. A phase-locked loop can be used as an FSK demodulator. The error voltage output will have distinctly different levels for each of the two frequencies. The output levels can be made directly compatible with TTL.

FSK is frequently used in modems (modulator/demodulators) for interfacing telephone lines with digital equipment. Since phone lines are limited to the range of speech frequencies, the two FSK frequencies in this case would be audio tones.

Another application for FSK is that of interfacing digital equipment to inexpensive audio-type cassette or reel-to-reel tape recorders. These recorders are coming into common use in microprocessor systems as a storage substitute for prohibitively expensive discs, drums, or computer-grade tape machines.

1-15 Bidirectional Bus Drivers

One of the recent innovations in computer architecture is the bidirectional bus organization system, which has bus organization arithmetic logic units, registers, and so on all connected to a common bus. Data is transferred from any device to any other device on the bus by opening and closing appropriate gates.

The device that interfaces units and the bus is called a *bus interface driver*. Usually bidirectional, it provides directional control *transmit*-to-the-bus or *receive*-from-the-bus and frequently has a high impedance (open-circuit) state. Open collector devices are sometimes used for this purpose, but the tristate output circuit is more nearly ideal. This output permits 100 or more devices on a single bus, with isolation in the off state that is far superior to that of open collector devices.

Bus (or line) drivers and receivers are available as transmitters only or receivers only, as well as in combination packages.

These interface units are used for a number of machine communication purposes and, consequently, the terminology varies somewhat with different applications. The following are pairs of synonyms that may be interchanged:

a. Driver—transmitter
b. Transmitter/receiver—bidirectional bus driver
c. Receiver—sense amplifier
d. Bus—line

The input circuit is normally a differential comparator with switching sensitivities of from +5 to 50 mV. Receivers are normally driven in a differential mode when the line length exceeds a very few feet. For longer lines, twisted pairs are preferred. Noise pickup is more nearly the same for each line in a twisted pair, allowing for the cancellation of the common mode noise signal in the differential amplifier. Many of the later IC data bus transmitter/receivers have input impedances high enough to be used with MOS. These devices can be used to interface bus lines where MOS will have to drive TTL inputs. They are commonly used for interfacing MOS memories to TTL.

Figure 11-37 shows how bidirectional driver/receivers are connected on a bus. By varying the control levels at the driver/receivers, any device—*A*, *B*, or *C*—can have its output connected to the input of any other device on the bus. The output of any device can also drive the inputs to more than one device on the bus. Bus organization provides a simple interconnection system and avoids the limitations of most hard-wired interconnections in which some combinations must be left out because of the impracticality of interconnecting all outputs to all other inputs. The wiring in the bus system can be altered simply by changing the control input levels. What might be very complex wiring can be reduced to a single bus (and few wires) with significant advantages in a high-frequency system.

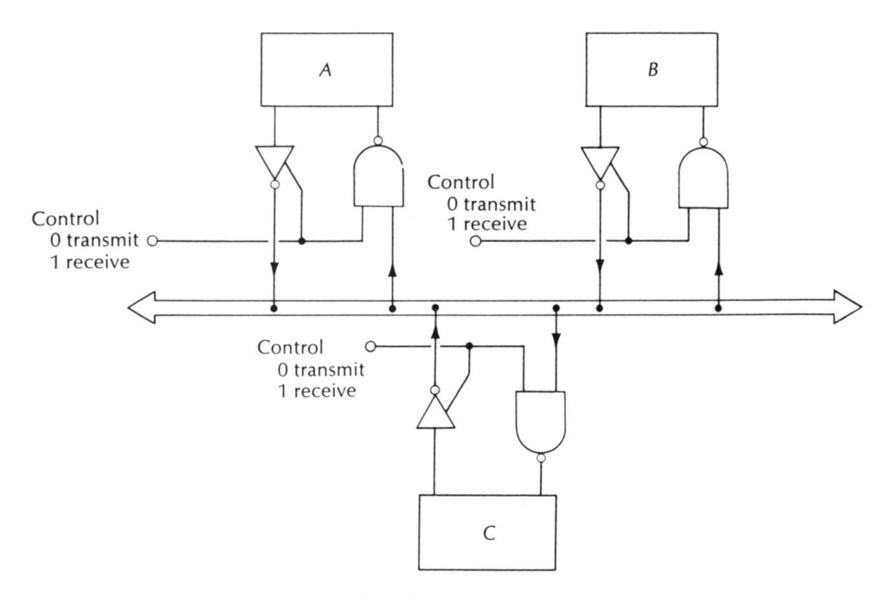

Figure 11-37 Bus Interfacing with Bidirectional Bus Driver/Receiver

When longer lines are used, a twisted pair is the most common transmission line, although coaxial lines can also be used. The line is often terminated in the characteristic impedance of the line, typically about 100 ohms for a 100-foot twisted pair.

The terminating resistor may have a capacitor in series with it to avoid loading the system for DC but still maintaining the proper termination impedance for digital pulses. Terminating the line serves the same purpose as it does in any RF transmission line—to minimize reflections. In digital circuits, reflected pulses could be accepted by some device on the bus as a regular, intended, pulse.

Figure 11-38 shows a *party line* system with terminations. The line drivers must supply the power dissipated by the terminating resistors in addition to the receiver input currents. Most modern drivers can drive as many as twenty receiver inputs and the terminating resistors. Several bus drivers have built-in D latches for temporarily holding data on a bus.

11-16 Data Communication Line Drivers and Receivers

Line drivers and receivers are required to transmit and receive data on a line between data communication equipment and terminals. The driver converts TTL levels to standard communications levels (usually EIA or MIL-STD). The receiver accepts data at line transmission levels, separates them from line noise, and converts them into standard TTL levels. Hysteresis is normally provided to improve the noise immunity. Because of this feature, receivers are sometimes used in applications that call for a Schmitt trigger.

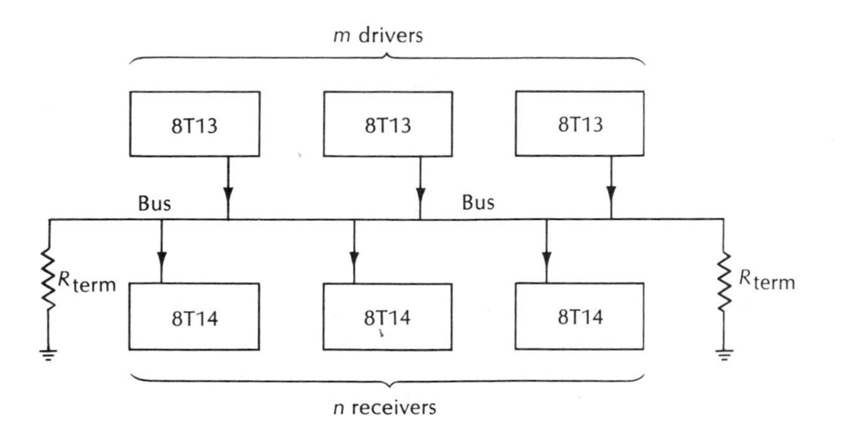

Figure 11-38 Drivers and Receivers on a Terminated Bus

Table 11-3 Partial Table of Data Transmission Standards

Specifications	MIL STD 188 B	EIA RS-232 B, C
Output Voltage: "0"	+6 ±1V	+5 to +15V
Output Voltage: "1"	−6 ±1V	−5 to −15V
Rise/Fall Time	±5% of pulse interval	±4% of pulse interval
Bit Rate	4 kHz typical	0 to 20 kHz

Table 11-3 compares MIL-STD and EIA communications standards. MIL standards provide for both single-ended and differential lines, but EIA standards apply only to a single-ended line. However, an additional wire that forms half of the twisted pair serves as a noise antenna to provide differential amplifier noise cancellation (see Figure 11-39).

Built-in hysteresis is required for many transmission lines because of inherently high noise levels. Figure 11-40 shows the hysteresis curves for MIL-STD, EIA standard, and EIA fail-safe modes. The EIA fail-safe operation enables the interface on a logical *zero*. In a bus-organized system, if the interface driver disable PC card is removed, it places the interface device at the high impedance state so that it does not affect other devices on the bus.

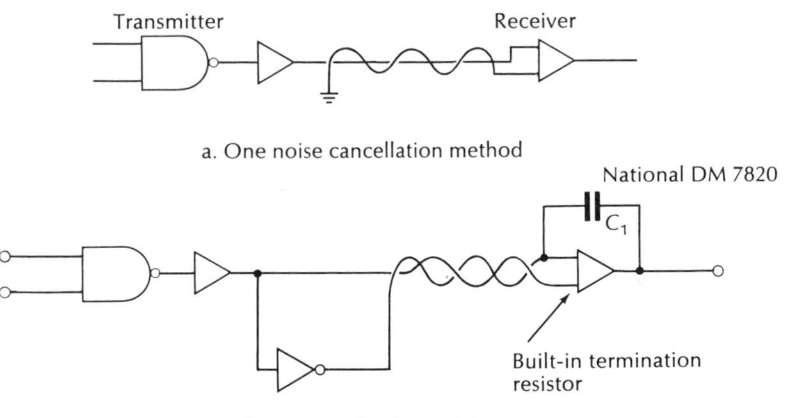

Figure 11-39 Noise Cancellation Methods

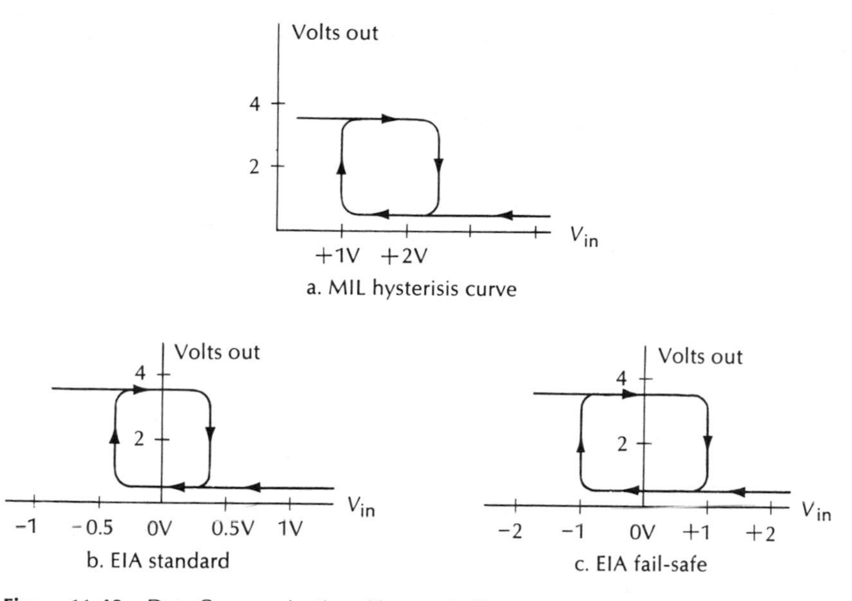

Figure 11-40 Data Communications Hysteresis Curves

1-17 Baud Rate

The rate at which data is transmitted over a line can be defined in bits per second, but the term *baud rate* is more often used. For a 50 percent duty cycle, the baud rate is twice the bit rate. Figure 11-41 shows a waveform that is not symmetrical (50 percent duty cycle). In Figure 11-41 T_2 shows the bit parameters and T_1 defines the baud parameters.

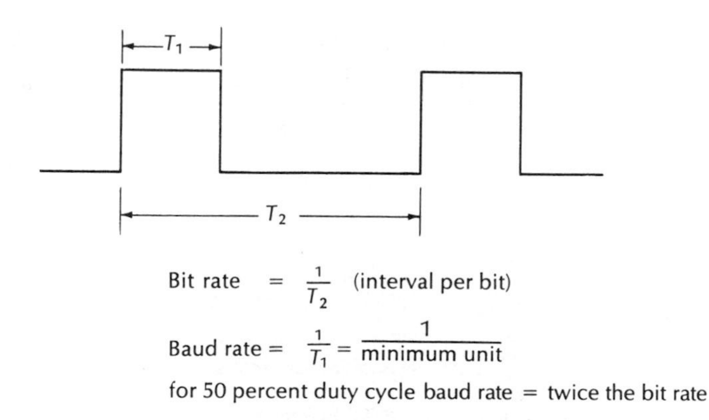

Figure 11-41 Baud Rate Defined

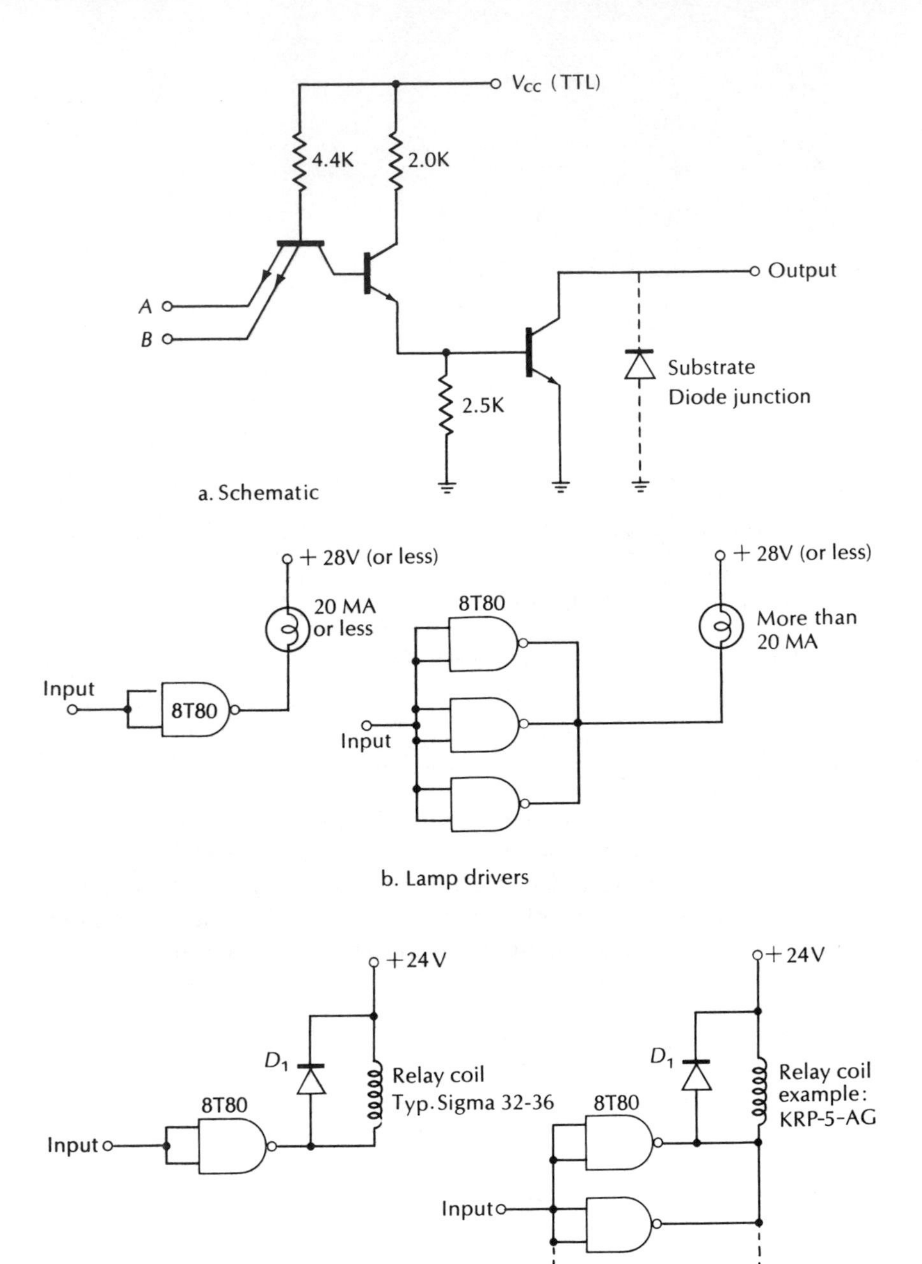

Figure 11-42 Typical Applications of the 8T80 Interface

$$\text{Bit rate} = \frac{1}{T_2} = \frac{1}{\text{interval per bit}}$$

$$\text{Baud rate} = \frac{1}{T_1} = \frac{1}{\text{minimum unit interval}}$$

It has become customary in some areas to equate baud rate and bit rate. The error in this assumption is not a problem as long as you are aware of how the terms are used in a given set of circumstances.

11-18 High-Level Interface Devices

It is sometimes necessary to interface TTL or MOS to devices that require higher voltage, current, or power levels than standard logic can drive. Typical of such devices are relays, lamps, and solenoids. The Signetics 8T18, 8T80, and 8T90 are intended for this kind of service. These devices typically convert the 5 V, 25 mW standard TTL level to a 28 V, 280 mW level. The 8T80, for example, is a Hex device with a two-input NAND gate input with an uncommitted (open) collector. Figure 11-42 shows the schematic and some typical applications for this device.

11-19 Contact Conditioning Interface

Mechanical contacts almost invariably bounce when contact is made. Ordinarily the bounce and multiple make-and-break action that results is no problem. However, when mechanical contacts are used with high-speed logic, the bounce is seen by the gate as multiple pulses. A simple NAND or NOR latch is often used as a switch debouncer (or bounceless pushbutton).

The set (or reset) action is initiated by the first contact closure pulse. Because of the regenerative behavior of the circuit, the switching action once initiated continues even with loss of switch closure.

The only pulse that counts is the first one—the rest are ineffective. Some contact conditioning circuits are shown in Figure 11-43.

There are several commercial IC contact conditioning circuits. One is the DM7544, a quad switch debouncer with tristate outputs and an enable/disable input. The 7544 debouncer is bus compatible. The device can store four switch closures to be held until released by a strobe pulse.

11-20 MOS-to-TTL and TTL-to-MOS Level Translators

The National Semiconductor DM7802 and 7806 are examples of packaged IC MOS-to-TTL level converters. The devices are intended to

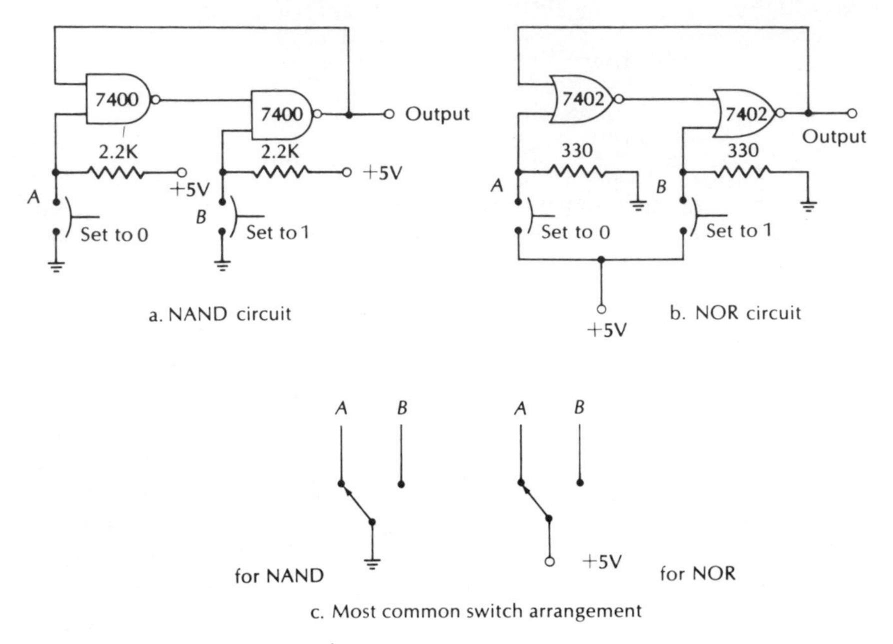

Figure 11-43 Contact Conditioning Circuits

interface MOS devices such as memory and shift-register circuits with TTL devices. They feature high drive capability, high input impedance, voltage level translation, tristate outputs, and built-in latch capability.

When TTL must drive MOS devices that are operating at logic voltage levels where a logic 1 is as much as +14 volts, interface devices such as the DM7810 and 7812 are available. The DM7810 is a quad two-input NAND-gate level translator. The 7812 is a Hex-inverter level translator. The output circuit is open-collector to provide a broad range of interfacing capabilities.

11-21 MOS to LED Interface Circuits

The DM75491 and 75492 are interface devices originally designed to interface MOS devices to 7-segment light-emitting diode (LED) displays. The 75491 consists of four Darlington pair transistors. Each transistor pair has both collector and emitter as IC pins. The 75492 consists of six Darlington pair transistors that all have their emitters returned to a common pin on the IC. The six collectors are available separately. The 75491 can source or sink 50 mA per driver, and the 75492 can sink up to 250 mA per driver.

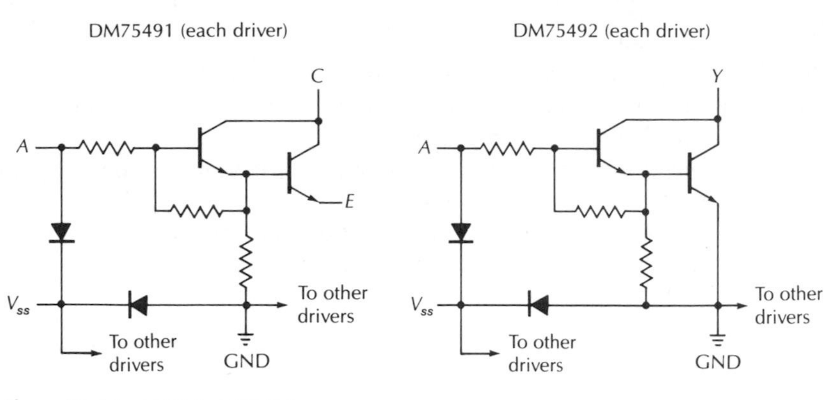

Figure 11-44 MOS to LED Drivers

The combination of high input impedance (low input current), high current gain, and high current output drive capability has made these devices a popular choice for a great many high-level interface circuits. Figure 11-44 shows the schematic diagram for the two devices.

SUMMARY

There are two specific classifications of interface circuits. The first involves logic devices that have different voltage or current levels but are still part of the logic system.

The second involves interfacing logic systems to outside-world devices that not only may have different current or voltage levels but also may use codes or operate at speeds that differ from those in the logic circuits.

A special case of interfacing logic to the outside world is that of converting analog data into digital data or digital data into an analog form.

There are a tremendous number of interface problems, but many of the most common ones can be solved by using off-the-shelf IC devices. This chapter merely surveys the common techniques. Nearly all of the circuits described are available as functional integrated circuits.

Problems

1. Explain the function of a transducer.
2. Explain the difference between analog and digital data.
3. Explain how the terms *step-size* and *resolution* are related.
4. Define the acronym *DAC*.

5. Define the acronym *ADAC*.
6. Define the terms *accuracy* and *resolution*.
7. Given the circuit in Figure 11-45, compute the voltage gain and give the approximate input impedance.
8. Given the circuit in Figure 11-46, compute the voltage gain and give the approximate input impedance.
9. Given the circuit in Figure 11-47, state the approximate voltage gain, input impedance, and the name of the circuit.
10. Identify the circuit in Figure 11-48.
11. Explain what the circuit in Figure 11-48 does.
12. Explain the purpose of a comparator.

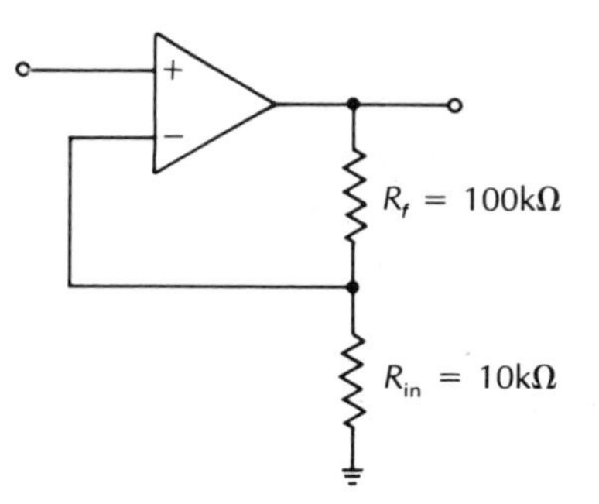

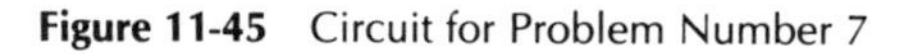

Figure 11-45 Circuit for Problem Number 7

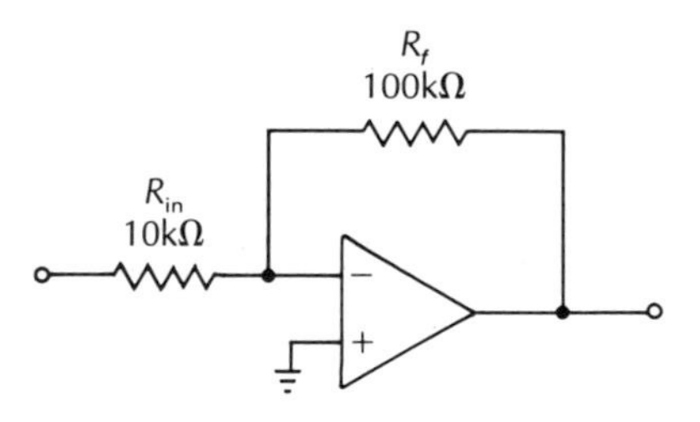

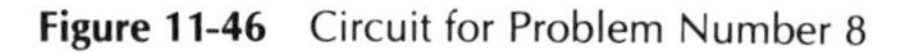

Figure 11-46 Circuit for Problem Number 8

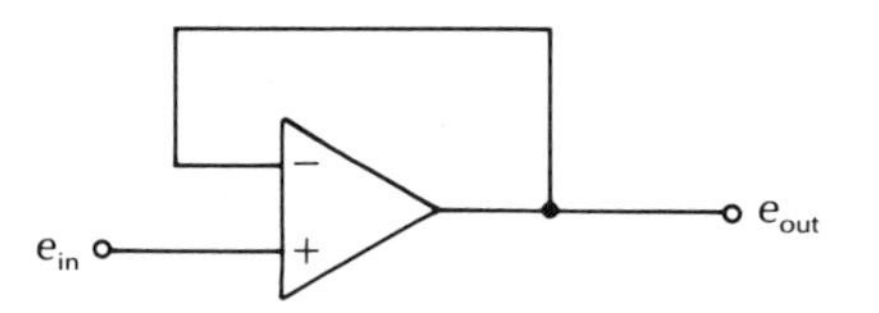

Figure 11-47 Circuit for Problem Number 9

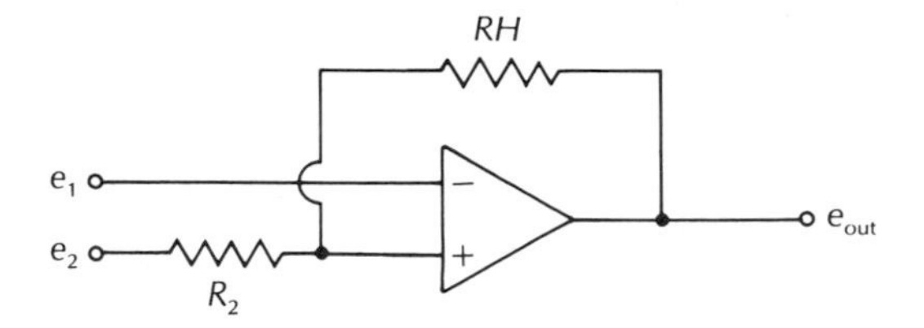

Figure 11-48 Circuit for Problems 10 and 11

13. Define the term *hysteresis* in terms of comparators.
14. Draw the block diagram of a 555 timer configured in a monostable mode. Show all external components.
15. Draw the block diagram of a 555 timer configured in an astable mode. Show all external components.
16. Go back to problem #14. Add component values to the monostable circuit to get a 10 millisecond monostable delay.
17. Define the term *monostable*.
18. Define the term *astable*.
19. What is the function of a digital clock?
20. What is a master clock?
21. What is a subordinate clock?
22. Why is the ladder network generally preferred over the simpler summing network in digital-to-analog converters?
23. If simultaneous (parallel) analog-to-digital converters are faster and just as accurate as sequential converters, why are sequential converters so often used?
24. Why is the successive approximation analog-to-digital converter generally preferred to the digital ramp converter? Give two reasons.
25. Given the successive approximation analog-to-digital converter in Figure 11-28, show how you would connect a 7-segment decoder and LED to display the digital equivalent of the analog input signal.
26. Make a table like Table 11-2 showing the register contents for each approximation as an analog input voltage of 13 volts is converted. Refer to Figure 11-28.
27. Special challenge: Given the block diagram of the analog-to-digital converter in Figure 11-49, explain how it works.
28. Why is time-sharing multiplexing often used in digital control systems that use DACs or ADACs?
29. What characteristic of many real-world physical systems makes time-sharing multiplexing possible?
30. What is a sample-and-hold amplifier?
31. What is the Gray code and where is it used? Explain.

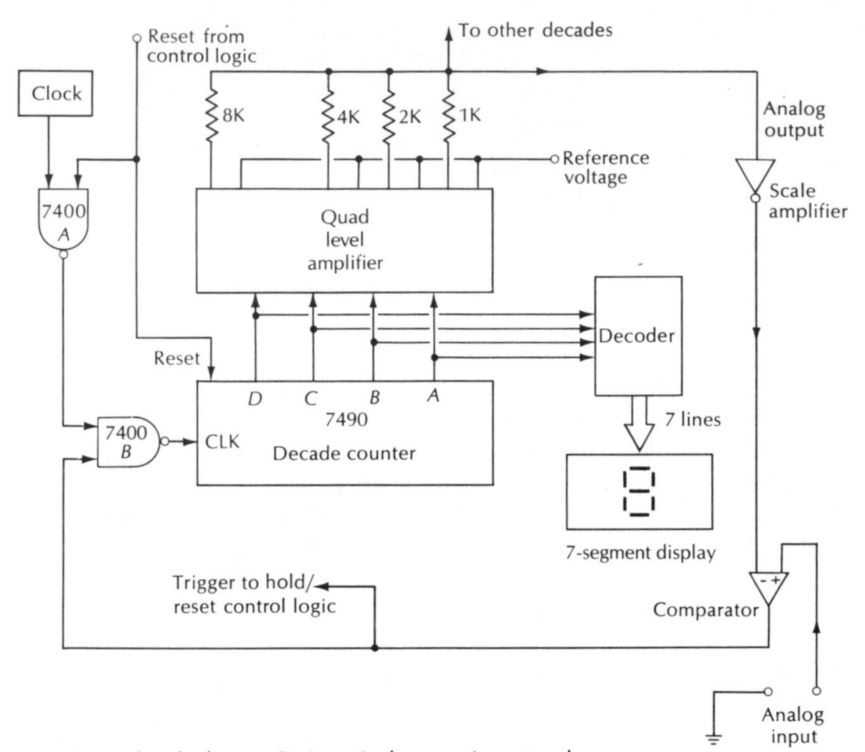

Note: Units decade shown. Resistors in the summing network are:
10K, 20K, 40K, 80K, for the 10's decade.

Figure 11-49 Digital Voltmeter Circuit

32. What is the function of an optical isolator?
33. Draw a block diagram of a phase-locked loop.
34. Explain how a phase-locked loop works. Use a block diagram.
35. Draw the block diagram of a phase-locked loop digital frequency multiplier.
36. Define frequency shift keying and describe one or more applications for it.
37. Describe the advantages of bidirectional driver/receivers in a bus-organized system.
38. What is the reason for terminating resistors in a data transmission line system?
39. What kind of line is most commonly used for distances of 100 feet or so?
40. Define baud rate.

41. Why is hysteresis used in data transmission line receivers?
42. Define the term *one-shot*.
43. How does non-retriggerable operation differ from retriggerable operation in a monostable?
44. Why are C-MOS devices preferable to TTL devices in crystal-controlled oscillator circuits?

Chapter 12

DIGITAL SYSTEMS

Learning Objectives. *Upon completing this chapter you should know:*

1. *The principles of multiplexing display devices.*
2. *The properties of a large-scale C-MOS counter circuit.*
3. *How to list the main building blocks of the microprocessor.*
4. *How to draw a simplified block diagram of a microcomputer.*
5. *How to explain the function of each register in the sample microprocessor.*
6. *How to describe the functions of a peripheral interface device.*
7. *That there are several levels of programming languages.*
8. *How to describe one-, two-, and three-byte instructions.*
9. *How to draw the block diagram of a simplified synchronous data communications system.*
10. *How to draw the block diagram of a simplified asynchronous data communications system.*

12-1 Multiplexing Seven-Segment Displays

Multiplexing is a common method used to minimize the number of packages and interconnecting lines in a digital system. In multiplexing operation a single decoder/driver is time-shared among the digits in the display. The digits are turned on one at a time in sequence at a 100 Hz or greater rate to insure against flicker.

Because segments are switched on and off rather than operated continuously, brightness suffers unless the pulse current is increased while a segment is on.

A typical multiplex scheme is illustrated in Figure 12-1a. In the simplified representation (Fig. 12-1a) the diodes are the LED segments, and mechanical switches are shown in place of electronics. All *a* segments are connected in parallel, all *b* segments are connected in parallel, and so on. The appropriate segment (anode) enable lines to form a given digit are activated. The scanner activates the first digit allowing appropriate segments to be activated for a time. It then moves on to the next digit, at which time a new combination of segments is selected. Because of the retention of the eye the digits appear to be continuously lit. Placing the current-limiting resistors in the segment enable lines instead of the common lines helps to avoid uneven current distribution among segments with the resulting uneven brightness.

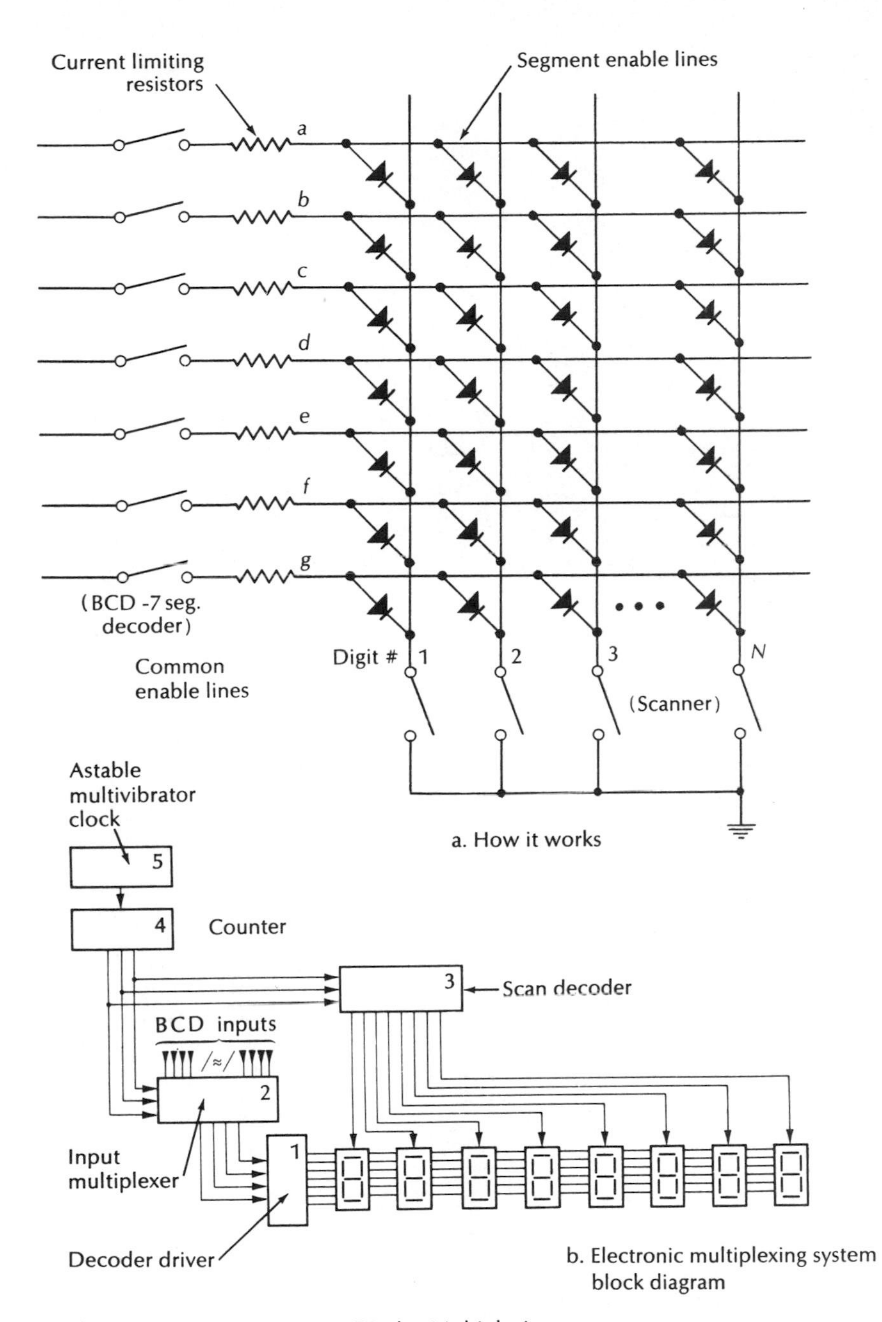

Figure 12-1 Seven-Segment Display Multiplexing

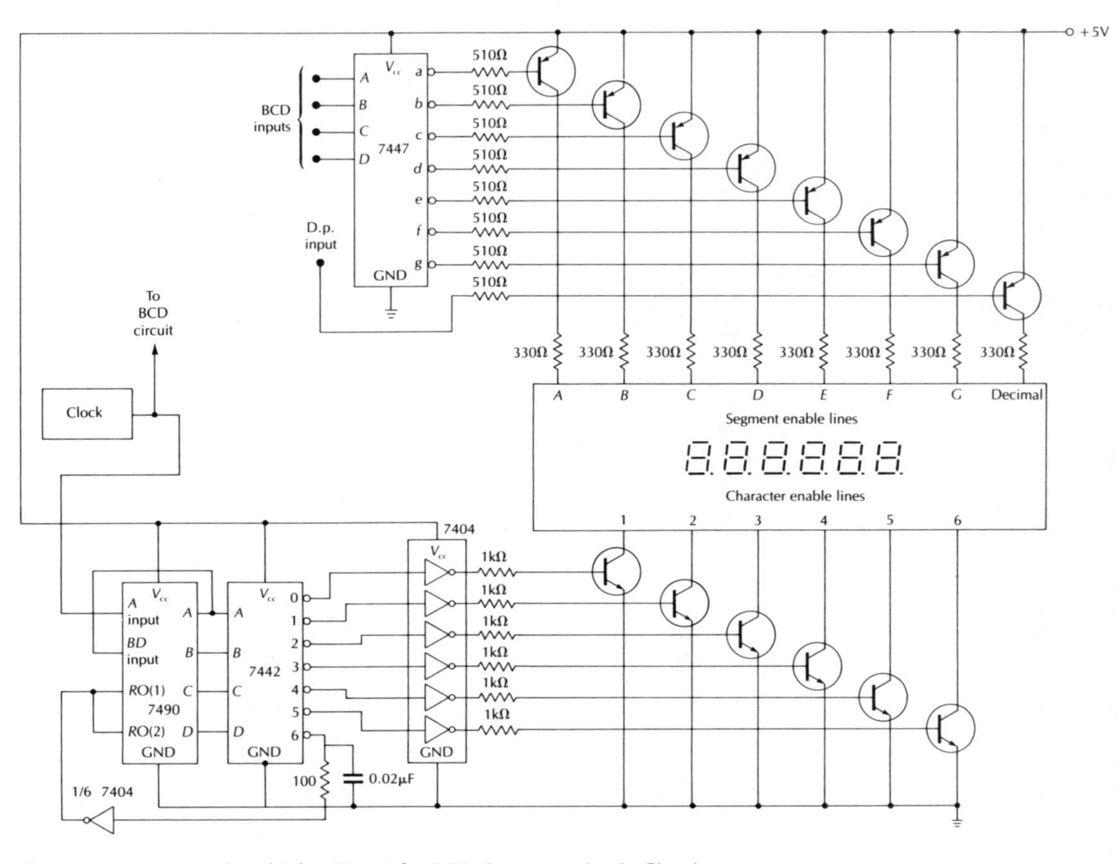

Figure 12-2 Practical Multiplex Circuit for BCD Counter or Logic Circuit

Figure 12-1b shows the functional block diagram (the electronic equivalent of Figure 12-1a) of an LED multiplex system. The following blocks fulfill the function (see Fig. 12-1):

1. Decoder/driver decodes BCD input data into the seven-segment display pattern code.
2. Input address selector multiplexer (or shift register) selects the particular BCD group to be displayed at a given time.
3. Scan decoder selects the LED digit to be activated. The input address multiplexer and the scan decoder together see that the proper BCD group is selected and that its corresponding LED digit is activated.
4. Scan counter sequences the input multiplexer and scan decoder through, sampling each BCD group in turn and routing that data to the proper digit in the display.
5. Clock drives the scan counter at the desired rate.

Figure 12-2 illustrates a practical circuit for the multiplexed display of a BCD logic circuit. The 7447 decodes the BCD output of a counter or other logic circuit to provide the drive for the appropriate segments for the required digits. Seven PNP transistors increase the segment drive current and an eighth transistor drives the decimal point.

A 7490 decade counter is decoded by a 7442 4-to-10-line decoder to provide scanning for the 6 digits. A feedback pulse is taken from the decoder to reset the 7490 counter after 6 counts (because there are 6 digits to scan). A 7404 Hex inverter and 6 NPN transistors provide drive for the *character enable* inputs to the display.

FOUR-DIGIT DECODER COUNTER WITH BUILT-IN MULTIPLEXING CIRCUITRY

Figure 12-3 shows the circuit for a complete 4-digit decade counter using a C-MOS MM74C925 integrated circuit. The chip contains the complete 4-digit decade counter, 7-segment decoder, and display multiplexing circuitry. Four-digit drive transistors and one decimal point driver transistor are required.

The display is a low-current common cathode unit. Larger, higher current, LED display units may require additional segment drive transistors as shown in Figure 12-2.

12-2 A Brief Introduction to Microcomputers

The microprocessor is probably the most exciting computer development in recent years. The basic idea behind large-scale, fully pro-

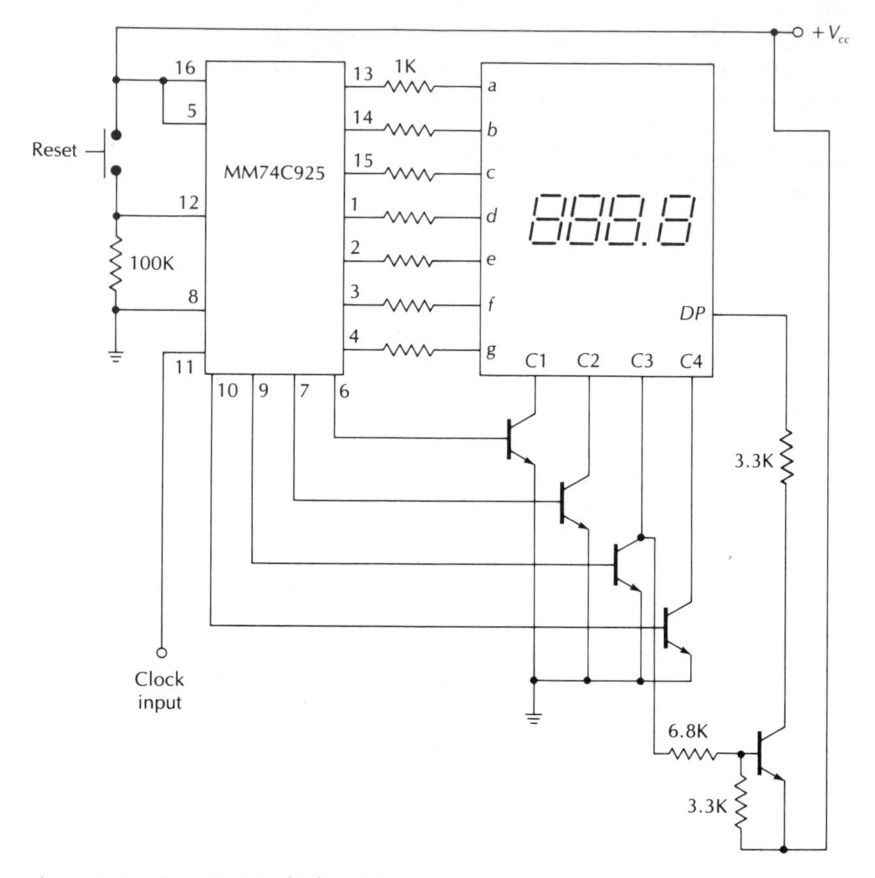

Figure 12-3 Four-Digit Multiplexed C-MOS Counter

grammable computers of the data-processing variety was to have a single basic machine that could be made to handle nearly any computing or data-processing task simply by providing a set of instructions called a *program*. The program and data could then be loaded into a memory, and the machine could take it from there at a fantastic rate of speed. Such a powerful machine could also be used for such dedicated (special-purpose) tasks as traffic-signal control, electronic instrumentation, electronic scales, automobile systems, and industrial control, but until recently the cost has been prohibitive.

Now the microprocessor is available and its cost is dropping the same way the price of pocket calculators did not long after their introduction. The microprocessor is a slower, less powerful, but very inexpensive equivalent of the central processing unit of the large-scale digital computer.

The microprocessor is the heart of a system called a *microcomputer*. An entire microcomputer can be put on a single chip, although until recently this practice has been limited to military applications. A large part of the computer is commonly put on a single LSI chip and called a *microprocessor*. The microprocessor unit (MPU) nearly always contains the ALU (arithmetic logic unit), minimum necessary working and control registers, and often some specifically committed ROM (read-only memory) along with counters and control logic.

The MPU is designed so that it has very little internal direction. It is virtually uncommitted to anything but the simplest operations common to any computer operation. Although the MPU is the central part and coordinator of the system, it is virtually a slave to an external set of instructions. Every action is dictated by external programming in the form of firmware or software. The firmware is a programmed ROM that is a kind of job description that dedicates the machine to a particular job. The ROMs are external to the MPU chip and can be programmed for the functions required of a general-purpose machine or of a dedicated (special-purpose) machine.

The following are some examples of dedicated machines:

Automotive ignition control
Automotive brake control
Automated gas pumps
Fast-food cash registers
Home-appliance controllers
Vending-machine control
Electronic scales
Specialty calculators
Adaptive control systems
Automotive analyzers
Medical instruments
Machine-tool control
Electronic games
Film-processing control
Communication-line controllers
Printer controllers
Traffic-light controllers
Photocopy-machine controllers
Point-of-sales inventory control
Electronic typesetting
Automatic drafting systems
Automatic elevator control
Chemical-process control

Once programmed, the firmware becomes permanent software that directs the activities of the processor and permits communication with keyboards, displays, and other peripheral devices. In dedicated systems, firmware may constitute the entire program (stored in the ROM). A plug-in ROM assembly is often called a *personality module*. These personality modules can be unplugged and replaced with a module that can change the computer's task. For example, the machine could be altered to cope with new traffic patterns in a traffic-signal controller. The same microcomputer might be used in any of the dedicated systems listed above by providing the proper personality module for each system.

Microcomputers can also be used for traditional data processing, such as payroll and inventory control, in smaller companies where the larger machines are too expensive.

Many industrial and medical instruments have built-in microprocessors that permit a relatively untrained person to use them. Thus physicians and highly trained technicians are freed for more vital tasks.

In aircraft and space-vehicle systems, microprocessors can be distributed throughout the control system to provide local computer-control under the direction of a central master-computer. A distributed system can provide a much faster response than a large, central-control computer.

Microprocessors are becoming so inexpensive that relatively simple logic circuits can be economically implemented with microprocessor hardware. The microprocessor is being used not only in computing-type applications; it is also fast becoming a popular programmable logic form.

Figure 12-4 is the block diagram of a simplified microcomputer. Real machines will have some additional registers. The condition or status register is used in nearly all machines to make decisions when an operation results in an overflow (carry), a negative value, a zero result, and so forth. Other special registers are often used, depending on the design of the specific computer. Aside from some refinements and a 4-bit limitation (only 16 memory addresses), the sample microcomputer in Figure 12-4 is reasonably typical. Clock circuits are omitted in the sample to keep it simple.

The following subsystems in Figure 12-4 would normally be on the integrated-circuit microprocessor chip:

1. Instruction register and bus driver
2. Instruction decoder
3. Program counter and bus driver
4. General register and bus driver
5. Accumulator and bus driver
6. Arithmetic logic unit and bus driver
7. Control circuitry
8. Condition (status) register

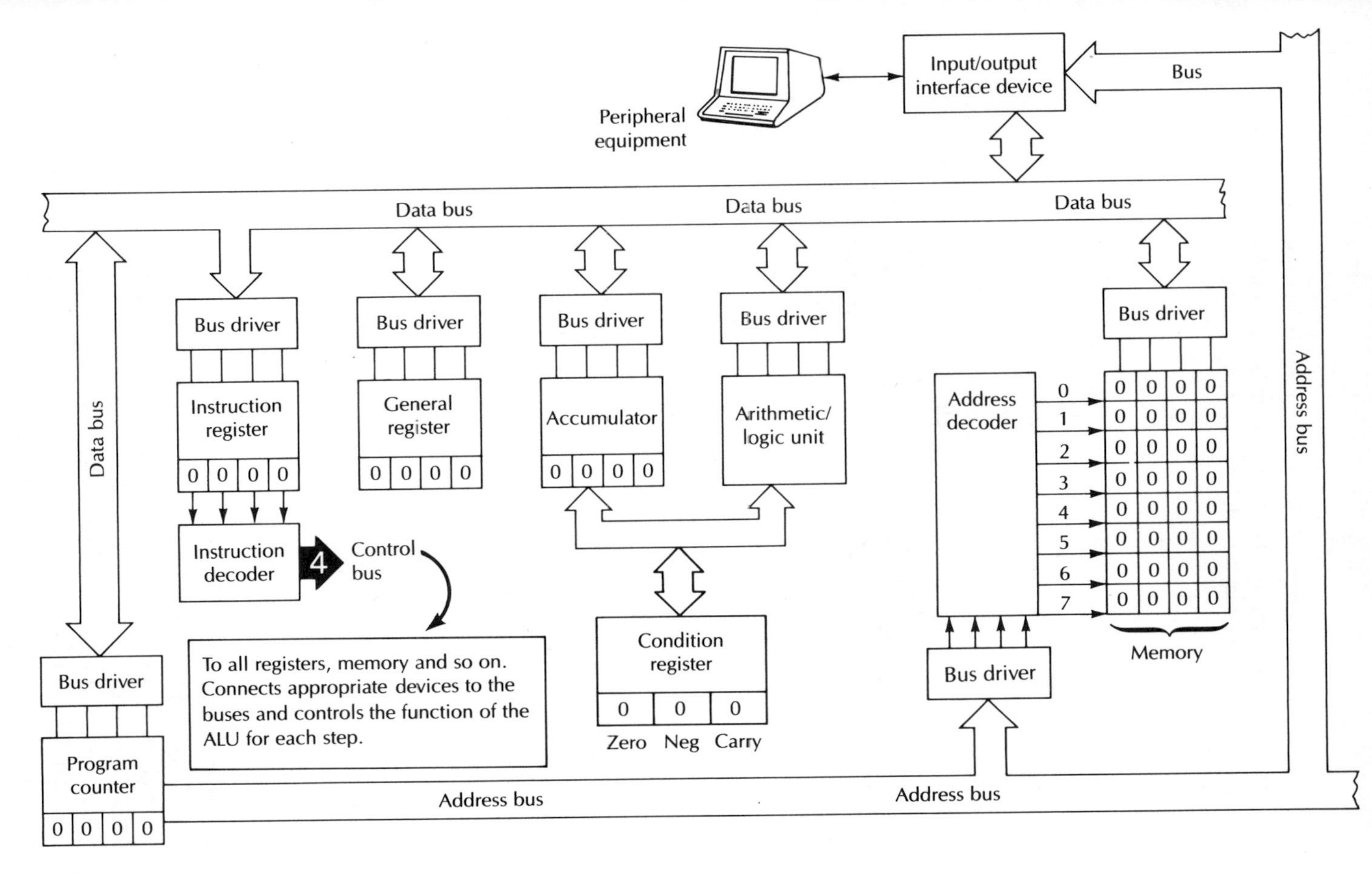

Figure 12-4 Sample Microcomputer Block Diagram

There may be a number of integrated-circuit memory chips. Each memory chip will contain the memory cells, the memory address decoder, and sometimes a bus driver for the decoder and one for the memory itself.

THE MEMORY

The memory is loaded with both instructions and data before any computing is done. The computer will fetch data and instructions from memory as they are needed. Results from the various computing operations are stored in the memory until needed. The memory is usually a combination of RAM (random-access memory) and ROM (read-only memory). Data in the RAM are frequently changing as they are processed. The ROM contains certain permanent instructions—often-used numerical values such as log tables and so on. Memory size can be expanded by connecting additional memory units to the bus. The microcomputer in our illustration processes 4 binary digits (bits) at a time. Most contemporary microcomputers process 8 to 16 bits simultaneously. A typical 8-bit microcomputer can address 65,536 memory locations, each containing one 8-bit (1 byte) word.

THE ADDRESS DECODER (ON THE MEMORY CHIP)

The address decoder decodes a 4-, 8-, or n-bit binary address (depending on the particular processor) and selects a particular word in memory to be read out of or written into. Most 8-bit microcomputers use 16-bit addresses; the decoders therefore have 65,536 individual word locations in memory from which to select. A word is 8 bits in an 8-bit machine, 16 bits in a 16-bit machine, and so on.

THE ARITHMETIC LOGIC UNIT

The ALU performs various arithmetic and logic operations under the command of the control unit. The ALU is called the *arithmetic section* in some machines.

THE CONDITION (STATUS) REGISTER

The condition or status register stores special data as a result (generally) of ALU operations. Each bit in the register is usually a special 1-bit register. For example, if the ALU produces a carry as the result of an addition, the carry bit would be stored in the carry (or overflow) register. A subtraction operation that yields a negative result would store that information in the negative 1-bit register. A subtraction of two equal values would be indicated by storing the information in the zero register.

Real machines often have other special-condition registers within the condition-register block. The number of bits in the condition register is not related to the word length of the processor because each bit is really an independent 1-bit register.

The condition register allows for *conditional instructions* that may tell the machine, for example, to do one thing if the result of a numerical operation is equal to or less than some specified value, and to do something different if the result is greater than that specific value. The microcomputer's flexibility and power are greatly enhanced through the use of conditional commands.

THE ACCUMULATOR

The accumulator is a temporary memory (register) composed of 4, 8, or 16 flip-flops. Processors are classified according to the word length. In an 8-bit machine the memory word will consist of 8 bits; the accumulator and other registers will also be 8-bits wide. The accumulator works directly with the arithmetic logic unit. It always stores one of the two operands for the ALU, and after each ALU operation it generally stores the results of that operation.

In most microprocessors the accumulator is the gateway to the outside world. Data from keyboards, displays, and so on pass through the accumulator on their way to memory or from the processor to the outside world.

THE GENERAL REGISTER

The general register is a one-word long group of flip-flops. It is used to store data that will be acted upon. It is often used to store the second operand for the arithmetic logic unit. The general register (and the accumulator) are much faster than main memory, and the data stored in registers are immediately available for processing. Some machines use the accumulator and general register as interchangeable registers and designate them as accumulator *A* and accumulator *B*.

THE PROGRAM COUNTER

The program counter (often called the memory-address register) is set to some binary number. Each time it is advanced (incremented) it addresses the next consecutive memory location. The binary number stored in the program-counter flip-flops represents a definite memory location. The program counter tells the memory-address decoder to select a particular memory address and set it up for reading the information stored there. The next step in the computer's operation will transfer the contents of that memory location along the bus to the appropriate register. The memory-address decoder decodes the contents of the program counter and selects the proper location in memory. The program counter can be instructed to *set* to any memory location and start its count at that location; thus the machine can jump from one group of memory locations to another. Much of the computer's flexibility depends on the ability to move around in the memory to find the required data and instructions.

THE INSTRUCTION REGISTER

Instructions stored in the memory are sent along the data bus into the instruction-register flip-flops. The instructions are in a binary code called the *operation code* or *op-code*. Each machine has a set of instructions. An instruction tells the machine what to do next. The instruction tells the machine which units to connect to the bus and governs transfers from memory to registers, addition operations, and so on.

THE INSTRUCTION DECODER

The instruction decoder decodes the binary instruction, and, through a group of AND and OR gates, places an enable voltage on the units that are to be connected to the bus or produces the proper code to tell the arithmetic logic unit what function to perform.

THE CLOCK

The clock pulse is generally divided into 4 discrete segments (4 phases), each occurring at a different time. No device can be connected to the bus without simultaneous commands from both the instruction decoder and the clock.

PERIPHERAL EQUIPMENT

The peripheral equipment is connected through an interface device to the bus. The peripheral devices form a communication link between the computer and the human operator or between the computer and the devices the computer is intended to control. Data from peripheral equipment are normally loaded into or fed out of the accumulator.

Peripheral Interface Devices

Peripheral interface devices serve several purposes. The computer itself operates at very low power levels, far too low to actuate relays, control a printer, or light indicator lamps. Interface devices often have a power-amplification function.

Peripheral devices may also operate in a code that is not standard binary. The peripheral interface device may be required to make the translation. If the peripheral devices are analog devices, the interface circuits must make digital-to-analog or analog-to-digital conversions.

Many peripheral devices are very slow in comparison to the computer. The interface device must often have some buffer (RAM) memory. The memory can store input data from the peripheral devices at whatever rate it is received, and discharge its contents at computer-compatible speeds. Buffer memory is also needed to record data at high computer speeds and unload it to a typewriter or other machine at a speed that it can keep up with.

12-3 A Computer Processing Example Using The Sample Microcomputer

Suppose we follow a sample microcomputer through all of its operations in the process of performing an addition.

Problem

Add: $0111_2 + 0001_2$

By hand we get:

```
  111    Carry
  0111
 +0001
 -----
  1000
```

In the decimal system:

8	4	2	1	Headings
0	1	1	1	Binary
(0×8) +	(1×4) +	(1×2) +	(1×1)	= 7 in base 10

8	4	2	1	Headings
0	0	0	1	Binary
(0×8)	(0×4)	(0×2)	(1×1)	= 1 in base 10

So, we are adding:

```
  111    Carry
  0111   (7 in base 10)
 +0001   (1 in base 10)
 -----
  1000 = 8 in base 10
```

Table 12-1 Partial Listing of Sample Microprocessor Instruction Set

Instruction	Operation Code
Clear all registers.	0000
Load the accumulator with the data stored in the next (in sequence) memory address.	0001
Load the general register with the data stored in the next (in sequence) memory address.	0010
Transfer the data in the accumulator to the designated memory address.	0011
Add the contents of the general register to the contents of the accumulator, and store the results in the accumulator.	0100

We need to know two things to start: how data and instructions must be entered into the memory and the operation codes for each of the instructions we will need. The instruction codes are shown in Table 12-1.

The instructions must be loaded into the memory in the following order:

1. First instruction
2. First word of data
3. Second instruction
4. Second word of data
5. Third instruction, and so on.

Now, we must write a formal program and load it into the computer memory. The program is shown in Table 12-2.

Now that we have written the program, Table 12-3 shows how it is stored in the computer's memory after the program has been loaded.

Table 12-2 Sample Microcomputer Addition Program Listing

Step	Description	Operation Code	Data	Memory Address Decimal	Memory Address Binary
1	Transfer the first instruction from memory address 0 (0000) to the instruction register.	0010		0	0000
2	Increment the program counter. Execute the instruction (0010): Load the general register from memory.		0111	1	0001
3	Increment the program counter. Transfer the second instruction from memory address 2 (0010) to the instruction register.	0001		2	0010
4	Increment the program counter. Execute the instruction (0001): Load the accumulator from memory.		0001	3	0011
5	Increment the program counter. Transfer the third instruction from memory address 4 (0100) to the instruction register.	0100		4	0100
6	Execute the instruction (0100): A. Add the contents of the general register to the contents of the accumulator. B. Store the sum in the accumulator.		1000 (in Acc)		
7	Increment the program counter. Transfer the fourth instruction from memory address 5 (0101) to the instruction register.	0011		5	0101
8	Increment the program counter. Execute the first part of the instruction (0011): set the program counter to the address stored in memory (1111).		1111	6	0110
9	Execute the second part of the instruction (0011): transfer the contents of the accumulator to memory. Note: In this case the data is transferred to memory address 15 (1111) which is the address of the input/output terminal.		1000	15	1111

Table 12-3 Contents of the Memory before the Microcomputer Starts

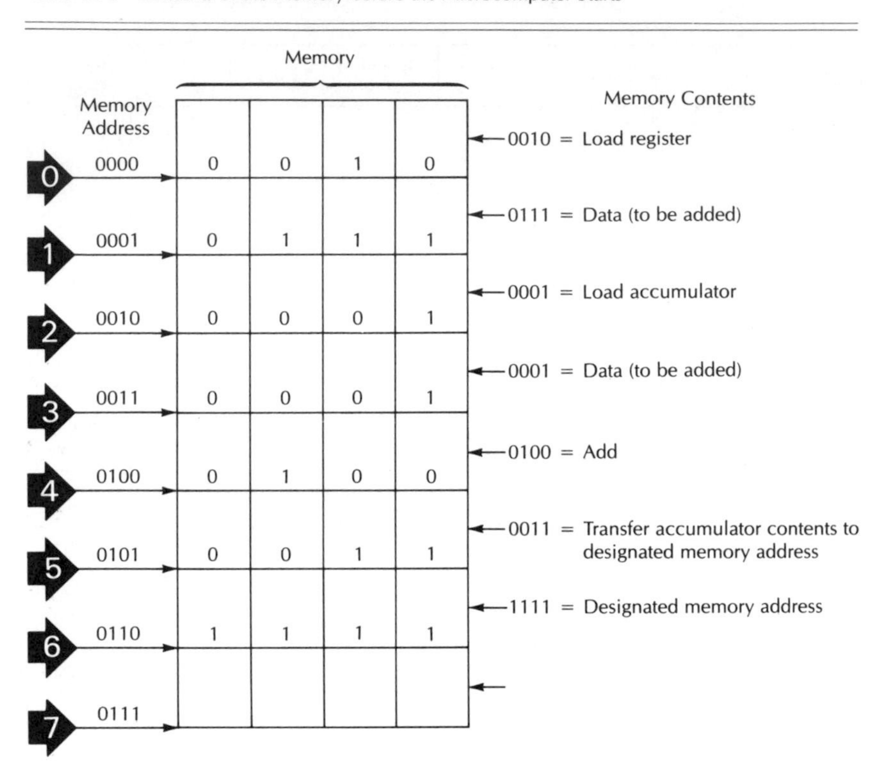

Suppose we follow the computer through its processing of the problem we have programmed. The step-by-step computer operation is: (see Fig. 12-5).

1. Fetch the first instruction.
   a. Select memory location 0000 (0) for reading. The program counter sends 0000 along the address bus to the memory address decoder.
   b. Because this machine cycle is an instruction cycle, the contents of the memory at address 0000 are automatically sent along the data bus to the instruction register. The data in memory address (0000) are loaded into the instruction register. The instruction is 0010, which in the machine's language, tells the instruction decoder to connect the memory and the general register to the bus to allow the transfer from memory to the general register.

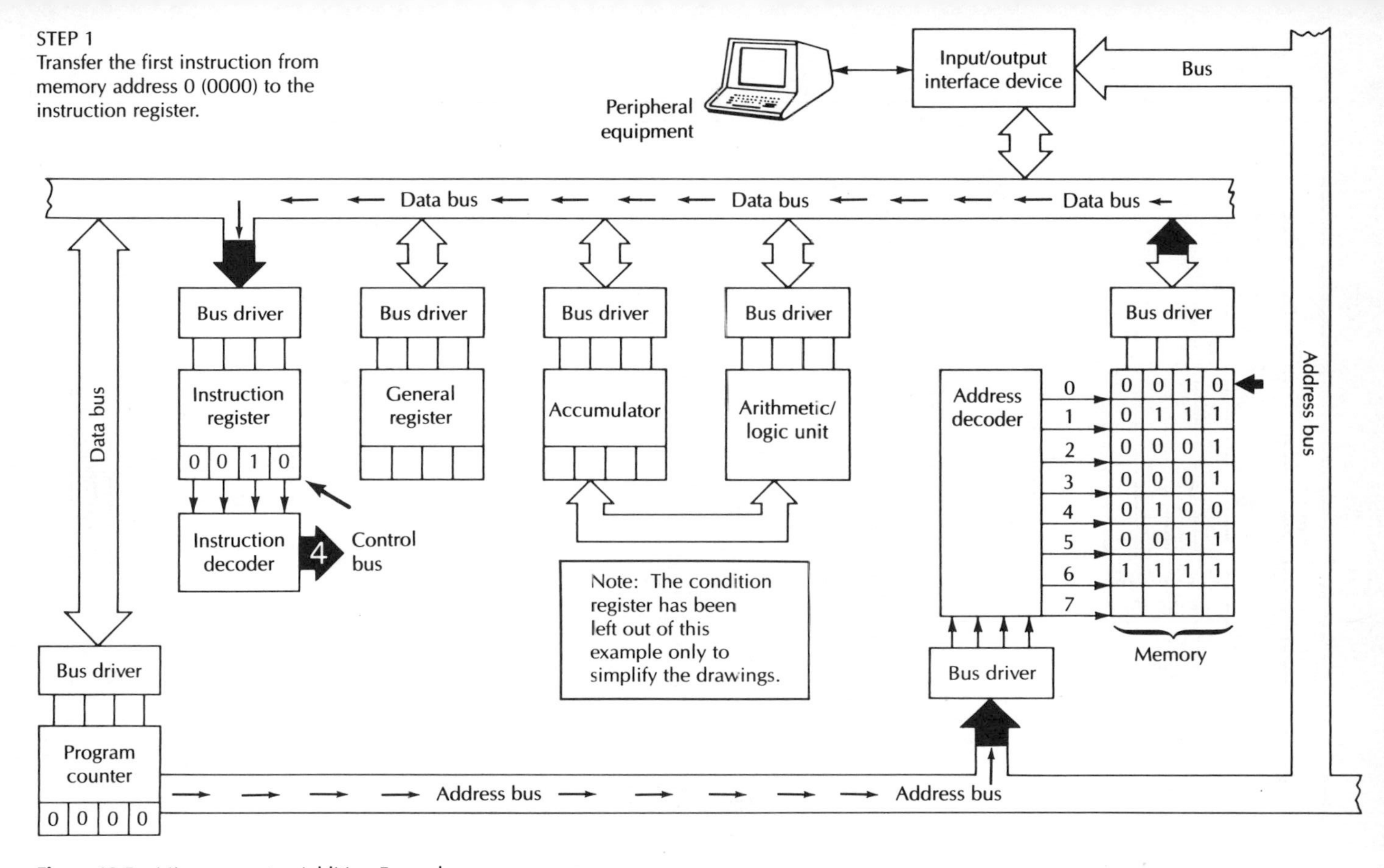

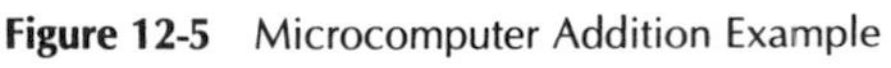

Figure 12-5 Microcomputer Addition Example

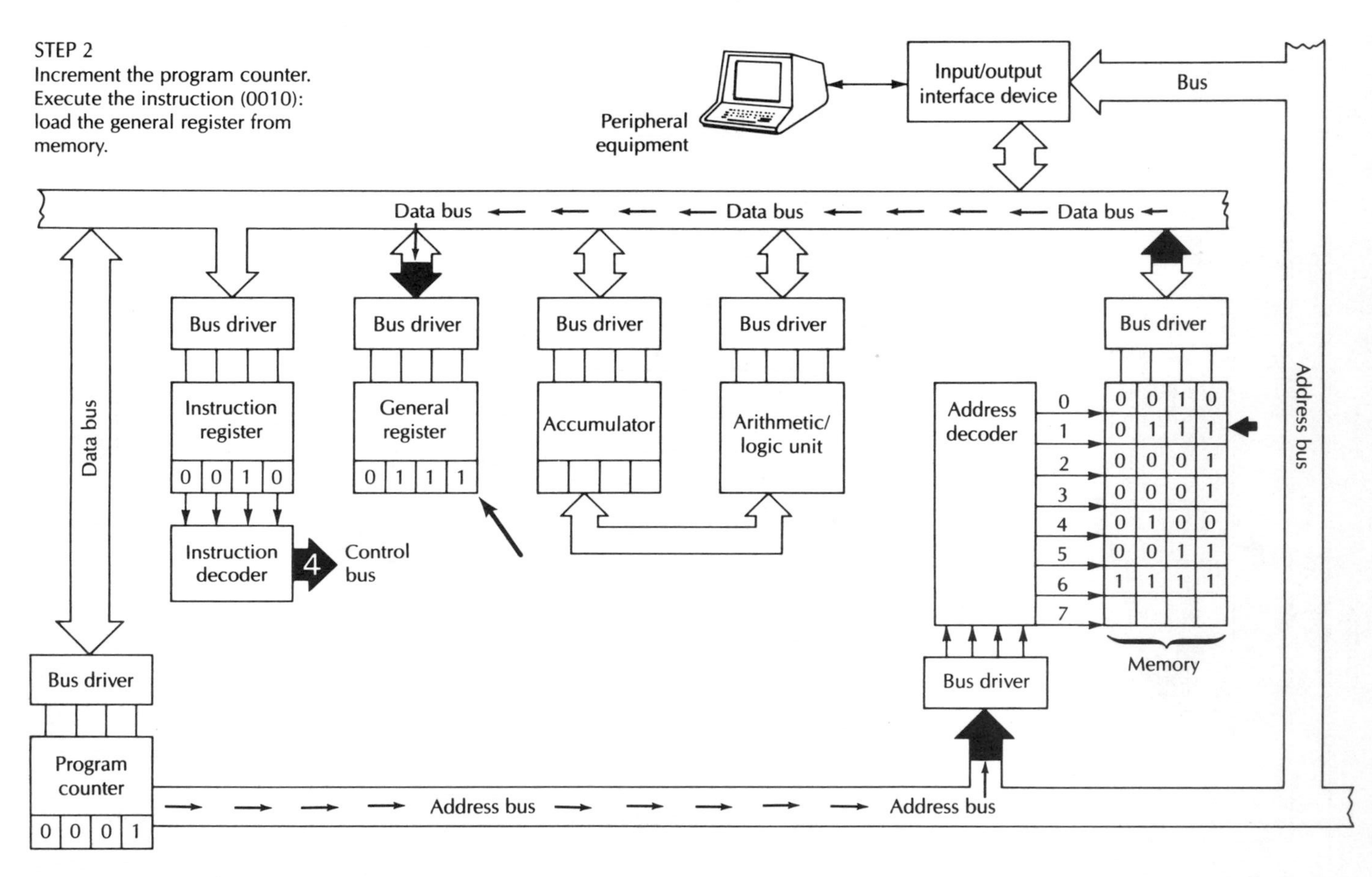

Figure 12-5 continued

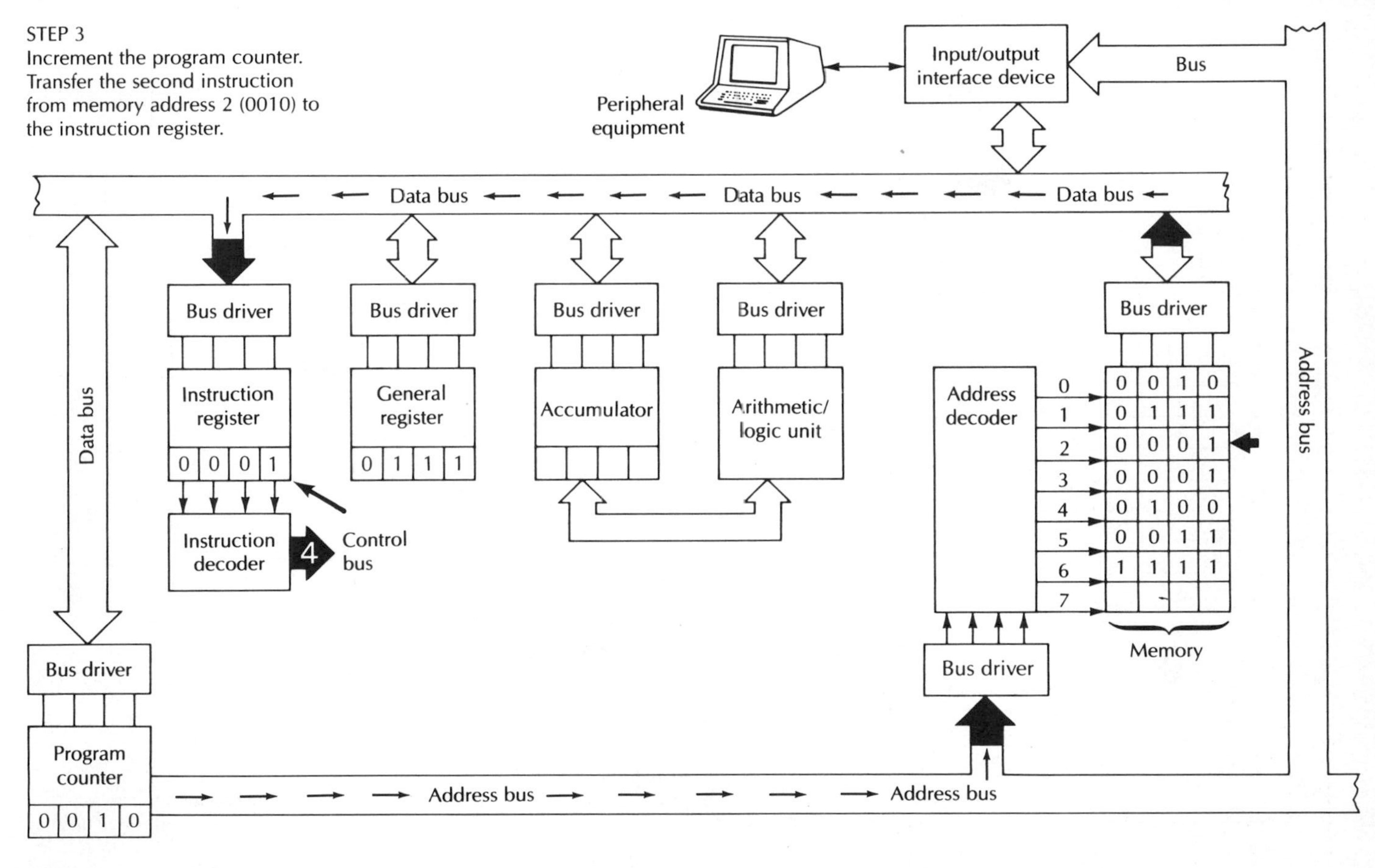

Figure 12-5 continued

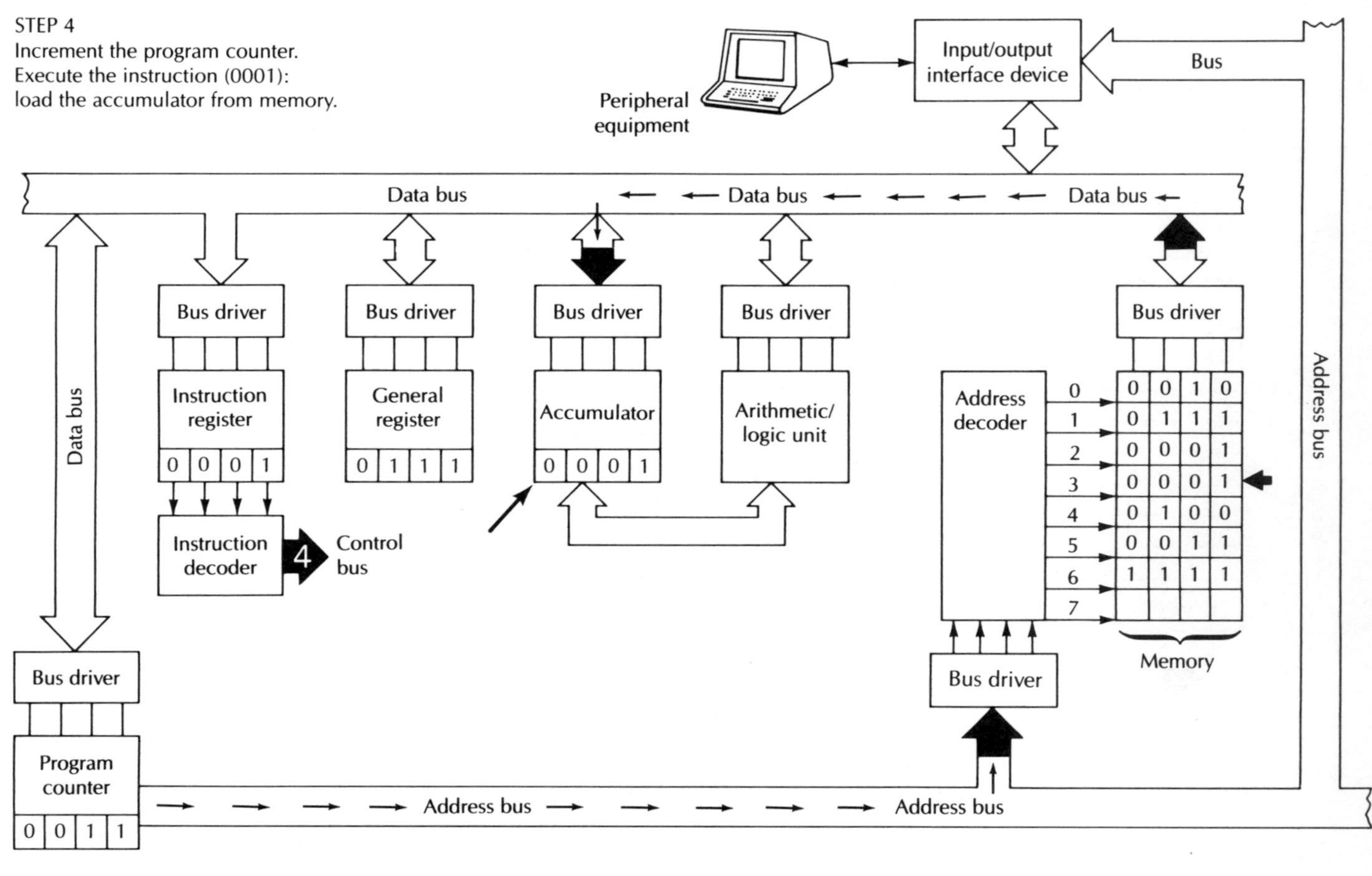

Figure 12-5 continued

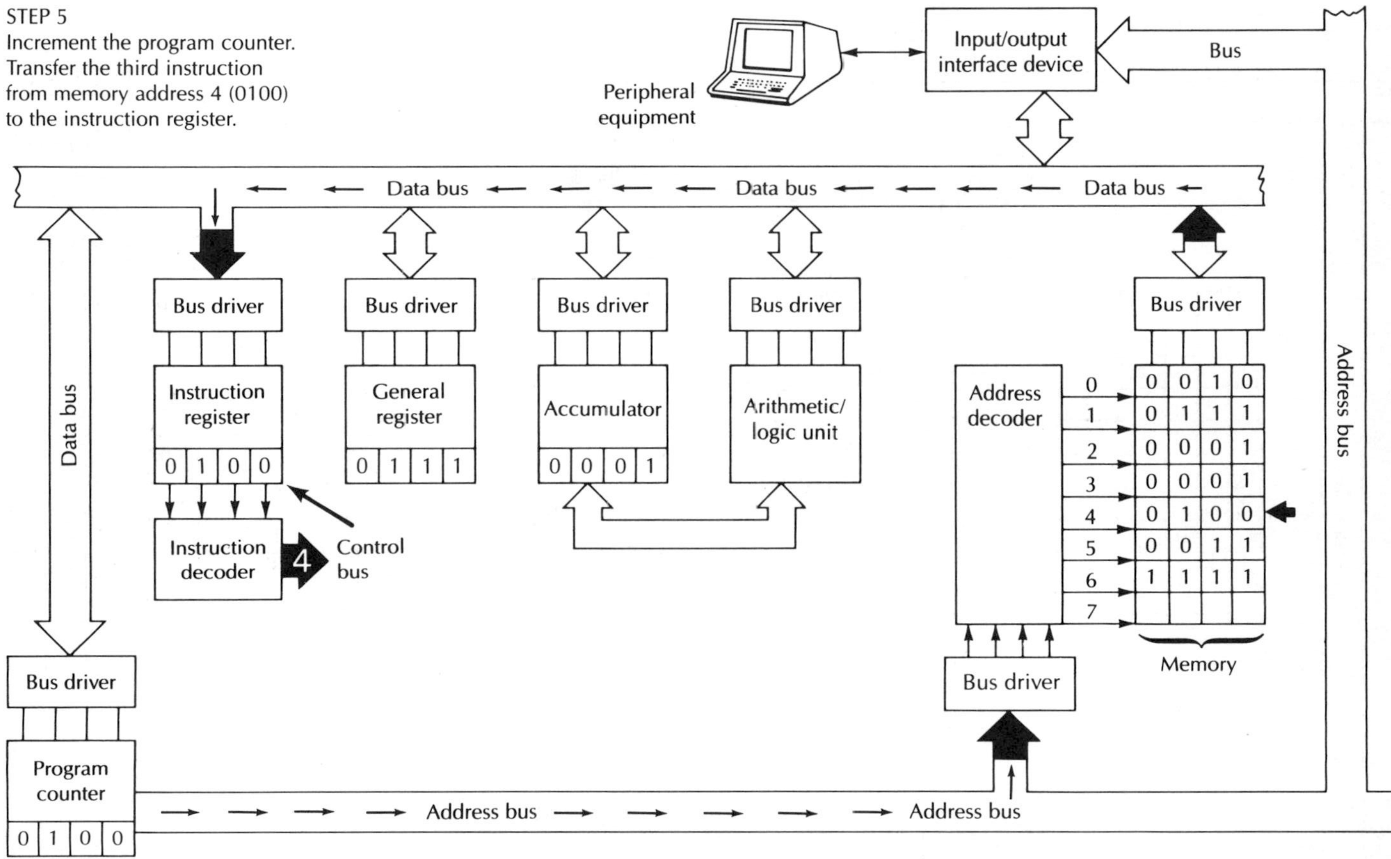

Figure 12-5 continued

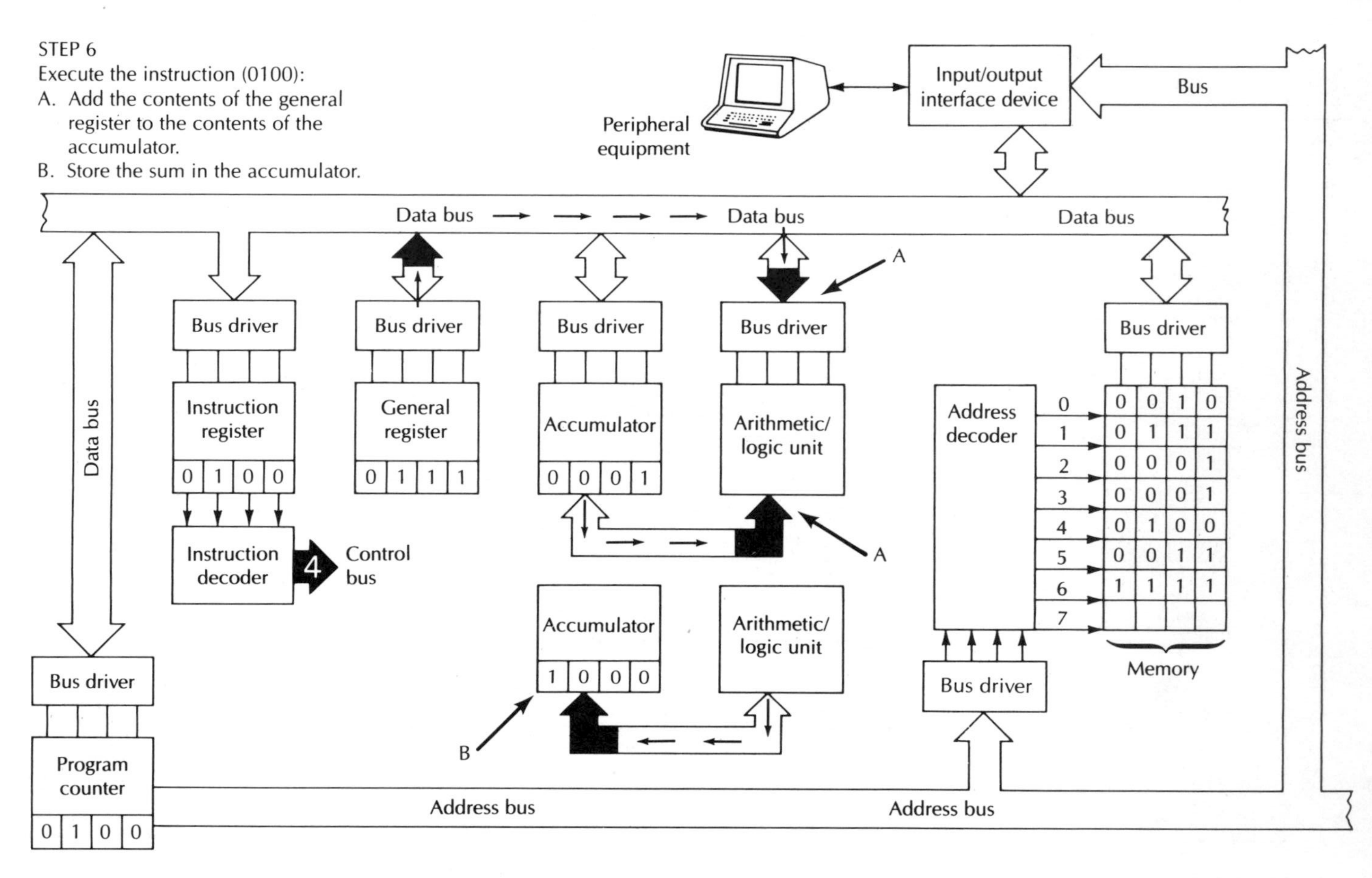

Figure 12-5 continued

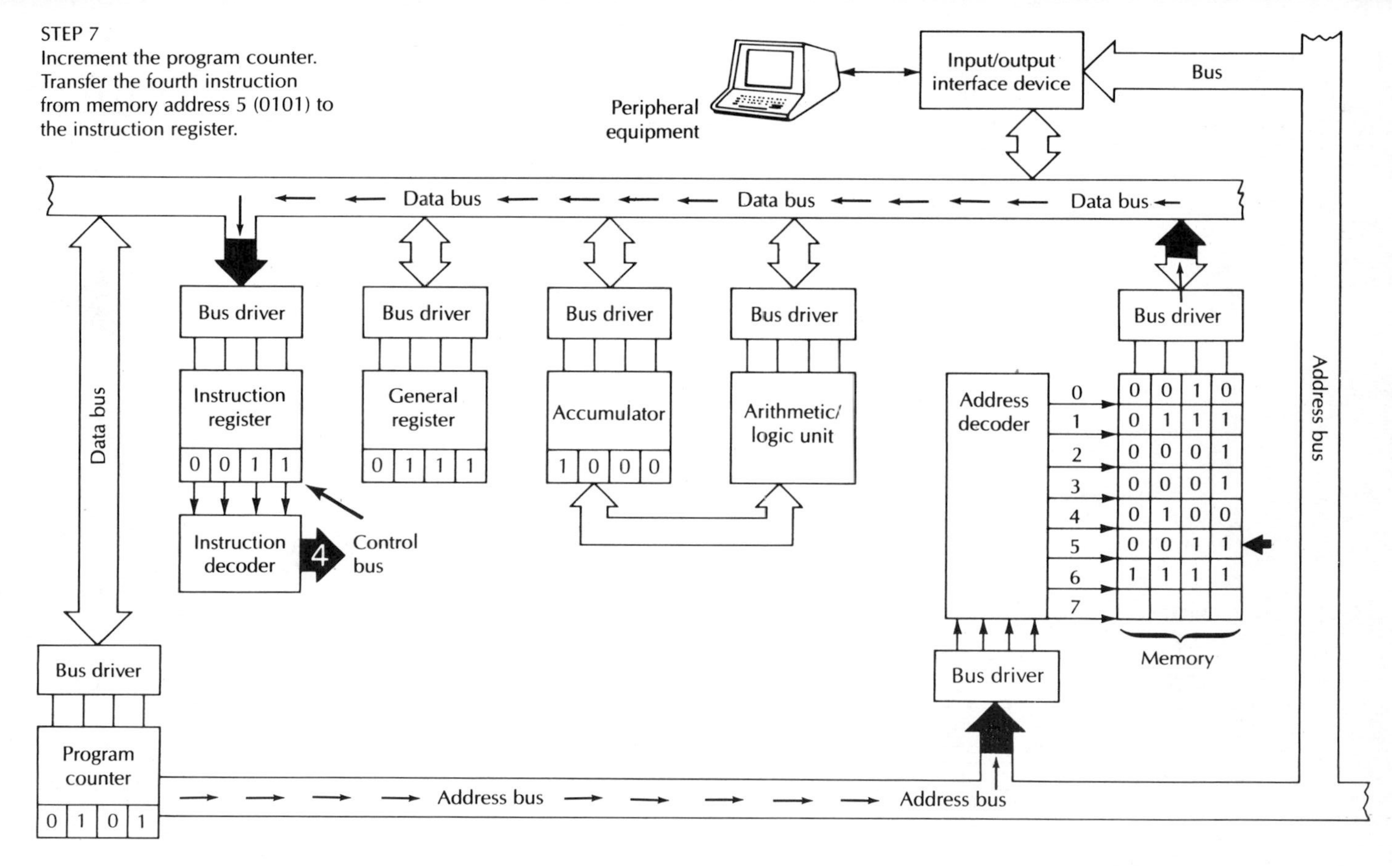

Figure 12-5 continued

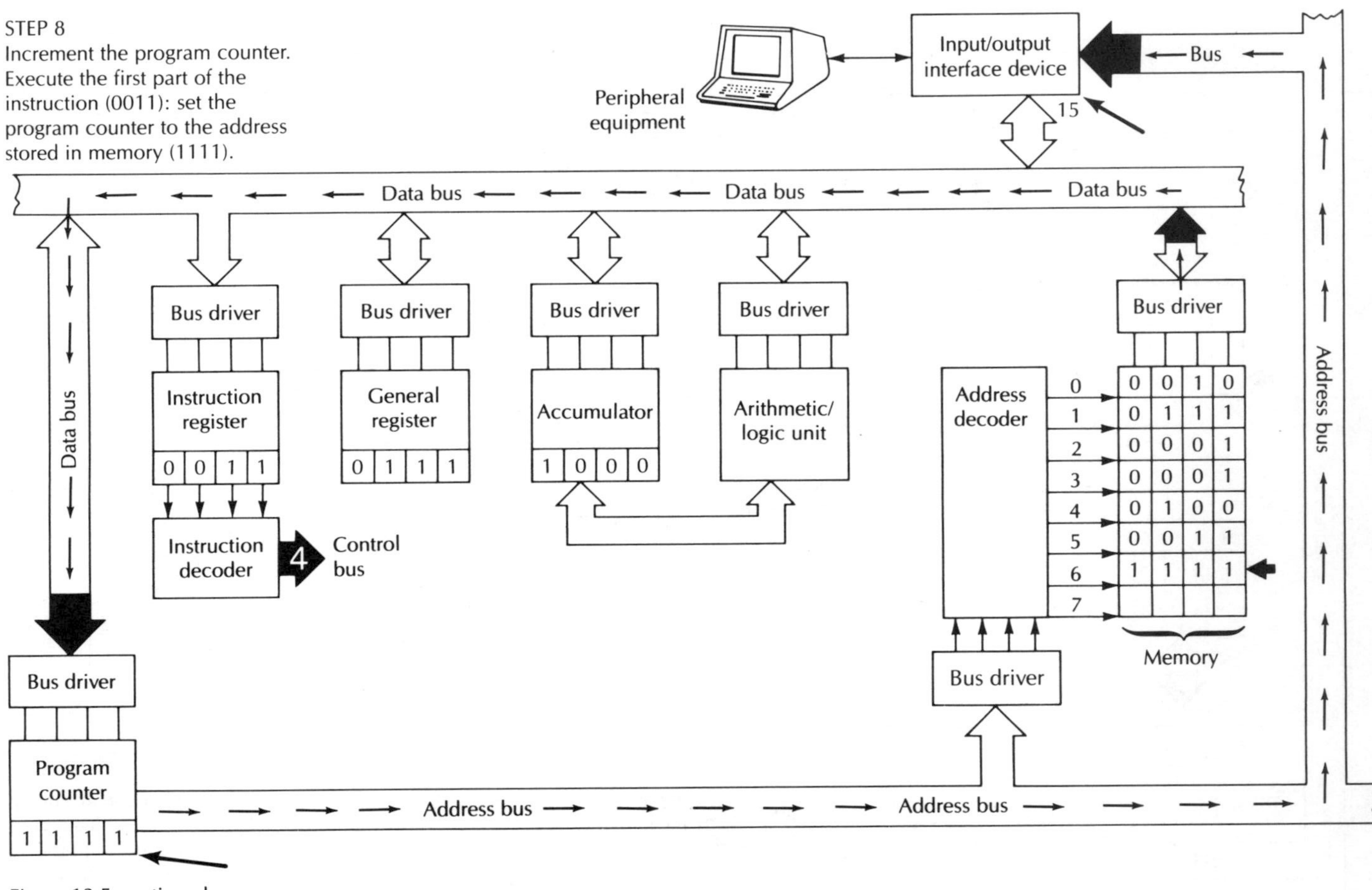

Figure 12-5 continued

STEP 9

Execute the second part of the instruction (0011): transfer the contents of the accumulator to memory. Note: In this case the data is transferred to memory address 15 (1111) which is the address of the input/output terminal.

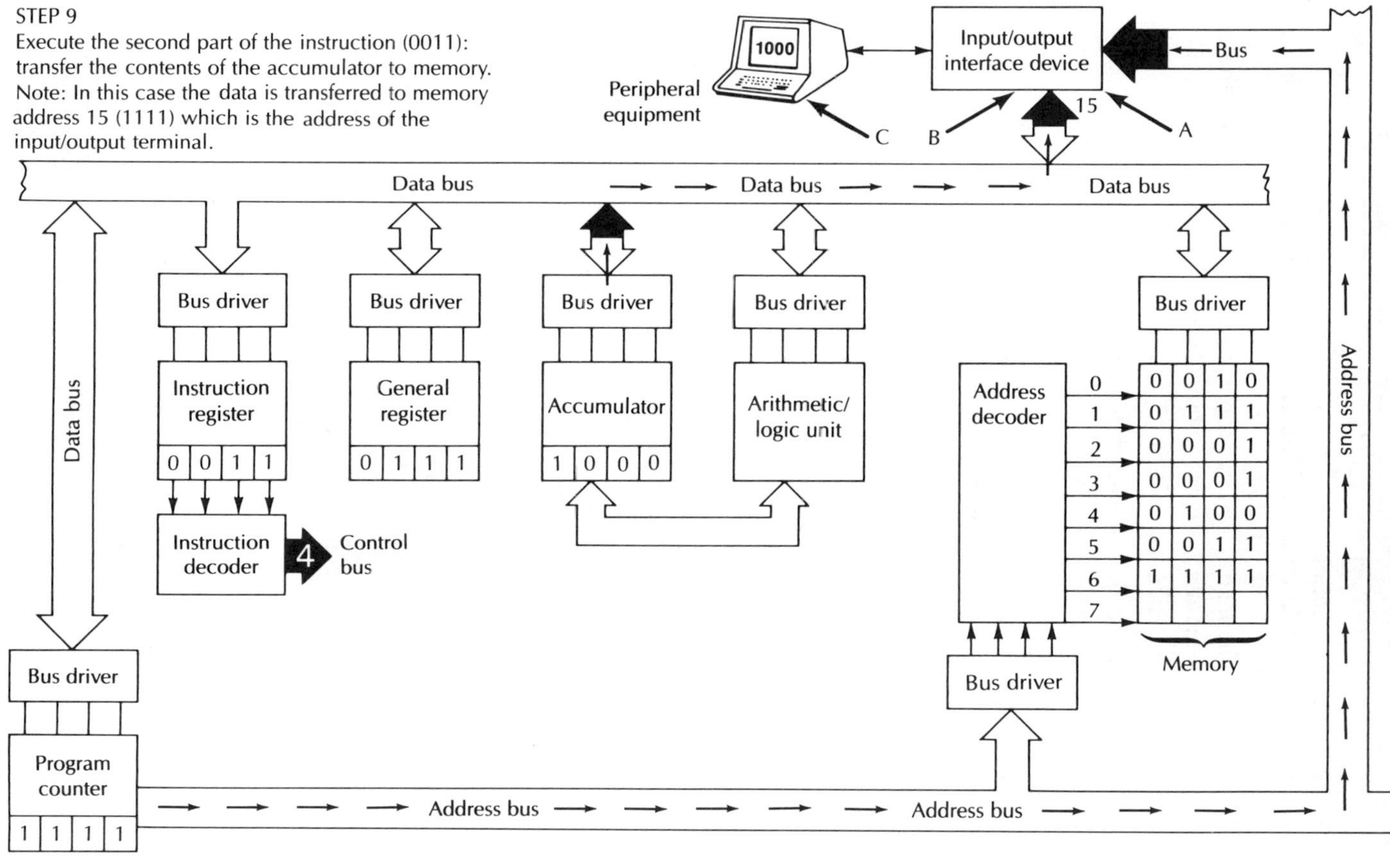

Figure 12-5 continued

2. Execute the instruction.
   a. Advance program counter (increment). Designate memory location 0001 (1) for reading.
   b. Carry out the instruction in Step 2. Load the register with the contents of memory location 0001 (1).

      Data: Load 0111 into the register.

   The data 0111 is the *addend* of this addition problem and is now stored in the general register.

3. Fetch the second instruction.
   a. Advance program counter (increment).
   b. Select memory location 0010 (2) for reading.
   c. Transfer the contents of memory location 0010 (2) into the instruction register.

      This is an instruction cycle, so the contents of memory address 0010 (2) are automatically loaded into the instruction register via the data bus.

   The instruction is: 0001 = load accumulator with the contents in the next memory location.

   Notice that memory addresses 2 (0010) and 3 (0011) both contain the binary digits 0001, but in one case 0001 is an instruction and in the other it is the augend of the addition problem. The microcomputer "knows" which 0001 is an instruction and which 0001 is data *only* as a result of *when* each occurs in the timing cycle. If the contents of a particular memory address are accessed during an instruction cycle, the computer will interpret the memory contents as an instruction. It is up to the programmer to see that the *order* of instructions and data is correct. The instruction decoder translates the instruction code (0001) into the appropriate enable signals to connect the memory and the accumulator to the data bus in preparation for the data transfer.

4. Execute the second instruction.
   a. Advance program counter (increment).
   b. Designate memory location 0011 (3) for reading.
   c. Carry out the instruction and transfer the contents of memory location 0011 (3) into the accumulator.

      The data loaded into the accumulator consist of the number 0001, which is the augend of the addition problem. The augend of the addition problem is now stored in the accumulator.

5. Fetch the third instruction.
   a. Advance program counter (increment).

b. Designate memory location 0100 (4) for reading.
c. Transfer the contents of memory location 0100 (4) into the instruction register.

The instruction is: 0100 = add the contents of the register to the contents of the accumulator and store the result in the accumulator.

6. Execute the third instruction.
   a. Carry out the instruction in Step 5: Add the contents of the register to the contents of the accumulator.
   b. The result of the addition in Step 6 is automatically stored in the accumulator. The result of the addition 0111 + 0001 is now stored in the accumulator.
7. Fetch the fourth instruction.
   a. Increment the program counter to read the contents of memory address 5 (0101).
   b. Transfer the contents of memory address 5 (0101) to the instruction register.

      Because this is an instruction cycle the memory contents are automatically destined for the instruction register.

      In this case, the instruction decoder not only connects the appropriate devices to the bus, but also recognizes that this particular instruction requires two separate actions for its execution.
8. Fetch the memory address to which the contents of the accumulator are to be transferred.
   a. Force the program counter to 1111, the "memory" address of the input/output terminal. The idea is to get the *sum* of the addition out of the accumulator and display it on the terminal.
9. Transfer the contents of the accumulator to the input/output terminal.

   The program counter selects memory address 15 (1111). The input/output interface device contains a decoder that recognizes that it is to receive data from the data bus. The interface device also makes any necessary code conversions, data rate changes, and so on, to allow the final answer to be displayed on the screen.

12-4 An Introduction to Programming

Programming is a broad term covering all methods of providing instructions for a computer or processor to follow. Some programs are hardwired and permanent, some are in the form of read-only memories (called firmware), and others are in the form of software, requiring no physical changes to change the program. A logic system designed for a specific job is hardware programmed and cannot be adapted for other tasks without physical modification and possibly a complete redesign. A logic system using only ROM devices is firmware programmed and can be transformed into a completely different system for an entirely different task by changing or reprogramming the ROM's. The most flexible of programs is software. The program is created on paper and transferred to the logic system by means of a keyboard, punched or magnetic tape, or punched cards. The machine stores the instructions in some kind of random-access memory and is dedicated to the performance of the special task directed by the instructions in the RAM. While a particular program is in the RAM, the machine is just as dedicated to a special task as a hard-wired logic system. However, all that is necessary to rededicate it to an entirely new task is to load in a new program. Once a library of programs has been built up, it takes but minutes to convert the machine from the performance of one task to another.

Many microprocessors are used in so-called dedicated systems rather than for general-purpose computing. For example, assume that a city is in the process of installing traffic signal controls. This process could involve activities such as sampling current traffic patterns and adjusting to them, varying signal behavior according to the season or time of day, and so on. The basic function of the controller for each signal is essentially the same, but each has its own unique problems. If the microprocessor controllers are used throughout the city, the controller at each signal can be adapted to its unique set of conditions by simply programming a ROM or a RAM for each signal. If a street is widened or other conditions changed, the controller for that signal does not have to be redesigned or modified. All that is required is inserting a ROM programmed for the new situation or loading a new software program into a RAM.

MACHINE-LANGUAGE PROGRAMMING

Machine language consists of groups of binary digits that direct the actions of the logic circuits controlling the transfer and handling of logic-level data within the machine. Machine-language instructions consist of two basic parts: the operation to be performed and the address or addresses of the data (operands) to be operated upon. Because machine language consists entirely of many zeros and ones, it is not very suitable for direct human use. A hierarchy of languages has therefore been developed to interface human language and machine language. Each successively higher level of language more closely approaches standard American English.

CHARACTERISTICS OF MACHINE LANGUAGE

In machine language all instructions must be expressed in binary code, generally octal or hexadecimal. An instruction set consisting of nothing but ones and zeros is difficult to remember, tedious to look up, and difficult to use without frequent errors.

The single advantage of machine language is its direct access to even the simplest of operations. The directness of machine language permits direct, efficient programming but makes the programmer's job difficult.

In addition, actual numerical addresses must be spelled out for every instruction and piece of data.

Editing or error correction in machine language is difficult because a single change usually requires that the entire program be rewritten.

ASSEMBLY LANGUAGE LEVEL

One step up in the language hierarchy is assembly language. In this language a single instruction is machine translated from a mnemonic (memory aid) code* such as ADD MOV CLA (clear and add) into the appropriate binary code groups. The human programmer can make statements in easily remembered mnemonic terms. A specially programmed ROM is often used to convert mnemonic codes into appropriate machine-language groups of ones and zeros.

Assembly language is closely related to the machine language of a specific machine except that mnemonic symbols are used instead of zeros and ones to specify operations and memory addresses. Both machine and assembly languages are machine dependent; that is, their construction is dependent upon the organization of the hardware of a particular machine.

PROCEDURE- (OR PROBLEM-) ORIENTED LANGUAGES

Higher-level languages—such as FORTRAN (FORmula TRANslation), a mathematically oriented language, COBOL (COmmon Business/Oriented Language), and BASIC—are not machine dependent. A translator interfaces these standard languages to any given machine. With an appropriate translator these high-level languages can be used without regard to the kind of machine involved. They are called procedure-oriented languages because they are concerned strictly with the problem, not with internal machine operation.

A characteristic of these procedure-oriented languages is that a single statement results in a number of machine-language instructions and the housekeeping functions of loading data into memory, keeping track of

* Example: STB might mean *store* in register B. STA would then mean *store* in register A.

data locations, and retrieving the data when required are handled automatically.

A sample BASIC statement might be similar to the following:

IF THEN/GO TO

This statement would finally be translated into control signals that cause addition and other operations to take place within the machine.

Many high-level languages contain a vocabulary of key or reserved words that cause the system to act, along with a set of optional words that can be used to make written statements more easily understood by humans. These optional words are ignored by the machine and are for human use only.

Programming languages are carefully constructed artificial languages. While we might debate about the use of commas in a term paper, there is no room for debate in programming languages. The syntax and punctuation must be correct or the computer will refuse to act on a statement and will call the programmer's attention to an error. At the machine-language level, an incorrect 1 or 0 entry would be executed by the machine and the final results would be in error. Because programming errors are inevitable, many mistakes are avoided by using high-level languages.

12-5 Machine-Language Instructions

Instructions for machine language consist of binary numbers, n-bits in length. The binary numbers are divided into sections called *fields*. One field is usually devoted to defining the operation to be performed and one or more fields give the memory location of the operands to be operated on. Instructions may contain from one to four addresses depending upon the type of machine involved. Some machines use two or more address formats.

FOUR-ADDRESS INSTRUCTIONS

Operation code	Source of operand #1	Source of operand #2	Destination	Location of next Instruction

The four-address instruction contains a field that specifies the address in memory of each of the two operands (sources), the address where the result of the operation is to be stored, and the address where the next instruction will be found.

THREE-ADDRESS INSTRUCTIONS

Operation code	Source of operand #1	Source of operand #2	Destination

When three-address instructions are used, a special register called a *program counter* keeps track of the address of the next instruction. In this case, it is not necessary to specify the next instruction as part of the initial instruction. The instructions are stored in sequence and branch locations are ordered. The program counter simply advances one location at a time or branches to another location.

TWO-ADDRESS INSTRUCTIONS

Operation code	Source of operand #1	Destination

If one of the operands is always in a known location in an accumulator register and the system includes a program counter to keep track of the location of the next instruction, a two-address instruction will be sufficient.

The two-address instruction makes it difficult to perform subtraction and division if the subtrahend or divisor happens to be stored in the accumulator. This difficulty can be overcome, however, by adding MOVE or LOAD and STORE instructions to the instruction set. The MOVE instruction transfers the memory contents from one location to another. Symbolically this MOVE can be indicated by the following form:

$$Y \leftarrow X$$

This is interpreted as: MOVE the contents of location X into location Y. The problem in division and subtraction occurs because neither is commutative, that is:

$$A - B \neq B - A \text{ and } A/B \neq B/A$$

Many computers have LOAD and STORE instructions instead of MOVE instructions. These are usually more convenient in one-address machines.

a. The LOAD instruction moves data to the accumulator.
b. The STORE instruction moves data out of the accumulator to some other memory location.

ONE-ADDRESS INSTRUCTIONS

Operation code	Address

In the one-address instruction, the counter is used, and one operand is always in the accumulator. The address of one operand is specified. The destination of the result is always the accumulator, so it does not have to be specified. Again, MOVE or LOAD and STORE instructions are often necessary for subtraction or division.

PUSHDOWN STACK INSTRUCTIONS

The pushdown stack memory permits zero-address instructions. The instruction contains only an operation code. A binary group is loaded on the top layer of the stack. It is then *pushed down* to the next lower level to make room for an additional group. The last information to be loaded is the first out. Loading a number into the stack is called *pushing* and extracting the data is called *popping*.

The principal disadvantage of the stack is that only the data resting on the top of the stack is accessible. Normally, the two operands destined for the next operation are resting in the top two positions in the stack.

MICROPROCESSOR INSTRUCTIONS

Microprocessor instructions consist of one, two, or three bytes. Data must follow instruction commands in successive memory locations.

One-Byte Instructions

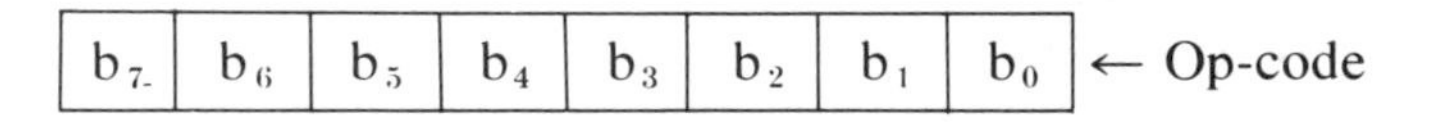

One-byte instructions, often called inherent instructions, are used primarily for manipulating accumulator registers. No address code for the operand needs to be specified because it is inherent in the instruction. For example, CLRA (clear accumulator register *A*) requires no definition of the data to the operated on, nor is it necessary to specify an address for the data to be operated on (operand). Clearing the register clears whatever data is stored in it.

Two-Byte Instructions

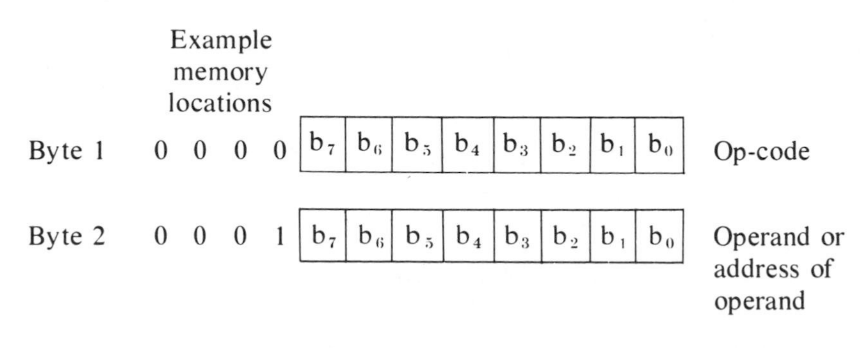

In two-byte instructions the operand must be in the memory location immediately following the op-code location.

Three-Byte Instructions

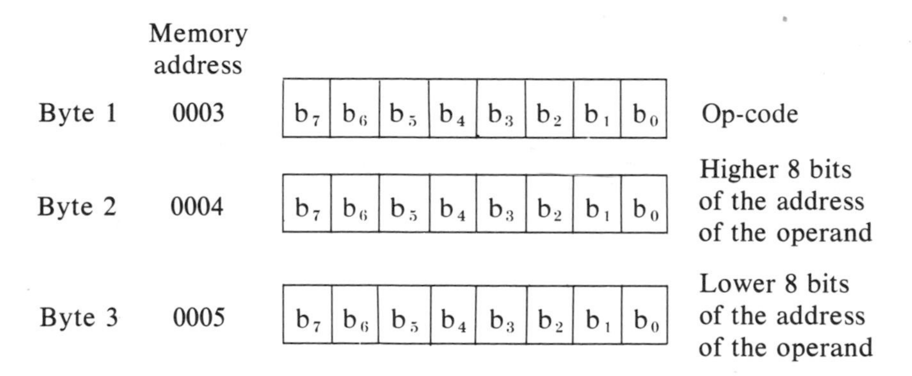

Three-byte instructions are used in the *extended* address mode. The three bytes are located sequentially in memory as shown above.

12-6 Synchronous and Asynchronous Data Communications Systems

Many logic systems and all microcomputers process data in parallel groups. When it is necessary to provide data communications between two distant points, it is not practical or economical to run multiwire cable. The solution to the problem is to convert the parallel data into a serial bit-stream, transmit the serial data along a transmission line, and convert the serial data back into parallel data at the receiver.

The conversion from parallel into serial data, and from serial into parallel, is accomplished by a shift register at each end of the line. Data are loaded in parallel into the transmit/shift register and clocked out as

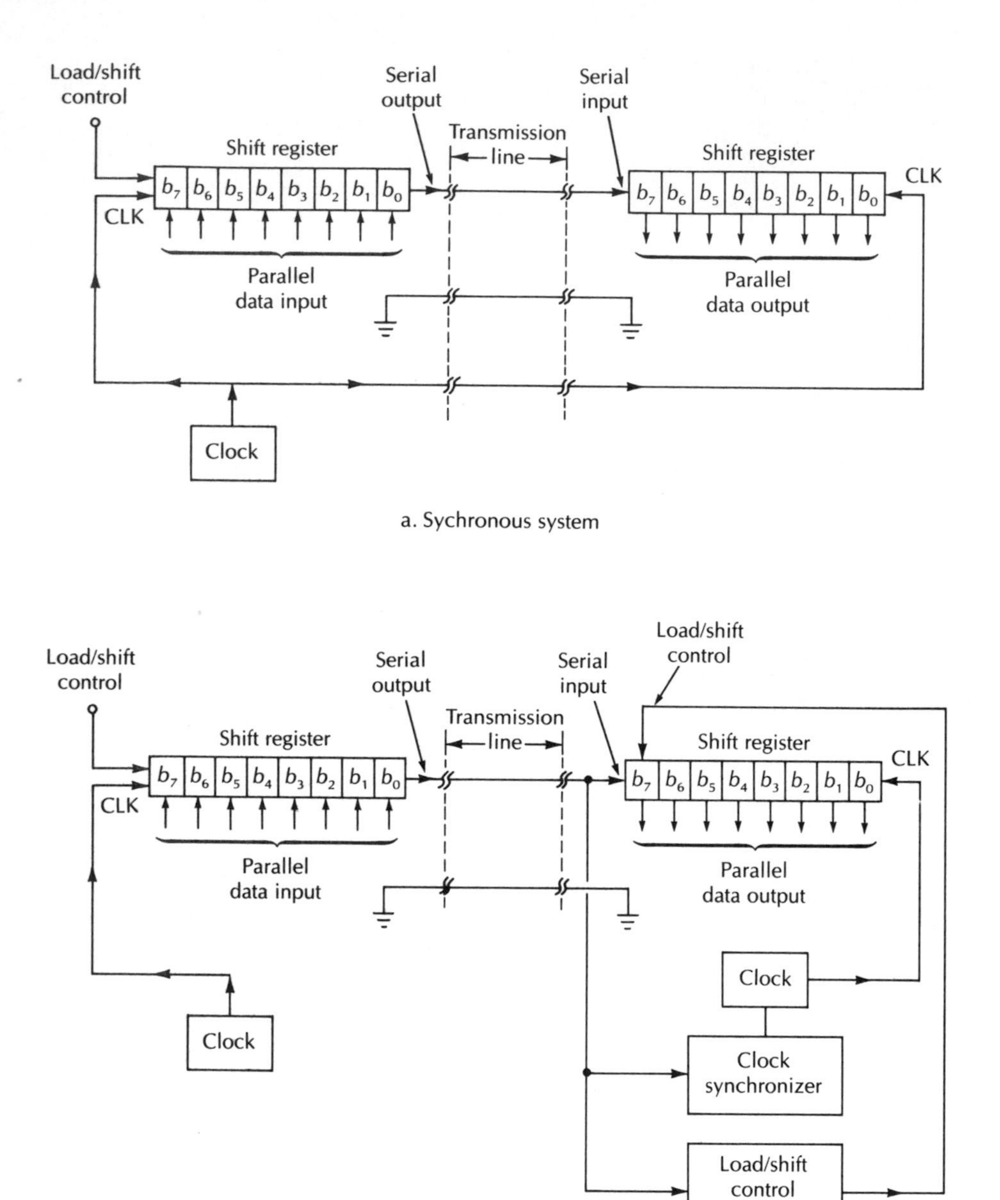

a. Sychronous system

b. Asynchronous system

Figure 12-6 Data Communications Systems

a serial bit-stream. At the receiving end, the data are clocked into the serial input of a shift register. As soon as the register is loaded, the output is available at the parallel outputs of the register.

A prime requirement for data communications systems is the synchronization of the received data with that of the transmitted data. There are two common methods of accomplishing the synchronization. The synchronous method uses a separate line to carry clock pulses from the transmitter to the receiver, whereas the asynchronous approach uses two clocks, one at the transmitting end and another at the receiving end.

The receiving clock will maintain the same frequency as the transmitting clock for short periods of time, but can be expected to drift away from that frequency with time. To avoid that problem, start and stop bits are added to the data stream to resynchronize the receiver clock for each transmitted data group.

Figure 12-6 shows simplified block diagrams of synchronous and asynchronous data transmission systems. Most data communications systems are bidirectional; they have a transmitter and a receiver at each end. Reliable data communications systems require complex control circuitry, the ability to add control bits and error-checking bits to the data stream, noise-immune receiving circuits, and, frequently, handshaking capability.

Handshaking allows the transmitter to ask the receiver to reply when it is ready to receive data. When the receiver replies *ready,* the data are transmitted. When the receiver has received all of that data group, it sends a *data received* signal back to the transmitter. The transmitter can then request the next *ready to receive* reply from the receiver. In some systems, called *full duplex,* the receiver transmits the data group back to the original transmitter where it is compared for accuracy.

All of these requirements demand very complex circuitry. Integrated circuits are available off-the-shelf, to satisfy these requirements. The USART (universal synchronous-asynchronous receiver-transmitter) provides most of the necessary circuitry on a chip. Several manufacturers produce USARTs. The UART (universal asynchronous receiver-transmitter) chip provides only asynchronous capability.

SUMMARY

Display Multiplexing

1. Display multiplexing reduces the number of decoder packages required to drive multidigit displays.
2. In larger-scale, integrated multidigit devices, muliplexing reduces the chip complexity.

Microcomputers

1. The arithmetic logic unit (ALU) performs arithmetic and logic operations on binary numbers.
2. The accumulator is a register composed of flip-flops. Data from the outside world generally pass through the accumulator.
3. Data and instructions from the outside world (or memory) enter the accumulator and are then transferred to specific memory locations, to the outside, or to special registers.
4. The accumulator is closely associated with the arithmetic logic unit. It provides temporary storage for data for the ALU and stores the results of ALU operations.
5. Data and instructions are usually loaded into memory before the computer begins its operation.
6. The computer then "fetches" instructions and data from memory at its own speed.
7. Instructions are transferred from memory to a group of flip-flops called the *instruction register,* where they are held until the instruction has been carried out.
8. The instruction decoder decodes the instruction and connects the proper units to the bus with the proper direction of data flow.
9. The program counter produces a binary signal output that specifies the memory location of the next instruction or data required. The program counter is also called the *memory-address register.*
10. The program counter's binary signal is decoded by the memory-address decoder.
11. The memory-address decoder places data at the defined memory location at the *ready.*
12. The next time the computer fetches from the memory, the data are ready and waiting.
13. The general register is similar to the accumulator. It provides temporary data storage for the ALU, but it is also free for special temporary storage needs.
14. Most of the circuits simply store data and instructions. Data and instructions are moved along the bus from unit to unit. The ALU operates on the data and places the results in the accumulator.

The control circuitry then transfers the contents of the accumulator to other units in the computer or to the outside world, according to the prewritten program.

Programming

1. A logic system designed for a specific job is known as a *hardwired logic system.* It must be redesigned for new applications.
2. In microcomputer systems, a nearly infinite number of different tasks can be accomplished by simply writing a new program.
3. A written program can be loaded into RAM. The written program is called *software.*
4. A *program* is a detailed set of instructions that defines a sequence of operations for the computer.
5. RAMs can be erased and a new program can be loaded in.
6. Programs that are intended to be more permanent can be recorded in ROM. The ROMs are plug-in units and can be exchanged for new ROMs with a new program. ROM programs are called *firmware*.
7. Machine language is the lowest-level programming language and consists of many binary 0's and 1's.
8. Machine-language programming is the most difficult programming language.
9. Machine language requires no extra memory and is the cheapest language when the machine must be programmed on a permanent basis.
10. When a machine must be reprogrammed often, higher-level languages become cheaper because of the fewer task-hours required to write the program.
11. The higher the program language level, the more nearly it resembles ordinary English.
12. A single statement in a high-level language results in several to many machine-language steps.
13. Ultimately, higher-level language statements must be converted into machine-language instructions.
14. The rules for translating from higher-level languages into machine language are stored in memory.

15. The memory space taken up by the translation rules subtracts from the total memory space available for the processing program.
16. There is always a trade-off between language level and available memory space.
17. Machine-language-level programs utilize memory space most efficiently. Higher-level languages utilize memory space less efficiently.

Data Communications

1. When data must be transmitted over a distance, parallel transmission is not practical.
2. Data communications systems convert data from parallel into a serial bit-stream, transmit it along the transmission line, and a receiver converts the data back into parallel form.
3. Synchronous systems require a separate clock line.
4. Asynchronous systems have independent clocks at the transmitting and receiving ends. The receiving clock is synchronized by adding extra control bits to the data stream. The receiving clock is resynchronized for each data group.
5. Reliable data transmission systems tend to become very complex.
6. Commercial LSI USART (universal synchronous-asynchronous receiver-transmitter) and UART (universal asynchronous receiver-transmitter) integrated circuits are available to solve most data transmission problems.

Problems

1. What is multiplexing as used with display devices?
2. What is the principal advantage of display multiplexing?
3. What is the function of the ALU?
4. When data are to be loaded into memory from a keyboard, what register stores the data temporarily before transferring them to the memory?
5. What is the function of the memory-address decoder?
6. What is the purpose of the program counter?
7. What is the function of the general register?
8. What is the function of the instruction register?
9. What is the function of the instruction decoder?
10. When the adder performs an operation on two binary numbers, where is the result first stored?

11. What is the function of the bus drivers?
12. What are some of the functions that might be performed by an interface device?
13. Photocopy Figure 12-4 and follow through the following addition problem in the same fashion as the example in Section 12-3. Microprocessors have a 1-bit overflow (carry) register as shown in Figure 12-4. Show how this 1-bit register is used in the following problem:

    Add: 0111 + 1001.

14. Define *dedicated* (committed) *system*.
15. In a dedicated system, how do microprocessors differ from hard-wired logic?
16. Define *programming*.
17. List the hierarchies of programming languages.
18. Contrast the following:
    a. Machine language
    b. Assembler language
    c. Procedure-oriented language
19. Match each of the descriptions below with either (1) hardware, (2) firmware, or (3) software.
    a. The least flexible program form
    b. The most flexible program form
    c. Requires *no* physical modification to change the machine's task
    d. Requires replacement or programming of a ROM
20. Contrast one-, two-, and three-byte instructions.
21. How is a software program used by the computer? Explain in detail.
22. Define the word *mnemonics* and explain how it applies to programming.
23. What are the advantages of assembler language over machine language?
24. What are the advantages of high-level procedure-oriented languages over assembler and machine languages?
25. How is assembler language converted into machine language?
26. Does it make any difference to the programmer what kind of machine COBOL, BASIC, or FORTRAN is used with?
27. Make a sketch of the fields for:
    a. Four-address instruction
    b. Three-address instruction
    c. Two-address instruction
    d. One-address instruction
28. In a three-address instruction, how is the location of the next instruction defined?

29. In a two-address instruction, the source of one operand is given; how is the location of the second operand specified?
30. In a one-address instruction, how is the destination specified?
31. What is the difference between synchronous and asynchronous data transmission systems?
32. Describe handshaking in data communications systems.

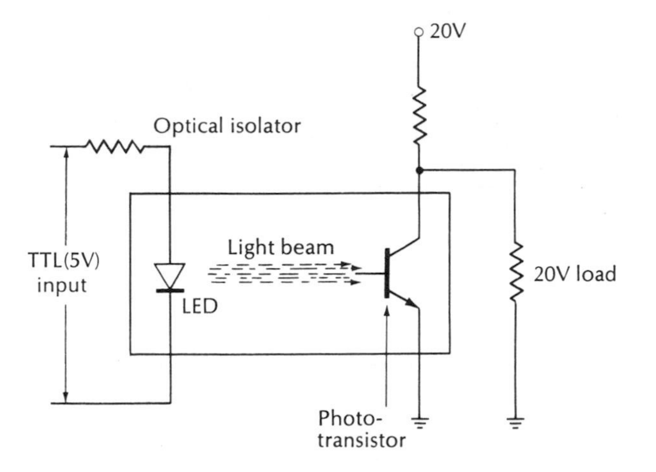

Figure 11-32 Optical Isolator

11-14 Phase-Locked Loop

The phase-locked loop (PLL) is a closed-loop feedback, self-regulating system. The basic system has been in use for many years, but until the implementation of IC technology, its overall complexity and cost prevented its use in many applications.

THEORY OF OPERATION

The phase-locked loop is a closed-loop electronic servo-mechanism. Figure 11-33 shows the block diagram of a closed-loop servo-mechanism. In the electromechanical servo in Figure 11-33b, R_1 and R_2 form a bridge. When the reference potentiometer and the error-detector potentiometer are in the same electrical position, the bridge is balanced, the amplifier sees zero voltage, and the gear motor remains off. Suppose that the reference potentiometer is rotated. Because the error-detector potentiometer is connected to the rotary antenna shaft (the load), movement of the reference potentiometer (only) unbalances the bridge. The bridge error signal is amplified to drive the motor. The motor drives the antenna and the error-detector potentiometer in the proper direction to rebalance the bridge. At balance, the error voltage vanishes and the motor stops.

The phase-locked loop consists of the following elements:

1. Load (value to be controlled): frequency
2. Error detector: comparator
3. Error corrector: voltage controlled (voltage variable) oscillator (VCO)

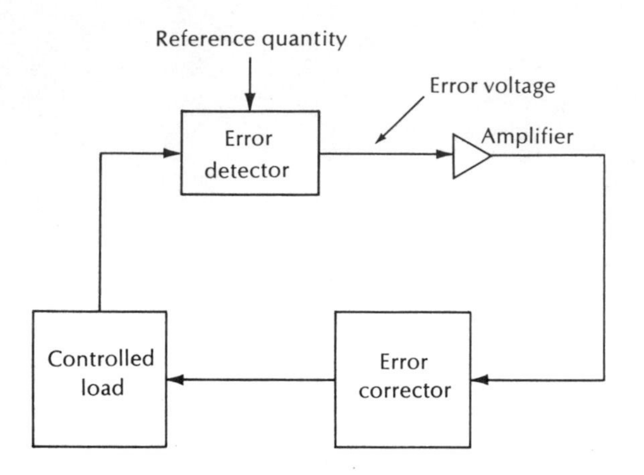

a. Generalized servo-mechanism block diagram

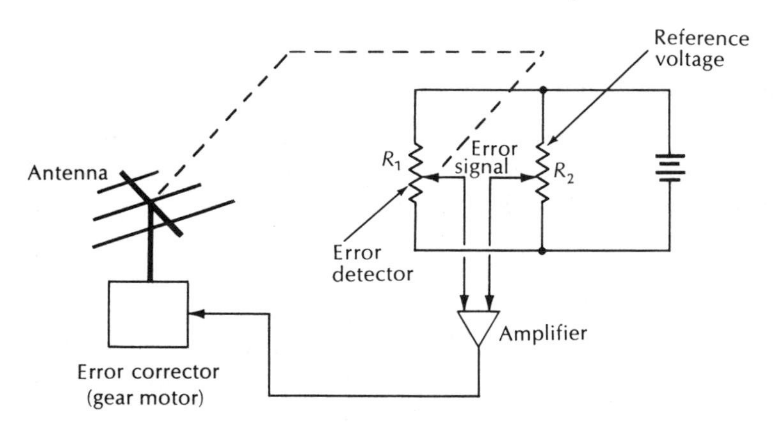

b. Electromechanical servo-mechanism (antenna rotator)

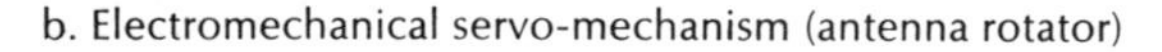

Figure 11-33 The Servo-Mechanism

4. Amplifier: electronic amplifier
5. Reference: input frequency

Figure 11-34 shows the block diagram of a phase-locked loop. The error detector compares the phase of the input signal with that of the voltage controlled oscillator (VCO). If the two are identical, no error voltage is generated and the VCO runs at its free-running frequency. When there is a difference between the phases of the input signal and the VCO, the two signals are mixed in the phase detector.

Assume that the input frequency is 100 kHz and the VCO is running at 101 kHz. The phase comparator mixes the 100 kHz and 101 kHz to

11-12 Electromechanical A-to-D Conversion

In machine and process control it is often necessary to translate the angular position of a rotating shaft into digital information. The most common method for accomplishing this task is to use an optically read binary-code disc mounted on the shaft in question. Figure 11-31 shows the arrangement and the code disc layout for Gray code representation.

The Gray code is used because only one bit in the group changes at a time, which improves resolution and minimizes conversion error. The Gray code is not suitable for processing purposes, and data are normally translated into normal binary as an integral part of the conversion process.

11-13 Optical Isolators

An increasingly important interface device consists of an LED and a phototransistor optically connected in a light-tight package. This device eliminates reflected loading because the only coupling is by way of a light beam. Variations in the load are not reflected back to the source as happens in nearly all other coupling devices. Optical isolators are particularly useful for coupling devices with incompatible voltage requirements. Figure 11-32 shows a typical optical isolator application. Optical isolators provide isolation between the LED and the phototransistor up to several kilovolts.

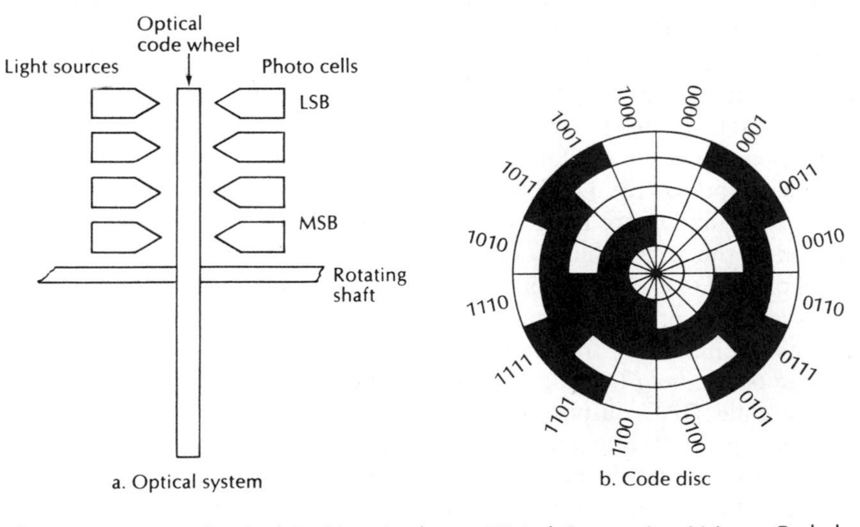

Figure 11-31 Mechanical Position Analog-to-Digital Conversion Using a Coded Optical Disc

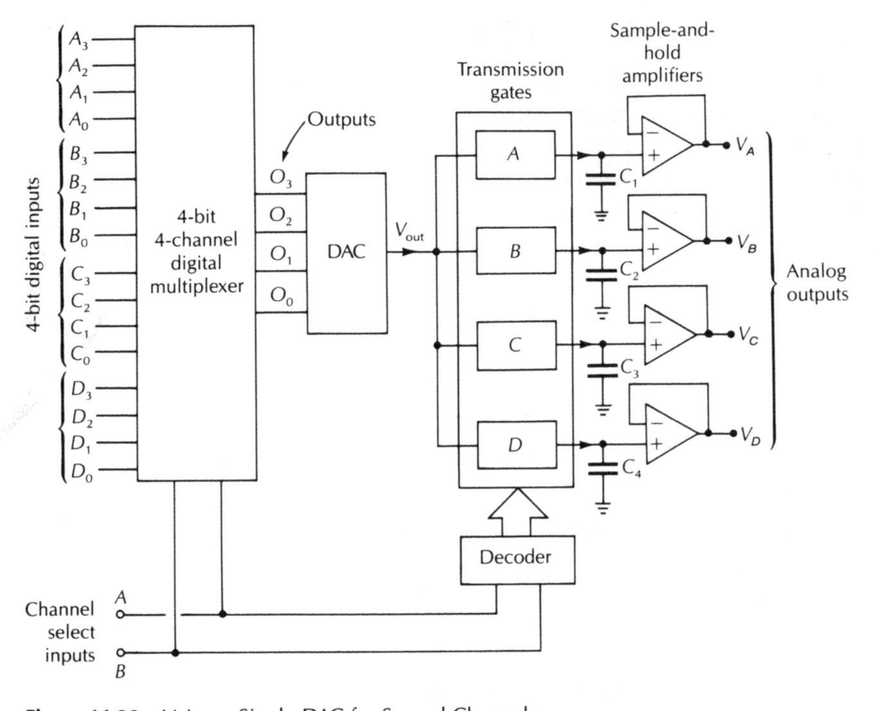

Figure 11-30 Using a Single DAC for Several Channels

The multiplexed DAC system in Figure 11-30 uses a 4-bit, 4-channel digital multiplexer to select the appropriate 4-bit digital channel. A single 4-bit DAC performs the conversion for all 4 channels.

The analog multiplexer connects the output of the DAC to the appropriate sample-and-hold amplifiers. Assume that the *channel select* inputs are set for the selection of channel A. The 4-channel digital multiplexer connects the digital input $A_0 = A_3$ to the DAC and channels the analog output of the DAC through transmission gate A. Transmission gates B, C, and D are in the open-circuit condition. The analog voltage output of the DAC quickly charges capacitor C_1 to the output voltage level of the DAC. Capacitor C_1 serves as an analog memory and allows the system to proceed to sample the other channels. When the system moves to channel B, transmission gate A becomes an open circuit. The charge on the *hold* (analog memory) capacitor C_1 continues to produce the original DAC output voltage at V_A even though the DAC is now converting a different channel. The input impedance of the amplifier must be very high to minimize capacitor discharge between samples. Samples must also be taken frequently enough that the inevitable *hold* capacitor discharge will be negligible and the analog quantity will not change too much between samples.

In the case of the divide-by-3, divide-by-2 configuration, the divide-by-3 counter goes through its complete counting sequence before the divide-by-2 is toggled. At the end of the second divide-by-3 cycle, the divide-by-2 circuit resets to zero. (See the truth table in Figure 7-19.).

TTL MSI COUNTERS

The 54/74 series provides a small but versatile collection of off-the-shelf counters to satisfy many common counting circuit requirements. Table 7-3 summarizes the common available types.

7-8 The 7493 Binary Counter

Figure 7-21 shows the logic diagram of the 7493 binary counter, a divide-by-2, divide-by-8 ripple counter. They may be used separately or connected externally to form a divide-by-16 counter. The counter has a gated reset-to-0 input. Either or both (set to 0) inputs to R_0 and R_0 must be at ground for normal counting. Both set-to-0 inputs are taken high (5 V) to reset the counter to zero.

The counter advances on the negative-going edge of the clock. The clock waveform must be TTL compatible and can have a maximum

Table 7-3 MSI Counters

Non-Presettable

Type	Mod.	Count Direction	Synchronous Max. Clock Rate MHz	Ripple Max. Clock Rate MHz	Configuration	Clock
7490	10	up		18	2 × 5	neg. edge
7492	12	up		18	2 × 6	neg. edge
7493	16	up		18	2 × 8	neg. edge

Presettable

Type	Mod.	Count Direction	Synchronous Max. Clock Rate MHz	Ripple Max. Clock Rate MHz	Configuration	Clock
74160	10	up	32		1 × 10	pos. edge
74161	16	up	32		1 × 16	pos. edge
74190	10	up/down	25		1 × 10	pos. edge
74191	16	up/down	25		1 × 16	pos. edge
74192	10	up/down	32		1 × 10	pos. edge
74177	16	up		35	2 × 8	neg. edge

truth table. An examination of the truth table in this figure indicates that the count for this configuration is a normal binary series count.

Divide-by-3, Divide-by-2 Counter

By placing the divide-by-3 section at the input of the chain, a mod 6 counter is obtained but the counting sequence is 0, 1, 2, 4, 5, 6, 0, (decimal equivalents) rather than the natural 0, 1, 2, 3, 4, 5, 0, binary sequence count. The block diagram and truth table are shown in Figure 7-19 and the logic diagram is shown in Figure 7-20.

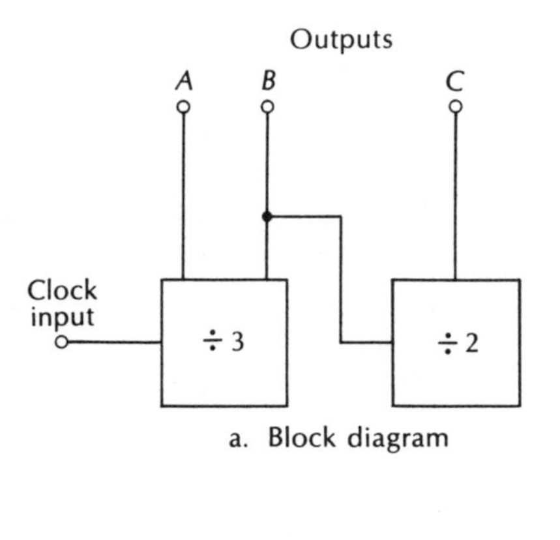

a. Block diagram

Count	C	B	A
0	0	0	0
1	0	0	1
2	0	1	0
3	1	0	0
4	1	0	1
5	1	1	0
6	0	0	0

b. Truth table

Figure 7-19 Divide-by-3, Divide-by-2 Mod 6 Counter and Truth Table

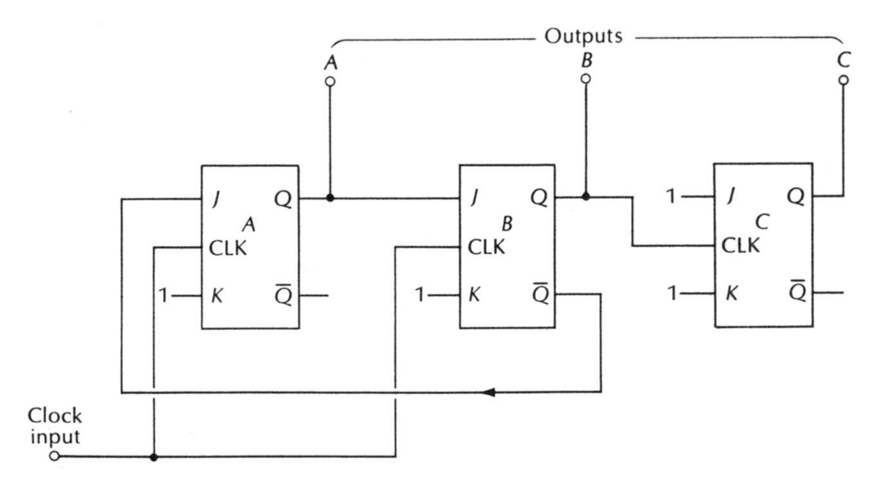

Figure 7-20 Logic Diagram for the Divide-by-3, Divide-by-2 Mod 6 Counter

MULTIPLEXING ANALOG-TO-DIGITAL CONVERTERS

Figure 11-29 illustrates how an analog multiplexer can be used to sample the outputs from four different transducers or other analog signal sources in some orderly sequence. One analog-to-digital converter can then convert to several sets of data. In some systems a single ADAC can be used with 50 or more sensing systems, depending on the speed and other characteristics of the analog devices and the speed of the digital system.

A high-resolution ADAC is fairly expensive. It may not be good economics to let it wait when it could just as well be converting other signals.

MULTIPLEXING DIGITAL-TO-ANALOG CONVERTERS

Digital-to-analog converters can also be time multiplexed, as illustrated in Figure 11-30. In process control systems you are likely to find a time-multiplexed ADAC and a time-multiplexed DAC sharing a single digital processor.

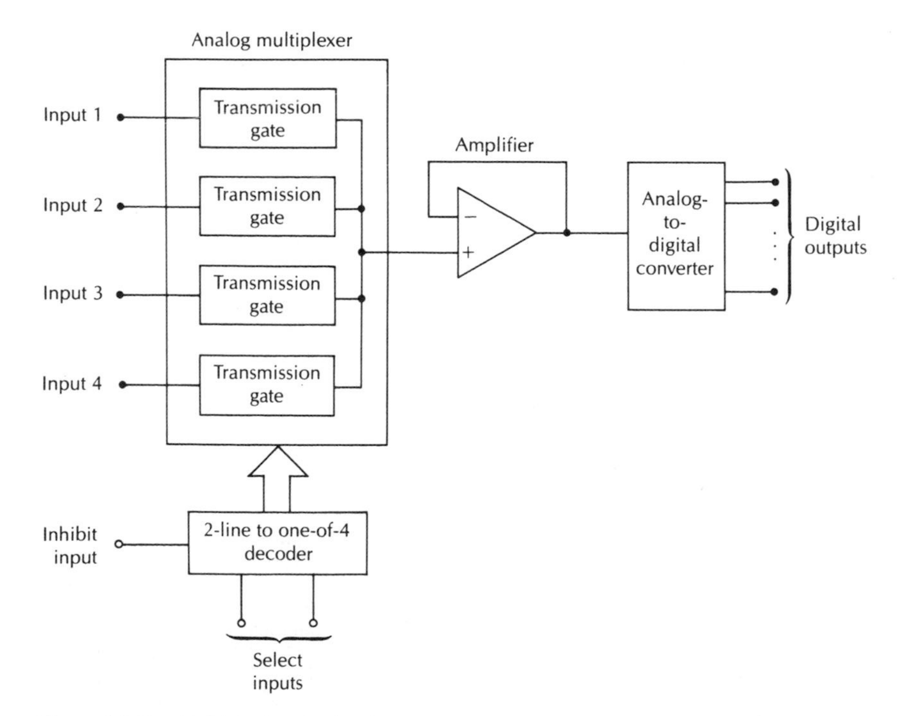

Figure 11-29 Multiplexing Analog Inputs

Table 11-2 Successive Approximation Example

Analog Input Voltage = 11V

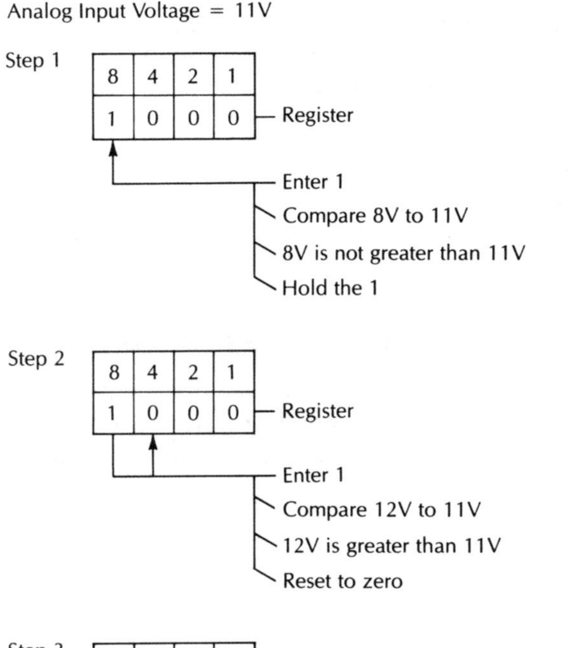

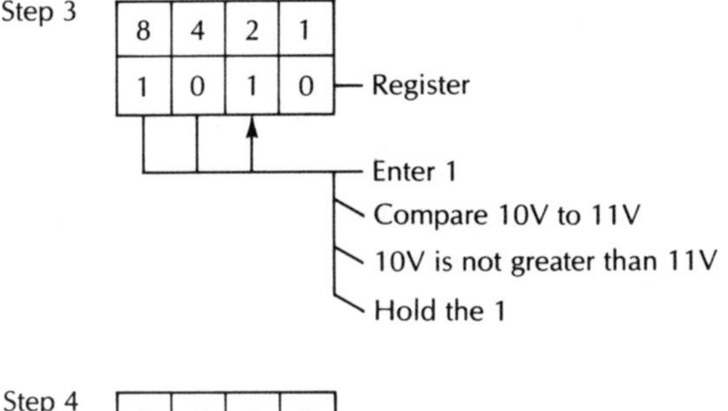

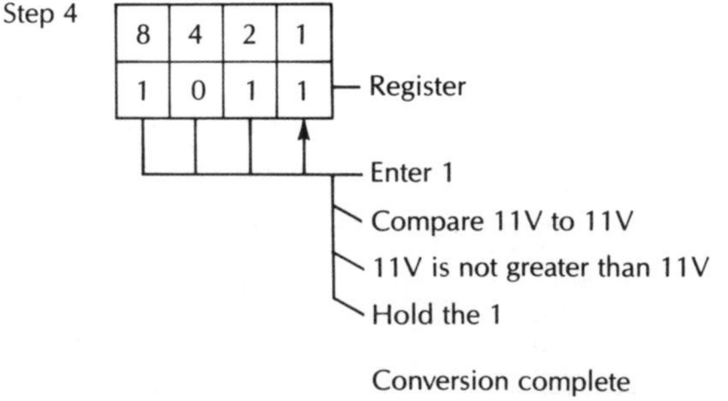

Conversion complete

   c. F-F C toggles to $Q = 1$
   d. F-F C returns a logical 0 from $\bar{Q}_c$ back to F-F A. This will prevent F-F A from toggling to $Q = 1$ on the fifth clock pulse (count 5)
   e. Q on F-F C places a logical 1 on K of F-F C. This will force F-F C to reset to 0 on count 5
6. Count 5
   a. F-F A remains at $Q = 0$ because of the logical 0 fed back from $\bar{Q}$ of F-F C
   b. F-F B remains at $Q = 0$ CBA
      The next natural count would be 1 0 1 , and nothing need be done to insure a $Qb = 0$ for count 5. It would not have toggled in any event
   c. F-F C has a logical 1 on the K input causing it to reset to $Q = 0$ on count 5

MOD 6 COUNTERS

Divide-by-2, Divide-by-3 Counter

Mod 6 counters can be formed by combining the mod 3 counter from Figure 7-16 with a single F-F (divide-by-2). If the binary (divide-by-2) is used as the input stage, each two clock pulses will provide one subordinate clock pulse to the divide-by-3 circuit. The binary stage will toggle on each clock pulse. The divide-by-3 stage will follow the count pattern for the mod 3 circuit in Figure 7-16, except that it will remain in each count state for two counts instead of making some change at every count as it does when used alone. Figure 7-18 shows a block diagram of the divide-by-2, divide-by-3 arrangement and its corresponding mod 6

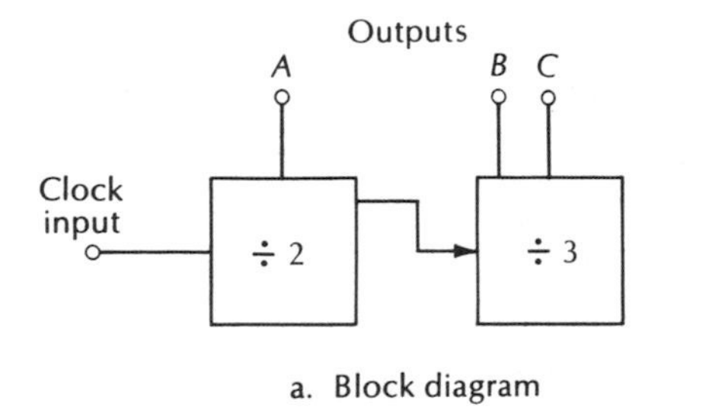

a. Block diagram

Count	C	B	A
0	0	0	0
1	0	0	1
2	0	1	0
3	0	1	1
4	1	0	0
5	1	0	1

b. Truth table

Figure 7-18 Divide-by-2, Divide-by-3 Mod 6 Counter and Truth Table

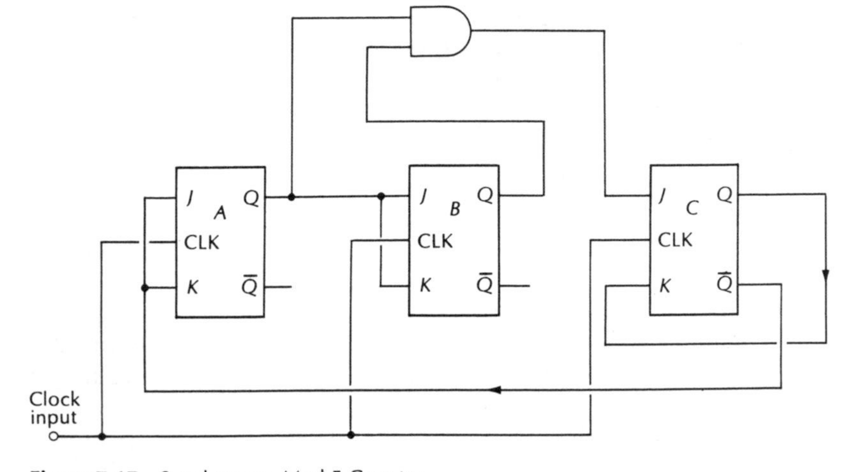

Figure 7-17 Synchronous Mod 5 Counter

Table 7-2 Truth Table for Mod 5 Hybrid Counter

Count	C	B	A
0	0	0	0
1	0	0	1
2	0	1	0
3	0	1	1
4	1	0	0
5	0	0	0

2. Count 1
   a. F-F *A* toggles to 1 ($Q = 1$)
   b. F-F's *B* and *C* remain at $Q = 0$
   c. $\bar{Q}$ of F-F $C = 1$ and places *J-K* of F-F *A* at logical 1
3. Count 2
   a. F-F *A* toggles to $Q = 0$
   b. F-F *B* toggles to $Q = 1$
   c. F-F *C* remains at $Q = 0$ ($\bar{Q} = 1$)
4. Count 3
   a. F-F *A* toggles to $Q = 1$
   b. F-F *B* remains at $Q = 1$
   c. F-F *C* remains at $Q = 0$
5. Count 4
   a. F-F *A* toggles to $Q = 0$
   b. F-F *B* toggles to $Q = 0$

INDEX